10th Anniversary of *Actuators*

10th Anniversary of *Actuators*

Editor

Zhuming Bi

MDPI • Basel • Beijing • Wuhan • Barcelona • Belgrade • Manchester • Tokyo • Cluj • Tianjin

Editor
Zhuming Bi
Civil and Mechanical Engineering
Purdue University Fort Wayne
Fort Wayne
United States

Editorial Office
MDPI
St. Alban-Anlage 66
4052 Basel, Switzerland

This is a reprint of articles from the Special Issue published online in the open access journal *Actuators* (ISSN 2076-0825) (available at: www.mdpi.com/journal/actuators/special_issues/LO28Y02719).

For citation purposes, cite each article independently as indicated on the article page online and as indicated below:

LastName, A.A.; LastName, B.B.; LastName, C.C. Article Title. *Journal Name* **Year**, *Volume Number*, Page Range.

ISBN 978-3-0365-7901-6 (Hbk)
ISBN 978-3-0365-7900-9 (PDF)

Contents

About the Editor

Zhuming Bi

Zhuming Bi is a Section Editor in Chief of *Actuators* in Advanced Manufacturing. He received a Ph.D. degree from the Harbin Institute of Technology, Harbin, China, and the University of Saskatchewan, Saskatoon, SK, Canada, in 1994 and 2002, respectively. Currently, he is a Professor at the Department of Civil and Mechanical Engineering, Purdue University Fort Wayne (PFW). Professor Bi is the 2023–2024 Fullbright-Nokia Distinguished Chair in Information and Communications Technologies; he is the Harris Chair of Wireless Communication and Applied Research at PFW. He has authored or co-authored four books, including three textbooks on "Finite Element Analysis Applications", "Computer Aided Design and Manufacturing", and "Digital Manufacturing". In addition, Professor Bi has published 10 book chapters, 168 journal articles, and over 60 articles in conference proceedings. His research interests are actuators, enterprise systems, digital manufacturing, finite element analysis, machine design, and robotics and automation.

Preface to "10th Anniversary of *Actuators*"

Actuators is an international, peer-reviewed, open-access journal on the science and technology of actuators and control systems; it was officially launched by MDPI in 2012. In its first ten years from 2012 to 2022, it has published a total of 629 articles that cover all aspects of the studies in the field of actuators including working principles, design optimization, performance evaluation, new applications, robotics, and system integration. *Actuators* has been included the in Science Citation Index Expanded (SCIE) since 2017, and its impact factor (IF) has steadily increased over years. The Special Issue entitled "10th Anniversary of *Actuators*" was prepared to celebrate the 10th anniversary of *Actuators*. It consists of 18 representative works with four main topics: systematic review for specific actuator types, performance evaluation, design and control, and new actuators' applications for readers to understand the challenges and the trends of research and development in advancing the theories, methodologies, and design tools of actuators.

Zhuming Bi
Editor

Review

Recent Developments on Dielectric Barrier Discharge (DBD) Plasma Actuators for Icing Mitigation

Frederico Rodrigues [1,*], Mohammadmahdi Abdollahzadehsangroudi [1], João Nunes-Pereira [1,2] and José Páscoa [1]

1   C-MAST, Centre for Mechanical and Aerospace Science and Technologies, Universidade da Beira Interior, Rua Marquês D'Ávila e Bolama, 6201-001 Covilhã, Portugal
2   CF-UM-UP, Centro de Física das Universidades do Minho e do Porto, Campus de Gualtar, 4710-057 Braga, Portugal
*   Correspondence: fmfr@ubi.pt

Abstract: Ice accretion is a common issue on aircraft flying in cold climate conditions. The ice accumulation on aircraft surfaces disturbs the adjacent airflow field, increases the drag, and significantly reduces the aircraft's aerodynamic performance. It also increases the weight of the aircraft and causes the failure of critical components in some situations, leading to premature aerodynamic stall and loss of control and lift. With this in mind, several authors have begun to study the thermal effects of plasma actuators for icing control and mitigation, considering both aeronautical and wind energy applications. Although this is a recent topic, several studies have already been performed, and it is clear this topic has attracted the attention of several research groups. Considering the importance and potential of using dielectric barrier discharge (DBD) plasma actuators for ice mitigation, we aim to present in this paper the first review on this topic, summarizing all the information reported in the literature about three major subtopics: thermal effects induced by DBD plasma actuators, plasma actuators' ability in deicing and ice formation prevention, and ice detection capability of DBD plasma actuators. An overview of the characteristics of these devices is performed and conclusions are drawn regarding recent developments in the application of plasma actuators for icing mitigation purposes.

Keywords: plasma actuators; dielectric barrier discharge; deicing; ice sensing; flow control

Citation: Rodrigues, F.; Abdollahzadehsangroudi, M.; Nunes-Pereira, J.; Páscoa, J. Recent Developments on Dielectric Barrier Discharge (DBD) Plasma Actuators for Icing Mitigation. *Actuators* **2023**, *12*, 5. https://doi.org/10.3390/act12010005

Academic Editor: Subrata Roy

Received: 9 November 2022
Revised: 15 December 2022
Accepted: 16 December 2022
Published: 21 December 2022

1. Introduction

Ice accretion on various component surfaces causes undesirable loads that pose a serious problem for aviation and wind turbines [1]. When aircraft fly in cold climates and pass through clouds containing supercooled water droplets, severe ice formation and subsequent icing on aircraft surfaces often occur [2,3]. This phenomenon of ice formation can negatively affect the normal operation of various aircraft components, for example, the aircraft wings. Ice accumulation on the leading edge of the aircraft wings disturbs the flow and degrades aerodynamic performance by increasing drag and reducing lift, which can seriously compromise flight safety [4,5]. It is estimated that 9% of aircraft safety accidents are originated by ice accretion [6].

Depending on the icing formation mechanism and conditions, different types of ice can form, usually defined as rime ice, glaze ice, or mixed ice. Rime ice occurs when droplets freeze almost immediately after impinging on the surface. This phenomenon usually occurs at low temperatures, low liquid water content, and low flow velocity [7]. On the other hand, glaze ice forms when the cooled water droplets freeze gradually after impinging on the surface. This phenomenon usually occurs at temperatures around 0 °C and with high liquid water content. Mixed ice, in turn, consists of a mixture of rime and glaze ice and occurs when both types of ice are formed [8,9]. Regardless of the type of ice, any ice accretion phenomenon can pose a hazard to the aircraft, but the formation of rime ice is usually less harmful to the aerodynamic performance of the wings [9]. The ice accretion on the surface, even in small amounts, increases the roughness of the surface and reduces maneuverability to a dangerous level [10]. In addition, ice can also build up in the aircraft's speedometers and pressure sensors, making them difficult to operate [10]. Furthermore, ice crystals can form at high altitudes and coexist with the supercooled water droplets. If ice crystals are continuously ingested into the engine's core compression system, they can provide sufficiently cold conditions to refreeze and accumulate on the engine surfaces, a phenomenon commonly referred to as "ice crystal icing". This phenomenon can lead to loss of thrust, compressor damage, and/or flameout [11,12].

As in the aviation industry, ice accretion is also a major problem in the wind power industry [13]. Considering the limited wind power near the ground, the idea of operating wind turbines at high altitudes has grown in recent years, because wind speed increases significantly with altitude [14]. In the first thousand meters of altitude, the wind speed usually increases by about 0.1 m/s per 100 m of altitude, which is a considerable increase in wind energy, because the wind energy varies with the cube of wind speed [15]. In addition, cold regions are very attractive for wind energy generation [16]. In these regions, the higher air density at lower temperatures provides about 10% more available wind energy than in other regions [17]. Therefore, cold climate regions at high altitudes provide optimal conditions for wind energy generation, and, for this reason, about a quarter of the world's wind turbine capacity is now installed in cold climate regions [18,19]. However, these regions also have favorable conditions for ice formation and accretion on wind turbines blades. The performance of wind turbines is affected by several factors, such as the blade design, the technical parameters of the turbine, the location, and the climate, in which blade icing is a critical factor that reduces the performance of the wind turbine [20]. It is estimated that the power loss of wind turbines due to icing can reach values up to 50% per year, depending also on the duration of icing [21]. In addition to the losses in electricity generation, the unbalanced loads caused by ice formation lead to deterioration of the components and shortening of their lifetime, which, in turn, significantly increases the maintenance costs of wind turbines [22]. Moreover, considering the possible use of small-scale vertical axis wind turbines in urban areas, ice accumulation increases the load and noise and can lead to ice shedding and ice throwing events that can generate dangerous projectiles [23].

Considering this background, and given the impact of ice accretion on the aviation and wind energy industries, several works have been carried out over the years to investigate and further develop different deicing systems [17,21]. One of the latest proposed deicing technologies is based on the use of surface dielectric barrier discharge (DBD) plasma actuators. These simple devices have been studied over the years for active flow control and aerodynamic efficiency improvement [24–27]. However, only recently have researchers started to investigate the possibility of using the thermal effects produced by these devices for deicing and anti-icing purposes [28,29]. Since these devices present a behavior similar to a capacitor, researchers have also investigated the possibility of using plasma actuators as capacitive ice sensors [30]. By combining all these functionalities, DBD plasma actuators can simultaneously detect ice formation and accumulation on the surface, perform deicing, and improve aerodynamic efficiency. Considering this enormous potential of DBD plasma actuators for anti-icing applications, the present work aims to review and critically exploit the recent developments in plasma actuators for icing mitigation applications. First, the traditional devices and techniques for deicing and ice sensing are discussed. Then, the design and operation of surface dielectric barrier discharge plasma actuators are explained, after which the ability of plasma actuators to induce thermal effects, perform deicing, and operate as ice sensors will be exploited in different subsections. To finalize the current work, several conclusions will be drawn about the potential of DBD plasma actuators for simultaneous deicing, ice sensing, and active flow control in aeronautics and wind turbine applications.

2. Traditional Techniques for Ice Protection Systems

Ice protection systems are designed to prevent ice accretion on wind turbine and aircraft surfaces (particularly leading edges) such as wings, propellers, rotor blades, engine intakes, and environmental control intakes. When the ice builds up to a significant thickness on the surfaces, it alters the shape of the airfoils and flight control surfaces, reduces wind turbine performance, and reduces the control or handling characteristics of the aircraft [31,32]. Ice protection systems can be based on deicing operations (removal of snow, ice, or frost from the surface), anti-icing operations (prevention of ice adhesion or delay of its formation or reformation), or combined deicing and anti-icing operations. Ice protection systems include the application of active and/or passive ice protection techniques. Active ice protection systems include the use of mechanical methods, the application of heat, the use of dry or liquid chemicals to lower the freezing point of water, or a combination of these various techniques. Passive ice protection systems, on the other hand, use coatings or modifications to surface structures to prevent or delay ice accretion on the surface. Figure 1 summarizes the different types of techniques used in ice protection systems that we will assess in the following subsections.

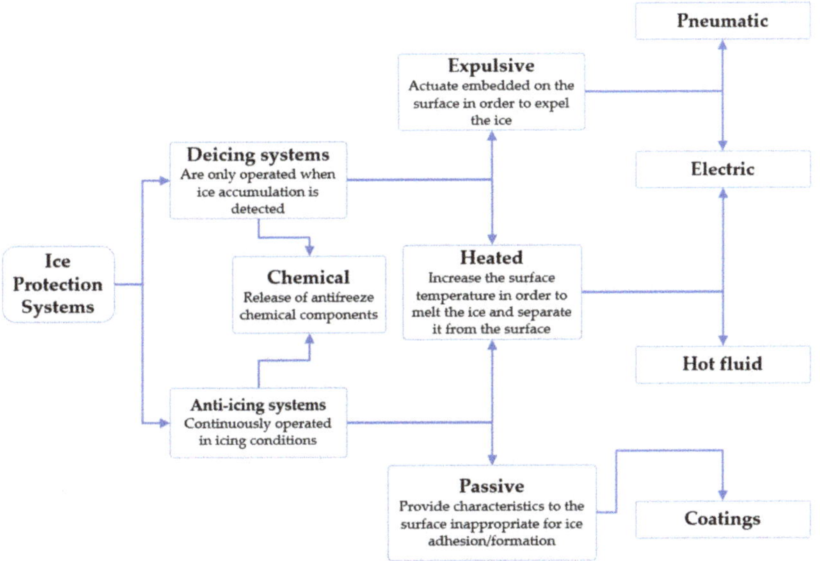

Figure 1. Different anti-icing and deicing techniques generally used in ice protection systems.

2.1. Traditional Deicing and Anti-Icing Techniques Used in Ice Protection Systems

The pneumatic deicing boot is a common mechanical technique for ice protection systems, generally used for in-flight deicing. This system consists of pneumatic boots generally installed on the leading edge of the wings and/or control surfaces, as ice accumulates particularly frequently in these areas. The pneumatic boot is commonly made of rubber layers with one or more air chambers between them so that they can be rapidly inflated and deflated to expel ice that has accumulated on the surface [33,34]. This rapid movement of the rubber breaks the adhesion of the ice, and it is removed and carried away by the airflow around the aircraft surface. This type of system is suitable for low and medium speed aircraft [17]. Ice protection systems based on pneumatic boots have the advantage of requiring low engine bleeding air, but the pneumatic boots need to be replaced frequently and also require frequent inspection and maintenance.

Another type of deicing system is based on the use of electrically resistive elements and is commonly referred to as an electro-thermal protection system [35–38]. A schematic illustrating the operation of an electro-thermal ice protection system is shown in Figure 2.

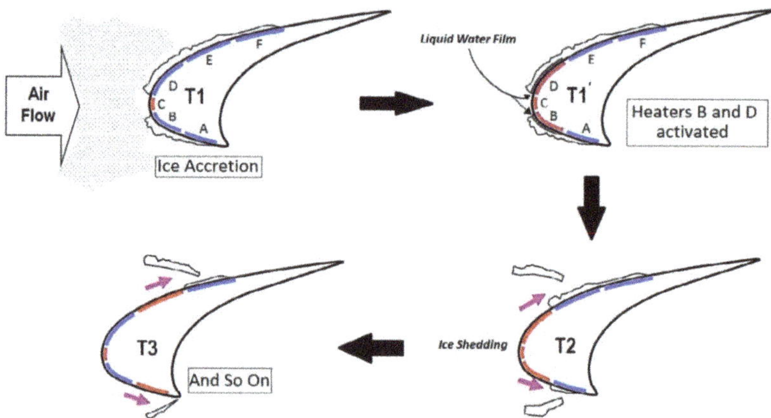

Figure 2. Illustration of the operation of an electro-thermal ice protection system (ETIPS) (reproduced from Huang et al. [39] with permission of Elsevier).

These elements generate heat when current is applied to them and are usually embedded in a rubber layer that is implemented in the leading edges of the surfaces that should be protected. These devices can operate continuously, providing a constant heat flux and preventing ice accretion, or intermittently, by removing ice only when it accretes. The second mode, in which the device operates as a deicer, involves much less power consumption because the system melts only the ice in contact with the surface to break the ice's adhesion and shed it through the flow around the surface. Normally, when ice accumulation is detected, these systems first operate as a deicing system to melt and remove the ice that has already accumulated, and then operate as an anti-icing system to prevent continuous ice accumulation.

Electro-mechanical ice removal devices form another group of ice protection systems. Often referred to as EMEDS (Electro-Mechanical Expulsion Deicing Systems) [40–42], this type of system contains actuators installed on the surface of the structure that produce a mechanical force to expel the ice from the surface [43]. A schematic of this type of system is shown in Figure 3.

The system generally consists of an electronic deicing control unit (DCU) that controls the apparatus and operating times, an energy storage bank (ESB) that supplies high-current electrical pulses to the system, and a leading-edge assembly (LEA) that contains the actuators that are moved to induce a shock wave on the protected surface and expel the ice. The system uses microsecond-scale high-current electrical pulses delivered to the actuators to generate opposing electromagnetic fields. The electromagnetic fields cause the actuators to rapidly change shape, which causes the leading-edge surface to vibrate at a very high frequency, dislodging the accumulated ice. This system can also be combined with an electrical heating element which prevents ice accretion while the actuators remove the ice that has accumulated in the area downstream of the heated part. This hybrid system is generally called a Thermo-Mechanical Expulsion Deicing System (TMEDS) [44].

A recent method, still under development, is based on the application of ultrasonic vibrations to break and delaminate the ice accumulated on the surface that should be protected [45–47]. In this type of system, ultrasonic waves with high energy are used to generate a shear stress on the surface in contact with the ice layer, which can break and remove the ice [48]. This shear stress is generated by the difference in wave propagation speed between the ice and the surface due to the different physical properties of the two media. Compared to the electro-thermal deicing systems, the ultrasonic deicing systems are not thermal and, therefore, require much less energy. Some works have reported the potential of this type of ice protection systems. Palacios et al. [49] demonstrated that certain ultrasonic modes generated by horizontal shear waves produce sufficient shear stress to remove the accreted ice layer from a host structure, and Budinger et al. [50] reported on the resonance modes and ultrasonic frequencies for low-power ultrasonic deicers.

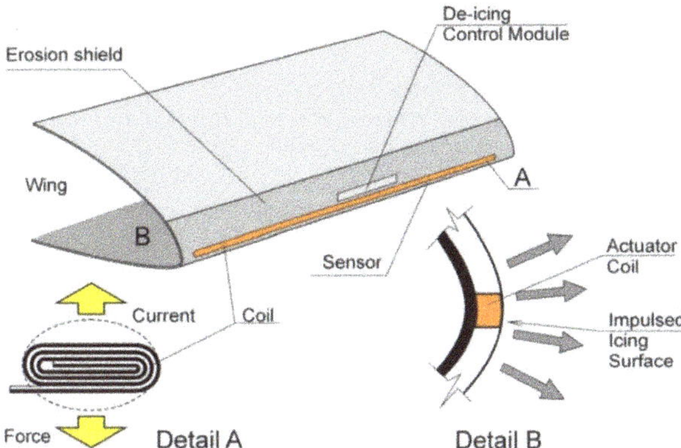

Figure 3. Schematic of the Electro-Mechanical Expulsion Deicing Systems (reproduced from Huang et al. [39] with permissions from Elsevier).

The deicing and ice removal operations can also be achieved using shape memory alloys. Shape memory alloys are used in this type of system because these materials can undergo a very large dimensional change induced by relatively small temperature variations. Considering this possible effect, the shape-changing properties of these materials are used in ice protection systems to promote ice removal [51–53].

Ice protection systems can also be based on very rapid electromagnetically induced vibrations. These systems are usually referred to as electro-expulsive systems and use an electric current passing through a parallel copper layer to create a repulsive magnetic field that induces small jumps of high acceleration on the upper conductor. This motion causes the ice to break into very small particles that are carried away by the adjacent airflow. This type of system is attractive for small aircraft, but it is not easy to install it on already existing wing structures [54,55]. On the other hand, there is also the electro impulse deicing system, which uses high-voltage capacitors that can be rapidly discharged by an electromagnetic coil, which in turn induces strong eddy currents in the metal surface that should be protected [56–59]. As a result, strong opposing electromagnetic forces are generated between the actuator coil and the metal surface, which is rapidly accelerated. This motion leads to the disruption of the ice and consequent shedding by the adjacent air stream [60]. The main disadvantage of this system is the electromagnetic interference and structural fatigue caused by its operation.

Another electrically operated ice protection system uses heating tapes made of thin graphite foil. This is a fast system that, when activated, instantaneously increases the temperature of the tape and melts the ice accreted on its surface, which is then carried away by the air flow. This system is very interesting because the thin graphite foil can be easily applied to the surface of various aircraft or wind turbine components. However, it has currently not been certified in aircraft flying under icing conditions [61,62].

The thermal air bleed anti-icing system is an additional ice protection system used primarily in modern larger jet aircraft and fixed wing transport aircraft [63,64]. In these systems, hot air is bled off the engine and transported by tubes in order to be exhausted through small holes in the lower surface of the wing. This hot air increases the surface temperature to a value above the freezing point of water in order to melt or evaporate the ice. In turbine-powered air vehicles, the hot air is usually extracted from the compressor section of the engine [65]. In contrast, for turbocharged and piston-powered aircraft, the bleed air is taken from the turbocharger. The work of Domingos et al. [66] is an example of a study that analyzed the implementation of a bleed air ice protection system. A few works have also reported the possible application of loop heat pipes (LHP) as an ice protection system. These loop heat pipes are highly efficient two-phase heat transfer systems that allow considerable amounts of heat to be transported over long distances. Studies have demonstrated their anti-icing ability and shown that they can be efficiently used as ice protection systems [67].

Another type of ice protection system is based on chemical deicing devices. In these systems, chemical antifreeze is pumped through small orifices on the surface to be protected to prevent ice accretion. Propylene glycol and ethylene glycol alcohol are examples of chemicals that can be used for this purpose. These components allow a protective layer to form on the surface of aircraft or wind turbine components that prevents the water from freezing at temperatures lower than the freezing point of water, thus delaying the possibility of ice formation and accumulation. However, the effectiveness of these systems in preventing ice formation is limited by the properties of the chemical and the prevailing weather conditions. In addition, under certain flow conditions, the rheology of the fluid may change and the thickness of the chemical layer may become too thin. In addition, the environmental impact of chemical deicing agents is still a major concern. Propylene glycol and ethylene glycol, although supposedly biodegradable, can pollute water resources, and they exert high levels of biochemical oxygen, which can adversely affect aquatic life. In addition, the system and liquid container are usually heavy and, thus, increase the weight of the aircraft [68–70].

All of the above-mentioned are systems that actively control ice formation and can be activated or deactivated. Another class of ice protection systems is the passive systems, which generally include techniques to modify the physicochemical properties of the surface to make it more difficult for ice to adhere to the surface [71]. The hydrophobic or super-hydrophobic coatings represent a passive ice protection system. These coatings have a high degree of water resistance and a self-cleaning effect that can repel water, making it difficult or impossible for ice to form [72]. Another type of passive ice protection system is the use of surfaces with micro- and nanoscale rough structure. This type of surface usually reduces the time that the water remains in contact with the surface to a period that is less than the time it needs for the water droplet to come into contact with the frozen material to freeze and adhere [73]. The implementation of icephobic coatings is also a topic that has been investigated as a new type of ice protection system. Icephobic surfaces are different from hydrophobic or superhydrophobic surfaces and do not require any special treatment or chemicals, but instead use icephobic materials such as carbon nanotubes and slippery liquid infused porous surfaces. Huang et al. [39] performed a study on icephobic coatings and the possibility of using them in a hybrid coating and active ice mitigation system. These icephobic and hydrophobic coatings can also be combined with electro-thermal and auxiliary photo-thermal performances as passive ice protection systems [74,75]. The schematic shown in Figure 4 illustrates this possible combination. Solar anti-icing/deicing surfaces based on photothermal materials constitute a recent method for ice protection systems. This type of surface has grabbed the attention of several researchers that recently have focused their works on the development of photothermal materials which can be used for solar anti-icing/frosting surfaces (SASs) [76,77]. These surfaces can absorb sunlight efficiently and convert the absorbed solar light energy into thermal energy, which, in turn, can be used for delaying or preventing ice formation [78]. Since solar energy is pollution free and a renewable energy source available in most areas, the use of photothermal materials for solar icing mitigation surfaces is a sustainable and environmentally friendly approach [79]. In addition, this method is associated with superior surface adaptability, long-term durability, and low-cost, which makes it a promising and beneficial option for a wide range of anti-icing applications [80]. Carbon-based, plasmonic-metal-based, and semiconductor-based methods are some of the most relevant approaches reported in the literature for SASs. The most studied photothermal conversion materials include carbon materials, conjugated polymers, two-dimensional nano-structural materials, and metallic particles [78]. The main disadvantages of this technique include the fact that solar radiation is inherently intermittent and, due to weather changes, it might become hard to receive sunlight even during the daytime. In addition, most of the reported SASs cannot remove the condensed water effectively, which significantly enhances reflectance and leads to reduced photothermal efficiency and decreased temperature [78,79]. Considering this, it is highly important to understand the phase transitions of water molecules and the nucleation and growth process of crystals during the water solidification process, and solar interfacial evaporation can be considered in order to promote the water evaporation from the surface. For that purpose, solar interfacial evaporation methods can be used in order to photothermally concentrate the solar radiation absorbed by the surface, increase the surface temperature, and promote the vapor generation [81].

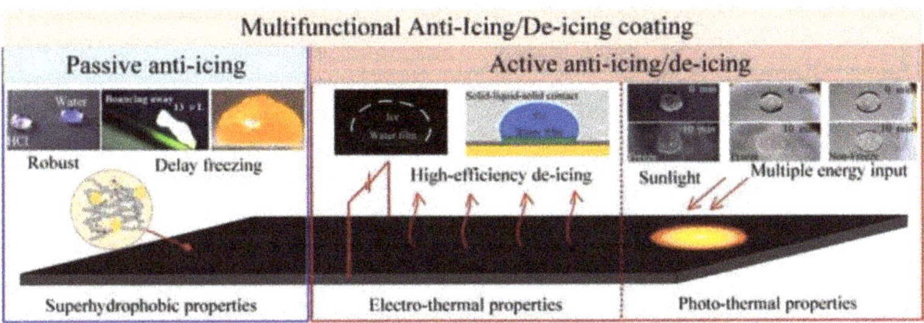

Figure 4. Combination of superhydrophobic coatings with electrothermal and photothermal properties (reproduced from Zhao et al. [74] with permissions from Elsevier).

As we have seen up to now, the existent ice protection systems still present some issues. In addition, the existent ice protection techniques only present the functionality of deicing and/or anti-icing. This means that, when the weather changes and there are no favorable icing conditions, the ice protection system becomes useless and adds weight to the wind turbine blade. Therefore, in the current work, we focus on the use of DBD plasma actuators as ice protection devices. The main advantage of using DBD plasma actuators instead of the existent deicing technologies consists of the fact that they allow deicing and ice prevention accumulation to be performed when required, and when it is not necessary to perform deicing, the plasma actuator can be operated in active flow control mode, improving the wind turbine's efficiency. In addition, we should emphasize that although there are some existing deicing technologies, such as superhydrophobic coatings or solar anti-icing/deicing surfaces, that do not lead to a significant weight increase, several other technologies, such as pneumatic deicing boots, electro-mechanical ice removal devices, or chemical deicing systems, are clearly associated with a considerable undesirable weight increase. Another advantage of plasma actuators is that they are very light, and the operation and control of plasma actuators can be achieved with circuit boards, including a full bridge converter and a transformer cascade board, that do not represent a considerable weight increase for the system.

2.2. Traditional Ice Sensors Used in Ice Protection Systems

Ice sensors can be integrated into anti/deicing systems to provide sufficient information to effectively operate ice mitigation devices [82]. In most aircraft ice detection systems, sensors cannot be placed directly on the airfoil surfaces that most need to be kept free. The addition of a protruding sensor would compromise the aerodynamics of the aircraft. While various attempts have been made to fabricate ice detectors, they have been limited by their accuracy, by their inability to distinguish between ice and water [83], and by their inability to measure the thickness of the ice.

Current existing ice detection methods can be divided into direct and indirect methods of ice detection [21]. Indirect methods use some weather conditions such as temperature, humidity, and visibility in conjunction with empirical deterministic models to predict icing events. These types of instruments are more expensive and not entirely reliable. Direct ice detection methods, on the other hand, measure the change in a property such as mass, conductivity, inductance, and dielectric constant to measure ice formation on the surface [84]. A direct ice detection method is based on ultrasonic damping [85]. The principle of this method is based on measuring the change in the transmission of sound waves. Some similar methods use piezoelectric sensors. This type of sensor detects the presence of ice by analyzing the wave packet energy of the signal. Some other direct ice detection methods use the principle of vibration to measure the presence of ice. These methods use a vibrating probe or a vibrating diaphragm. When ice forms on the probe or diaphragm, the vibration frequency of these devices changes [86]. Some ice sensing techniques are also based on measuring the electrical charge change on the ice surface. When ice is present on a surface, the capacitance of the surface changes, which can be measured as an indication of the presence of ice [87,88]. An illustration of the implementation of capacitive ice sensors can be seen in Figure 5.

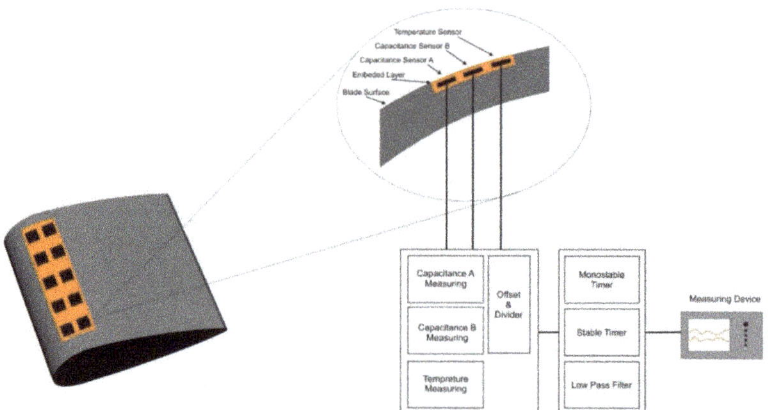

Figure 5. Schematic diagram of the implementation of capacitive ice sensing (reproduced with permission from Madi et al. [89]).

In another technique, temperature sensors are used in the stack. In this device, one temperature sensor is kept at room temperature and another at icing. Both sensors are heated, and the difference in the change in sensor temperature over time signals the presence of ice. The other method of detecting ice is based on an optical measurement. The principle of this technique is based on changing the optical properties of the surface, such as the reflection of light and the change in emissivity. As we have seen in this subsection, different techniques can be used for ice detection. However, these techniques still have some drawbacks, which are summarized in Table 1. This explains the need for new optimized technologies, such as DBD plasma sensors/actuators, for ice protection systems.

As we can see, the conventional ice protection systems require the additional implementation of ice sensors in order to detect the presence of ice and operate the deicing device efficiently. Considering this, an additional advantage of plasma actuators against the existent ice protection methods consists of the fact that plasma actuators can simultaneously operate as an ice sensor and deicing system and, thus, with the same device, it is possible to detect the presence of ice and then remove it.

Table 1. Summary of the main disadvantages of the different ice sensing techniques.

Drawbacks of Various Conventional Ice Sensing Techniques	
Ultrasonic damping	Lack of practical application experience; feasibility under practical conditions not proven.
Piezoelectric sensor	Degradation of aerodynamic performance of blade surface, associated with significant measurement error.
Resonance frequency measurement	Associated with large measurement error that does not allow accurate determination of ice accumulation. In addition, ice accumulation detection is affected by the shape of the surface and the velocity of the object to which it is applied.
Vibration diaphragm	Lack of practical application.
Electrical change	Allows only monitoring of icing conditions in the vicinity of the instrument. Monitoring instruments can affect the aerodynamic performance of the rotor blades.
Temperature change	Unable to detect ice formation on the surface of the blades in a timely manner.
Optical measurement technology	Significant deviation between the calculation result and the actual situation, the observation period is limited, the ice accumulation may change the position and shape of the projection aperture, and it is difficult to use during the day. In addition, the installation of the light source may affect the aerodynamic performance of the surface.

3. Dielectric Barrier Discharge Plasma Actuators for Ice Mitigation

Dielectric barrier discharge plasma actuators are very simple electronic devices that have attracted great interest from the scientific community in recent years [90–92]. These devices are capable of imparting a momentum to the adjacent flow that can be used to manipulate the surrounding flow field. For this reason, they have been studied primarily for active flow control. The conventional configuration shown in Figure 6 is based on two electrodes placed asymmetrically and separated by a dielectric layer. One of the electrodes is located on top of the dielectric layer, while the other is located at the bottom. As a result, the first electrode is fully exposed to the adjacent atmosphere and is referred to as the exposed electrode, while the second electrode is embedded in the dielectric layer and is commonly referred to as the embedded, covered, or encapsulated electrode. The embedded electrode is electrically isolated from the exposed electrode and the surrounding atmosphere and is connected to ground. The exposed electrode, on the other hand, is excited with a sinusoidal voltage signal with a frequency on the order of kilohertz [93–95]. The dielectric layer can be made of any dielectric material with good insulating properties and high dielectric strength [96,97].

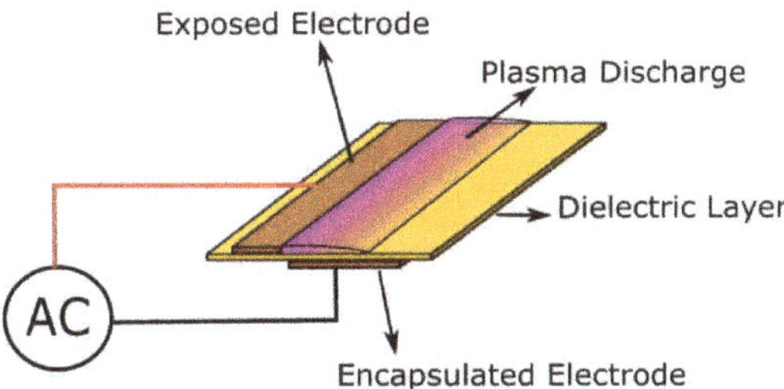

Figure 6. Schematic of the conventional surface dielectric barrier discharge plasma actuator (reproduced from Rodrigues et al. [98] with permission of Elsevier).

When the actuator is supplied with a sufficiently high voltage amplitude, the adjacent layer is ionized due to the fast electron movement between the exposed electrode and the dielectric surface. These electrically charged particles are accelerated downstream in the presence of the strong electric field and exert an impulse on the adjacent neutral air molecules. As a result, an ionic wind is generated, tangential to the surface to which the actuator is attached, which can be used to influence the adjacent air flow field [99,100]. In other words, the generation of the plasma discharge exerts a body force on the adjacent air, which is pulled toward the surface and accelerated downstream, tangentially to the actuator surface. The performance of plasma actuators can be affected by various factors and parameters. Various works have been performed in order to improve the actuator's performance by, for example, changing its configuration or the type of dielectric material [101,102]. The most important parameters affecting the performance of the DBD plasma actuator are summarized in Table 2.

Table 2. Summary of the main parameters that influence the DBD plasma actuators performance.

Parameters That Influence the Plasma Actuator Performance	
Input signal characteristics	➢ Voltage amplitude (1–80 kV$_{pp}$) [103,104] ➢ Frequency (1–60 kHz) [105,106] ➢ Waveform type (sinusoidal, quadratic, or triangular) [107]
Geometrical parameters	➢ Exposed electrode width (1–10 mm) [92,108] ➢ Embedded electrode width (8–20 mm) [109,110] ➢ Gap between the electrodes (0–3 mm) [109] ➢ Dielectric thickness (0.3–4 mm) [110,111]
Dielectric materials	➢ Kapton [112] ➢ Teflon [109] ➢ PMMA [110] ➢ PIB rubber [111] ➢ Macor [113] ➢ Cirlex [92] ➢ PVDF [114]
Actuator configuration	➢ Single DBD Plasma actuator [115] ➢ Micro DBD Plasma actuator [108] ➢ Nano second pulsed plasma actuator [116] ➢ Multiple encapsulated electrode actuator [112] ➢ Sliding DBD plasma actuator [110] ➢ Stair-shaped DBD plasma actuator [117] ➢ Segmented electrode plasma actuator [118] ➢ Plasma heat knife actuator [119] ➢ Plasma synthetic jet actuator (linear or annular) [120,121] ➢ Curved plasma actuators (horseshoe or serpentine) [122,123]

One of the reasons these devices have become so popular in aviation is that they are fully electronic and do not use moving mechanical parts, which usually add significant weight to the aircraft. They are also instantaneous in response, very lightweight, robust, easy to manufacture and implement, and consume little power [124–126]. In addition, several numerical approaches are now available that allow investigation under conditions that are difficult to reproduce experimentally [127–131]. Considering all these features, plasma actuators have been studied over the years for a variety of applications in the fields of active flow control and heat transfer. Table 3 provides an overview of the possible applications of DBD plasma actuators.

Initially, DBD plasma actuators were investigated only for active flow control applications. However, over the years, several authors realized that these devices produce significant thermal effects and began to investigate the thermal properties of these devices. After realizing the potential of plasma actuators to produce thermal effects, several authors began to focus their research on the use of DBD plasma devices for anti-icing or deicing purposes. In addition, since plasma actuators are capacitive devices, the authors discovered the possibility of using them as ice detection sensors. In the following subsections, the various works carried out considering these new functionalities of DBD plasma actuators are evaluated.

Table 3. Summary of the possible applications of DBD plasma actuators.

DBD Plasma Actuators' Applications	
Active Flow Control Field	➢ Flow separation control [132–134] ➢ Wake control [135] ➢ Aircraft noise reduction [136,137] ➢ Modification of velocity fluctuations [138–140] ➢ Drag reduction [94] ➢ Lift coefficient enhancement [103,141] ➢ Flow boundary layer modification [142] ➢ Turbulence reduction [143,144]
Heat Transfer Field	➢ Film cooling efficiency enhancement [145,146] ➢ Surface cooling [147] ➢ Deicing and anti-icing [28,148,149] ➢ Ice sensing [30,111]

3.1. Thermal Effects Induced by Dielectric Barrier Discharge Plasma Actuators

Although most of the initial studies mainly considered the aerodynamic plasma effect and neglected the heat dissipation phenomenon, we now know that a large percentage of the power applied to the DBD plasma device is dissipated in the form of heat. In fact, only a very small fraction of the power applied to the actuator is transferred to the adjacent air as kinetic energy. As explained by Kriegseis et al. [150] and Roth et al. [151,152], and illustrated in Figure 7, the total power delivered to the DBD plasma actuator is composed of the reactive power, the power consumed in dielectric heating, the power expended to maintain and promote the plasma discharge, and, finally, the fluid mechanic power delivered to accelerate the adjacent air.

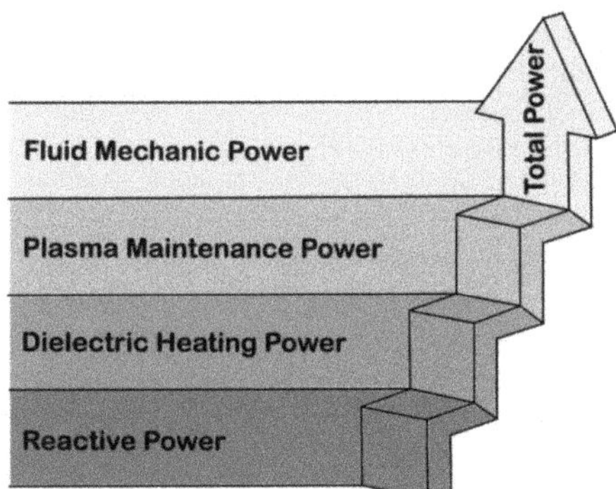

Figure 7. Power-flow diagram of a dielectric barrier discharge.

As we know, the reactive power is the complex power that represents the energy stored and retrieved by the load. Thus, it consists of the power that is continuously absorbed by the actuator and returned to the energy source. When dielectric materials are under the action of direct current, the DC conduction current is negligible if they are good insulators. However, this is not the case when a dielectric is under the action of an alternating electric field, as is the case with DBD plasma actuators. When an appreciable AC current is in phase with the electric field, part of the power is dissipated as dielectric heating due to the phenomenon of dielectric hysteresis, which is similar to hysteresis in ferromagnetic materials [153,154]. Due to this effect, the dielectric layer of the plasma actuator can consume a considerable amount of heat, usually referred to as dielectric power dissipation, which can be estimated as follows:

$$P_l = U^2 \frac{2\pi f A}{d} \varepsilon_R \varepsilon_0 \tan(\delta). \tag{1}$$

where P_l is the power loss in the dielectric materials, U is the voltage, f is the frequency, A is the area, d is the distance of the electrodes, meaning that in a plasma actuator will be given by the dielectric layer thickness, ε_R is the relative permittivity of the dielectric, ε_0 is the permittivity of vacuum, and $\tan g(\delta)$ is the dielectric loss factor.

During plasma discharge, electrons move from the exposed electrode to the di-electric surface in the first half of the AC cycle and back in the next half. During this movement, the electrodes collide with the neighboring particles and generate two plasma micro-discharges per AC cycle [155,156]. Due to this repetitive motion, a significant percentage of energy is also dissipated by elastic collisions of electrons, vibrational excitations, collisions between ions and neutral molecules, and thermal energy transferred from electrons to neutral particles [98,157]. This power consists of the power required to continuously maintain the plasma discharge. Finally, some of the power is transferred to the adjacent air as fluid mechanical power, which imparts momentum to the adjacent particles, accelerating them downstream in a tangential direction toward the surface. In general, this power is quantified by considering it as the equilibrium of the flow rate of the kinetic energy density [158–160] and is expressed by:

$$P_m = \int_0^\infty \frac{1}{2} \rho v(Y)^3 l dY. \tag{2}$$

where P_m is the mechanical power of the fluid, ρ is the air density, $v(Y)$ is the velocity profile, and l the length of the plasma discharge.

The works of Roth et al. [109,151] can be considered the first works to show the various sources of power loss, with emphasis on dielectric heating. Later, in the study of Dong et al. [161], although the focus of the work was not on a thermal analysis of DBD plasma actuators but on the study of the aerodynamic performance of these devices under subsonic flow, the authors calculated the energy lost in the dielectric and experimentally measured the vibrational and rotational temperatures of the plasma using spectroscopic emission measurements. In the same year, Jukes et al. [162] studied the jet flow induced by dielectric barrier discharge plasma actuators and, additionally, applied thermal imaging techniques to estimate the surface temperature of the dielectric material due to plasma operation. By analyzing the results obtained and considering various assumptions, an analytical formula for estimating the plasma gas temperature was derived. However, in a later work by Joussot et al. [163], it was shown that the formula derived by Jukes et al. [162] did not provide sufficiently accurate results for the plasma gas temperature. Joussot et al. [163] performed thermal infrared measurements to determine the dielectric surface temperature when the plasma was turned on and after the discharge was turned off. Similar to the work of Jukes et al. [162], but with much greater emphasis on analyzing the gas temperature distribution, Stanfield et al. [164] experimentally determined the rotational and vibrational temperature distributions as a function of voltage. They concluded that the rotational temperatures for N_2 and N_2^+ decrease in the induced flow direction and increase with the increase in the applied voltage.

Later, various authors began to focus more on the use of infrared thermal cameras to characterize the surface temperature of plasma actuators in order to analyze in depth the thermal effects of DBD plasma actuators. Tirumala et al. [165] performed infrared temperature measurements for plasma actuators with different dielectric thicknesses and for different glow states provided by different input voltage waveforms. In turn, Rodrigues et al. [166] experimentally investigated the influence of dielectric thickness and type of dielectric material on the surface temperature field of plasma actuators operated at different voltages. In a follow-up work, Rodrigues et al. [167] additionally studied the influence of an external flow on the surface temperature field of plasma actuators with different dielectric thicknesses and different dielectric materials.

A typical surface temperature contour of a DBD plasma actuator is shown in Figure 8. This figure shows the surface temperature field of a plasma actuator operated at quiescent conditions and under the influence of an external flow. As can be seen in this figure, the temperature of the top surface of the plasma actuator increases significantly due to the plasma discharge. The highest temperatures are measured at the onset of plasma formation near the exposed electrode. At quiescent conditions, the area of the exposed electrode has a similarly high temperature as the area where the plasma starts, due to the good thermal conductivity of the material of the exposed electrode. Under the influence of an external flow, however, the temperature values decrease along the entire actuator surface, with the area of the exposed electrode showing a significantly greater decrease.

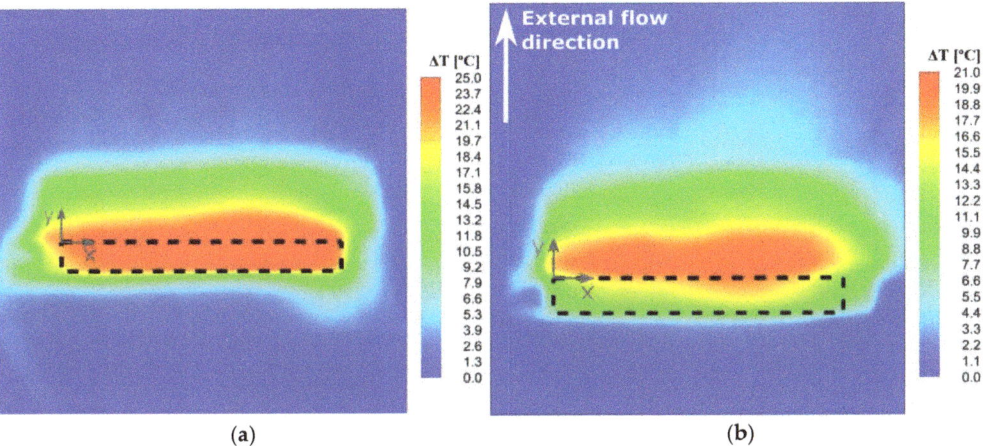

(a)   (b)

Figure 8. Typical surface temperature field of a dielectric barrier discharge plasma actuator with a dielectric thickness made of Kapton, 1 mm thick, operating at 8 kVpp and 24 kHz: (**a**) quiescent conditions (**b**) under external flow influence (reproduced from Rodrigues et al. [167] with permission of Elsevier).

In addition, Figure 9 shows the spatial temperature variation along the top surface of the plasma actuator for different input voltage levels. As we can see, the temperature along the x-axis is higher and almost constant between $x/l = 0$ and $x/l = 1$, which corresponds to the region of plasma formation. The temperature increases with increasing applied voltage, and at higher voltages, the temperature profile starts to oscillate more, which means that the plasma discharge becomes more and more filamentary. In the y-direction, the highest temperatures are found near the exposed electrode edge and decrease in the y-direction.

Later, Abbasi et al. [168] studied the thermal properties of plasma actuators in a turbulent boundary layer using infrared thermography and presented transient thermal results. They also investigated the effect of duty cycle on the thermal properties of plasma actuators and concluded that they exhibit strong oscillatory temperature variations at low frequencies in the range of 5–10 Hz, while at higher frequencies, a similar effect to continuous operation occurs. Recently, Kaneko et al. [169] also studied the thermal effects of plasma actuators using infrared thermal imaging to understand the effects of the shape of the exposed electrode. They demonstrated that the topology of the discharge differs between plate and wire electrodes, and, therefore, the surface temperature fields also show differences. The surface temperature measurements and thermal property analysis performed in the above studies have shown that plasma actuators cause a significant increase in the surface temperature of the dielectric. The temperatures determined in these studies are significantly higher than the solidification temperature of water, demonstrating the potential of plasma actuators to prevent ice formation and accumulation on the surface on which they are used.

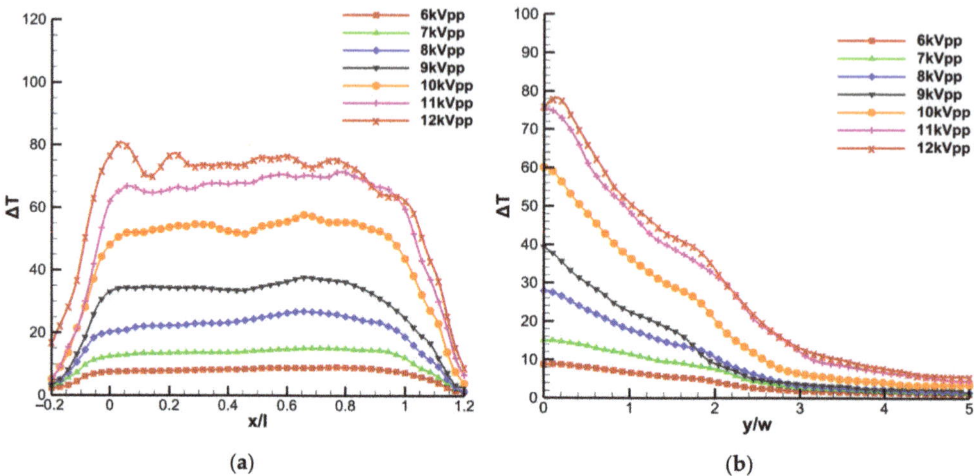

Figure 9. Spatial temperature variation along the *x*-axis and *y*-axis for a plasma actuator made of Kapton, 1 mm thick, operating at different voltage levels and 24 kHz: (**a**) variation along the *x*-axis and (**b**) variation along the *y*-axis (reproduced from Rodrigues et al. [166]).

Due to the nature of the plasma discharge, which results from the use of high voltage and the generation of a high electric field, several conventional measurement techniques, such as thermocouples, cannot be used directly because of the strong electromagnetic interference. In view of this, Rodrigues et al. [98] proposed a new calorimetric technique to quantify the thermal power generated by DBD plasma devices. In this technique, a thermally isolated duct with a fan at the inlet is used to generate a constant axial airflow. The plasma actuator is placed in the airflow calorimeter and the temperature is measured without plasma discharge and after a certain operating time with plasma discharge. Using the basic calorimetric law and applying it to this particular case, the thermal power generated by the dielectric barrier discharge can be estimated by the following equation:

$$P_t = \rho \Phi C_p \Delta T. \tag{3}$$

where P_t is the thermal power generated by the plasma device, ρ is the air density, Φ is the air flow rate, C_p is the specific heat of the air at constant pressure, and ΔT the temperature variation in the air due to the plasma discharge operation. This experimental technique was used in the works of Rodrigues et al. [98,114] to estimate the thermal power generated by DBD plasma actuators made of different dielectric materials and different dielectric layer thicknesses. In addition, they also estimated the dielectric power loss using the analytical formula previously published in Roth et al. [151]. Figure 10 shows the ratio of active power, thermal power, and dielectric power loss for actuators made of 0.3 mm Kapton and 1.02 mm Kapton operated at different voltages.

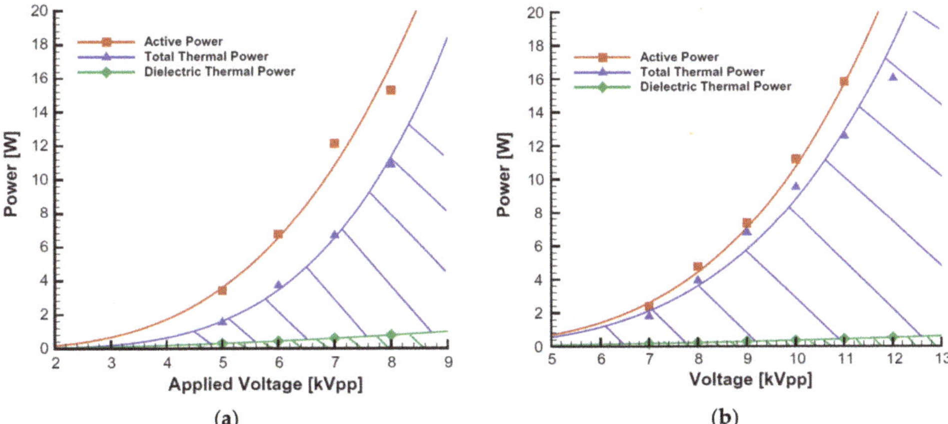

Figure 10. Comparison between active power, total thermal power generated, and power dissipated as dielectric heating in plasma actuators operating at different voltage levels and 24 kHz: (**a**) actuator made of Kapton with 0.3 mm thickness and (**b**) actuator made of Kapton with 1.02 mm thickness (reproduced from Rodrigues et al. [98]).

Figure 10 shows that plasma actuators dissipate a large percentage of power as thermal energy, regardless of dielectric thickness. Rodrigues et al. [114] studied the thermal power generated by actuators with different dielectric materials and concluded that depending on the dielectric material, 60% to 95% of the power applied to the actuator is released as thermal energy. This proves once again the potential of plasma actuators to prevent ice formation or promote local ice melt. Most studies dealing with the thermal effects of dielectric barrier discharge have been performed experimentally, but some numerical studies can also be found in the literature. Aberoumand et al. [170] numerically investigated the effects of various DBD plasma actuator arrangements on the temperature field in a channel flow, and Benmoussa et al. [171] numerically investigated the phenomenon of gas heating in dielectric barrier discharges due to the Joule heat effect for Ne–Xe gas mixtures. Recently, Zhang et al. [172] conducted a detailed review of recent developments in the thermal properties of dielectric barrier discharge surface plasmas and demonstrated the potential of these devices to produce significant thermal effects that can be used for deicing and ice prevention applications.

3.2. Plasma Actuators for Deicing and Ice Formation Prevention

Prevention of ice formation or deicing is an extremely important process in various industries. In particular, in the global aviation sector, ice formation on the surface of aircraft causes various types of alarms, ranging from environmental damage, as it affects aerodynamics and leads to an increase in fuel consumption and pollutant emissions, to catastrophic accident situations, as it can cause mechanical and electrical malfunctions [173,174]. Similarly, in the wind energy market, ice accumulation on the rotor blades leads to large loads on the wind tower, causing significant losses in power generation, dangerous ice throw around the turbine, and even structural failure [175]. Considering that a significant percentage of the energy supplied to DBD plasma actuators is converted into heat dissipated from their surface, as described in the previous section [98,114], these devices have been proposed as a viable alternative to prevent ice formation and/or deicing of aerodynamic components in various applications. Recent literature has reported on various configurations of DBD plasma actuators used in different systems and tested under different conditions.

The study by Cai et al. [29] is one of the first experimental reports on the use of DBD plasma actuators for deicing and anti-icing. The authors installed handmade plasma actuators on a cylinder model surface subjected to supercooled flow in a wind tunnel. The

DBD plasma actuators consisted of a 0.33 mm thick Kapton dielectric layer (six 0.056 mm thick stacked) and two 0.03 mm thick asymmetric copper electrodes with 3.3 and 10 mm wide exposed and encapsulated electrodes, respectively. The actuators were coupled to a Teflon cylinder (45 mm diameter and 220 mm length), without spacing, in a multiple actuator configuration covering ≈80 mm of the cylinder length, as shown in Figure 11.

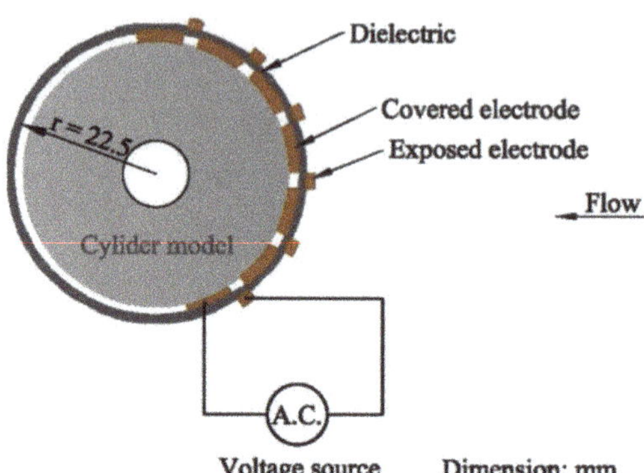

Figure 11. Multi-actuator configuration (adapted from Cai et al. [29] on with the permission of Springer).

The deicing and anti-icing tests were performed in a closed-circuit icing wind tunnel, where the airflow velocity (U) could range from 5 to 18 m/s and the air temperature (T) from −25 to 30 °C. The anti-icing tests, conducted at a wind speed of 15 m/s and with the actuator supplied at 15 kV$_{pp}$ and 13.4 kHz, showed that after 16 min of actuation, no ice accretion occurred on the cylinder surface covered by the DBD, while clear ice formed in the uncovered zone (Figure 12a–c). Deicing tests, conducted after the wind tunnel was operated for 15 min with the spray of supercooled droplets, showed complete removal of the 5 mm thick ice layer after 150 s of plasma actuation at 15 kVpp and 13.4 kHz and at constant airflow velocity and temperature (Figure 12d–f).

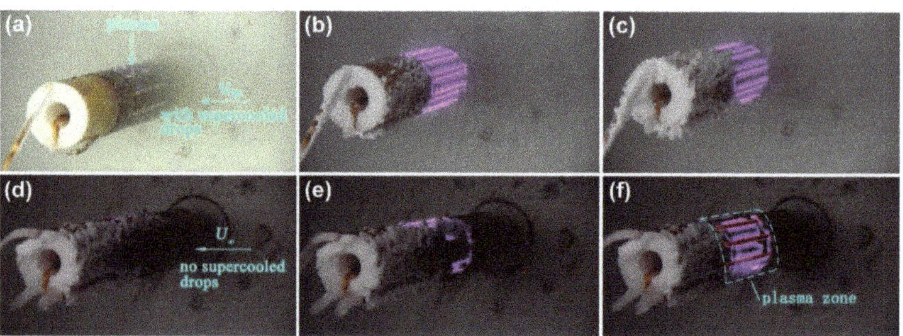

Figure 12. Anti-icing (**a–c**) and deicing (**d–f**) DBD plasma multi-actuator model at V = 15 m/s and T = −10 °C after: (**a**) t = 0 s, (**b**) t = 210 s, (**c**) t = 930 s, (**d**) t = 0 s, (**e**) t = 120 s, (**f**) t = 180 s (adapted from Cai et al. [29] with the permission of Springer).

Liu et al. [149] investigated the use of thermal effects induced by a DBD plasma actuator for aircraft icing mitigation compared to conventional electrical heating. Both systems were installed in parallel in a NACA 0012 airfoil with a chord length of 150 mm (Figure 13a), made of hard-plastic material through 3D printing rapid prototyping, and tested in an icing research tunnel with a wind speed capability of 60 m/s and airflow temperature as low as −25 °C, equipped with atomizing spray nozzles capable of injecting water droplets with mean volume diameter (MVD) ≈20 µm, allowing the liquid water content (LWC) in the tunnel to be adjusted. The DBD plasma actuator consisted of four encapsulated copper electrodes (350 mm length × 10.0 mm width) and five exposed copper electrodes (96 mm length × 3.0 mm width) with a thickness of 70 mm and zero overlap gap, separated by three layers of Kapton 130 µm thick. The four encapsulated electrodes were evenly distributed over ≈27% of the chord length of the airfoil and were spaced 3 mm apart each (Figure 13b).

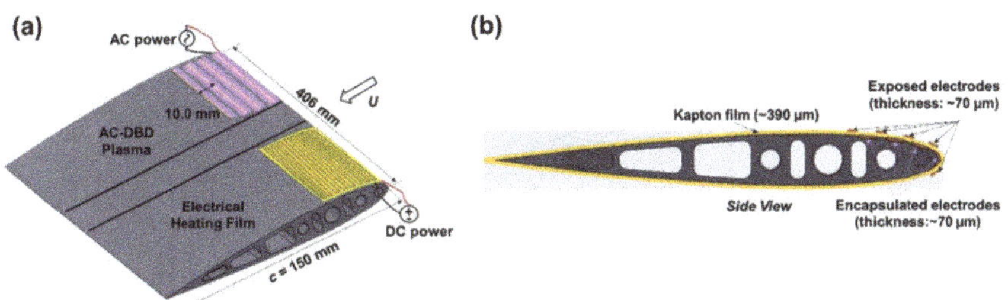

Figure 13. (**a**) Top view and (**b**) side view of NACA 0012 airfoil model with the DBD plasma actuator and electrical film heater side-by-side on the surface: (**a**) top view and (**b**) cross section view (adapted from Liu et al. [149] with the permission of Elsevier).

The first exposed electrode near the leading edge had a smaller width of 5.0 mm to generate more plasma in this area. The conventional electrical film heater consisted of an etched foil element with a thickness of 0.013 mm encapsulated between two layers of 0.05 mm thick Kapton film and 0.025 mm thick FEP adhesive tape over a total area of 50.8 × 101.6 mm. Prior the start of the tests, the icing tunnel was operated at −5 °C for 60 min to ensure a thermal steady state. The tests were performed at U = 40 m/s, T = −5 °C, and LWC = 1.0 g/cm^3. The power density of the two devices was kept constant and equal to 15.6 kW/m^2. The thermal profiles of both systems in operation were obtained through an infrared thermal imaging camera. In the first test phase, the DBD and the electric film heater were turned on for 60 s. After 10 s, the temperature of the DBD electrodes increased by about 10 °C, while the temperature of the dielectric layer (in the spaces between the electrodes) remained below 0 °C (Figure 14a). After 60 s of operation, thermal equilibrium was reached for both devices (Figure 14b). From 60 s (instant t0 in Figure 14c,d), the water spray system was connected, and it was found that after 25 s, the impingement of the supercooled water droplets on the airfoil caused a significant temperature drop in the electric film heater, while the temperature of the DBD plasma actuator dropped only slightly (Figure 14c). After 200 s, the temperature of both devices practically did not change, indicating that thermal equilibrium was reached (Figure 14d), which in turn means that the energy supplied to the devices was sufficient to prevent ice accumulation on the airfoil.

In the context of this study, another report by Liu et al. [176] compared the actuation of DBD plasma actuators with conventional electrical heating using the same experimental setup and conditions as reported in [149]. The reported results are consistent with the previous information. Only after 200 s of operation did ice begin to form at a nearby plasma discharge site. In view of these results, the authors considered that the DBD plasma actuator-based method had a more promising performance compared with the conventional electrical heating method, because the temperature drop on the DBD surface was much lower. This difference was explained by the fact that the water droplets were heated not only by heat conduction but also by heat convection as they moved through the hot air above the DBD (Figure 15).

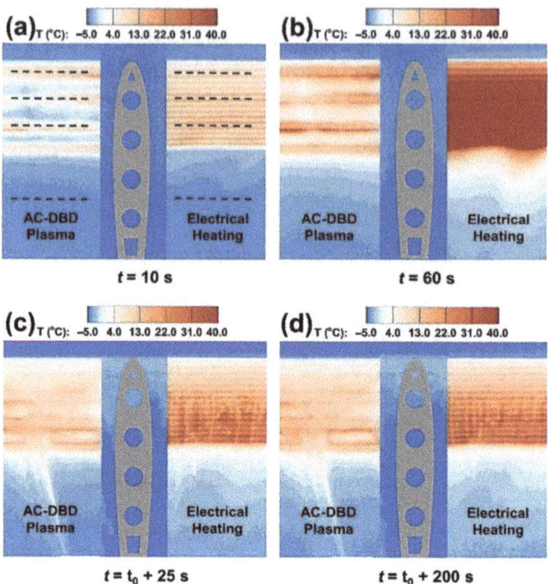

Figure 14. Time evolution of the temperature distributions over the DBD plasma actuator and the electrical film heater airfoil surfaces sides after: (**a**) t = 10 s, (**b**) t = 60 s, (**c**) t = t0 + 25 s, and (**d**) t = t0 + 200 s (adapted from Liu et al. [149] with the permission of Elsevier).

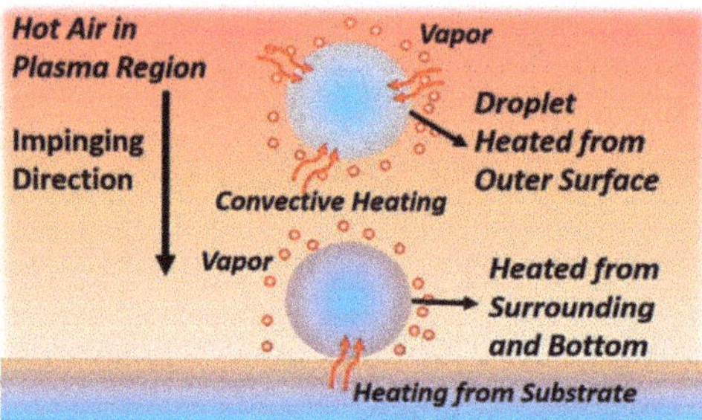

Figure 15. Heating mechanism of an impinging droplet on a DBD plasma actuator surface (adapted from Liu et al. [149] with the permission of Elsevier).

This half–half configuration (similar to [149]) was previously used by Zhou et al. [28] in a study designed to demonstrate the effectiveness of DBD plasma actuators as deicing and anti-icing devices. DBD plasma actuators, based on three layers of Kapton with a thickness of 130 μm and multiple exposed and encapsulated copper electrodes with a thickness of 70 μm, were symmetrically placed in the middle of a NACA 0012 airfoil (half on and half off). The system was exposed to typical glaze ice conditions (U = 40 m/s, T = −5 °C; LWC = 1.5 g/m^3) and rime ice conditions (U = 40 m/s, T = −15 °C; LWC = 1.0 g/m^3). In the half where the DBD plasma actuators were turned off, ice formed shortly after the start of the experiments, while in the half where the DBD were turned on, ice formed only in the region downstream of the covered area. This is a common problem in thermal based anti-icing and deicing systems, for which some authors propose a hybrid solution combining DBD with hydrophobic/icephobic coatings, which will be discussed later.

Chen et al. [177] evaluated the anti-icing performance of a nanosecond surface dielectric barrier discharge (NS-DBD) and investigated the effect of pulse frequency and voltage amplitude on actuation performance. The NS-DBD plasma actuator is a device driven by repetitive high voltage pulses with fast rise times on the order of nanoseconds [178]. The NS-DBD consisted of exposed (2 mm wide) and buried (10 mm wide) electrodes with thickness of 0.027 mm separated by three layers of Kapton tape as a dielectric layer (0.24 mm thick) (Figure 16a). The device was installed on an aluminum airfoil model NACA 0012 with a chord length of 280 mm and a spanwise length of 145 mm, as shown in Figure 16b. The airfoil was covered with a film 2.5 mm thick PTFE (polytetrafluoro-ethylene) to prevent heat transfer between NS-DBD and the airfoil structure.

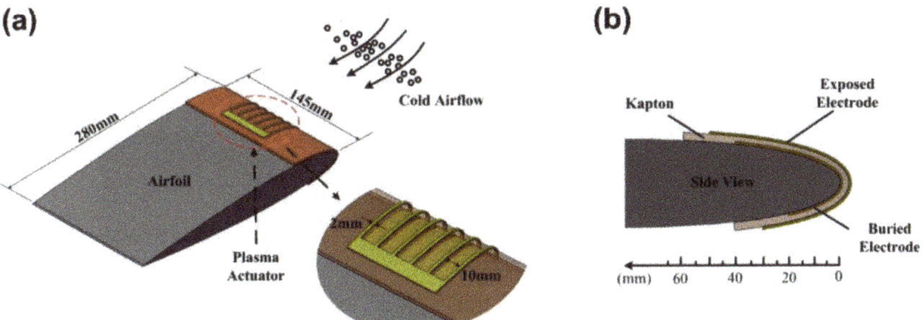

Figure 16. (**a**) Schematic of NACA 0012 airfoil with NS-DBD plasma actuator attached. (**b**) Side view of the airfoil with the NS-DBD (adapted from Chen et al. [177] under the Creative Commons Attribution License).

Experiments were performed in an icing research tunnel under conditions of U = 65 m/s, T = −10 °C, LWC = 0.5 g/cm^3, and MVD = 25 μm. The wind tunnel was turned on 1 h before the experiments to reach a steady state, then, the NS-DBD plasma actuator was turned on for ≈100 s (t = 0 s) to reach thermal equilibrium, and finally, the water sprayer was activated. To understand the effects of the discharge conditions, experiments with higher voltage amplitude and lower pulse frequency (HV-LF) and discharges with lower voltage amplitude with higher pulse frequency (LV-HF) were performed. Dynamic evaluation of anti-icing, with continuous impact of supercooled water droplets, showed that for the same input voltage, the LV-HF discharge performed better than the HV-LF discharge, as shown in Figure 17.

In the LV-HF discharge (Figure 17a), it was found that the supercooled droplets that hit the airfoil on the surface of the NS-DBD melted and flowed backward and, in turn, ice was formed in the areas of the airfoil that were not covered by the plasma actuator. In the discharge HV-LF (Figure 17b), ice accretion was detected even in the NS-DBD zone, first between the electrodes and at the end along the entire leading edge. The authors suggested the existence of a threshold frequency corresponding to the voltage amplitude of actuation signal and the incoming flow condition, which determined the anti-icing performance.

Following NS-DBD, Kolkbair et al. [179] reported the investigation of a hybrid system for deicing and anti-icing based on the combination of a NS-DBD plasma actuator and a superhydrophobic surface (SHS) coating on the airfoil surface. The exposed (≈95 mm wide) and grounded electrodes of the NS-DBD plasma actuator were made of copper tape ≈ 70 μm thick separated by a PVC (polyvinyl chloride) dielectric layer ≈ 0.3 mm thick. The device was coupled to a 3D-printed NACA 0012 airfoil with a chord length of 150 mm, in a configuration similar to that presented in [177], and, finally, was painted with enamel. To test the SHS, some prototypes were sprayed with the superhydrophobic coating Hydrobead on the top of enamel, according to the procedure described in [180]. The hydrophobized surfaces showed a significant increase in the static contact angle from ≈65° to ≈157° for the Enamel and Hydrobead surfaces. The experiments were conducted in an icing research tunnel (the same as in [149]), and the temperature map during ice accretion was obtained using an infrared thermal imaging system. Tests were conducted at U = 40 m/s, T = −5 °C, and LWC = 0.8 g/cm^3, and it was found that ice formed on the leading edge of the airfoil when the actuator was off, despite the hydrophobic treatment. In turn, when the NS-DBD was turned on at V = 14 kV, f = 2 kHz, and P = 175 W/m, the supercooled water droplets formed ice at the leading edge only after 60 s, but the backward-flowing water froze on the airfoil surface and formed rivulets from ≈30 s of experimental time. For the actuation conditions of V = 14 kV, f = 4 kHz, and P = 350 W/m, an ice-free leading edge was observed, but ice rivulets appeared on the back surface after ≈30 s. Finally, at a plasma actuation of V = 14 kV, f = 6 kHz, and P = 525 W/m, there was no ice formation at either the leading edge or at the trailing edge of the airfoil. It should be emphasized that for untreated airfoil surfaces, ice accretion was observed at the leading edge and at the trailing edge for all input conditions tested. The authors concluded that the combination of NS-DBD plasma actuation and SHS coating effectively prevents ice accretion on the entire structure of the airfoil.

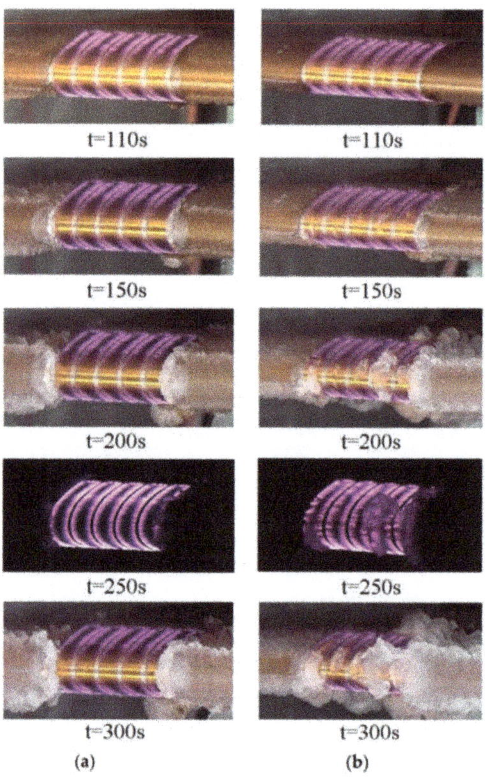

Figure 17. Dynamic anti-icing process of NS-DBD plasma actuator: (**a**) LV-HF discharge and (**b**) HV-LF discharge (adapted from Chen et al. [177] under the Creative Commons Attribution License).

Actuator surface wettability was also addressed in the study by Zheng et al. [181], where the effect of a hydrophobic coating on the dielectric layer of a SDBD plasma actuators was studied. Two types of dielectric layers were prepared, one of ordinary quartz (1 ± 0.01 mm thick) and another of SiC treated quartz glass, both with exposed electrodes (5 × 50 mm) and buried electrodes (15 × 40 mm) in copper foil (0.03 mm thick) arranged in an asymmetric configuration. The measured contact angle of the SiC-treated surface was 119.5°, about five times higher than that of the ordinary quartz glass surface. The static deicing experiments at 10 kV and 6 kHz showed that the SiC-coated actuator melted an area of 7.1 cm² of ice after 60 s of experiment time, while the quartz glass actuator melted 9.2 cm², with the estimated power consumption of the SiC-coated actuator being 7.35 W, while the uncoated quartz glass actuator consumed 14.7 W. Therefore, the efficiency of the

SiC-coated actuator after 60 s was 0.996 cm^2/W, which was 54.31% higher than that of the uncoated actuator (0.626 cm^2/W).

Returning to the scope of the NS-DBD, Liu et al. [182] also evaluated the anti-icing and deicing performance of these devices for in-flight ice mitigation of aircraft. The experiments were conducted in the same icing research tunnel used in [149], and the plasma actuators were fabricated with five layers of PVC as the dielectric layer ($\approx$100 µm thick per layer), one exposed electrode, and nine encapsulated electrodes, all with the same thickness and length of about 70 µm and 125 mm, respectively. The device was coupled to a NACA 0012 airfoil model (150 mm chord length) fabricated by 3D printing from hard plastic in a half–half configuration of the NS-DBD plasma actuator with plasma on versus plasma off, as shown in Figure 18.

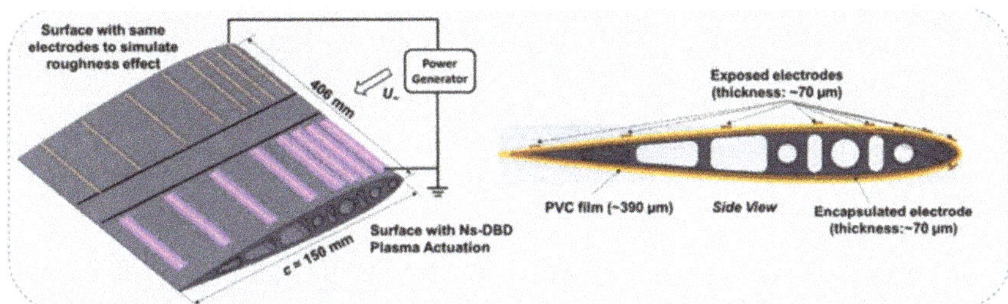

Figure 18. Schematic of the airfoil model used for anti-icing and deicing studies (adapted from Liu et al. (2019) based on [11] with the permission of IOP Publishing).

Ice accretion tests were performed under different temperature (-15 to $-5\,^\circ$C) and frequency (2 to 6 kHz) conditions, keeping U = 40 m/s and LWC = 1.0 g/cm^3 constant. As in the previous study, the authors found that increasing the operating frequency of the NS-DBD plasma actuator significantly improved the anti-icing and deicing performance. As shown in Figure 19a, the supercooled water droplets impinging on the wing surface quickly formed an ice layer around the leading edge at 2 kHz, while almost no ice formed around the leading edge of the wing airfoil at 6 kHz (Figure 19b), largely due to the higher thermal energy generation at higher operating frequency. Regarding the temperature effect, the authors found that the plasma actuation showed better anti-icing and deicing performance at warmer air temperatures of T = $-5\,^\circ$C for the same frequency input signal.

In another study by Liu el al. [183], the use of a DBD plasma actuator operating in duty cycle mode was tested for aircraft ice mitigation and compared with a DBD operating in continuous mode. The authors used a NACA 0012 airfoil, fabricated by 3D printing from hard plastic, as the wing profile, with DBD plasma actuators installed in a half–half configuration, similar to Figure 19 from [182], with each half used for the respective type of actuation. The DBD consisted of four encapsulated electrodes and five exposed electrodes (70 µm thick) separated by three layers of Kapton film (130 µm thick). The exposed electrodes were evenly distributed on both halves of the wing profile with 3 mm spacing. The airfoil was tested in an icing research tunnel at U = 40 m/s, LWC = 1.0 g/cm^3, and T = $-5\,^\circ$C, and it was turned on 60 min (t0) before testing to ensure steady-state conditions. The results showed that using duty cycle mode with a modulation frequency of 1 Hz exhibited better anti-icing performance at the same power input. Figure 20 shows that the DBD operated in duty cycle mode was ice-free by the end of the 140 s of test, while the conventional actuator had an ice film in the plasma zone after about 20 s of testing. The authors also demonstrated that increasing the duty cycle modulation frequency increased the heat dissipation of the DBD, thus improving its anti-icing and deicing capabilities.

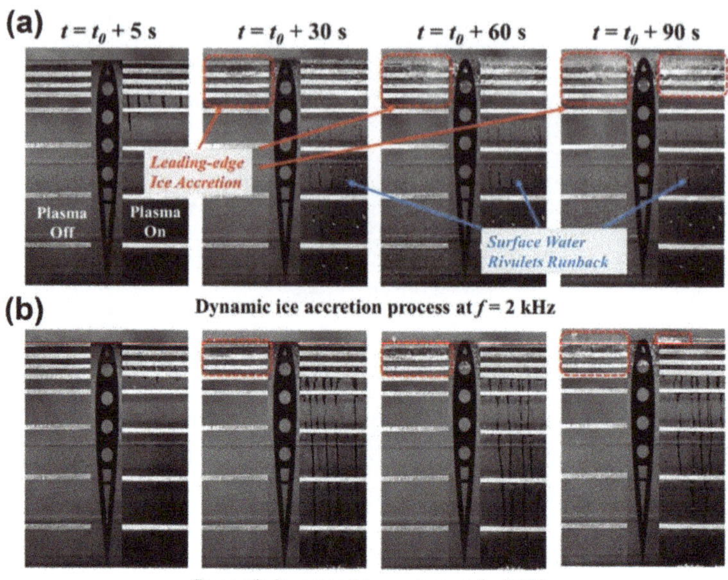

Figure 19. Time-evolution of the dynamic ice accretion experiments over the airfoil surface with the NS-DBD plasma actuator being operated at (**a**) 2 kHz and (**b**) 6 kHz (adapted from Liu et al. [182] with the permission of IOP Publishing).

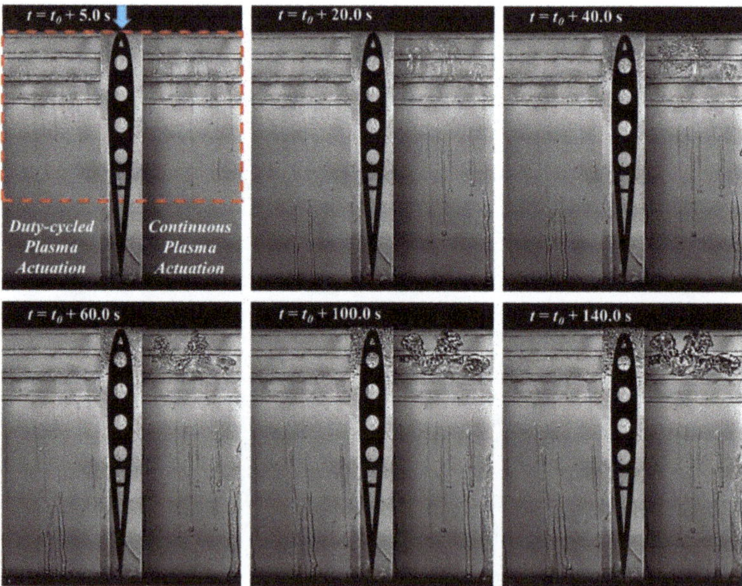

Figure 20. Ice accretion over the airfoil in duty-cycled plasma actuation (left side) versus conventional continuous plasma actuation (right side) (adapted from Liu et al. [183] with the permission of Elsevier).

An interesting anti-icing approach, based on the so-called "heat knife", was proposed by Wei et al. [119]. The authors developed a device based on a series of surface dielectric barrier discharge (SDBD) plasma actuators in a specific configuration, which they called the "stream-wise plasma heat knife" (see Figure 21).

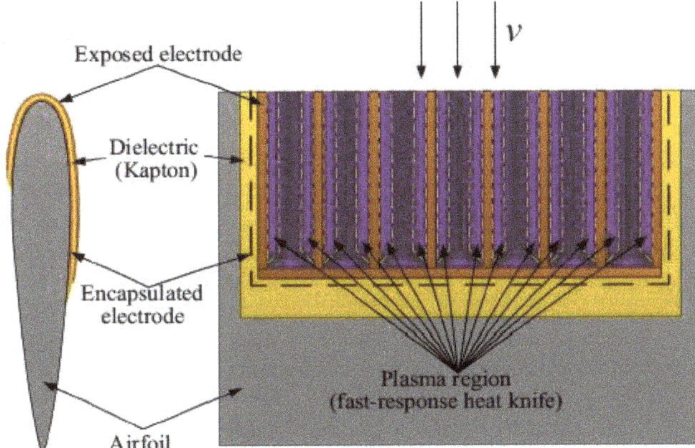

Figure 21. Side and top view of the "stream-wise plasma heat knife" (adapted from Wei et al. [119] with the permission of Elsevier).

According to the authors, in this configuration, when the device was powered by a high-voltage source, the generated plasma rapidly gave off heat around the discharge, allowing a fast thermal response that quickly heated the water droplets impinging on the "stream-wise plasma heat knife". The device was installed in the NACA 0012 airfoil model (280 mm chordwise length and 300 mm spanwise length), wrapped with a 2-mm thick polyimide film for thermal and electrical insulation, and tested in an icing research tunnel at LWC = 0.5 g/cm^3, MVD = 25 μm, U = 65 m/s, and two temperatures of −5 and −15 °C. The "stream-wise plasma heat knife", which consisted of a dielectric layer of 0.15-mm-thick Kapton tape separating the exposed (2-mm-wide) and encapsulated (10-mm-wide) electrodes made of 0.06-mm-thick copper foil, was operated with a nanosecond pulsed power with a peak voltage of 7.7 kV and frequency of 6 kHz. The typical snapshots of the time evolution of the dynamic ice accretion experiments (Figure 22) showed that at T = −5 °C, there was no ice formation in the plasma region after 180 s. At T = −15 °C, a small ice layer appeared between two exposed electrodes after 90 s, but it did not grow further or disappear over time. Under both conditions, ice accumulated in the plasma-free region.

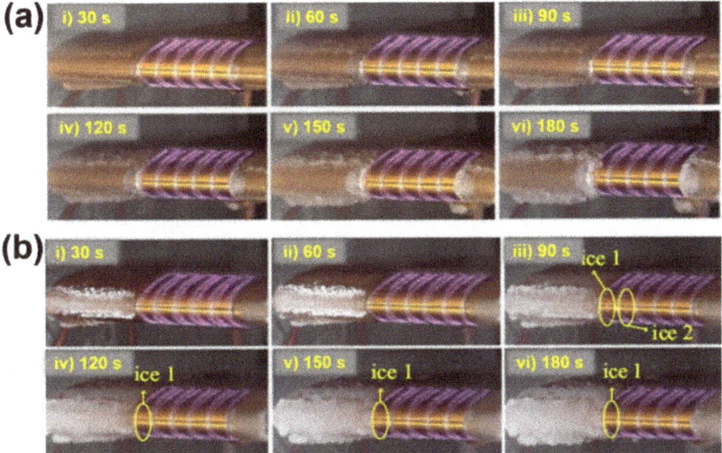

Figure 22. Time evolution of the dynamic anti-icing process at (**a**) T = −5 °C and (**b**) T = −15 °C (adapted from Wei et al. [119] with the permission of Elsevier).

In an identical report, Wei et al. [184] studied the removal of 3 mm thick ice accumulated on the surface of a nanosecond pulsed surface dielectric barrier discharge (nSDBD). The apparatus and experimental setup were similar to that used in [119], and tests were performed under typical glaze icing conditions: LWC = 0.5 g/cm^3, MVD = 25 μm, U = 65 m/s, and T = −5 °C. The nSDBD was connected after 90 s of ice formation on its surface (≈3 mm thick), and it was found that melting of the ice started immediately after the plasma discharge started, and the whole process took about 4 s. Moreover, ice accretion in the plasma region did not repeat during the rest of the experiment. The authors concluded that the thermal effects together with the aerodynamic force contributed to the good deicing performance of the nSDBD plasma actuator.

Fang et al. [185] also discussed the concept of the "heat knife" and proposed the plasma streamwise heat knife approach for anti-icing purposes. Two types of configurations were experimented with the same number of streamwise exposed electrodes (3 mm wide) spaced 10 mm apart and all connected by a spanwise electrode strip (5 mm wide). In configuration 1, the connecting strip separated the streamwise electrodes symmetrically, while in configuration 2, the connecting strip was placed at the end of the streamwise electrodes, as shown in Figure 23.

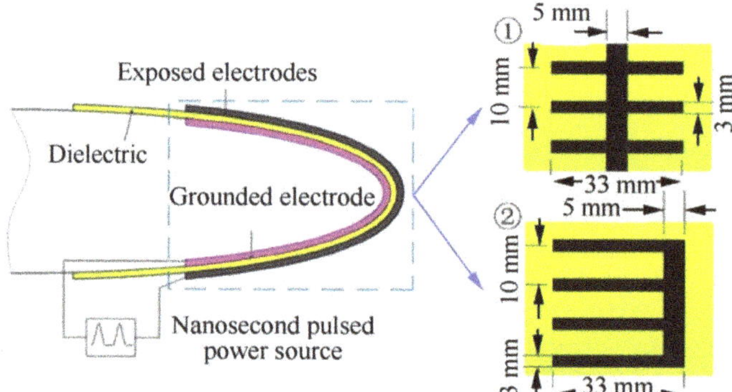

Figure 23. Two different configurations of streamwise plasma heat knife mounted at the leading edge of the airfoil from Su et al. [186] (with permission under a Creative Commons license).

The electrodes were made of a 0.06-mm-thick copper foil and a 0.18-mm-thick dielectric layer of Kapton tape. The devices were attached to the leading edge of a NACA 0012 organic glass airfoil (Figure 23) and tested in an icing wind tunnel with constant ice conditions of U = 20 m/s, T = −15 °C, LWC = 1 g/m^3, and MVD = 40 μm and electrical conditions of 9 kV$_{pp}$ and 6 kHz. Configuration 2 showed better anti-icing performance than configuration 1, because it was ice-free on most of the surface covered by the actuator, unlike configuration 1, where an ice ridge formed on the spanwise electrode. Based on these results, Su et al. [186] proposed a three-level electrode configuration by reformulating configuration 2 to improve heating under severe icing conditions. Anti-icing tests were conducted in an icing wind tunnel (U = 65 m/s, T = −15 °C, LWC = 0.5 g/cm^3, MVD = 25 μm) with a power consumption of 70 W. It was found that the reconfigured three-level heat knife achieved better anti-icing results than the original configuration at the leading edge of the airfoil, but a significant ice ridge was formed at the trailing edge.

Meng et al. [187] compared the anti-icing efficacy of three types of SDBD plasma actuators, each designed for a different type of actuation: type-1 to generate induced flow in the same direction as the incoming flow, type-2 to generate induced flow in the opposite direction of the in-coming flow, and type-3 to generate induced jets in the vertical direction. All actuators consisted of 0.07 mm thick copper electrodes separated by three 0.13 mm thick Kapton layers. The arrangement and size of the electrodes varied depending on the configuration: four exposed electrodes (5 mm wide) and four buried electrodes (5 mm wide) for type-1 and type-2, and five exposed electrodes (3 mm wide) and four buried electrodes (5 mm wide for the first and 10 mm wide for the others) for type-3, as shown in the cross-sectional images of the airfoils shown in Figure 24. The SDBD actuators were coupled to a model of a NACA 0012 airfoil (0.15 m chord length and 0.4 m span length) in a half–half configuration, similar to that used in [182], with two separated plasma on and plasma off zones. The device was tested in an icing wind tunnel under the conditions of U = 40 m/s, LWC = 1.0 g/cm^3, and T = −5 °C. The configuration that induced perpendicular flows (type-3) achieved the best anti-icing performance by ensuring no ice accumulated on the entire underside of the airfoil, while the configurations with induced flows in the same (type-1) and opposite (type-2) directions were 57% and 81% of the airfoil chord length ice-free (Figure 24).

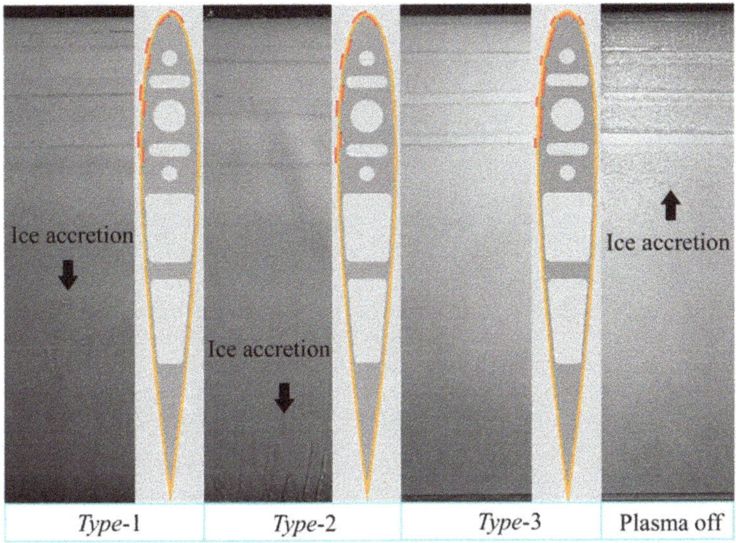

Figure 24. Ice accretion on the airfoil surface after 112 s (adapted from Meng et al. [187] with the permission of AIP Publishing).

Kolbakir et al. [188] also studied and compared the performance of different configurations of anti-icing DBD plasma actuators, arranged in different orientations and varying the number and width of exposed electrodes. The actuators were constructed with a PVC film of ≈400 μm thickness on the dielectric layer and a copper tape of ≈70 μm thickness on the electrodes. The exposed electrodes were all the same length (72 mm) and the width varied from 4.00 to 60.0 mm depending on the test case. The DBD were tested in an icing research tunnel (U = 40 m/s, T = −5 °C, LWC = 1.5 g/m^3) coupled to a NACA 0012 airfoil model (150 mm chord length and 400 mm span length) in streamwise and spanwise layouts. The streamwise layout resulted in higher plasma-induced surface heating and, thus, better anti-icing performance than the spanwise layout. Streamwise actuators prevented ice accretion not only at the leading edge, but also on the trailing edge of the airfoil, as the surface heating delayed the runback ice formation.

Lindner et al. [189] investigated the effect of electrode type and electrode configuration on the anti-icing performance of SDBD actuators. The authors developed SDBD plasma actuators using microelectromechanical systems (MEMS) technology and compared the performance with SDBD fabricated by printed circuit board (PCB) technology. The anti-icing and deicing experiments were conducted in an icing wind tunnel at U = 27 to 50 m/s, T = −18 to −20 °C, LWC = 3 g/m^3, and MVD = 20 μm. Different materials were used for the dielectric layer of the SDBD actuators: FR4 TG135 (500 μm thick) in PCB and zirconia

(150 μm thick) and borofloat glass (500 μm thick) in MEMS. It was concluded that a smaller thickness of MEMS SDBD electrodes (0.3 μm) compared to the 35 μm thickness of PCB SDBD favored the anti-icing performance, as shown in Figure 25. Moreover, the anti-icing effect is independent of the ionic wind generated, so the thickness of the substrate can be reduced. Titanium was also identified as the most suitable material for electrode fabrication. The authors suggested that the density of the electrodes and the thickness of the dielectric layer play a crucial role in the effectiveness of the device.

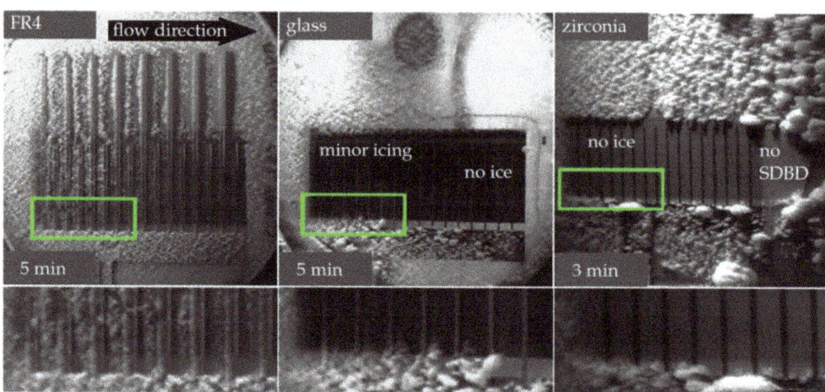

Figure 25. Anti-icing experiments of the three studied materials. Images below depict a magnification of the area marked in green (adapted from Lindner et al. [189] with permission under the Creative Commons Attribution License).

The ice shape modulation method on the leading-edge airfoil was studied by Jia et al. [190] using nSDBD. The actuators were constructed with a dielectric Kapton layer (0.18 mm thick), and exposed (3 mm wide) and encapsulated (5 mm wide) copper foils (0.027 mm thick) electrodes were installed on a NACA 0012 plexiglass airfoil (800 mm spanwise length and 200 mm chord length) and tested in an icing wind tunnel under glaze ice conditions (LWC = 1.5 g/m^3, MVD = 25 μm, T = −5 °C, U = 65 m/s) and under frost ice conditions (LWC = 0.5 g/m^3, MVD = 25 μm, T = −15 °C, U = 65 m/s). Using nSDBD ice shape modulation, the continuous ice at the leading edge of the airfoil was periodically modulated into segmented ice pieces. Under glaze ice conditions, the ice on both sides of the deicing zone began to coalesce after 360 s, while under frost ice conditions, this occurred after 270 s. The authors concluded that the ice shape modulation method can reduce energy consumption by more than 50% compared to full deicing.

Abdollahzadeh et al. [111] developed a parametric optimization of DBD plasma actuators for ice sensing and deicing performance. The DBD were built with a 0.3 mm thick Kapton layer, a 20 mm wide embedded copper electrode, a 5 mm wide exposed copper electrode, and a 0 mm gap between the electrodes. For deicing tests, the DBD surface was covered with an 8 mm thick ice layer, and the actuator was operated at 4 kV$_{pp}$ and 24 kHz (Figure 26a). After 1550 s, the ice layer detached from the DBD surface, and a large hole was formed. The authors considered that the melting process was complete at this point (Figure 26b). Soft ice (6 cm × 10 cm × 1.7 cm, Figure 26c) was also tested in deicing experiments, and it was observed that the frost layer was completely melted after about 780 s (Figure 26d).

Gao et al. [191] presented an innovative plasma synthetic jet actuator (PSJA) concept for ice mitigation, as an alternative to the usual SDBD-based deicing and anti-icing systems. Deicing experiments were conducted at a room temperature of 10 ± 2 °C, and the system was found to be effective in removing free columnar ice layers of 200 mm diameter and 3, 8, and 10 mm thickness in the different failure modes of radial and circumferential crushing, radial and circumferential cracking, and radial cracking. For improving the deicing efficiency of plasma actuators over an airfoil, Hu et al. [192] performed an optimization study for the implementation of plasma actuators on a realistic configuration of the NACA0012 airfoil model and emphasized that the biggest advantage of the AC-SDBD plasma actuator is that it can be simultaneously used for flow control and ice mitigation using the same device, meaning that the actuators can be used for icing control in icing conditions and flow control in the non-icing environment. Recently, Tanaka et al. [193] performed an experimental study about snowfall flow control using a high-durability designed plasma electrode, and Lilley et al. [194] studied the effects of water adhesion from droplets directly sprayed onto a plasma actuator and its plasma glow recovery.

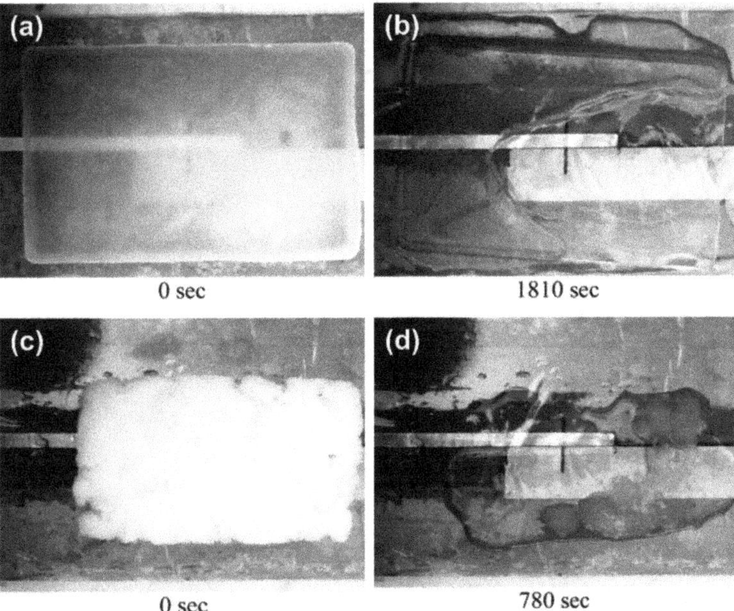

Figure 26. Time evolution of the deicing process of ice (**a,b**) and soft ice (**c,d**) layers (adapted from Abdollahzadeh et al. [111] with permission under Creative Commons license).

Various forms of deicing and anti-icing actuation based on dielectric barrier discharge were presented and discussed. The prevention of ice accretion is becoming a leading application for DBD plasma actuators; however, the fact that these devices are multifunctional and have a prominent application in active flow control to suppress boundary layer separation and/or delay airfoil stall, which improves aerodynamic performance, should not be underestimated. At the same time, they can be used as effective deicing and/or anti-icing devices to prevent ice accumulation on the airfoil surface and ensure safer and more efficient operation [188].

3.3. Ice Sensing by Dielectric Barrier Discharge Plasma Actuators

In addition to the possible use of plasma actuators as active flow control and deicing devices, Abdollahzadeh et al. [195] recently disclosed the possibility of using these devices as ice sensors. In this work, the authors explain that since plasma actuators behave similarly to a capacitor, they can also be operated as capacitive ice sensors and detect the presence of water, air, or ice on their top surface. The properties of water are different in different states of matter, i.e., the dielectric constant of ice is different from that of liquid water. When we apply an alternating current to the plasma actuator, an electric field is generated on the top of the dielectric layer, in the region between the exposed electrode and the covered electrode. This electric field is affected by the medium on top of the dielectric layer; since air, water, and ice are media with different properties and different dielectric permittivity, the presence of water or ice on the surface of the actuator changes the electric field and, consequently, the different electric parameters of the actuator, such as charge or capacitance. Therefore, by monitoring the electrical properties of the DBD plasma actuator, it is possible to detect the presence of air, ice, or water on the top surface of the plasma device and, in this respect, to operate the device as an ice sensor. Abdollahzadeh et al. [30] experimentally investigated the operation of a DBD plasma actuator as a simultaneous deicing and ice sensing device and demonstrated that their electrical properties change significantly when the medium adjacent to the actuator surface is changed. Figure 27 shows the change in the electrical properties of a single DBD plasma actuator due to the change in the properties of the adjacent medium.

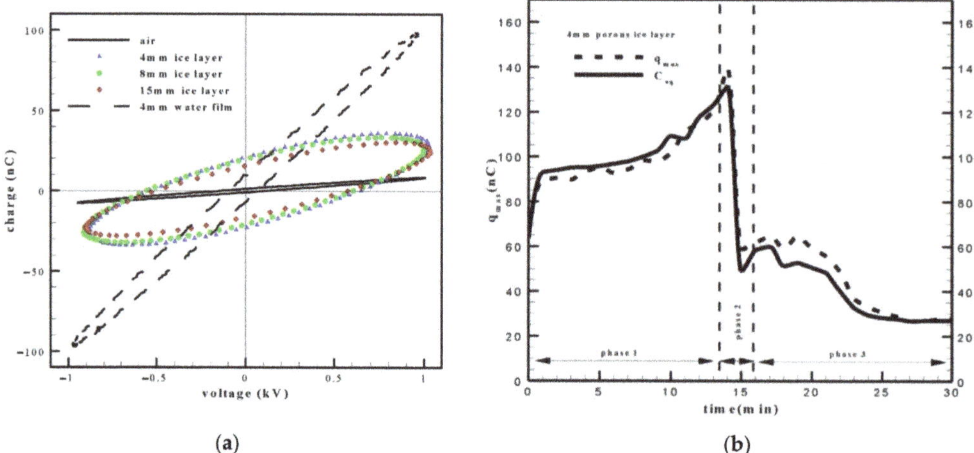

Figure 27. Variation in the plasma actuator electrical properties in the presence of different adjacent medium: (**a**) Variation in the DBD sensor/actuator charge in the presence of 4 mm, 8 mm, and 15 mm thick ice layer and a 4 mm thick water film. (**b**) Temporal variation in the DBD sensor/actuator capacitance and charge in the presence of an ice layer (reproduced from Abdollahzadeh et al. [30] with permissions of Elsevier).

Figure 27a shows that the charge variation along the voltage cycle changes significantly due to the presence of air, ice, or water on the top of the actuator. In the presence of air, the voltage–charge curve is narrower, and the maximum and minimum charges have lower absolute values. The absolute charge values increase in the presence of a layer of ice and become even larger in the presence of a thin film of water. In Figure 27b, we see that DBD plasma actuators can also be used as ice sensors to monitor the melting process of a thin ice layer over time. Figure 27b shows that in the first phase, both the charge and the capacitance increase, which means that the ice is melted and the fraction of liquid water increases with time. In phase 2, a sudden decrease in the values of charge and capacitance is observed, which means that the water between the ice and the actuator has been drained and the amount of ice/water on the actuator's surface has been significantly reduced. In phase 3, a slower decrease in capacitance and charge values is still observed, which means that smaller pieces of ice are still removed. Abdollahzadeh et al. [30] demonstrated not only the potential of plasma actuators for ice sensing, but also that by using networks of multiple plasma actuators, it is possible to detect the position of the ice, since each actuator provides an individual signal. Later, in the study of Rodrigues et al. [118], this possibility was optimized by developing a new plasma actuator configuration characterized by a plasma actuator with multiple encapsulated electrodes that allows for detection of the position of the ice, or where there is a major ice accumulation. This configuration is shown in Figure 28 along with the monitoring voltage signals when the ice is placed on the left segment.

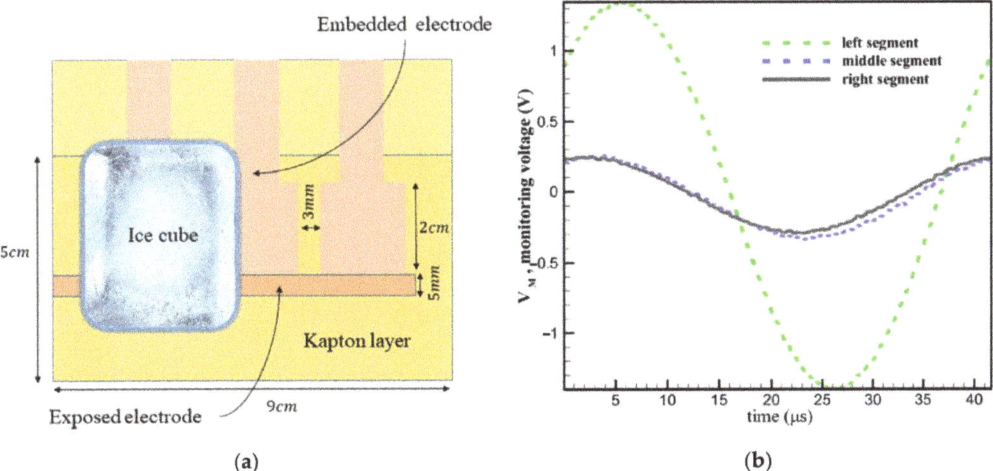

Figure 28. Segmented encapsulated electrode configuration for ice location detection: (**a**) Schematic of the actuator configuration with three segmented electrodes. (**b**) Variation of the monitoring voltage at each segmented electrode with the ice cube on the left side (reproduced from Rodrigues et al. [118]).

Figure 28b shows that when an ice cube is placed over one segment of the actuator while the other segments are kept ice-free, the monitoring voltage of that segment increases significantly, indicating that it is covered with ice. Since the monitoring signal of each segment is independent of the other segments, each segment operates as an independent sensor, and by knowing its position, it is possible to detect the specific position where ice accumulates. Recently, Abdollahzadeh et al. [111] performed a parametric study to understand the effect of dielectric material, dielectric thickness, exposed electrode width, embedded electrode width, and gap on the performance of the plasma actuator as an ice sensor. Rodrigues et al. [196] and Xie et al. [197] have conducted further studies to investigate the suitability of DBD plasma actuators for multipurpose applications and demonstrated that these devices are capable of performing ice sensing, deicing, and flow control simultaneously.

4. Conclusions

Ice accretion on aircraft and wind turbine surfaces is a major problem for both the aviation and wind power industries. For aircraft flying in cold conditions, the ice buildup phenomenon can be harmful for the normal operation of several aircraft components, including, for example, the aircraft wings. Ice accumulation on the leading edge of aircraft wings disturbs the flow and degrades aerodynamic performance by increasing drag and decreasing lift, which may seriously threaten flight safety. In a similar way, ice accretion is also a major problem for the wind power industry. Cold regions are very attractive for wind power generation, since these regions are usually associated with higher air density and higher wind speeds. However, these regions are also associated with favorable conditions for ice formation that affect and reduce wind turbine performance. In addition to the reduction in performance, the unbalanced loads caused by ice accumulation lead to deterioration of the components and a shortening of service life, which, in turn, considerably increases the maintenance costs of wind turbines. With this in mind, several works have been carried out over the years to study and further develop different ice protection systems. Ice protection systems include passive and active deicing and anti-icing techniques as well as ice sensing methods.

Active ice protection systems include the use of mechanical methods, the application of heat, the use of dry or liquid chemicals to lower the freezing point of water, or a combination of these different techniques. These methods include techniques such as pneumatic deicing boots, electrically resistive elements, electro-mechanical expulsion systems, thermo-mechanical expulsion devices, ultrasonic vibration methods, shape memory alloys, electromagnetically-induced vibration systems, thermal air bleed anti-icing systems, or chemical deicing devices. Passive ice protection systems, on the other hand, involve coatings or modifications to the surface structures to prevent or delay the ice accretion on the surface. These types of systems include hydrophobic or superhydrophobic coatings, micro- and nanoscale rough structures, and the combination of icephobic and hydrophobic coatings with electrothermal and auxiliary photothermal effects. Ice protection systems also involve the use of ice detection techniques which can be categorized in direct and indirect methods of ice detections. Indirect methods utilize some weather conditions such as temperature, humidity, and visibility combined with empirical deterministic models to predict icing events, while direct ice detection methods measure a change in a property such as mass, conductivity, inductance, and dielectric constants to estimate ice formation on the surface.

Recently, plasma actuators have been introduced as attractive devices for ice protection systems. These devices are capable of imparting momentum to the adjacent flow, which can be used to manipulate the local flow field. In addition, they have an instantaneous response time, are very lightweight, robust, easy to fabricate and implement, and consume low power levels. As demonstrated in the literature, these devices produce significant thermal effects which can be used for anti-icing or deicing purposes. The total power delivered to the DBD plasma actuator is composed of the reactive power, the power dissipated in dielectric heating, power spend to maintain and promote the plasma discharge, and, finally, the fluid mechanic power that is delivered to the adjacent air accelerating it. Different authors studied the actuator surface temperature field induced by the plasma discharge and concluded that the surface temperature increases significantly, especially at the onset of the plasma formation. At quiescent conditions, the exposed electrode region has a similar high temperature to the region where the plasma is generated due to the good thermal conductivity of the exposed electrode material. However, under the influence of an external flow, the temperature magnitudes decrease along the actuator surface, with the exposed electrode region showing a much larger decrease. In addition, researchers quantified the thermal power generated and concluded that plasma actuators dissipate a large percentage of power as thermal energy, regardless of the dielectric thickness. This thermal power released by the actuator is approximately 60 to 95% of the power supplied to the device, depending on the dielectric material. These results prove the potential of plasma actuators to avoid ice formation or promote local ice melting. Considering that a significant percentage of the power supplied to DBD plasma actuators is converted to heat dissipated from their surface, these devices have been proposed as a viable alternative to prevent ice formation and/or deicing in aerodynamic components in various applications. Various configurations of DBD plasma actuators have been reported in recent literature, applied to different systems, and tested under diverse conditions. Plasma actuators have been studied as deicing devices, and their ability to prevent ice accumulation on different objects, such as airfoils or cylinders, has been demonstrated. The performance of using plasma actuators as ice prevention systems was also compared to conventional electric heaters commonly used in ice protection systems. The studies showed that the DBD plasma actuator-based method presented a more promising performance compared to the conventional electrical heating method, because the temperature drop on the DBD surface was much lower. This difference was explained by the fact that the water droplets were heated not only by thermal conduction but also by thermal convection as they travel through the hot air above the DBD. In addition, various authors have also demonstrated the potential of using nanosecond plasma actuators for deicing and removal of ice from the surface. In line with these studies, it has also been shown that these devices can be combined with superhydrophobic surface coatings to improve the efficiency of the ice protection system. In addition, to improve the deicing efficiency of these devices, a new configuration called a "stream-wise plasma heat knife" has been proposed. Various works were developed considering this new configuration to improve its performance. Other works also addressed the study of different layouts, actuators positioning, and different types of electrodes and dielectric materials. In all the works, authors agreed on the ability of plasma actuators to efficiently perform deicing and ice prevention operations.

In addition to the potential use of plasma actuators as active flow control and deicing systems, these devices have recently been proposed as ice sensors. Since plasma actuators exhibit a behavior similar to a capacitor, they can also be operated as an ice capacitive sensor and detect the presence of water, air, or ice on their surface. When an alternating current is applied to the plasma actuator, an electric field is generated on the top of the dielectric layer in the region between the exposed electrode and covered electrode. This electric field is affected by the medium on top of the dielectric layer. Thus, the presence of water or ice on the actuator's surface will alter the electric field in comparison to the electric field produced in air and, consequently, will also modify the different electrical parameters of the actuator, such as the charge or the capacitance. Therefore, by monitoring the electrical characteristics of the DBD plasma actuator, it is possible to detect the presence of air, ice, or water on the top surface of the plasma device and operate the device as an ice sensor. Considering all the works and findings explored in this paper, we may conclude that plasma actuators are ideal devices for ice protection systems, because the same device is able to perform anti-/deicing operations and detect ice accumulation. This eliminates the need for two different techniques for deicing and detecting ice accumulation. Furthermore, they still have the advantage of controlling flow and improving the aerodynamic performance, even when no deicing or ice prevention measures need to be implemented.

Author Contributions: Conceptualization, F.R. and M.A.; methodology, F.R., M.A. and J.N.-P.; investigation, F.R., M.A. and J.N.-P.; writing—original draft preparation, F.R., M.A. and J.N.-P.; writing—review and editing, F.R., M.A., J.N.-P. and J.P.; supervision, J.P.; project administration, J.P.; funding acquisition, F.R., M.A., J.N.-P. and J.P. All authors have read and agreed to the published version of the manuscript.

Funding: This research was funded by C-MAST (Center for Mechanical and Aerospace Science and Technology), Research Unit No. 151, project grant number UID/00151/2020, by FCT Project "Smart blade based on printed sliding dielectric barrier discharge plasma actuators and higher harmonic motion for ice removal of cycloidal wind turbine", grant number CEECIND/03347/2017, and by Fundação para a Ciência e a Tecnologia (FCT), Fundo Social Europeu (FSE), and Programa Operacional Regional Centro 2020 and Norte 2020 for the Strategic Funding grant number UID/FIS/04650/2020 and grant SFRH/BPD/117838/2016 (J.N.-P.).

Data Availability Statement: Not applicable.

Conflicts of Interest: The authors declare no conflict of interest.

References

1. Gohardani, O.; Hammond, D.W. Ice adhesion to pristine and eroded polymer matrix composites reinforced with carbon nanotubes for potential usage on future aircraft. *Cold Reg. Sci. Technol.* **2013**, *96*, 8–16. [CrossRef]
2. Bagherzadeh, S.A.; Asadi, D. Detection of the ice assertion on aircraft using empirical mode decomposition enhanced by multi-objective optimization. *Mech. Syst. Signal Process.* **2017**, *88*, 9–24. [CrossRef]
3. Zilio, C.; Patricelli, L. Aircraft anti-ice system: Evaluation of system performance with a new time dependent mathematical model. *Appl. Eng.* **2014**, *63*, 40–51. [CrossRef]
4. Zhang, X.; Wu, X.; Min, J. Aircraft icing model considering both rime ice property variability and runback water effect. *Int. J. Heat Mass Transf.* **2017**, *104*, 510–516. [CrossRef]
5. Cao, Y.; Huang, J.; Yin, J. Numerical simulation of three-dimensional ice accretion on an aircraft wing. *Int. J. Heat Mass Transf.* **2016**, *92*, 34–54. [CrossRef]
6. Hu, T.; Lv, H.; Tian, B.; Su, D. Choosing Critical Ice Shapes on Airfoil Surface for the Icing Certification of Aircraft. *Procedia Eng.* **2014**, *80*, 456–466. [CrossRef]

7.   Liu, T.; Qu, K.; Cai, J.; Pan, S. A three-dimensional aircraft ice accretion model based on the numerical solution of the unsteady Stefan problem. *Aerosp. Sci. Technol.* **2019**, *93*, 105328. [CrossRef]
8.   Zhang, X.; Min, J.; Wu, X. Model for aircraft icing with consideration of property-variable rime ice. *Int. J. Heat Mass Transf.* **2016**, *97*, 185–190. [CrossRef]
9.   Cao, Y.; Ma, C.; Zhang, Q.; Sheridan, J. Numerical simulation of ice accretions on an aircraft wing. *Aerosp. Sci. Technol.* **2012**, *23*, 296–304. [CrossRef]
10. Deng, H.; Chang, S.; Song, M. The optimization of simulated icing environment by adjusting the arrangement of nozzles in an atomization equipment for the anti-icing and deicing of aircrafts. *Int. J. Heat Mass Transf.* **2020**, *155*, 119720. [CrossRef]
11. Rodríguez-Sanz, Á.; Valdés, R.A.; Comendador, F.G.; Ayra, E.S.; Cancela, J.C. Total air temperature anomalies as a metric for detecting high-altitude ice crystal events: Development of a failure indicator heuristic. *Eng. Fail Anal.* **2019**, *105*, 982–1005. [CrossRef]
12. Bucknell, A.; McGilvray, M.; Gillespie, D.R.H.; Jones, G.; Collier, B. A thermodynamic model for ice crystal accretion in aircraft engines: EMM-C. *Int. J. Heat Mass Transf.* **2021**, *174*, 121270. [CrossRef]
13. Wang, Q.; Yi, X.; Liu, Y.; Ren, J.; Li, W.; Wang, Q.; Lai, Q. Simulation and analysis of wind turbine ice accretion under yaw condition via an Improved Multi-Shot Icing Computational Model. *Renew. Energy* **2020**, *162*, 1854–1873. [CrossRef]
14. Ali, Q.S.; Kim, M.H. Design and performance analysis of an airborne wind turbine for high-altitude energy harvesting. *Energy* **2021**, *230*, 120829. [CrossRef]
15. Fortin, G.; Perron, J.; Ilinca, A. Behaviour and Modeling of Cup Anemometers under Icing Conditions. In Proceedings of the 11th International Workshop on Atmospheric Icing of Structures (IWAIS), Montreal, QC, Canada, 13–16 June 2005.
16. Manatbayev, R.; Baizhuma, Z.; Bolegenova, S.; Georgiev, A. Numerical simulations on static Vertical Axis Wind Turbine blade icing. *Renew. Energy* **2021**, *170*, 997–1007. [CrossRef]
17. Parent, O.; Ilinca, A. Anti-icing and de-icing techniques for wind turbines: Critical review. *Cold Reg. Sci. Technol.* **2011**, *65*, 88–96. [CrossRef]
18. Tao, T.; Liu, Y.; Qiao, Y.; Gao, L.; Lu, J.; Zhang, C.; Wang, Y. Wind turbine blade icing diagnosis using hybrid features and Stacked-XGBoost algorithm. *Renew. Energy* **2021**, *180*, 1004–1013. [CrossRef]
19. Stoyanov, D.B.; Nixon, J.D.; Sarlak, H. Analysis of derating and anti-icing strategies for wind turbines in cold climates. *Appl. Energy* **2021**, *288*, 116610. [CrossRef]
20. Cheng, X.; Shi, F.; Liu, Y.; Liu, X.; Huang, L. Wind turbine blade icing detection: A federated learning approach. *Energy* **2022**, *254*, 124441. [CrossRef]
21. Wei, K.; Yang, Y.; Zuo, H.; Zhong, D. A review on ice detection technology and ice elimination technology for wind turbine. *Wind Energy* **2020**, *23*, 433–457. [CrossRef]
22. Yirtici, O.; Tuncer, I.H. Aerodynamic shape optimization of wind turbine blades for minimizing power production losses due to icing. *Cold Reg. Sci. Technol.* **2021**, *185*, 103250. [CrossRef]
23. Baizhuma, Z.; Kim, T.; Son, C. Numerical method to predict ice accretion shapes and performance penalties for rotating vertical axis wind turbines under icing conditions. *J. Wind Eng. Ind. Aerodyn.* **2021**, *216*, 104708. [CrossRef]
24. Benmoussa, A.; Páscoa, J.C. Enhancement of a cycloidal self-pitch vertical axis wind turbine performance through DBD plasma actuators at low tip speed ratio. *Int. J. Thermofluids* **2023**, *17*, 100258. [CrossRef]
25. Benmoussa, A.; Páscoa, J.C. Performance improvement and start-up characteristics of a cyclorotor using multiple plasma actuators. *Meccanica* **2021**, *56*, 2707–2730. [CrossRef]
26. Rodrigues, F.; Páscoa, J.C.; Dias, F.; Abdollahzadeh, M. Plasma Actuators for Boundary Layer Control of Next Generation Nozzles. In Proceedings of the ASME International Mechanical Engineering Congress and Exposition. Proceedings (IMECE), Phoenix, AZ, USA, 11–17 November 2016. [CrossRef]
27. Rodrigues, F.; Mushyam, A.; Pascoa, J.; Trancossi, M. A new plasma actuator configuration for improved efficiency: The stair-shaped dielectric barrier discharge actuator. *J. Phys. D Appl. Phys.* **2019**, *52*, 385201. [CrossRef]
28. Zhou, W.; Liu, Y.; Hu, H.; Hu, H.; Meng, X. Utilization of Thermal Effect Induced by Plasma Generation for Aircraft Icing Mitigation. *AIAA J.* **2018**, *56*, 1097–1104. [CrossRef]
29. Cai, J.; Tian, Y.; Meng, X.; Han, X.; Zhang, D.; Hu, H. An experimental study of icing control using DBD plasma actuator. *Exp. Fluids* **2017**, *58*, 1–8. [CrossRef]
30. Abdollahzadeh, M.; Rodrigues, F.; Pascoa, J.C. Simultaneous ice detection and removal based on dielectric barrier discharge actuators. *Sens. Actuators A Phys.* **2020**, *315*, 112361. [CrossRef]
31. Cao, Y.; Wu, Z.; Su, Y.; Xu, Z. Aircraft flight characteristics in icing conditions. *Prog. Aerosp. Sci.* **2015**, *74*, 62–80. [CrossRef]
32. Cao, Y.; Tan, W.; Wu, Z. Aircraft icing: An ongoing threat to aviation safety. *Aerosp. Sci. Technol.* **2018**, *75*, 353–385. [CrossRef]
33. Khadak, A.; Subeshan, B.; Asmatulu, R. Studies on de-icing and anti-icing of carbon fiber-reinforced composites for aircraft surfaces using commercial multifunctional permanent superhydrophobic coatings. *J. Mater. Sci.* **2021**, *56*, 3078–3094. [CrossRef]
34. Drury, M.D.; Szefi, J.T.; Palacios, J.L. Full-Scale Testing of a Centrifugally Powered Pneumatic De-Icing System for Helicopter Rotor Blades. *J. Aircr.* **2016**, *54*, 220–228. [CrossRef]
35. Shinkafi, A.; Lawson, C. Enhanced Method of Conceptual Sizing of Aircraft Electro-Thermal De-icing System. *Int. J. Aerosp. Mech. Eng.* **2014**, *8*, 1073–1080. [CrossRef]

36. Mohseni, M.; Amirfazli, A. A novel electro-thermal anti-icing system for fiber-reinforced polymer composite airfoils. *Cold Reg. Sci. Technol.* **2013**, *87*, 47–58. [CrossRef]

37. Ibrahim, Y.; Kempers, R.; Amirfazli, A. 3D printed electro-thermal anti- or de-icing system for composite panels. *Cold Reg. Sci. Technol.* **2019**, *166*, 102844. [CrossRef]

38. Kaj, O.; Larsen, F.M.; Grabau, P.; Jespersen, J.E. Wind Turbine Blade with a System for Deicing and Lightning Protection. U.S. Patent 6,612,810, 2 September 2003.

39. Huang, X.; Tepylo, N.; Pommier-Budinger, V.; Budinger, M.; Bonaccurso, E.; Villedieu, P.; Bennani, L. A survey of icephobic coatings and their potential use in a hybrid coating/active ice protection system for aerospace applications. *Prog. Aerosp. Sci.* **2019**, *105*, 74–97. [CrossRef]

40. Olson, R.A.; Loyal, M.; Hanson, M. Electro-Expulsive De-Icing System for Aircraft and Other Applications. U.S. Patent 9,108,735, 5 February 2010.

41. Ingram, R.; Codner, G.; Gerardi, J. Electro-Magnetic Expulsion De-Icing System. U.S. Patent 5,782,435, 21 July 1997.

42. Shin, J.; Bond, T. Surface roughness due to residual ice in the use of low power deicing systems. In Proceedings of the 31st Aerospace Sciences Meeting, Reno, NV, USA, 11–14 January 1993. [CrossRef]

43. Goraj, Z. An Overview of the Deicing and Antiicing Technologies with Prospects for the Future. In Proceedings of the 24th International Congress of the Aeronautical Sciences, Yokohama, Japan, 29 August–3 September 2004.

44. Al-Khalil, K. Thermo-mechanical expulsion deicing system-TMEDS. In Proceedings of the Collection of Technical Papers—45th AIAA Aerospace Sciences Meeting, Reno, NV, USA, 8–11 January 2007; Volume 12, pp. 8562–8574. [CrossRef]

45. Palacios, J.; Smith, E.; Rose, J.; Royer, R. Ultrasonic De-Icing of Wind-Tunnel Impact Icing. *J. Aircr.* **2012**, *48*, 1020–1027. [CrossRef]

46. Wang, Y.; Xu, Y.; Huang, Q. Progress on ultrasonic guided waves de-icing techniques in improving aviation energy efficiency. *Renew. Sustain. Energy Rev.* **2017**, *79*, 638–645. [CrossRef]

47. Palacios, J.L.; Zhu, Y.; Smith, E.C.; Rose, J.L. Ultrasonic shear and lamb wave interface stress for helicopter rotor de-icing purposes. In Proceedings of the 47th Structures, Structural Dynamics and Materials Conference, Newport, RI, USA, 1–4 May 2006; Volume 11, pp. 8131–8142. [CrossRef]

48. Wang, Z. Recent progress on ultrasonic de-icing technique used for wind power generation, high-voltage transmission line and aircraft. *Energy Build.* **2017**, *140*, 42–49. [CrossRef]

49. Palacios, J.L.; Gao, H.; Smith, E.C.; Rose, J.L. Ultrasonic shear wave anti-icing system for helicopter rotor blades. In Proceedings of the Annual Forum Proceedings-AHS International. III, Phoenix, AZ, USA, 9–11 May 2006; pp. 1492–1502. [CrossRef]

50. Budinger, M.; Pommier-Budinger, V.; Napias, G.; da Silva, A.C. Ultrasonic Ice Protection Systems: Analytical and Numerical Models for Architecture Tradeoff. *J. Aircr.* **2016**, *53*, 680–690. [CrossRef]

51. Myose, R.Y.; Horn, W.J.; Hwang, Y.; Herrero, J.; Huynh, C.; Boudraa, T. *Application of Shape Memory Alloys for Leading Edge Deicing*; SAE Technical Papers; SAE: Warrendale, PA, USA, 1999. [CrossRef]

52. Gerardi, J.J.; Ingram, R.B.; Catarella, R.A. A shape memory alloy based de-icing system for aircraft. In Proceedings of the 33rd Aerospace Sciences Meeting and Exhibit, Reno, NV, USA, 9–12 January 1995; pp. 1–8. [CrossRef]

53. Sullivan, D.B.; Righi, F.; Hartl, D.J.; Rogers, J. Shape memory alloy rotor blade deicing. In Proceedings of the 54th AIAA/ASME/ASCE/AHS/ASC Structures, Structural Dynamics, and Materials Conference, Boston, MA, USA, 8–11 April 2013. [CrossRef]

54. Ilinca, A.; Ilinca, A. Analysis and Mitigation of Icing Effects on Wind Turbines. *Wind Turbines* **2011**. [CrossRef]

55. Dalili, N.; Edrisy, A.; Carriveau, R. A review of surface engineering issues critical to wind turbine performance. *Renew. Sustain. Energy Rev.* **2009**, *13*, 428–438. [CrossRef]

56. Zumwalt, G.; Friedberg, R. Designing an electro-impulse de-icing system. In Proceedings of the 24th Aerospace Sciences Meeting, Reno, NV, USA, 6–9 January 1986. [CrossRef]

57. Sommerwerk, H.; Horst, P.; Bansmer, S. Studies on electro impulse de-icing of a leading edge structure in an icing wind tunnel. In Proceedings of the 8th AIAA Atmospheric and Space Environments Conference, Washington, DC, USA, 13–17 June 2016. [CrossRef]

58. Jiang, X.; Wang, Y. Studies on the Electro-Impulse De-Icing System of Aircraft. *Aerospace* **2019**, *6*, 67. [CrossRef]

59. Goehner, R.; Glover, N.; Hensley, D. Electro-Impulse De-Icing System for Aircraft. U.S. Patent 4,678,144, 7 July 1984.

60. Endres, M.; Sommerwerk, H.; Mendig, C.; Sinapius, M.; Horst, P. Experimental study of two electro-mechanical de-icing systems applied on a wing section tested in an icing wind tunnel. *CEAS Aeronaut. J.* **2017**, *8*, 429–439. [CrossRef]

61. Sørensen, K.L.; Helland, A.S.; Johansen, T.A. Carbon nanomaterial-based wing temperature control system for in-flight anti-icing and de-icing of unmanned aerial vehicles. In Proceedings of the IEEE Aerospace Conference Proceedings, Big Sky, MT, USA, 7–14 March 2015. [CrossRef]

62. Vertuccio, L.; de Santis, F.; Pantani, R.; Lafdi, K.; Guadagno, L. Effective de-icing skin using graphene-based flexible heater. *Compos. B Eng.* **2019**, *162*, 600–610. [CrossRef]

63. Pellissier, M.P.C.; Habashi, W.G.; Pueyo, A. Optimization via FENSAP-ICE of Aircraft Hot-Air Anti-Icing Systems. *J. Aircr.* **2012**, *48*, 265–276. [CrossRef]

64. Zhang, F.; Deng, W.; Nan, H.; Zhang, L.; Huang, Z. Reliability analysis of bleed air anti-icing system based on subset simulation method. *Appl. Eng.* **2017**, *115*, 17–21. [CrossRef]

65. Federal Aviation Administration. Aviation Maintenance Technician Handbook-Airframe Volume 2. Available online: https://www.faa.gov/handbooksmanuals/aviation/aviation-maintenance-technician-handbook-airframe-volume-2 (accessed on 2 November 2022).

66. Domingos, R.H.; Papadakis, M.; Zamora, A.O. Computational methodology for bleed air ice protection system parametric analysis. In Proceedings of the AIAA Atmospheric and Space Environments Conference, Toronto, ON, Canada, 2–5 August 2010. [CrossRef]

67. Su, Q.; Chang, S.; Zhao, Y.; Zheng, H.; Dang, C. A review of loop heat pipes for aircraft anti-icing applications. *Appl. Eng.* **2018**, *130*, 528–540. [CrossRef]

68. Hem, L.J.; Weideborg, M.; Schram, E. Degradation and toxicity of additives to aircraft de-icing fluids; the effect of discharge of such fluids to municipal wastewater treatment plants. *Proc. Water Environ. Fed.* **2000**, *2000*, 419–433. [CrossRef]

69. Wang, Y.; Hudson, N.E.; Pethrick, R.A.; Schaschke, C.J. Poly(acrylic acid)–poly(vinyl pyrrolidone)-thickened water/glycol de-icing fluids. *Cold Reg. Sci. Technol.* **2014**, *101*, 24–30. [CrossRef]

70. Louchez, P.R.; Bernardin, S.; Laforte, J.L. Physical properties of aircraft de-icing and anti-icing fluids. In Proceedings of the 36th AIAA Aerospace Sciences Meeting and Exhibit, Reno, NV, USA, 12–15 January 1998. [CrossRef]

71. Fakorede, O.; Feger, Z.; Ibrahim, H.; Ilinca, A.; Perron, J.; Masson, C. Ice protection systems for wind turbines in cold climate: Characteristics, comparisons and analysis. *Renew. Sustain. Energy Rev.* **2016**, *65*, 662–675. [CrossRef]

72. Fortin, G. *Super-Hydrophobic Coatings as a Part of the Aircraft Ice Protection System*; SAE Technical Papers; SAE: Warrendale, PA, USA, 2017. [CrossRef]

73. Latthe, S.S.; Sutar, R.S.; Bhosale, A.K.; Nagappan, S.; Ha, C.S.; Sadasivuni, K.K.; Liu, S.; Xing, R. Recent developments in air-trapped superhydrophobic and liquid-infused slippery surfaces for anti-icing application. *Prog. Org. Coat.* **2019**, *137*, 105373. [CrossRef]

74. Zhao, Z.; Chen, H.; Zhu, Y.; Liu, X.; Wang, Z.; Chen, J. A robust superhydrophobic anti-icing/de-icing composite coating with electrothermal and auxiliary photothermal performances. *Compos. Sci. Technol.* **2022**, *227*, 109578. [CrossRef]

75. Guo, H.; Liu, M.; Xie, C.; Zhu, Y.; Sui, X.; Wen, C.; Li, Q.; Zhao, W.; Yang, J.; Zhang, L. A sunlight-responsive and robust anti-icing/deicing coating based on the amphiphilic materials. *Chem. Eng. J.* **2020**, *402*, 126161. [CrossRef]

76. Liu, G.; Xu, J.; Chen, T.; Wang, K. Progress in thermoplasmonics for solar energy applications. *Phys Rep.* **2022**, *981*, 1–50. [CrossRef]

77. Liu, G.; Xu, J.; Wang, K. Solar water evaporation by black photothermal sheets. *Nano Energy* **2017**, *41*, 269–284. [CrossRef]

78. Zhang, H.; Zhao, G.; Wu, S.; Alsaid, Y.; Zhao, W.; Yan, X.; Liu, L.; Zou, G.; Lv, J.; He, X.; et al. Solar anti-icing surface with enhanced condensate self-removing at extreme environmental conditions. *Proc. Natl. Acad. Sci. USA* **2021**, *118*, e2100978118. [CrossRef]

79. Sheng, S.; Zhu, Z.; Wang, Z.; Hao, T.; He, Z.; Wang, J. Bioinspired solar anti-icing/de-icing surfaces based on phase-change materials. *Sci. China Mater.* **2021**, *65*, 1369–1376. [CrossRef]

80. Wu, C.; Geng, H.; Tan, S.; Lv, J.; Wang, H.; He, Z.; Wang, J. Highly efficient solar anti-icing/deicing via a hierarchical structured surface. *Mater. Horiz.* **2020**, *7*, 2097–2104. [CrossRef]

81. Liu, G.; Chen, T.; Xu, J.; Yao, G.; Xie, J.; Cheng, Y.; Miao, Z.; Wang, K. Salt-Rejecting Solar Interfacial Evaporation. *Cell Rep. Phys. Sci.* **2021**, *2*, 100310. [CrossRef]

82. Virk, M. Atmospheric Icing Sensors—An insight. In Proceedings of the Seventh International Conference on Sensor Technologies and Applications, Barcelona, Spain, 25–31 August 2013. [CrossRef]

83. Codner, G.W.; Pruzan, D.A.; Rauckhorst, R.L., III; Reich, A.D.; Sweet, D.B. Impedance Type Ice Detector. U.S. Patent 5,955,887, 20 December 1996.

84. Mughal, U.N.; Virk, M.S.; Mustafa, M.Y. Dielectric based sensing of atmospheric ice. *AIP Conf. Proc.* **2013**, *1570*, 212. [CrossRef]

85. Wang, P.; Zhou, W.; Bao, Y.; Li, H. Ice monitoring of a full-scale wind turbine blade using ultrasonic guided waves under varying temperature conditions. *Struct. Control Health Monit.* **2018**, *25*, e2138. [CrossRef]

86. Homola, M.C.; Nicklasson, P.J.; Sundsbø, P.A. Ice sensors for wind turbines. *Cold Reg. Sci. Technol.* **2006**, *46*, 125–131. [CrossRef]

87. Mughal, U.N.; Virk, M.S.; Mustafa, M. State of the Art Review of Atmospheric Icing Sensors. *Sens. Transducers* **2016**, *198*, 2–15. Available online: https://munin.uit.no/handle/10037/11740 (accessed on 2 November 2022).

88. Jiang, J.H.; Wu, D.L. Ice and water permittivities for millimeter and sub-millimeter remote sensing applications. *Atmos. Sci. Lett.* **2004**, *5*, 146–151. [CrossRef]

89. Madi, E.; Pope, K.; Huang, W.; Iqbal, T. A review of integrating ice detection and mitigation for wind turbine blades. *Renew. Sustain. Energy Rev.* **2019**, *103*, 269–281. [CrossRef]

90. Páscoa, J.C.; Rodrigues, F.F.; Das, S.S.; Abdollahzadeh, M.; Dumas, A.; Trancossi, M.; Subhash, M. Exit Flow Vector Control on a Coanda Nozzle Using Dielectric Barrier Discharge Actuator. In Proceedings of the ASME International Mechanical Engineering Congress and Exposition Proceedings (IMECE), Houston, TX, USA, 15–18 November 2015. [CrossRef]

91. Mushyam, A.; Rodrigues, F.; Pascoa, J.C. A plasma-fluid model for EHD flow in DBD actuators and experimental validation. *Int. J. Numer. Methods Fluids* **2019**, *90*, 115–139. [CrossRef]

92. Nunes-Pereira, J.; Rodrigues, F.F.; Abdollahzadehsangroudi, M.; Páscoa, J.C.; Lanceros-Mendez, S. Improved performance of polyimide Cirlex-based dielectric barrier discharge plasma actuators for flow control. *Polym. Adv. Technol.* **2022**, *33*, 1278–1290. [CrossRef]

93. Corke, T.C.; Post, M.L.; Orlov, D.M. Single dielectric barrier discharge plasma enhanced aerodynamics: Physics, modeling and applications. *Exp. Fluids* **2009**, *46*, 1–26. [CrossRef]

94. Font, G.I.; Enloe, C.L.; McLaughlin, T.E. Plasma Volumetric Effects on the Force Production of a Plasma Actuator. *AIAA J.* **2012**, *48*, 1869–1874. [CrossRef]

95. Benmoussa, A.; Páscoa, J.C. Cycloidal rotor coupled with DBD plasma actuators for performance improvement. *Aerosp. Sci. Technol.* **2021**, *110*, 106468. [CrossRef]

96. Ferry, J.W.; Rovey, J.L. Thrust measurement of dielectric barrier discharge plasma actuators and power requirements for aerodynamic control. In Proceedings of the 5th Flow Control Conference, Chicago, IL, USA, 28 June–1 July 2010. [CrossRef]

97. Houser, N.M.; Gimeno, L.; Hanson, R.E.; Goldhawk, T.; Simpson, T.; Lavoie, P. Microfabrication of dielectric barrier discharge plasma actuators for flow control. *Sens. Actuators A Phys.* **2013**, *201*, 101–104. [CrossRef]

98. Rodrigues, F.; Pascoa, J.; Trancossi, M. Heat generation mechanisms of DBD plasma actuators. *Exp. Fluid Sci.* **2018**, *90*, 55–65. [CrossRef]

99. He, C.; Corke, T.C.; Patel, M.P. Plasma Flaps and Slats: An Application of Weakly Ionized Plasma Actuators. *J. Aircr.* **2012**, *46*, 864–873. [CrossRef]

100. Mertz, B.E.; Corke, T.C. Time-dependent dielectric barrier dishcharge plasma actuator modeling. In Proceedings of the 47th AIAA Aerospace Sciences Meeting Including the New Horizons Forum and Aerospace Exposition, Orlando, FL, USA, 5–8 January 2009. [CrossRef]

101. Durscher, R.; Roy, S. Novel multi-barrier plasma actuators for increased thrust. In Proceedings of the 48th AIAA Aerospace Sciences Meeting Including the New Horizons Forum and Aerospace Exposition, Orlando, FL, USA, 4–7 January 2010. [CrossRef]

102. Wilkinson, S.P.; Siochi, E.J.; Sauti, G.; Xu, T.B.; Meador, M.A.; Guo, H. Evaluation of dielectric-barrier-discharge actuator substrate materials. In Proceedings of the 45th AIAA Plasmadynamics and Lasers Conference, Atlanta, GA, USA, 16–20 June 2014. [CrossRef]

103. Benard, N.; Jolibois, J.; Moreau, E. Lift and drag performances of an axisymmetric airfoil controlled by plasma actuator. *J. Electrostat.* **2009**, *67*, 133–139. [CrossRef]

104. Zhao, G.Y.; Li, Y.H.; Liang, H.; Han, M.H.; Hua, W.Z. Control of vortex on a non-slender delta wing by a nanosecond pulse surface dielectric barrier discharge. *Exp. Fluids* **2015**, *56*, 1–9. [CrossRef]

105. Jayaraman, B.; Cho, Y.C.; Shyy, W. Modeling of dielectric barrier discharge plasma actuator. *J. Appl. Phys.* **2008**, *103*, 053304. [CrossRef]

106. Xiao, D.; Borradaile, H.; Choi, K.S.; Feng, L.; Wang, J.; Mao, X. Bypass transition in a boundary layer flow induced by plasma actuators. *J. Fluid Mech.* **2021**, *929*, A6. [CrossRef]

107. Pescini, E.; Suma, A.; de Giorgi, M.G.; Francioso, L.; Ficarella, A. Optimization of Plasma Actuator Excitation Waveform and Materials for Separation Control in Turbomachinery. *Energy Procedia* **2017**, *126*, 786–793. [CrossRef]

108. De Giorgi, M.G.; Ficarella, A.; Marra, F.; Pescini, E. Micro DBD plasma actuators for flow separation control on a low pressure turbine at high altitude flight operating conditions of aircraft engines. *Appl. Eng.* **2017**, *114*, 511–522. [CrossRef]

109. Roth, J.R.; Dai, X. Optimization of the aerodynamic plasma actuator as an electrohydrodynamic (EHD) electrical device. In Proceedings of the Collection of Technical Papers-44th AIAA Aerospace Sciences Meeting, Reno, NV, USA, 9–12 January 2006; Volume 19, pp. 14604–14631. [CrossRef]

110. Moreau, E.; Sosa, R.; Artana, G. Electric wind produced by surface plasma actuators: A new dielectric barrier discharge based on a three-electrode geometry. *J. Phys. D Appl. Phys.* **2008**, *41*, 115204. [CrossRef]

111. Abdollahzadeh, M.; Rodrigues, F.; Nunes-Pereira, J.; Pascoa, J.C.; Pires, L. Parametric optimization of surface dielectric barrier discharge actuators for ice sensing application. *Sens. Actuators A Phys.* **2022**, *335*, 113391. [CrossRef]

112. Erfani, R.; Erfani, T.; Utyuzhnikov, S.V.; Kontis, K. Optimisation of multiple encapsulated electrode plasma actuator. *Aerosp. Sci. Technol.* **2013**, *26*, 120–127. [CrossRef]

113. Huang, J.; Corke, T.C.; Thomas, F.O. Unsteady Plasma Actuators for Separation Control of Low-Pressure Turbine Blades. *AIAA J.* **2012**, *44*, 1477–1487. [CrossRef]

114. Rodrigues, F.F.; Pereira, J.N.; Abdollahzadeh, M.; Pascoa, J.; Mendez, S.L. Comparative Evaluation of Dielectric Materials for Plasma Actuators Active Flow Control and Heat Transfer Applications. In Proceedings of the ASME 2021 Fluids Engineering Division Summer Meeting, Virtual, 10–12 August 2021; Volume 3. [CrossRef]

115. Mertz, B.E.; Corke, T.C. Single-dielectric barrier discharge plasma actuator modelling and validation. *J. Fluid Mech.* **2011**, *669*, 557–583. [CrossRef]

116. Roupassov, D.V.; Nikipelov, A.A.; Nudnova, M.M.; Starikovskii, A.Y. Flow Separation Control by Plasma Actuator with Nanosecond Pulsed-Periodic Discharge. *AIAA J.* **2012**, *47*, 168–185. [CrossRef]

117. Rodrigues, F.F.; Pascoa, J.C. Implementation of stair-shaped dielectric layers in micro- and macroplasma actuators for increased efficiency and lifetime. *J. Fluids Eng. Trans. ASME* **2020**, *142*, 104502. [CrossRef]

118. Rodrigues, F.; Abdollahzadeh, M.; Pascoa, J.C.; Oliveira, P.J. An experimental study on segmented-encapsulated electrode dielectric-barrier-discharge plasma actuator for mapping ice formation on a surface: A conceptual analysis. *J. Heat Transfer* **2021**, *143*, 011701. [CrossRef]

119. Wei, B.; Wu, Y.; Liang, H.; Zhu, Y.; Chen, J.; Zhao, G.; Song, H.; Jia, M.; Xu, H. SDBD based plasma anti-icing: A stream-wise plasma heat knife configuration and criteria energy analysis. *Int. J. Heat Mass Transf.* **2019**, *138*, 163–172. [CrossRef]

120. Santhanakrishnan, A.; Jacob, J.D. Flow control with plasma synthetic jet actuators. *J. Phys. D Appl. Phys.* **2007**, *40*, 637. [CrossRef]
121. Segawa, T.; Furutani, H.; Yoshida, H.; Jukes, T.; Choi, K.S. Wall normal jet under elevated temperatures produced by surface plasma actuator. In Proceedings of the Collection of Technical Papers-45th AIAA Aerospace Sciences Meeting, Reno, NV, USA, 8–11 January 2007; Volume 14, pp. 9611–9618. [CrossRef]
122. Roy, S.; Wang, C.-C. Bulk flow modification with horseshoe and serpentine plasma actuators. *J. Phys. D Appl. Phys.* **2008**, *42*, 032004. [CrossRef]
123. Riherd, M.; Roy, S. Serpentine geometry plasma actuators for flow control. *J. Appl. Phys.* **2013**, *114*, 083303. [CrossRef]
124. Cattafesta, L.N.; Sheplak, M. Actuators for Active Flow Control. *Annu. Rev. Fluid Mech.* **2011**, *43*, 247–272. [CrossRef]
125. Kotsonis, M.; Veldhuis, L. Experimental study on dielectric barrier discharge actuators operating in pulse mode. *J. Appl. Phys.* **2010**, *108*, 113304. [CrossRef]
126. Corke, T.C.; Enloe, C.L.; Wilkinson, S.P. Dielectric Barrier Discharge Plasma Actuators for Flow Control. *Annu. Rev. Fluid Mech.* **2009**, *42*, 505–529. [CrossRef]
127. Pendar, M.-R.; Páscoa, J.C. Numerical Investigation of Plasma Actuator Effects on Flow Control Over a Three-Dimensional Airfoil with a Sinusoidal Leading Edge. *J. Fluids Eng.* **2022**, *144*, 081208. [CrossRef]
128. Pendar, M.R.; Pascoa, J. Study of the Plasma Actuator Effect on the Flow Characteristics of an Airfoil: An LES Investigation. *SAE Int. J. Adv. Curr. Pract. Mobil.* **2021**, *3*, 1206–1215. [CrossRef]
129. Abdollahzadeh, M.; Rodrigues, F.; Pascoa, J.C.; Oliveira, P.J. Numerical design and analysis of a multi-DBD actuator configuration for the experimental testing of ACHEON nozzle model. *Aerosp. Sci. Technol.* **2015**, *41*, 259–273. [CrossRef]
130. Abdollahzadeh, M.; Páscoa, J.C.; Oliveira, P.J. Modified split-potential model for modeling the effect of DBD plasma actuators in high altitude flow control. *Curr. Appl. Phys.* **2014**, *14*, 1160–1170. [CrossRef]
131. Amanifard, N.; Abdollahzadeh, M.; Moayedi, H.; Pascoa, J.C. An explicit CFD model for the DBD plasma actuators using wall-jet similarity approach. *J. Electrostat.* **2020**, *107*, 103497. [CrossRef]
132. Bouremel, Y.; Li, J.M.; Zhao, Z.; Debiasi, M. Effects of AC Dielectric Barrier Discharge Plasma Actuator Location on Flow Separation and Airfoil Performance. *Procedia Eng.* **2013**, *67*, 270–278. [CrossRef]
133. Kelley, C.L.; Bowles, P.; Cooney, J.; He, C.; Corke, T.C.; Osborne, B.; Silkey, J.; Zehnle, J. High Mach number leading-edge flow separation control using AC DBD plasma actuators. In Proceedings of the 50th AIAA Aerospace Sciences Meeting Including the New Horizons Forum and Aerospace Exposition, Nashville, TN, USA, 9–12 January 2012. [CrossRef]
134. Durasiewicz, C.; Singh, A.; Little, J.C. A Comparative Flow Physics Study of Ns-DBD vs Ac-DBD Plasma Actuators for Transient Separation Control on a NACA 0012 Airfoil. In Proceedings of the 2018 AIAA Aerospace Sciences Meeting, Kissimmee, FL, USA, 8–12 January 2018. [CrossRef]
135. Naghib-Lahouti, A.; Hangan, H.; Lavoie, P. Distributed forcing flow control in the wake of a blunt trailing edge profiled body using plasma actuators. *Phys. Fluids* **2015**, *27*, 035110. [CrossRef]
136. Huang, X.; Zhang, X. Streamwise and spanwise plasma actuators for flow-induced cavity noise control. *Phys. Fluids* **2008**, *20*, 037101. [CrossRef]
137. Huang, X.; Zhang, X. Plasma Actuators for Noise Control. *Int. J. Aeroacoustics* **2010**, *9*, 679–703. [CrossRef]
138. Grundmann, S.; Tropea, C. Experimental transition delay using glow-discharge plasma actuators. *Exp. Fluids* **2007**, *42*, 653–657. [CrossRef]
139. Grundmann, S.; Tropea, C. Active cancellation of artificially introduced Tollmien-Schlichting waves using plasma actuators. *Exp. Fluids* **2008**, *44*, 795–806. [CrossRef]
140. Grundmann, S.; Tropea, C. Experimental damping of boundary-layer oscillations using DBD plasma actuators. *Int. J. Heat Fluid Flow* **2009**, *30*, 394–402. [CrossRef]
141. Feng, L.H.; Jukes, T.N.; Choi, K.S.; Wang, J.J. Flow control over a NACA 0012 airfoil using dielectric-barrier-discharge plasma actuator with a Gurney flap. *Exp. Fluids* **2012**, *52*, 1533–1546. [CrossRef]
142. Joussot, R.; Hong, D.; Weber-Rozenbaum, R.; Leroy-Chesneau, A. Modification of the laminar-to-turbulent transition on a flat plate using a DBD plasma actuator. In Proceedings of the 5th Flow Control Conference, Chicago, IL, USA, 28 June–1 July 2010. [CrossRef]
143. Kozlov, A.V.; Thomas, F.O. Active noise control of bluff-body flows using dielectric barrier discharge plasma actuators. In Proceedings of the 15th AIAA/CEAS Aeroacoustics Conference (30th AIAA Aeroacoustics Conference), Miami, FL, USA, 11–13 May 2009. [CrossRef]
144. Thomas, F.O.; Corke, T.C.; Iqbal, M.; Kozlov, A.; Schatzman, D. Optimization of Dielectric Barrier Discharge Plasma Actuators for Active Aerodynamic Flow Control. *AIAA J.* **2012**, *47*, 2169–2178. [CrossRef]
145. Yu, J.; Wang, Z.; Chen, F.; Yan, G.; Wang, C. Large eddy simulation of the elliptic jets in film cooling controlled by dielectric barrier discharge plasma actuators with an improved model. *J. Heat Transfer.* **2018**, *140*, 122001. [CrossRef]
146. Li, G.; Chen, F.; Li, L.; Song, Y. Large Eddy Simulation of the Effects of Plasma Actuation Strength on Film Cooling Efficiency. *Plasma Sci. Technol.* **2016**, *18*, 1101. [CrossRef]
147. Uehara, S.; Takana, H. Surface cooling by dielectric barrier discharge plasma actuator in confinement channel. *J. Electrostat.* **2020**, *104*, 103417. [CrossRef]

148. Rodrigues, F.; Abdollahzadeh, M.; Pascoa, J.; Pires, L. Influence of Exposed Electrode Thickness on Plasma Actuators Performance for Coupled Deicing and Flow Control Applications. In Proceedings of the ASME 2021 Fluids Engineering Division Summer Meeting, Virtual, 10–12 August 2021; Volume 3. [CrossRef]
149. Liu, Y.; Kolbakir, C.; Hu, H.; Hu, H. A comparison study on the thermal effects in DBD plasma actuation and electrical heating for aircraft icing mitigation. *Int. J. Heat Mass Transf.* **2018**, *124*, 319–330. [CrossRef]
150. Kriegseis, J.; Möller, B.; Grundmann, S.; Tropea, C. On Performance and Efficiency of Dielectric Barrier Discharge Plasma Actuators for Flow Control Applications. *Int. J. Flow Control.* **2012**, *4*, 125–132. [CrossRef]
151. Roth, J.R.; Dai, X.; Rahel, J.; Shermann, D.M. The physics and phenomenology of paraelectric One Atmosphere Uniform Glow Discharge Plasma (OAUGDPTM) actuators for aerodynamic flow control. In Proceedings of the 43rd AIAA Aerospace Sciences Meeting and Exhibit-Meeting Papers, Reno, NV, USA, 10–13 January 2005; pp. 14057–14068. [CrossRef]
152. Roth, J.R.; Rahel, J.; Dai, X.; Sherman, D.M. The physics and phenomenology of One Atmosphere Uniform Glow Discharge Plasma (OAUGDPTM) reactors for surface treatment applications. *J. Phys. D Appl. Phys.* **2005**, *38*, 555. [CrossRef]
153. Kraus, J. *Electromagnetics*, 4th ed.; McGraw-Hill Inc.: New York, NY, USA, 1991.
154. Rodrigues, F.F.; Pascoa, J.C.; Trancossi, M. Analysis of Innovative Plasma Actuator Geometries for Boundary Layer Control. In Proceedings of the ASME 2016 International Mechanical Engineering Congress and Exposition, Phoenix, AZ, USA, 11–17 November 2016. [CrossRef]
155. Enloe, C.L.; McLaughlin, T.E.; VanDyken, R.D.; Kachner, K.D.; Jumper, E.J.; Corke, T.C. Mechanisms and Responses of a Single Dielectric Barrier Plasma Actuator: Plasma Morphology. *AIAA J.* **2012**, *42*, 589–594. [CrossRef]
156. Enloe, C.L.; McLaughlin, T.E.; VanDyken, R.D.; Kachner, K.D.; Jumper, E.J.; Corke, T.C.; Post, M.; Haddad, O. Mechanisms and Responses of a Dielectric Barrier Plasma Actuator: Geometric Effects. *AIAA J.* **2012**, *42*, 595–604. [CrossRef]
157. Abdollahzadeh, M.; Páscoa, J.C.; Oliveira, P.J. Two-dimensional numerical modeling of interaction of micro-shock wave generated by nanosecond plasma actuators and transonic flow. *J. Comput. Appl. Math.* **2014**, *270*, 401–416. [CrossRef]
158. Pons, J.; Moreau, E.; Touchard, G. Asymmetric surface dielectric barrier discharge in air at atmospheric pressure: Electrical properties and induced airflow characteristics. *J. Phys. D Appl. Phys.* **2005**, *38*, 3635. [CrossRef]
159. Léger, L.; Moreau, E.; Touchard, G. Electrohydrodynamic airflow control along a flat plate by a DC surface corona discharge-Velocity profile and wall pressure measurements. In Proceedings of the 1st Flow Control Conference, St. Louis, MO, USA, 24–26 June 2002. [CrossRef]
160. Rodrigues, F.F.; Pascoa, J.C.; Trancossi, M. Experimental Analysis of Alternative Dielectric Materials for DBD Plasma Actuators. In Proceedings of the ASME International Mechanical Engineering Congress and Exposition. Proceedings (IMECE), Salt Lake City, UT, USA, 11–14 November 2019. [CrossRef]
161. Dong, B.; Bauchire, J.M.; Pouvesle, J.M.; Magnier, P.; Hong, D. Experimental study of a DBD surface discharge for the active control of subsonic airflow. *J. Phys. D Appl. Phys.* **2008**, *41*, 155201. [CrossRef]
162. Jukes, T.N.; Choi, K.S.; Segawa, T.; Yoshida, H. Jet flow induced by a surface plasma actuator. *Proc. Inst. Mech. Eng. Part I J. Syst. Control. Eng.* **2008**, *222*, 347–356. [CrossRef]
163. Joussot, R.; Hong, D.; Rabat, H.; Boucinha, V.; Weber-Rozenbaum, R.; Leroy-Chesneau, A. Thermal characterization of a DBD plasma actuator: Dielectric temperature measurements using infrared thermography. In Proceedings of the 40th AIAA Fluid Dynamics Conference, Chicago, IL, USA, 28 June–1 July 2010. [CrossRef]
164. Stanfield, S.A.; Menart, J.; DeJoseph, C.; Kimmel, R.L.; Hayes, J.R. Rotational and Vibrational Temperature Distributions for a Dielectric Barrier Discharge in Air. *AIAA J.* **2012**, *47*, 1107–1115. [CrossRef]
165. Tirumala, R.; Benard, N.; Moreau, E.; Fenot, M.; Lalizel, G.; Dorignac, E. Temperature characterization of dielectric barrier discharge actuators: Influence of electrical and geometric parameters. *J. Phys. D Appl. Phys.* **2014**, *47*, 255203. [CrossRef]
166. Rodrigues, F.F.; Pascoa, J.C.; Trancossi, M. Experimental Thermal Characterization of DBD Plasma Actuators. In Proceedings of the ASME International Mechanical Engineering Congress and Exposition. Proceedings (IMECE), Pittsburgh, PA, USA, 9–15 November 2018. [CrossRef]
167. Rodrigues, F.F.; Pascoa, J.C.; Trancossi, M. Experimental Analysis of Dielectric Barrier Discharge Plasma Actuators Thermal Characteristics under External Flow Influence. *J. Heat Transfer.* **2018**, *140*, 102801. [CrossRef]
168. Abbasi, A.A.; Li, H.; Weiwei, H.; Meng, X. Thermal characteristics of plasma actuators in turbulent boundary layer. In Proceedings of the AIAA AVIATION 2020 FORUM, Virtual Event, 15–19 June 2020. [CrossRef]
169. Kaneko, Y.; Nishida, H.; Tagawa, Y. Visualization of the Electrohydrodynamic and Thermal Effects of AC-DBD Plasma Actuators of Plate- and Wire-Exposed Electrodes. *Actuators* **2022**, *11*, 38. [CrossRef]
170. Aberoumand, S.; Jafarimoghaddam, A.; Aberoumand, H. Numerical Investigation on the Impact of DBD Plasma Actuators on Temperature Enhancement in the Channel Flow. *Heat Transfer—Asian Res.* **2017**, *46*, 497–510. [CrossRef]
171. Benmoussa, A.; Belasri, A.; Harrache, Z. Numerical investigation of gas heating effect in dielectric barrier discharge for Ne-Xe excilamp. *Curr. Appl. Phys.* **2017**, *17*, 479–483. [CrossRef]
172. Zhang, X.; Zhao, Y.; Yang, C. Recent developments in thermal characteristics of surface dielectric barrier discharge plasma actuators driven by sinusoidal high-voltage power. *Chin. J. Aeronaut.* **2022**, *36*, 1–21. [CrossRef]
173. Preventing the Dangerous Formation of Ice on Aircraft | PHOBIC2ICE Project | Results in Brief | H2020 | CORDIS | European Commission. Available online: https://cordis.europa.eu/article/id/386841-preventing-the-dangerous-formation-of-ice-on-aircraft (accessed on 6 November 2022).

174. Volpe, A.; Gaudiuso, C.; Ancona, A. Laser Fabrication of Anti-Icing Surfaces: A Review. *Materials* **2020**, *13*, 5692. [CrossRef]
175. Alsabagh, A.S.Y.; Tiu, W.; Xu, Y.; Virk, M.S. A Review of the Effects of Ice Accretion on the Structural Behavior of Wind Turbines. *Wind. Eng.* **2013**, *37*, 59–70. [CrossRef]
176. Liu, Y.; Kolbakir, C.; Hu, H.; Hu, H. A comparison study on AC-DBD plasma and electrical heating for aircraft icing mitigation. In Proceedings of the AIAA Aerospace Sciences Meeting, Kissimmee, FL, USA, 8–12 January 2018. [CrossRef]
177. Chen, J.; Liang, H.; Wu, Y.; Wei, B.; Zhao, G.; Tian, M.; Xie, L. Experimental Study on Anti-Icing Performance of NS-DBD Plasma Actuator. *Appl. Sci.* **2018**, *8*, 1889. [CrossRef]
178. Little, J.; Takashima, K.; Nishihara, M.; Adamovich, I.; Samimy, M. Separation Control with Nanosecond-Pulse-Driven Dielectric Barrier Discharge Plasma Actuators. *AIAA J.* **2012**, *50*, 350–365. [CrossRef]
179. Kolbakir, C.; Hu, H.; Liu, Y.; Hu, H. A hybrid anti-/de-icing strategy by combining ns-dbd plasma actuator and superhydrophobic coating for aircraft icing mitigation. In Proceedings of the AIAA Scitech 2019 Forum, San Diego, CA, USA, 7–11 January 2019. [CrossRef]
180. Waldman, R.M.; Li, H.; Guo, H.; Li, L.; Hu, H. An experimental investigation on the effects of surface wettability on water runback and ice accretion over an airfoil surface. In Proceedings of the 8th AIAA Atmospheric and Space Environments Conference, Washington, DC, USA, 13–17 June 2016. [CrossRef]
181. Zheng, X.; Song, H.; Bian, D.; Liang, H.; Zong, H.; Huang, Z.; Wang, Y.; Xu, W. A hybrid plasma de-icing actuator by using SiC hydrophobic coating-based quartz glass as barrier dielectric. *J. Phys. D Appl. Phys.* **2021**, *54*, 375202. [CrossRef]
182. Liu, Y.; Kolbakir, C.; Starikovskiy, A.Y.; Miles, R.; Hu, H. An experimental study on the thermal characteristics of NS-DBD plasma actuation and application for aircraft icing mitigation. *Plasma Sources Sci. Technol.* **2019**, *28*, 014001. [CrossRef]
183. Liu, Y.; Kolbakir, C.; Hu, H.; Meng, X.; Hu, H. An experimental study on the thermal effects of duty-cycled plasma actuation pertinent to aircraft icing mitigation. *Int. J. Heat Mass Transf.* **2019**, *136*, 864–876. [CrossRef]
184. Wei, B.; Wu, Y.; Liang, H.; Chen, J.; Zhao, G.; Tian, M.; Xu, H. Performance and mechanism analysis of nanosecond pulsed surface dielectric barrier discharge based plasma deicer. *Phys. Fluids* **2019**, *31*, 091701. [CrossRef]
185. Fang, X.; Su, Z.; Song, H.; Liang, H.; Xie, L.; Liu, X. Parametric Investigation of High-voltage Plasma Discharge for Aircraft Icing Mitigation. In Proceedings of the 2021 IEEE 2nd China International Youth Conference on Electrical Engineering, CIYCEE 2021, Chengdu, China, 15–17 December 2021. [CrossRef]
186. Su, Z.; Liang, H.; Zong, H.; Li, J.; Fang, X.; Wei, B.; Chen, J.; Kong, W. Geometrical and electrical optimization of NS-SDBD streamwise plasma heat knife for aircraft anti-icing. *Chin. J. Aeronaut.* 2022; *in press*. [CrossRef]
187. Meng, X.; Hu, H.; Li, C.; Abbasi, A.A.; Cai, J.; Hu, H. Mechanism study of coupled aerodynamic and thermal effects using plasma actuation for anti-icing. *Phys. Fluids* **2019**, *31*, 037103. [CrossRef]
188. Kolbakir, C.; Hu, H.; Liu, Y.; Hu, H. An experimental study on different plasma actuator layouts for aircraft icing mitigation. *Aerosp. Sci. Technol.* **2020**, *107*, 106325. [CrossRef]
189. Lindner, M.; Pipa, A.V.; Karpen, N.; Hink, R.; Berndt, D.; Foest, R.; Bonaccurso, E.; Weichwald, R.; Friedberger, A.; Caspari, R.; et al. Icing Mitigation by MEMS-Fabricated Surface Dielectric Barrier Discharge. *Appl. Sci.* **2021**, *11*, 11106. [CrossRef]
190. Jia, Y.; Liang, H.; Zong, H.; Wei, B.; Xie, L.; Hua, W.; Li, Z. Ice shape modulation with nanosecond pulsed surface dielectric barrier discharge plasma actuator towards flight safety. *Aerosp. Sci. Technol.* **2022**, *120*, 107233. [CrossRef]
191. Gao, T.X.; Luo, Z.B.; Zhou, Y.; Yang, S.K. A novel de-icing strategy combining electric-heating with plasma synthetic jet actuator. *Proc. Inst. Mech. Eng. Part G J. Aerosp. Eng.* **2020**, *235*, 513–522. [CrossRef]
192. Hu, H.; Meng, X.; Cai, J.; Zhou, W.; Liu, Y.; Hu, H. Optimization of Dielectric Barrier Discharge Plasma Actuators for Icing Control. *J. Aircraft* **2020**, *57*, 383–387. [CrossRef]
193. Tanaka, T.; Matsuda, H.; Takahashi, T.; Chiba, T.; Watanabe, N.; Sato, H.; Takeyama, M. Experimental Study on the Snow-fall Flow Control of Backward-Facing Steps Using a High-Durability Designed Plasma Electrode. *Actuators* **2022**, *11*, 313. [CrossRef]
194. Lilley, A.J.; Roy, S.; Michels, L.; Roy, S. Performance recovery of plasma actuators in wet conditions. *J. Phys. D Appl. Phys.* **2022**, *55*, 155201. [CrossRef]
195. Abdollahzadehsangroudi, M.; Pascoa, J.; Rodrigues, F. System for Ice Detection/Prevention and Flow Control Based on the Impression of Sliding Plasma Actuators with Dielectric Discharge Barrier. U.S. Patent WO2018060830A1, 25 September 2017.
196. Rodrigues, F.F.; Abdollahzadeh, M.; Pascoa, J. Dielectric Barrier Discharge Plasma Actuators for Active Flow Control, Ice Formation Detection and Ice Accumulation Prevention. In Proceedings of the ASME 2020 Fluids Engineering Division Summer Meeting collocated with the ASME 2020 Heat Transfer Summer Conference and the ASME 2020 18th International Conference on Nanochannels, Microchannels, and Minichannels, Virtual, 13–15 June 2020; Volume 2. [CrossRef]
197. Xie, L.; Liang, H.; Zong, H.; Liu, X.; Li, Y. Multipurpose distributed dielectric-barrier-discharge plasma actuation: Icing sensing, anti-icing, and flow control in one. *Phys. Fluids* **2022**, *34*, 071701. [CrossRef]

Review

Development of Composite Hydraulic Actuators: A Review

Marek Lubecki [1], Michał Stosiak [1], Paulius Skačkauskas [2,*], Mykola Karpenko [2], Adam Deptuła [3] and Kamil Urbanowicz [4]

1 Faculty of Mechanical Engineering, Wrocław University of Science and Technology, 7/9 Łukasiewicza St., 50-371 Wrocław, Poland
2 Faculty of Transport Engineering, Vilnius Gediminas Technical University, Saulėtekio al. 11, LT-10223 Vilnius, Lithuania
3 Faculty of Production Engineering and Logistics, Opole University of Technology, 76 Prószkowska St., 45-758 Opole, Poland
4 Faculty of Mechanical Engineering and Mechatronics, West Pomeranian University of Technology, 70-310 Szczecin, Poland
* Correspondence: paulius.skackauskas@vilniustech.lt

Abstract: With the development of engineering materials, as well as the growing requirements for weight reduction and the reduction of energy consumption by mechanical systems, attempts have been made to utilize composite materials in the design of hydraulic cylinders. In many cases, the reduction in the weight of the actuators may lead to a reduction in the values of bending moments acting on the booms of working machines, as well as leading to a reduction in the power demand in drive systems. The use of composite materials can also increase the reliability of cylinders in corrosive environments and places with strong electromagnetic fields. This paper presents the development of hydraulic actuators made of composite materials, presenting both the achievements of research centers and commercial companies. The main research and engineering problems are presented along with the methods of solving them resulting from the literature available. The directions for further research that should be undertaken in order to increase reliability, improve efficiency, and reduce weight are also outlined.

Keywords: CFRP; filament winding; weight reduction; hydraulic drive; composite; actuators

Citation: Lubecki, M.; Stosiak, M.; Skačkauskas, P.; Karpenko, M.; Deptuła, A.; Urbanowicz, K. Development of Composite Hydraulic Actuators: A Review. *Actuators* **2022**, *11*, 365. https://doi.org/10.3390/act11120365

Academic Editor: Zhuming Bi

Received: 26 October 2022
Accepted: 5 December 2022
Published: 6 December 2022

Publisher's Note: MDPI stays neutral with regard to jurisdictional claims in published maps and institutional affiliations.

1. Introduction

A hydrostatic drive is one of the most popular types of drives in mechanical engineering. It consists of a pump that converts kinetic energy (most often shaft rotation) into the energy of fluid pressure, a set of valves whose task is to control the flow parameters (flow rate, direction, and pressure), receivers (motors and actuators) converting fluid pressure energy into mechanical energy, and other elements such as hoses or accumulators. The popularity of this type of systems is a result of a number of advantages, such as a high ratio of power transmitted to the mass of the system, the possibility of arranging the elements on the machine, or the possibility of obtaining almost any functional structure as well as the ease of automation [1].

Hydraulic cylinders are actuators whose task is to convert the energy of fluid pressure into the mechanical energy of a reciprocating movement. The fluid acts on the movable element of the actuator (piston or plunger) causing a force directly proportional to the pressure and the area of the element. A double-rod hydraulic cylinder is shown on Figure 1. It consists of a front joint (1), piston rod (2), front-end cap (3), piston (4), barrel (5), rear-end cap (6), and a tail joint (7) [2]. Standard hydraulic cylinders are designed to operate at working pressures up to 25 MPa; versions with higher working pressures are also available on request. Diameter and pitch ranges vary from a few millimeters [3,4] to several meters [5]. Many years of perfecting the design of actuators has allowed a general efficiency close to 100% [6,7] to be obtained.

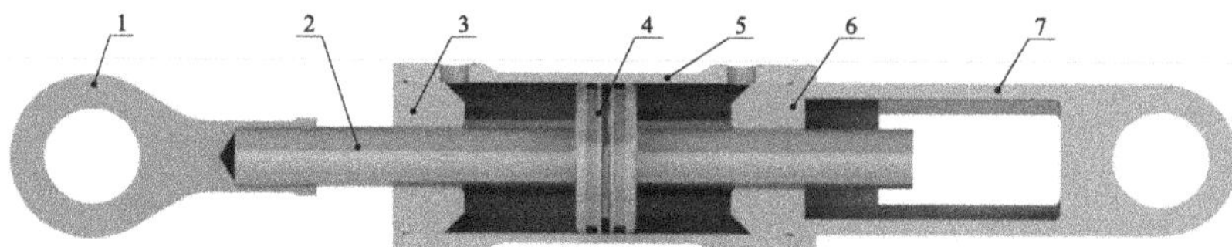

Figure 1. Double tie-rod hydraulic cylinder: 1—front joint, 2—piston rod, 3—front-end cap, 4—piston, 5—barrel, 6—rear-end cap, and 7—tail joint [2].

When choosing a material, the issues of strength, durability, manufacturing technology, and cost must be taken into account [8]. Environmental issues are becoming more and more important, forcing the reduction or elimination of harmful materials, emissions, and the amount of waste produced. Efforts are also being made to design structures in such a way that they consumes as little energy as possible during operation. Weight reduction is a desirable step in this context, as it is usually associated with the increased energy efficiency of the machine [9,10]. Conventional hydraulic cylinders are mostly made of steel or aluminum alloys, and the seals are made of plastics and bronze. The most popular alloys include structural steel S355J2G3 [11], austenitic stainless steel AISI 304, and aluminum alloy Al7075 [12,13].

A composite is a material obtained by combining two or more base materials with (most often) radically different properties. The resulting material has better and (or) new properties compared to the components used separately or resulting from their simple summation. In most cases, one of the materials takes the role of a matrix (continuous, bonding medium), while the rest becomes the filler (reinforcement) [14,15].

Among the fibrous composites (in which the reinforcement is realized by fibers), we can distinguish two main types: those reinforced with short fibers and those reinforced with continuous fibers. Fibers (most often glass, carbon, or aramid) are characterized by a high modulus of elasticity along their axis and high tensile strength [16]. There are many types of fibers on the market that differ in terms of parameters and the materials from which they are made and thus also the price.

The second component of the composite material is the matrix, the purpose of which is to bond the reinforcement material and to enable load transfer between the fibers. It can also stop or slow down the propagation of cracks initiated in the reinforcement and protect the fibers from adverse environmental conditions [17]. The most commonly used matrix materials include polyester, vinyl ester, or epoxy resins, as well as thermoplastics (polyethylene, polypropylene, and polyamide).

Year by year, there is growing interest in the use of fiber-reinforced composite materials. This is due to their high strength, low weight, and corrosion resistance. There are many methods of manufacturing fiber-reinforced polymeric composite materials, such as lamination (hand, spray, vacuum bag), infusion molding, winding, weaving, and pultrusion. In the production of high-pressure cylinders and tanks, mainly winding and weaving methods have been used [18].

2. Design and Research Issues

When designing a hydraulic cylinder, it is necessary to take into account the differences resulting from the nature of the composite material. A number of problems that the designer faces during the design process of such an element are presented below.

2.1. Material Anisotropy

The use of fiber-reinforced composites in a load-bearing structure entails the need to take into account the significant anisotropy of such materials. A single unidirectional layer shows high strength and a high Young's modulus along the fibers, but much lower parameters in the direction perpendicular to the reinforcement fibers. The classical methods of calculating the barrel wall thickness taken from conventional actuator designs cannot be used [19]. Designers must therefore use computational methods developed for multilayer elements, often combining anisotropic composite layers with isotropic steel or aluminum.

2.2. Piston–Barrel Interface

It is crucial to ensure the proper frictional cooperation of the piston seal with the inner surface of the barrel. The phenomena occurring at the contact of polymer seals and the steel or aluminum inner layer of a classic cylinder are already well known. Changing the material, however, makes it necessary to describe these phenomena anew. Material and technological solutions are also sought to make the inner layer as resistant to wear as possible and with the lowest possible coefficient of friction [20,21]. One of the solutions is the use of a thin-walled steel pipe (so-called liner) on which a composite overwrap is made.

2.3. Connecting the Barrel with the End Caps

In conventional hydraulic actuators, threaded or tie-rod connections are most often used to connect the barrel with the end caps [22]. While the latter solution is applicable to composite cylinders, cutting threads in the composite material is not advised [23]. Designing an actuator without tie-rods requires an adhesive or form-fit connection. An alternative solution is to cut threads in the liner material.

3. Research Work on the Development of Composite Cylinders

The scientific literature offers various approaches to the issue of designing a hydraulic actuator with a composite barrel.

In as early as 1986, Hashimoto et al. [24] presented an innovative composite hydraulic cylinder with an integrated piston position sensor. They presented a method of connecting composite parts made by winding with steel elements with the use of pins, with which the fiber was wrapped during manufacturing. The advantage of this form-fit connection is the continuity of the fibers and the avoidance of mechanical processing of the composite as well as the elimination of the adhesive connection. This method was tested both statically and with impulse loads. A magnetostrictive piston position sensor was integrated in the actuator barrel. Such a sensor was characterized by high accuracy and repeatability of reading, and the wound barrel structure was well suited to integrating this solution. The described prototype had an internal diameter of 50 mm, a piston rod diameter of 28 mm, a stroke of 500 mm, and a weight of 4.7 kg. This was a 2/3 reduction in the weight of the element compared to a conventional steel actuator. A similar design was presented by Sumali et al. [25]. The sensor consisted of a coil wound up during the manufacturing of the barrel and thus embedded in its wall, and a steel piston rod that also served as the core of the coil. The sliding of the piston rod into the coil changes the position of the core in the coil, which changes the inductance and thus the impedance of the coil. The impedance measurement allowed the position of the actuator piston to be determined. The resulting cylinder had an inside diameter of 38.1 mm, an outside diameter of 44.45 mm, and a stroke of 203.2 mm.

Mantovani et al. [26,27] presented a method of designing a composite cylinder with a steel liner in terms of strength and buckling. The analysis presented by the authors was performed for an actuator with a working pressure of 35 MPa, an internal diameter of 300 mm, and a stroke of 1960 mm. For this purpose, the authors used the solution of the Lame problem modified for anisotropic materials. Design constraints have been adopted to ensure the appropriate immediate strength of the cylinder as well as the appropriately low values of the circumferential deformation of the cylinder and the axial deformation of the liner. Due to the fact that the considered barrel would also transmit axial forces, the authors designed layers in it, in which the fibers would be arranged along the axis of the element. Finally, a modal analysis was carried out, which did not show any significant differences in the natural frequencies between the composite barrel and the reference (steel) one. The designed barrel had a mass of 80 kg compared to 407 kg of the reference barrel, which translates into a weight reduction of approx. 80%.

Ritchie et al. [28] performed the FEM analysis and manufactured and tested the piston and the front cap from a composite material reinforced with carbon fiber. The elements were made using subtractive methods from ready-made pipes, which is not recommended in the case of composite materials. The authors defined the machinability of the composite as medium and the obtained surface quality and dimensional accuracy as low. The conducted corrosion tests as well as the comparison of the weights of both elements showed the superiority of the composite material over steel. The paper does not present the tests of the complete actuator.

Scholz and Kroll [20] considered removing the steel liner from the inside of the barrel and replacing it with a nanocomposite coating with properties ensuring appropriate tribological conditions at the barrel–piston interface. The authors presented methods of calculating stresses in individual composite layers, as well as a method of producing nanocomposite coatings (Figure 2). However, the process of selecting and optimizing the laminate structure is not presented in detail. The manufactured cylinders were tested, the deformation as a function of internal pressure and the abrasive wear of the nanocomposite layer were determined (Figure 3). The prototypes had an internal diameter of 85.5 mm, a barrel length of 250 mm, and a barrel thickness of 2.8 mm. The operating pressure was 35 MPa with a burst pressure of 700 MPa. The authors confirmed the suitability of a selected coating as a sliding material in a cylinder.

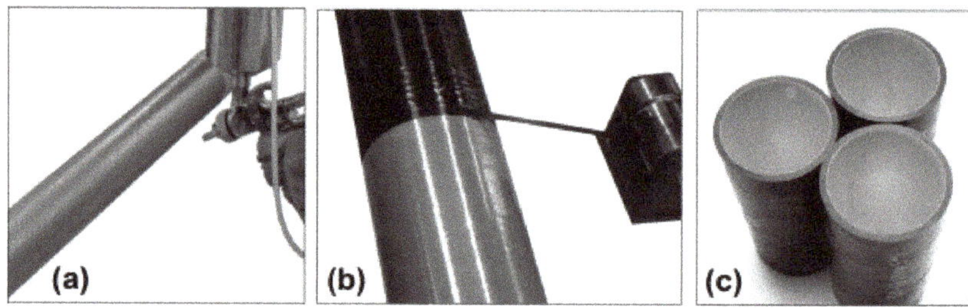

Figure 2. Manufacturing of CFRP barrels with internal nano-gelcoat: (**a**) sprayed application of particle–resin mixture; (**b**) filament-winding process; and (**c**) cylinders after curing and demolding [20].

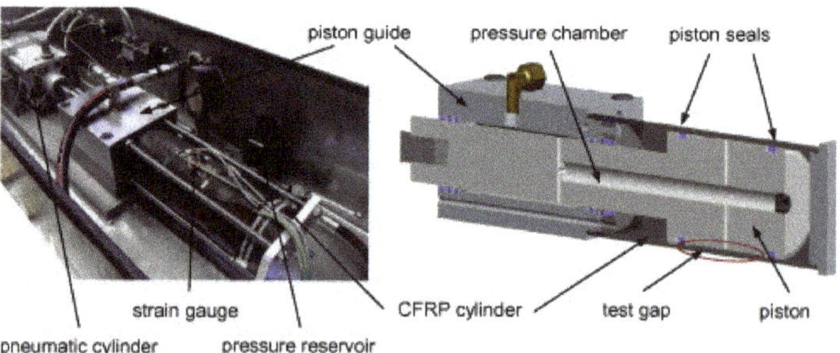

Figure 3. Test configuration (left) and cross-sectional view (right) of the test set-up [20].

Nowak and Schmidt [29–31] conducted an in-depth theoretical analysis of composite pipes with a steel liner, deriving analytical formulas allowing the stress distribution to be calculated, taking into account the loads of internal pressure, axial force, and temperature. They also conducted experimental verification with the use of strain gauges and acoustic emission. The analysis was made for barrels with a length of 420 mm, with a liner with an internal diameter of 185 mm, and an external diameter of 193 mm and 204 mm. The thickness of the composite winding ranged from 7 mm to 15 mm. The internal pressure up to 75 MPa was used in the tests. The authors showed that as long as the liner works in the range of elastic deformations, the composite overwrap does not transfer significant loads.

However, when the liner is in the range of plastic deformation, the composite provides the structural integrity with the necessary strength.

Various methods of connecting the end caps to the composite barrel were considered by Zhang et al. [32]. The authors proposed a proprietary method called inlaid connection consisting of connecting the front-end cap, liner, and the rear-end cap using a composite overwrap. According to the authors, this ensures a more secure connection of elements, easier service, and more economical use of materials. The paper presents laminate strength calculations assuming a constant winding angle of 63.43° and a laminate thickness of 5 mm. The calculations were carried out for a cylinder with an internal diameter of 50 mm, a piston rod diameter of 28 mm, a stroke of 500 mm, a working pressure of 25 MPa, and a maximum pressure of 37.5 MPa. However, the authors did not take into account the strength of the steel liner, as it was assumed that the entire load would be carried by the composite.

Lightweight hydraulic actuators can also be used in robotics. El Asswad et al. [33,34] designed and manufactured a prototype, and they subjected a hydraulic cylinder with a composite barrel to bench tests and intended for it to be used as the drive of a humanoid robot. The process of barrel design was shown taking into account the material anisotropy using the Tsai–Hill criterion. The genetic algorithm was used to find the optimal geometric dimensions of individual actuator elements while minimizing the final mass of the assembly. Two prototypes of actuators were made one with tie-rods and the other, where the end caps were connected to the barrel using adhesive. The friction coefficient was determined, which turned out to be quite high (0.61), which was related to the properties of the composite surface. The presented design had an internal diameter of 25 mm, a piston rod diameter of 12 mm, and a stroke of 110 mm. It was tested at a supply pressure of 4 MPa. It is worth noting the use of a piston was made possible by 3D printing.

In a series of publications, Solazzi and colleagues presented the design process of a composite hydraulic cylinder along with its implementation and testing. Basic strength calculations of the barrel, front, and end caps, as well as the piston rod, were presented. The cylinder calculations took into account the material anisotropy, and 0°, ±45°, and 90° were assumed as the allowed fiber angles [35]. The criterions of Tresca and Huber–Mises were used as the strength criteria. The authors provided several variants of the individual elements (made of steel, aluminum alloy, aluminum alloy, and an epoxy–carbon composite) with a weight comparison. A prototype of a barrel, consisting of an inner steel liner and a carbon–epoxy overwrap, manufactured using the filament winding method was also pressure tested. Winding was performed by hand and the element was post-treated to obtain a uniform outside diameter. Next, the authors presented a process of designing a telescopic actuator made of a composite material [36]. The material selection and the basic strength calculations for an isotropic material and also partially for an anisotropic material, as well as finite element analysis, were presented. Particular emphasis was placed on preventing buckling of the element. Once again, the Huber–Mises hypothesis, commonly used for metallic materials, was used to assess the strength of the composite material. The authors took into account the difficulties in joining the actuator parts resulting from the fact that they were made of different materials and proposed the adhesive joint as the best suited. The next paper presents strength calculations for the cylinder, taking into account its layering (the presence of an internal aluminum liner and an external composite reinforcement) [37]. The Huber–Mises hypothesis was used to assess the strength. Both the end caps and the liner were made of aluminum alloy and then joined together by welding (Figure 4). A composite reinforcement made of carbon fabric and epoxy resin was placed on the prepared element by means of hand laminating with the help of a vacuum bag (Figure 5). Additionally, the paper presents a piston rod made of a composite material with a method of connecting an aluminum piston and a front joint with a composite piston rod by means of a form fit. The last part of the article presents experimental tests of a finished actuator. The tested actuator had a liner with an internal diameter of 50 mm and a wall thickness of 3 mm with a 4 mm composite overwrap. The piston rod diameter was 40 mm and the stroke diameter was 500 mm. Next, the problem of stress variability in a cylinder consisting of an aluminum liner reinforced with an epoxy–glass laminate was described [38]. The variability of the stress will result from the variability of operating parameters, such as internal pressure, as well as from the geometrical variability (manufacturing inaccuracies) and material properties. The coefficient of variation was introduced, which is a measure of the spread of actual stress values around the mean value, which increased with the increase in the input parameters tolerance. Solazzi and Buffoli in their paper from 2021 [39] described the design process of a composite cylinder taking into account the action of fatigue loads. A membrane model of the material modified with anisotropy was adopted for strength calculations. The Tsai–Hill criterion was adopted as the failure hypothesis, which is better than the previously used criteria of Tresca or Huber–Mises to assess the strength of composites. The fatigue strength was estimated using from available S-N curves for composite materials.

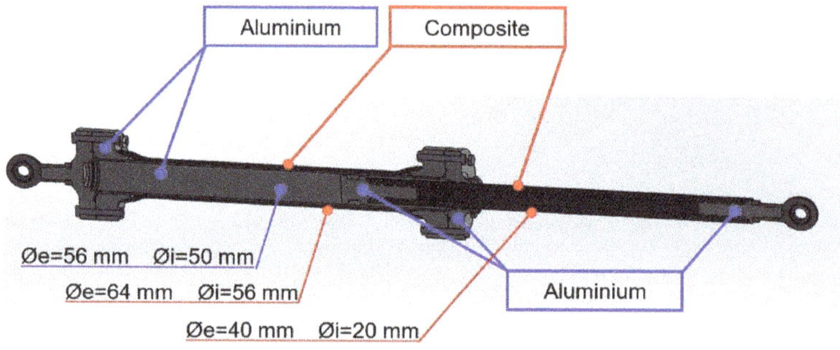

Figure 4. Geometrical model of an actuator designed by Solazzi [37].

Figure 5. An actuator prototype designed by Solazzi [37].

Praveen Kumar and Lee [40] developed hybrid piston rods combining a steel core with an external composite reinforcement. They performed buckling analysis using the finite element method as well as experimental tests for a wide range of elements with a length of 650 mm and outer diameters ranging from 7.5 mm to 30 mm. The CFRP content ranged from 0% for solid steel rods to 100% for all-composite designs. It has been shown that the increase in the proportion of composites in the element increases its resistance to buckling. This was also influenced by the angle of the fibers, where the reinforcement at 0° worked best; however, it should not appear alone, but rather in combination with the layers of ±45° and 90°. The authors suggest that the use of hybrid piston rods will allow for a significant reduction in the mass of the cylinder without sacrificing its buckling resistance, which in the case of these elements is one of the key parameters taken into account in the selection process.

A different approach to multi-material composite elements was presented by Siegfarth et al. [41]. The authors proposed two designs of a monolithic piston integrated with the piston rod and seal manufactured using the MMAM (multi material additive manufacturing) method. The components shown had a piston diameter of 3.6 mm, a seal outer diameter of approximately 4.5 mm, and a piston rod diameter of 2.5 mm. It is worth noting that the entire element was made together with sealing in one process using

PolyJet 3D printing technology (Figure 6). The paper presents experimental studies of water absorption, friction, and wear. Despite achieving lower operational parameters than conventional designs, the authors indicate a high potential for the further development of the presented solutions.

Figure 6. Three-dimensional-printed pistons integrated with the piston rod and seal [41].

The complete design, manufacturing, and testing process was presented by Li et al. [2]. The authors described a double-rod actuator with a metallic liner reinforced with a CFRP overwrap (Figure 7). The end caps were connected to the barrel liner with screws, which ensured the possibility of servicing. The paper includes a liner material selection process and FEM calculations for a multi-material barrel, as well as a description of the manufacturing process. The prototype consisted of an aluminum alloy liner on which composite layers were applied by means of filament winding and hand laminating, and it had an internal diameter of 75 mm, a piston rod diameter of 30 mm, and a barrel length of 80 mm. The design pressure was 21 MPa with a maximum pressure of 31.5 MPa. The designed laminate had layers arranged at an angle of 0 and 90 to the axis of the cylinder. The FEM calculations did not use the strength hypotheses specific to composite materials, and the stresses were presented according to the Huber–von Mises hypothesis. The authors performed leakage and friction tests, as well as deformation measurements. The proposed design approach managed to reduce the barrel weight by more than 56% while maintaining the current operating parameters.

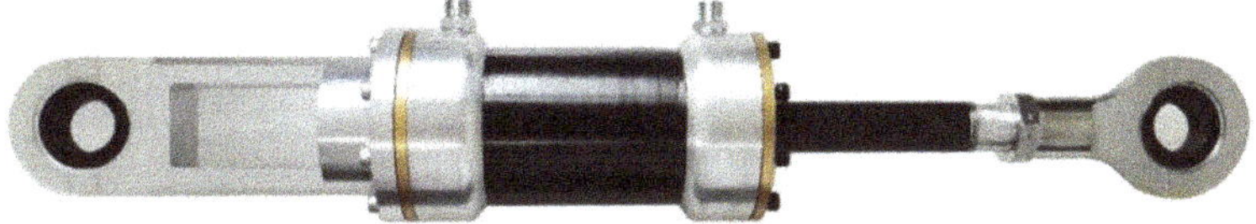

Figure 7. CFRP hydraulic cylinder prototype designed by Li et al. [2].

Coskun and Sahin presented a novel approach to the design of a composite hydraulic cylinder [42]. This solution was inspired by composite pressure vessels, in which the composite winding is made not only on the cylindrical part, but also on the domes (Figure 8). In the case of an actuator, this solves the issue of connecting the end caps to the barrel. The authors focused on the computational part in which they optimized the structure of the barrel material using FEM and steepest ascent methods. The described solution allows the mass of the element to be significantly reduced and the problem of connecting the barrel with the end caps to be eliminated. On the other hand, in the event of failure, it significantly hinders the servicing of such an actuator.

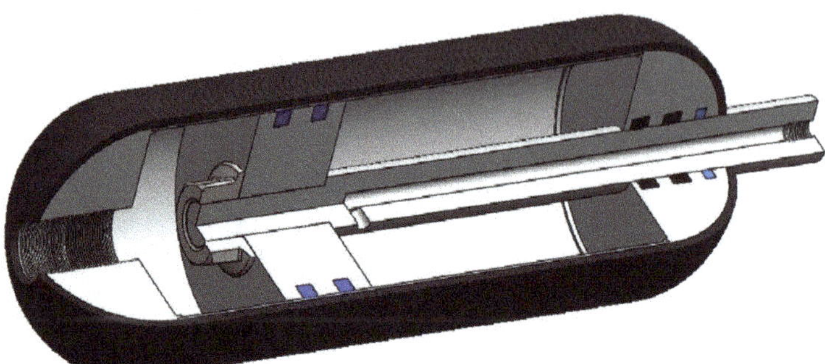

Figure 8. Schematic visualization of the composite hydraulic cylinder design by Coskun and Sahin [42].

Lubecki et al. conducted tribological tests using the ball on disk method [43,44] and adhesive tests using the pull-off method [45,46] of potential materials for the inner coating of the composite barrel. The aim was to replace the steel liner used so far in most designs. From these tests, it was concluded that the best material was thermosetting polyurethane, which achieved the best adhesion results to the composite substrate, as well as a low coefficient of friction and low wear. The team also conducted thermomechanical tests on PET plastic confirming its suitability for the end caps of a hydraulic cylinder [47]. The authors presented the FEM analysis taking into account the non-linearity of the material and experimental validation on the completed end cap prototype (Figure 9). The presented end cap was intended for use in a cylinder with an internal diameter of 40 mm. Prototype tests were carried out at an internal pressure of 25 MPa.

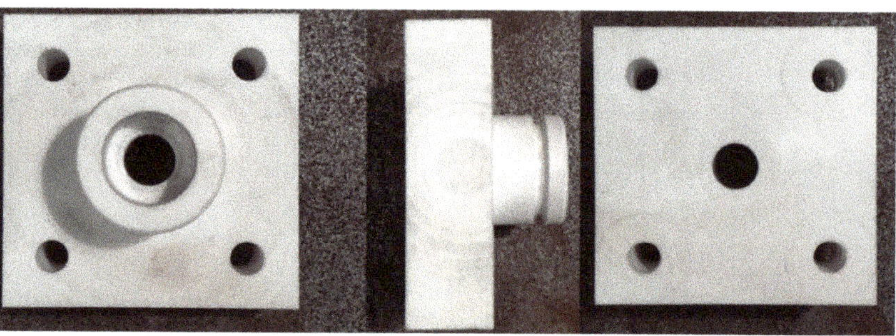

Figure 9. End cap made of PET plastic tested by Lubecki et al. in [47].

The next paper presents theoretical calculations and an experimental analysis of the deformation of a barrel of a tie-rod hydraulic cylinder [48]. The authors indicated that the load distribution in the real cylinder differs from that usually assumed in theoretical calculations. An example of calculating the strength of a composite cylinder using the classical laminate theory method was also presented. The authors pointed out that during the strength calculations of the composite barrel, it is crucial to correctly identify the loads. Even small deviations from the actual distribution may result, in the case of a strongly anisotropic material, in obtaining incorrect values of the stresses and strains in the layers.

4. Patents

Searching the patent databases, one can find a number of solutions of hydraulic and pneumatic actuators using composite materials in their design. Patent US4685384A [49] presents a hydraulic composite actuator, the barrel of which would consist of an internal liner ensuring appropriate conditions for cooperation with the piston and an external composite reinforcement. The integrity of the element is ensured by an additional longitudinal overwrap that connects the front cap with a tail joint and is fixed on the supporting ring by means of a form fit and an additional circumferential overwrap (Figure 10).

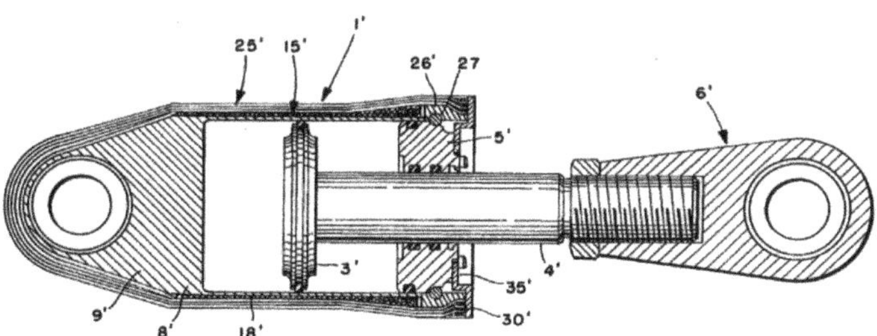

Figure 10. The connection of actuator elements proposed in patent US4685384A [49]: 1—barrel, 3—piston, 4—piston rod, 5—front cap, 6—front joint, 8—end cap, 9—tail joint, 15—hoop winding, 25—longitudinal winding, 26—retainer ring, 27—retainer wire, 30—circumferential winding, and 35—end plate.

A different approach is proposed in patent US5415079A [50]. Here, the connection is made by specially shaping both ends of the liner. The helical winding is made in such a way that the turn takes place on the outer parts of the diameter change, which results in a permanent connection between the liner and the composite. End caps are connected to the cylinder by a threaded connection (Figure 11).

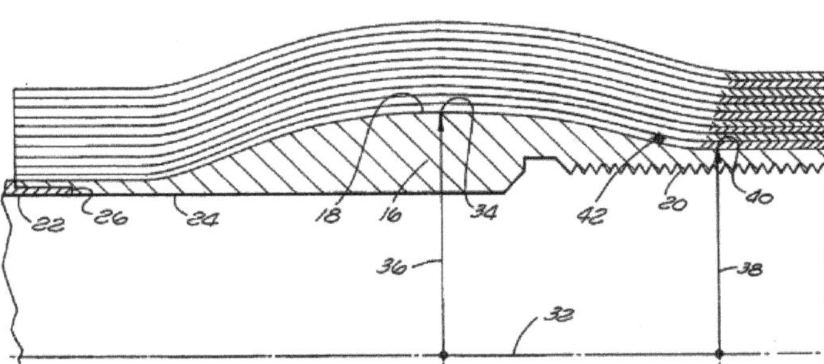

Figure 11. A form-fit connection of liner and composite overwrap shown in US5415079A [50].

Patent US5435868A [51] describes a method of manufacturing composite barrels with internal threads at the ends using the filament winding method (Figure 12). It uses a cylindrical core with threaded inserts (36) placed on it. The barrels (10), with smooth parts (12) and threaded parts (20), (22) are produced by a winding process and separated from each other by distances (58). After the process, the core is pulled from the center and the threaded inserts are unscrewed from the barrels. In this way, the outer thread contour of the inserts is transferred to the inner thread of the composite barrel. The advantage of this method is that it can produce multiple elements on a single core.

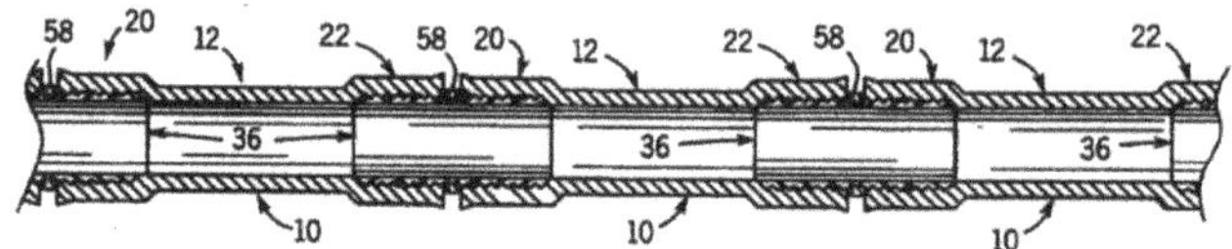

Figure 12. A method of producing composite barrels with internal threads at the ends [51].

Patent US5465647A [52] shows a method of connecting a composite barrel with an end cap (Figure 13) by making a groove (34) inside the barrel (10) cooperating with a barb (38) of the cover (14). In order to increase the stiffness of the connection, after its completion, an expanding ring (42) is used, whose task is to press the skirt (36) of the cover against the inner wall of the barrel. As a rule, such a connection is inseparable.

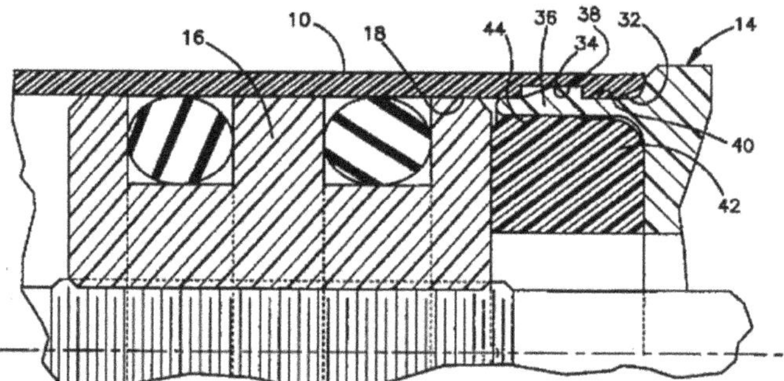

Figure 13. The method of connecting the composite barrel with the end cap [52].

Another approach to detachably connect the barrel with the end cap was presented in US7240607B2 [53] (Figure 14). The barrel (12) has a groove (28) made on its outer surface (26) in which holes (30) are drilled radially. The groove (28) is designed to receive the split ring (20) also with holes (44). An end cap (16) with tapped holes (40) matching the holes in the ring (44) and barrel (30) is placed inside the cylinder. The connection is made by means of screws (22). The advantages of this solution are the ease of connecting and disconnecting elements using popular tools, the possibility of using pipes available on the market, and improving the load distribution through the use of a ring. Due to the machining of the composite cylinder, the immediate and fatigue strength of such a connection may be reduced.

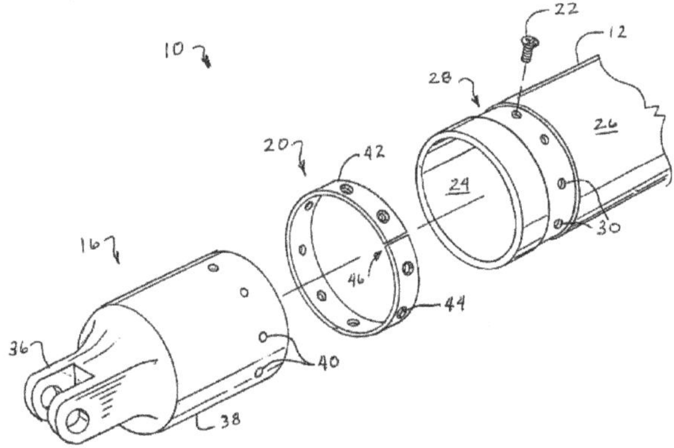

Figure 14. The method of connecting the cylinder and end cap by means of a split ring and screws [53].

5. Commercial Designs

In 2014, Parker Hannifin introduced a line of Lightraulics actuators consisting of a whole range of elements in which the barrel is reinforced with a composite overwrap. According to the manufacturer, these cylinders can work at operating pressures up to 70 MPa with a simultaneous weight reduction of up to 60%, increased corrosion resistance, reduced susceptibility to vibration, and high resistance to fatigue loads [54,55]. Inside the barrel, there is a steel or aluminum liner that ensures tightness, appropriate conditions for cooperation with the piston, and transmitting axial forces. The composite overwrap is made only circumferentially (Figure 15a). In most designs, the end caps are connected to the steel liner by means of a threaded connection. The exceptions are the cylinders with a very short stroke (the ratio of the piston diameter to the stroke is close to 1:1), where there is a variant of connecting the end caps using tie-rods (d). Short stroke cylinders (referred to as heavy duty) come in piston diameters from 160 mm to 310 mm and a stroke from 150 mm to 300 mm. More slender solutions, mainly intended for marine applications, can be ordered for a piston diameter up to 125 mm and a stroke up to 2500 mm.

Figure 15. Examples of Parker Lightraulics actuator designs: (**a**) a cross-section showing the cylinder structure, (**b**) a cylinder with a high stroke-to-diameter ratio, (**c**) a cylinder with a low stroke-to-diameter ratio, and (**d**) a cylinder with a low stroke-to-diameter ratio with tie-rods [55].

Composite cylinder hydraulic cylinders called PolySlide are manufactured by Polygon [56,57]. The manufacturer declares increased resistance to impact and corrosion as well as improved conditions of cooperation with the seal in relation to steel surfaces. The manufacturer offers solutions with several variants of the end caps connection with the cylinder (Figure 16): press-fit, using snap rings, pins, adhesive connections, and even threaded connections.

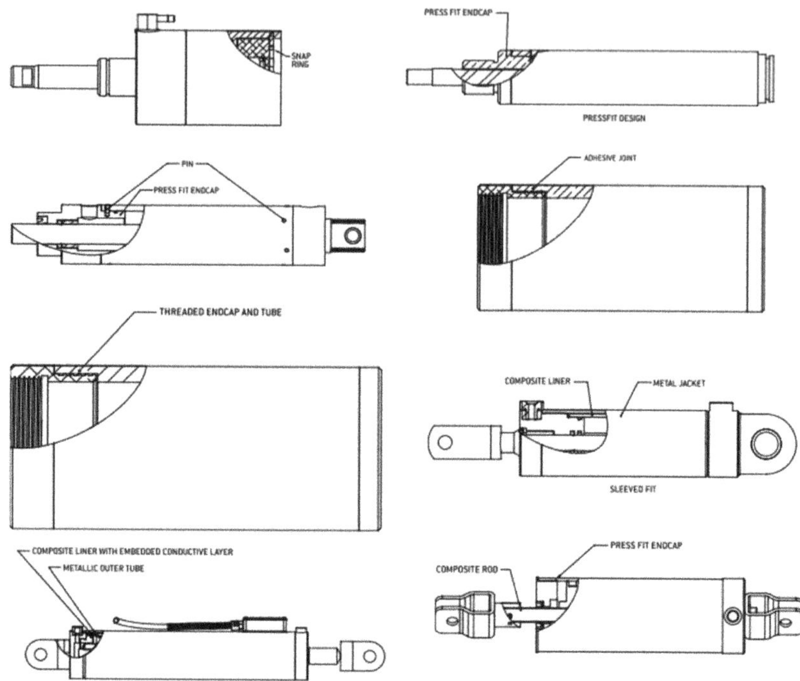

Figure 16. Various designs offered by Polygon [56].

Composite actuators are also offered by Liebherr [58]. The company describes advantages such as high strength and stiffness, low weight, good fatigue properties, and corrosion resistance.

6. Conclusions and Outlook

The development of composite hydraulic cylinders has gained momentum in recent years. This is indicated by the growing number of publications in scientific journals, patents, and the appearance of commercial structures on the market. The patents focus mainly on the methods of producing composite cylinders and their combination with end caps. The authors of scientific papers also deal with the issues of strength calculations of composite cylinders and the methods of eliminating the steel liner and replacing it with polymer materials. Creative ways to use the properties of composite materials to embed the piston position sensors in the structure of the element are also indicated.

Development work has been carried out on a wide spectrum of sizes and working pressures from micro scale [41] to conventional sized elements [20,26,27,38,54]. With more conservative designs, it is already possible to achieve satisfactory performance results today [54], but for more innovative designs, more tests are needed [33,34,41].

Despite the laying of the foundations for the working structures of composite hydraulic cylinders, it seems necessary to carry out work in the following directions:

- Material selection and testing for a non-metallic liner. There is no doubt that there are many promising materials that should be subjected to this test. They can be both homogeneous materials as well as composites reinforced with particles in macro-, micro-, or nano-scales.
- Development and validation of an algorithm that optimizes the structure of the cylinder material (type of material and number and angle of layers) using the already published calculation methods.

- Conducting long-term tests of the developed prototypes to determine the structural integrity of the designs as well as liner-braid adhesion.
- An attempt to further eliminate metallic materials from the actuator structure (piston, piston rod, and end caps), which would improve its resistance to environmental conditions and eliminate the influence of the magnetic field on the operation of the element.

Author Contributions: Conceptualization, M.L., M.S., P.S., M.K., A.D., K.U.; writing—original draft preparation, M.L., M.S., P.S., M.K., A.D., K.U.; writing—review and editing, M.L., M.S., P.S., M.K., A.D., K.U. All authors have read and agreed to the published version of the manuscript.

Funding: This research received no external funding.

Data Availability Statement: Not applicable.

Conflicts of Interest: The authors declare no conflict of interest.

References

1. Rabie, M.G. *Fluid Power Engineering*; McGraw Hill: New York, NY, USA, 2009; ISBN 978-0-07-162246-2.
2. Li, Y.; Shang, Y.; Wan, X.; Jiao, Z.; Yu, T. Design and Experiment on Light Weight Hydraulic Cylinder Made of Carbon Fiber Reinforced Polymer. *Compos. Struct.* **2022**, *291*, 115564. [CrossRef]
3. Li, N. Multi-Degree-of-Freedom Manipulator Driven by Micro Hydraulic System. In Proceedings of the 2016 7th International Conference on Education, Management, Computer and Medicine (EMCM 2016), Shenyang, China, 29–31 December 2016; Atlantis Press: Paris, France, 2016.
4. Xia, J.; Durfee, W.K. Analysis of Small-Scale Hydraulic Actuation Systems. *J. Mech. Des.* **2013**, *135*, 091001. [CrossRef]
5. Truong, V.T.; Hwang, Y.L.; Cheng, J.K.; Tran, K.D. Dynamics Analysis and Numerical Simulation of Large-Scale Hydraulic Cylinder Actuators. In *Engineering Tribology and Materials IV*; Trans Tech Publications Ltd.: Zurich, Switzerland, 2020; Volume 900, pp. 14–19.
6. Xia, J.; Durfee, W.K. Experimentally Validated Models of O-Ring Seals for Tiny Hydraulic Cylinders. In Proceedings of the ASME/BATH 2014 Symposium on Fluid Power and Motion Control, Bath, UK, 10–12 September 2014; American Society of Mechanical Engineers: New York, NY, USA, 2014.
7. Kollek, W. *Microhydraulic Components and Systems—Fundamentals of Design, Modelling and Operation*; Oficyna Wydawnicza Politechniki Wrocławskiej: Wrocław, Poland, 2011; ISBN 978-83-7493-617-0.
8. Ashby, M.F. Chapter 2—The Design Process. In *Materials Selection in Mechanical Design*, 4th ed.; Ashby, M.F., Ed.; Butterworth-Heinemann: Oxford, UK, 2011; pp. 15–29, ISBN 978-1-85617-663-7.
9. Joost, W.J. Reducing Vehicle Weight and Improving U.S. Energy Efficiency Using Integrated Computational Materials Engineering. *JOM* **2012**, *64*, 1032–1038. [CrossRef]
10. Kaluza, A.; Kleemann, S.; Fröhlich, T.; Herrmann, C.; Vietor, T. Concurrent Design & Life Cycle Engineering in Automotive Lightweight Component Development. *Procedia CIRP* **2017**, *66*, 16–21. [CrossRef]
11. Marczewska, I.; Bednarek, T.; Marczewski, A.; Sosnowski, W.; Jakubczak, H.; Rojek, J. Practical Fatigue Analysis of Hydraulic Cylinders and Some Design Recommendations. *Int. J. Fatigue* **2006**, *28*, 1739–1751. [CrossRef]
12. Moore, R.L.; Nordmark, G.E.; Kaufman, J.G. Fatigue and Fracture Characteristics of Aluminum Alloy Cylinders under Internal Pressure. *Eng. Fract. Mech.* **1972**, *4*, 51–63. [CrossRef]
13. Cerrini, A.; Beretta, S. Failure Investigation and Design Improvements of Al 7075 Piston for Hydraulic Actuators. *Eng. Fail. Anal.* **2006**, *13*, 18–31. [CrossRef]
14. Kaw, A.K. *Mechanics of Composite Materials*; CRC Press: Boca Raton, FL, USA, 2005; ISBN 9780429125393.
15. Gay, D.; Hoa, S.V.; Tsai, S.W. *Composite Materials: Design and Applications*; CRC Press: Boca Raton, FL, USA, 2002; ISBN 9781420031683.
16. Datoo, M.H. *Mechanics of Fibrous Composites*; Elsevier: Amsterdam, The Netherlands, 1991.
17. Vasiliev, V.V.; Morozov, E.V. *Advanced Mechanics of Composite Materials*; Elsevier: Amsterdam, The Netherlands, 2007; ISBN 9780080453729.
18. Peters, S.T. *Composite Filament Winding*; ASM International: Materials Park, OH, USA, 2011; ISBN 978-1-61503-722-3.
19. Xia, M.; Takayanagi, H.; Kemmochi, K. Analysis of Multi-Layered Filament-Wound Composite Pipes under Internal Pressure. *Compos. Struct.* **2001**, *53*, 483–491. [CrossRef]
20. Scholz, S.; Kroll, L. Nanocomposite Glide Surfaces for FRP Hydraulic Cylinders—Evaluation and Test. *Compos. Part B Eng.* **2014**, *61*, 207–213. [CrossRef]
21. Pan, Q.; Zeng, Y.; Li, Y.; Jiang, X.; Huang, M. Experimental Investigation of Friction Behaviors for Double-Acting Hydraulic Actuators with Different Reciprocating Seals. *Tribol. Int.* **2021**, *153*, 106506. [CrossRef]
22. Mogbei, O.R. *Mechanical Design—Part 5 Materials Review And Selection*; Butterworth-Heinemann: Sunderland, UK, 2015.

23. Skowrońska, J.; Zaczyński, J.; Kosucki, A.; Stawiński, Ł. Modern Materials and Surface Modification Methods Used in the Manufacture of Hydraulic Actuators. In *Lecture Notes in Mechanical Engineering*; Springer Science and Business Media Deutschland GmbH: Berlin/Heidelberg, Germany, 2021; Volume 24, pp. 427–439, ISBN 9783030595081.
24. Hashimoto, H.; Tamura, M.; Ichiryu, K. Newly Developed Carbon Fiber Reinforced Plastics (Cfrp) Hydraulic Cylinder Incorporated Stroke Sensor. In *JFPS International Symposium on Fluid Power*; The Japan Fluid Power System Society: Tokyo, Japan, 1986.
25. Sumali, H.; Bystrom, E.P.; Krutz, G.W. A Displacement Sensor for Nonmetallic Hydraulic Cylinders. *IEEE Sens. J.* **2003**, *3*, 818–826. [CrossRef]
26. Mantovani, S.; Costi, D.; Strozzi, A.; Bertocchi, E.; Dolcini, E. Double Acting Composite Tube Cylinder for Fluid Power Applications: A Design Procedure. In Proceedings of the International Conference on Mechanical, Automotive and Aerospace Engineering, Kuala Lumpur, Malaysia, 17–19 May 2011.
27. Mantovani, S. Feasibility Analysis of a Double-Acting Composite Cylinder in High-Pressure Loading Conditions for Fluid Power Applications. *Appl. Sci.* **2020**, *10*, 826. [CrossRef]
28. Ritchie, J.; Mumtahina, U.; Rasul, M.; Sayem, A. Alternative Materials in Hydraulic Cylinder Design—Application Of Carbon Fibre Components. *Mech. Eng. Res. J.* **2013**, *9*, 43–47.
29. Nowak, T.; Schmidt, J. Non-Linear Mechanical Analysis of the Composite Overwrapped Cylinder for Hydraulic Applications. *Adv. Manuf. Sci. Technol.* **2014**, *37*, 31–48. [CrossRef]
30. Nowak, T.; Schmidt, J. Prediction of Elasto-Plastic Behavior of Pressurized Composite Reinforced Metal Tube by Means of Acoustic Emission Measurements and Theoretical Investigation. *Compos. Struct.* **2014**, *118*, 49–56. [CrossRef]
31. Nowak, T.; Schmidt, J. Theoretical, Numerical and Experimental Analysis of Thick Walled Fiber Metal Laminate Tube under Axisymmetric Loads. *Compos. Struct.* **2015**, *131*, 637–644. [CrossRef]
32. Zhang, J.; Bao, J.; Zhang, D.; Xu, B.; Chao, Q. Inlaid Connection of Carbon Fibre Reinforced Plastic Cylinder. In Proceedings of the 2016 IEEE International Conference on Aircraft Utility Systems (AUS), Beijing, China, 10–12 October 2016; pp. 1024–1029.
33. El Asswad, M.; AlFayad, S.; Khalil, K. Experimental Estimation of Friction and Friction Coefficient of a Lightweight Hydraulic Cylinder Intended for Robotics Applications. *Int. J. Appl. Mech.* **2018**, *10*, 1850080. [CrossRef]
34. Elasswad, M.; Tayba, A.; Abdellatif, A.; Alfayad, S.; Khalil, K. Development of Lightweight Hydraulic Cylinder for Humanoid Robots Applications. *Proc. Inst. Mech. Eng. Part C J. Mech. Eng. Sci.* **2018**, *232*, 3351–3364. [CrossRef]
35. Solazzi, L. Feasibility Study of Hydraulic Cylinder Subject to High Pressure Made of Aluminum Alloy and Composite Material. *Compos. Struct.* **2019**, *209*, 739–746. [CrossRef]
36. Solazzi, L.; Buffoli, A. Telescopic Hydraulic Cylinder Made of Composite Material. *Appl. Compos. Mater.* **2019**, *26*, 1189–1206. [CrossRef]
37. Solazzi, L. Design and Experimental Tests on Hydraulic Actuator Made of Composite Material. *Compos. Struct.* **2020**, *232*, 111544. [CrossRef]
38. Solazzi, L. Stress Variability in Multilayer Composite Hydraulic Cylinder. *Compos. Struct.* **2021**, *259*, 113249. [CrossRef]
39. Solazzi, L.; Buffoli, A. Fatigue Design of Hydraulic Cylinder Made of Composite Material. *Compos. Struct.* **2021**, *277*, 114647. [CrossRef]
40. S P, P.K.; Lee, S.-S. Design and Experimental Analyses of Hybrid Piston Rods Used in Hydraulic Cylinders under Axial Load. *Appl. Sci.* **2021**, *11*, 8552. [CrossRef]
41. Siegfarth, M.; Pusch, T.P.; Pfeil, A.; Renaud, P.; Stallkamp, J. Multi-Material 3D Printed Hydraulic Actuator for Medical Robots. *Rapid Prototyp. J.* **2020**, *26*, 1019–1026. [CrossRef]
42. Coskun, T.; Sahin, O.S. Design of the Composite Hydraulic Cylinder with Geodesic Dome Trajectory: A Numerical Study. *Polym. Compos.* **2022**, *43*, 5894–5907. [CrossRef]
43. Lubecki, M.; Stosiak, M.; Leśniewski, T. Comparative Studies of Tribological Properties of Selected Polymer Resins for Use in Hydraulic Systems. *Tribologia* **2019**, *288*, 31–37. [CrossRef]
44. Lubecki, M.; Stosiak, M.; Leśniewski, T. Antiwear Coatings for Multi-Material Composite Hydraulic Cylinder. A Tribological Study. In *TRANSBALTICA XII: Transportation Science and Technology*; Prentkovskis, O., Yatskiv (Jackiva), I., Skačkauskas, P., Junevičius, R., Maruschak, P., Eds.; Springer International Publishing: Cham, Switzerland, 2022; pp. 184–193.
45. Mayer, P.; Lubecki, M.; Stosiak, M. An Influence of the Surface Treatment of the Composite Substrate on the Pull-Off Strength of the Aged Polyurea And Polyurethane Coatings. In Proceedings of the 4th Polish Congress of Mechanics and 23rd International Conference on Computer Methods in Mechanics PCM-CMM-2019, Krakow, Poland, 8–12 September 2019.
46. Mayer, P.; Lubecki, M.; Stosiak, M.; Robakowska, M. Effects of Surface Preparation on the Adhesion of UV-Aged Polyurethane Coatings. *Int. J. Adhes. Adhes.* **2022**, *117*, 103183. [CrossRef]
47. Lubecki, M.; Stosiak, M.; Gazińska, M. Numerical and Experimental Analysis of the Base of a Composite Hydraulic Cylinder Made of PET. In *Lecture Notes in Mechanical Engineering*; Springer Science and Business Media Deutschland GmbH: Berlin/Heidelberg, Germany, 2021; Volume 24, pp. 396–405.
48. Lubecki, M.; Michał, S.; Banaś, M.; Stryczek, P.; Urbanowicz, K. Experimental and Theoretical Analysis of Hydraulic Cylinder Loads. In *Fatigue and Fracture of Materials and Structures*; Lesiuk, G., Duda, S., Correia, J.A.F.O., De Jesus, A.M.P., Eds.; Springer International Publishing: Cham, Switzerland, 2022; pp. 85–91.
49. Dirkin, W.; Douglass, D.; Tootle, J.; Benton, T. Fluid Actuator Including Composite Cylinder Assembly. U.S. Patent 4,685,384, 1986.
50. Ching, F. Composite Cylinder for Use in Aircraft Hydraulic Actuator. U.S. Patent 5,415,079, 1992.

51. Yu, X.; Waldenstrom, C. Method of Winding a Fiber-Resin Composite Pressure Fluid Cylinder. U.S. Patent 5,435,868, 1993.
52. Fish, E. Fluid Cylinder End Cap Assembly. U.S. Patent 5,465,647, 1994.
53. Fish, E. Removable End Plug. U.S. Patent 7,240,607, 2007.
54. Stelling, O.; Otte, B.; Petker, J. Composite High Pressure Hydraulic Actuators for Lightweight Applications. In Proceedings of the 9th International Fluid Power Conference (IFK), Aachen, Germany, 24–26 March 2014.
55. Parker Hannifin Corporation Lightraulics® Composite Hydraulic Cylinders; Catalogue HY07-1410/UK. 2017. Available online: https://www.parker.com/literature/Cylinder%20Europe/Cylinder%20Europe%20-%20English%20Literature/Composites/Composite%20Cylinders_1410-UK.pdf (accessed on 1 December 2022).
56. Polygon Composites Technology. *Composite Finished Cylinders Design Guide*; Polygon Composites Technology: Walkerton, ON, Canada, 2012.
57. Polygon Composites Technology Cylinder Tubing. Available online: https://polygoncomposites.com/tailored-solutions/cylinder-tubing/ (accessed on 23 September 2021).
58. Liebherr Liebherr Hybrid Cylinder. Available online: https://www.liebherr.com/en/deu/products/components/hydraulics/hybrid-cylinders/hybrid-cylinder.html#soentstehtihrhybridzylinder (accessed on 6 December 2021).

Review

Gelatin Soft Actuators: Benefits and Opportunities

Sandra Edward [1] and Holly M. Golecki [2,*]

[1]   Department of Mechanical Engineering, University of Illinois at Urbana-Champaign, Urbana, IL 61801, USA
[2]   Department of Bioengineering, University of Illinois at Urbana-Champaign, Urbana, IL 61801, USA
*   Correspondence: golecki@illinois.edu

Abstract: Soft robots are being developed as implantable devices and surgical tools with increasing frequency. As this happens, new attention needs to be directed at the materials used to engineer these devices that interface with biological tissues. Biocompatibility will increase if traditional materials are replaced with biopolymers or proteins. Gelatin-based actuators are biocompatible, biodegradable, versatile, and tunable, making them ideal for biomedical and biomechanical applications. While building devices from protein-based materials will improve biocompatibility, these new materials also bring unique challenges. The properties of gelatin can be tuned with the addition of several additives, crosslinkers, and plasticizers to improve mechanical properties while altering the characteristic fluid absorption and cell proliferation. Here, we discuss a variety of different gelatin actuators that allow for a range of actuation motions including swelling, bending, folding, and twisting, with various actuation stimulants such as solvent, temperature, pneumatic pressure, electric field, magnetic field, or light. In this review, we examine the fabrication methods and applications of such materials for building soft robots. We also highlight some ways to further extend the use of gelatin for biomedical actuators including using fiber-reinforced gelatin, gelatin cellular solids, and gelatin coatings. The understanding of the current state-of-the-art of gelatin actuators and the methods to expand their usage may expand the scope and opportunities for implantable devices using soft hydrogel robotics.

Keywords: gelatin; actuator; soft robotics; biocompatible actuator; hydrogels

Citation: Edward, S.; Golecki, H.M. Gelatin Soft Actuators: Benefits and Opportunities. *Actuators* **2023**, *12*, 63. https://doi.org/10.3390/act12020063

Academic Editor: Federico Carpi

Received: 1 November 2022
Revised: 22 January 2023
Accepted: 26 January 2023
Published: 31 January 2023

1. Introduction

Gelatin is a heterogeneous mixture of peptides and proteins derived from collagen through partial hydrolysis that destroys cross-linkages between polypeptide chains and breaks polypeptide bonds [1,2]. In this hydrolyzed state, gelatin is versatile and has many benefits. Gelatins are readily derived without the need for synthesis, are harmless to the environment due to rapid degradation rates, and allow for the incorporation of water-soluble additives [3]. Gelatin is used extensively in the food [2,4,5], pharmaceutical [6,7], cosmetic [8], and photography [9,10] industries. In food industries, gelatin is used for emulsification, texturization, and stabilization [11]. In pharmaceutical industries, gelatin is used for drug encapsulation [12], drug stabilization [13], wound dressing [14,15], plasma expansion [16], ointment filling [17], and emulsification [18]. After the long historical use of gelatin in these industries, applications of gelatin are now also used in diverse biomechanical fields from drug delivery [19] to bone tissue engineering [20,21].

Gelatin has more recently been studied for use in non-conventional active roles such as actuators in soft robotic devices [22,23], and wearables [3,24,25]. Soft actuators are sensitive to environmental fluctuations that convert chemical and physical energy into mechanical work [22]. However, hydrogels such as gelatin are considered to be mechanically weak materials [26], they are soft and brittle and cannot withstand large deformations from environmental fluctuations due to their polymer networks containing inhomogeneities such as dangling chains and loops [26]. Gelatin gels are also easily broken into fragments under a modest compressive load [27]. Gelatin-based gels have a moderate performance when strained and rapidly dry when operated in air, causing stiffening and limiting the stability and durability of the gelatin elements [3]. The mechanical properties of conventional hydrogels such as gelatin are also evaluated by shearing or compression rather than by stretching because of poor deformability [26].

Although alone they are characterized as mechanically weak materials, the physical properties of gelatin gels can be easily tuned with additives, nanocomposites, or plasticizers [28]. Gelatin hydrogels can also serve as a platform for building actuators for biomedical applications because of their biocompatibility [29,30], ability to promote cell adhesion and proliferation [31], tissue-like stiffness [32,33] and other attractive properties such as fluid absorbance [34]. With the use of various material additives and fabrication techniques that enhance gelatin's mechanical properties, the benefits of gelatin hydrogel materials can be exploited and they can be applied as actuators. Many soft fluidic actuators have been modeled and controlled through various methods, including differential simulations, topology optimizations, data-driven modeling and control methods, hardware control boards, nonlinear estimations, and analytical and numerical modeling methods, such as dynamic modeling and finite element modeling [35,36]. Finite element modeling methods were found to be especially effective in capturing the strong nonlinearities and complex geometries of soft actuators [37]. Since gelatin can be both hyperelastic and viscoelastic [38,39], finite element modeling methods are useful to predict gelatin actuation performance and optimize designs.

In this review, we classify gelatin actuators based on their kinetic motion, i.e., linear type, bending type, and twisting type (Figure 1). We identify for each of these different motions the different ways in which gelatin can be engineered to be stimulated by environmental fluctuations in pH [40,41], solvent [22], temperature [42], humidity [43], electric field [44], magnetic field [45], or light [46]. The actuation motion and stimuli diversity of gelatin actuators is dictated by how the gelatin is modified with additives, such as crosslinkers, reinforcements, and plasticizers, and by actuator structural design [47]. The structural designs of these gelatin actuators can be fabricated through several techniques, such as mold casting [23,48], spin-coating [43], and 3D printing [49,50]. In classifying the gelatin actuators according to kinetic motions, it can be seen how currently widely used non-biocompatible biomedical actuators of a particular kinetic motion can be replaced with gelatin actuators that carry out the same motions, through different fabrication methods, depending on the stimulations, the additives, and the structural designs.

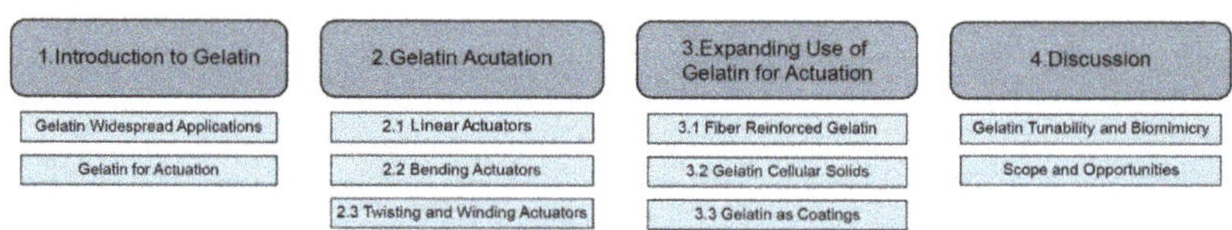

Figure 1. Overview schematic of gelatin actuators.

While gelatin has been demonstrated to be useful and functional as an actuator in the biomechanical field, there are many ways that evolving gelatin applications can be extended within soft robotics. This can be achieved by improving the material properties of gelatin with the use of reinforced fibers or with the use of gelatin cellular solids or foams. Gelatin coatings may also serve to enhance the functional biocompatibility and efficiency of synthetic actuators. With these methods, there is an opportunity to widen the scope for innovation with gelatin actuators and medical device design.

2. Versatile Actuation Using Gelatin

2.1. Linear Actuators

The majority of small gelatin linear actuators are stimulated by swelling. The swelling of hydrogels such as gelatin can be used for actuating and sensing applications [51]. The swelling of gelatin actuators is stimulated through solvents creating environmental fluctuations in terms of pH, compositions, and concentrations [52]. The swelling also depends on the substrate and solvent material synthesis as well as on the time spent in the solvent. Swelling-stimulated hydrogel actuators work as drug delivery systems by initially trapping molecules in their crosslinked polymer networks. When the hydrogel is in contact with a solution that causes it to swell, the distance between the polymer chains in the gel increase, and the drug is released into the bloodstream [53]. Gelatin is a hydrophilic material with a water uptake of above 100% when immersed for periods as short as 5 min, and above 500% when immersed for 15 min [54]. In general, a high temperature, low gelatin concentration, and deviation from the isoelectric point can promote the swelling or disintegration of the gelatin polymer networks [46,55].

Some linear actuation due to solvent swelling depends on the solvent's pH. Ooi et al., studied the effects of cellulose nanocrystals (CNCs) on the swelling ratio of a CNC-reinforced gelatin hydrogel crosslinked with glutaraldehyde [53]. They found that the hydrogels were highly sensitive to changes in pH and had a maximum swelling ratio at pH 3. This ability to vary swelling ratios depending on the stimulant pH allows for the control of swelling with a viable pH in the body and makes it a good candidate for drug carriers. Hajikarimi and Sadeghi studied the effect of pH on the swelling of nanocomposite hydrogels of N-vinyl pyrrolidone via acrylic acid (gelatin-g-NVP-AA) with and without the addition of nanoclay crosslink sodium montmorillonite (gelatin-g-NVP-AA/MMT) [56]. The maximum drug absorption occurred at pH 8 and the rate of swelling increased with decreasing particle size. The hydrogel with the nanoclay significantly increased the absorption capacity of the nanocomposite.

Some linear actuation due to solvent swelling depends on the solvent's and gelatin's concentrations and compositions. Microbial transglutaminase (m-TG) is a common non-toxic crosslinking agent for gelatin [57,58]. A study by Besser et al., used a bi-layer gelatin-laminin hydrogel crosslinked with microbial transglutaminase that swells in a high glucose content solution [59]. The substrate with both laminin and gelatin more authentically recapitulated the composition and mechanical properties of the extracellular matrix. Using a gelatin concentration between 4% and 10% for the hydrogels allows for an elastic modulus of about 1–25 kPa which corresponds to the stiffnesses appropriate for multiple cell types. The crosslinked hydrogel acts as a substrate for human stem cells, Schwann cells, and skeletal muscle cells to adhere to and proliferate and the degree of swelling is inversely proportional to gelatin concentration. Li et al., developed a PHEMA-gelatin actuator with a ductile gelatin micro-disc embedded in a 3D rigid PHEMA polymeric network that can contract (swell) in high-concentration salt solutions and relax (de-swell) in dilute solutions [60]. This behavior is due to the synergy of phase separation and the Hofmeister effect of the hydrogel. Tattanon et al., studied the swelling of a gelatin hydrogel with hydroxyapatite crosslinked with genipin, a low toxicity, water-soluble bi-functional crosslinking reagent, to increase material strength [52,61]. The material has superior swelling behavior and degradation rate in deionized water compared to that in a phosphate-buffered saline solution; this is due to the hydroxyl groups of deionized water that react with gelatin's hydrophilic functional groups.

Gelatin is also used to create actuators that disintegrate or dissolve when stimulated photothermally. Wu et al., developed a self-propelled biodegradable multilayer rocket for anticancer drug delivery to cancer cells [46]. The layers of the rocket are bovine serum albumin (BSA), poly-l-lysine (PLL), and heat-sensitive gelatin hydrogel with gold nanoparticles, doxorubicin, or catalase. Near-infrared irradiation is absorbed as heat into the gelatin which melts the actuator down and rapidly releases the drugs into cancer cells.

2.2. Bending Actuators

Bending actuators convert actuation energy into a bending motion. Although gelatin has a reputation for having relatively poor mechanical properties, it can produce a wide range of bending motions [3,44,62]. Gelatin is also extremely versatile and can be actuated by several stimuli, including pH [41], temperature [63], solvent [22,43], mechanically [64], pneumatically [3,23,28,48,49,65,66], or electrically [44,67–71]. The following section describes a variety of efforts to build bending actuators from gelatin-based materials.

A mechanically actuated gelatin finger (Figure 2A) was proposed by Harris et al. [64]. Its components were fabricated by crosslinking gelatin with microbial transglutaminase which allows for tuning the stiffness of the material and prevents the material from turning brittle by increasing the Young's Modulus [72]. The resulting gelatin actuator compliance matches in vivo tissues and may be a favorable chemical interface for host tissues.

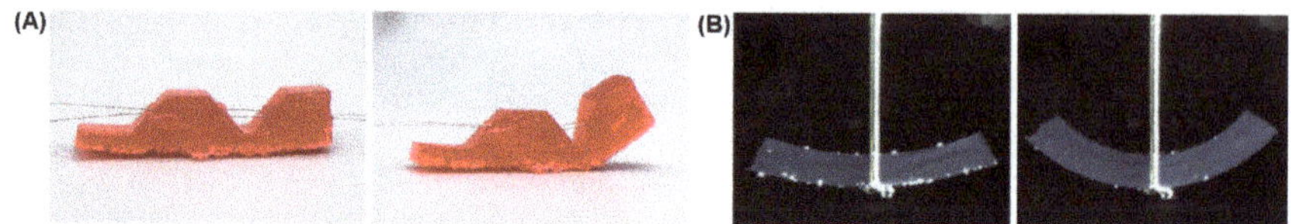

Figure 2. Bending actuators. (**A**) Mechanically actuated gelatin bending actuators with Nitinol wire and crosslinked with microbial transglutaminase ([64], original image). (**B**) pH-sensitive swelling bilayer gelatin actuator in water before and after deformation [41]. (Images (**A**,**B**) are licensed under CC BY 4.0 (http://creativecommons.org/licenses/by/4.0/ (accessed on 30 October 2022))).

A temperature-responsive reversible bi-layer film actuator was proposed by Stroganov et al., with one of the layers being gelatin and the other being polycaprolactone (PCL) [63]. The actuator can be both folded and unfolded. The gelatin layer swells in water applying compressive stress on the hydrophobic PCL layer and the PCL layer bends and folds at high temperatures and unfolds at room temperature. This actuator has applications in cell encapsulation and release, and scaffold design [63].

Solvent-stimulated bending actuators based on solvent or gelatin concentrations have different forms. Self-folding robots were developed to undergo geometric transformations due to the varying thickness and stiffness of the gelatin layer crosslinked with microbial transglutaminase and chemically coupled to an aluminum-nylon tri-layer [62]. The constructs fold when dehydrated in a polyethylene glycol solution, where water diffuses out, and unfolds when rehydrated in a NaCl solution, where water diffuses into the system. The folding actuation is affected by the changes in hydrogel mechanics, thickness, and crosslinking density. The bilayer actuators produce a simple bending motion when stimulated to swell [22]. Hanzly et al., proposed a solvent-stimulated gelatin bi-layer actuator formed with two different gelatin layers of different stiffnesses [22]. Crosslinking gelatin with glutaraldehyde (GTA) increases cross-link density, increasing the stiffness of the gelatin layer. The higher modulus layer acts as a strain-limiting layer when the actuator swells in response to a soluble starch, causing bending. The actuator can then return to a relaxed state when the starch is hydrolyzed with the enzyme, α-amylase. The actuator can be used as a shape-shifting hydrogel in soft machines or to deliver nutraceuticals, flavors, or drugs upon ingestion. The actuators can also create folding motions.

A swelling-stimulated bending actuator based on pH sensitivity to the solvent was proposed by Riedel et al., in the form of a bilayer gelatin actuator [41]. It was appropriately modified with energetic electron beams to control the switching behavior with environmental conditions and programmable bio-absorbability (Figure 2B). They found that there is a strong dependence of the swelling on the pH, which is correlated to the isoelectric point, and that the strongest swelling occurs at a pH of 2–3. Varying material properties such as gel concentration, irradiation dose, pH value, and salt concentration can tune the stimuli' responsiveness. This actuator has a wide range of applications in bio-sensors and self-expanding bio-actuators.

A magnetic field-stimulated bending actuator was designed by Helminger et al. [45]. They developed a gelatin-based ferrogel by embedding iron ions into the gelatin hydrogel. They do this by bonding gelatin molecules to ferrous metal cations which act as a template for the co-precipitation of the magnetic nanoparticles. These magnetic nanoparticles are stimulated by an external magnetic field to create a bending actuator. The study also found that the adsorption of these magnetic nanoparticles onto the polymer matrix limits the swelling of the hydrogel and makes them lose their thermoreversible properties.

The pneumatically-stimulated bending gelatin actuators have more variety in design and form. Some benefits of soft pneumatic actuation are that they are more lightweight, flexible, and versatile with actuation modes [73], and offer less pollutive and hazardous complications than actuating with non-sustainable chemicals or electric fields [74]. One of the trepidations of using soft natural biomaterials as a pneumatic actuator is that they may fail under the high, pressurized forces required of a pneumatic actuator to carry out the desired actuation; hydrogel actuators can typically handle around 15 psi actuation pressure [75]. However, by leveraging gelatin composites, a strong elastomeric material can be made that is comparable to silicone [48] or India rubber [76]. These new materials have then been used to fabricate effective pneumatic actuators from gelatin analogs.

When gelatin was mixed with glycerol and water in a ratio of 1:1:8, the resulting elastomer can be cast to create a new type of edible, biodegradable pneumatic actuator [23]. The same material was also used to fabricate a tube-like actuator that bends due to a strain-limiting layer on one side (Figure 3A) [66]. An "elephant trunk" actuator that can actuate into an s-shape or a u-shape guided by a crocheted cotton yarn exoskeleton was created using a modified mixture of these three ingredients that included citric acid [3]. This created an acidic environment and increased the safety of the material by preventing bacterial growth. They found that using a shellac resin as a biocompatible coating on the actuator delays dissolution even in acidic, stomach fluid-like environments. They also added other sugars and food additives to enhance the extensibility and structural properties of the actuator without sustainability and edibility being compromised. PneuNets (pneumatic network actuators) are a class of pneumatic soft actuators known for their ability to make sophisticated motions with simple controls [77]. It was found that modifying the gelatin, glycerol, and water mixture with citric acid and sodium chloride additives can mold room-temperature self-healing pneumatic actuators such as the pneuNets (Figure 3B) [28].

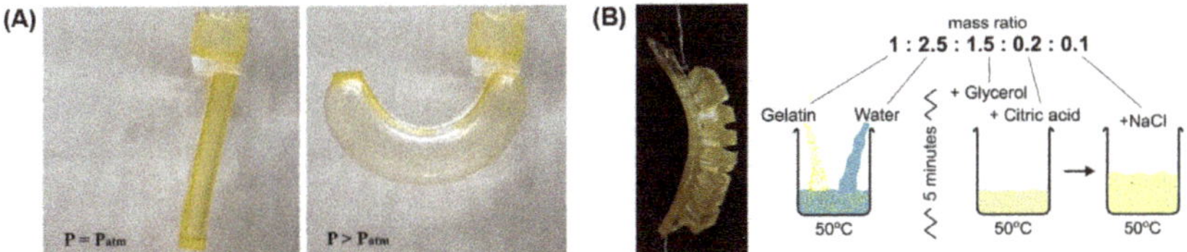

Figure 3. Pneumatically-stimulated bending actuators. (**A**) Biodegradable actuator under non-pressurized and pressurized states [66]. (**B**) PneuNet actuators made of gelatin and glycerol are used as grippers to apply a force that picks up objects [23,28]. (Images (**A**,**B**) are licensed under CC BY 4.0 (http://creativecommons.org/licenses/by/4.0/ (accessed on 30 October 2022))).

The combination of materials was also used to create pneumatic pouch-based actuators that were edible and biodegradable while also being untethered and self-actuating due to the food-safe cyclic chemical reactions between citric acid and sodium bicarbonate filling the pouch actuator with gases that release through a valve when pressurized [48]. The ratios of gelatin, glycerol, and water were altered to increase the glycerol content in a ratio of 1:6:5; in addition, the actuator was biodegradable in warm water. These characteristics allow the actuator to be used as deployable agricultural robots that roam freely, acting as sources of nutrients to various flora and fauna before degrading.

Fiber-reinforced elastic enclosure (FREE) gelatin actuators were also built to perform an omnidirectional movement at rapid response times of less than 1 s [49]. The FREE gelatin actuator is a cylindrical body with three chambers arranged in 120° rotation and was 3D printed based on the fused deposition modeling method. The 3D-printed stretchable waveguides were integrated with this soft robotic actuator to combine optical sensor networks to enable integrated curvature, direction, and force sensing with high precision. It was then reinforced with cotton fiber obliquely wound around the actuator body to improve bending performance. The robots were capable of real-time control and could be used to detect and remove obstacles.

Pneumatic gelatin actuators can also be made to be consumed. Since edible gummy candies use gelatin as a main ingredient, it was found that they could be melted and recast into edible pneuNet molds [65]. Although the resulting actuator was edible, it takes about three days to cure recast gummy bears, and the actuators had reduced elasticity. Thus, the authors cast an alternate mixture, made of gelatin, water, and corn syrup, that was more elastic and appealing to consume. A single-pour mold was used for more robust designs of the gelatin candy actuators [78].

Gelatin-based bending actuators can also be electromechanically stimulated. Electroactive polymers (EAPs) are electrically simulated materials with many unique appealing properties, including low weight, flexibility, and high energy density [44]. Gelatin alone is not an EAP because it has a low water resistance [67], limited mobility of polarized groups due to hydrogen bonds [44], and reduced mechanical properties when treated with an acid or base at high temperatures [67]. However, gelatin can be used for EAP applications by reinforcing the gelatin with filler materials or through chemical cross-linking [67].

Elhi et al., developed tri-layer electroactive polymer actuators out of polypyrrole and a gelatin hydrogel with choline acetate and choline isobutyrate ionic liquids [79]. The actuator was robust due to the high elasticity of the membrane and the ionic conductivity of the choline electrolytes. Another study used a gelatin-water hydrogel film to create a simple ionic actuator by immersing it in a 0.1 M NaOH solution with two contactless steel electrodes [69]. They assessed the electro-mechanical performance of the film and demonstrated the bending behavior of the swollen material in response to the electric field.

Graphene as a filler material in gelatin was studied for its electrical, thermal, and electromechanical properties under electric fields through copper electrodes [67]. It was found that the increased surface area and concentration of graphene led to increased storage modulus responses. A 0.1% volume graphene/gelatin hydrogel delivered the greatest deflection distance and di-electrophoresis force, making it a good candidate for actuation. Graphene additives were also found to exhibit high tensile strength [68]. Graphene oxide (GO) is easier to disperse in composite solvents than graphene [68,70]. Thus, GO is found to be a useful additive for electroactive gelatin polymers as it increases the tensile strength, Young's modulus, and energy at the break of gelatin [80]. The electro-responsive properties of the GO gelatin composite were investigated under external electric fields through copper electrodes. It was found to have larger magnitude responses as compared to gelatin hydrogels and graphite-gelatin composites (Figure 4A) [44]. GO has also been incorporated in gelatin methacrylate (GelMA) due to the excellent photo-patternable properties of GelMA in the fabrication of bio-compatible microscale structures [70]. The resulting hydrogel had tunable mechanical strength and enhanced electrical properties. However, GO does not have as strong electrical properties as graphene does, so chemically reducing GO to graphene to form reduced graphene oxide (RGO) restores its properties, but requires reductants and suspension agents that are toxic, making it unsuitable for biomedical applications [68].

One study used multi-walled carbon nanotubes (MWCNTs) in gelatin composites because carbon nanotubes have excellent electrical properties [71]. The composite was dispersed with ionic surfactant and then placed in an aqueous medium of NaCl solution between two platinum electrodes generating a DC electric field. The resulting actuator bent due to the osmotic pressure difference at the solution gel interface and showed good reversible behavior compared to gelatin alone, which eroded significantly on continuous exposure to DC currents.

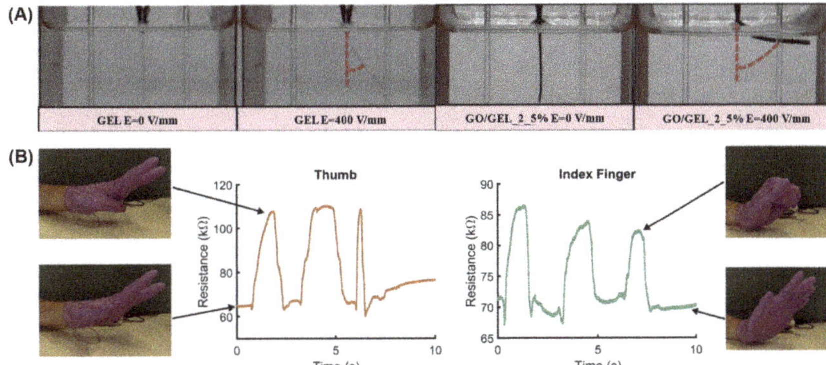

Figure 4. Electromechanically-stimulated gelatin actuators. (**A**) The degree of deflection and the electrode distance of 4 cm for gelatin actuators or graphite/gelatin actuators with increasing electric field strength [44]. (**B**) Gelatin can also be used as a sensor. All five fingers of the glove are 3D printed with gelatin sensor networks and can return separate strain responses [28]. (Images (**A**,**B**) are licensed under CC BY 4.0 (http://creativecommons.org/licenses/by/4.0/ (accessed on 30 October 2022))).

Alternatively, gelatin could also be used to build sensors, and electrodes, or even generate electric signals. Hardman et al., used a mixture of gelatin, glycerol, water, and citric acid in the ratio of 1:1.5:2.5:0.2, along with an additive of NaCl, that they extruded to 3D print ionic strain sensors that have highly linear responses to strain with impressive electrical properties (Figure 4B) [28]. NaCl reduces the baseline resistance of the material for strain sensing. They printed the sensors on a glove to read strain data while flexing their fingers. Choe et al., used electrohydrodynamic printing to fabricate gelatin-based electrodes with self-healing capabilities that could almost fully recover performance even if the electrodes were damaged due to tannic acid in the mixture forming hydrogen bonds [81]. These electrodes could be attached with elastomers to create dielectric elastomer actuators with good actuator operation and could also be used as strain sensors. Liu et al., created an actuator that generates a piezoelectric signal while being actuated by changes in humidity [43]. Gelatin was doped with poly(3,4-ethylenedioxythiophene)-poly(styrenesulfonate) (PEDOT:PSS) to improve mechanical integrity. The gelatin mixture was then spin-coated and combined with piezoelectric poly(vinylidene fluoride) (PVDF) film that can work as an alternating current generator when actuated. Energy from water vapor is absorbed into the water-responsive material and is converted to mechanical motion which then generates an electric signal. The actuator had good mechanical and humidity-responsive properties and may have applications in self-powering biomedicine robotic systems and sensors.

2.3. Twisting and Winding Actuators

While soft actuators have previously been widely studied for bending, expanding, and shrinking movements, there is now also more recent literature on twisting or torsional soft actuators [36,82]. However, there is a need to further explore twisting actuators in soft robotics, especially those made from biocompatible materials such as gelatin. These actuation motions are necessary to simulate the bio-morphology of the left ventricle of the heart, for example [83]. Twisting and winding gelatin actuators were inspired by self-winding and shape-changing mechanisms in nature such as seed dispersal units, climbing plants, carnivorous plants, and plant tendrils (Figure 5A,B) [84–86]. Helical shapes in nature are usually formed by competition between bending and in-plane stretching energy due to internal or external forces [47].

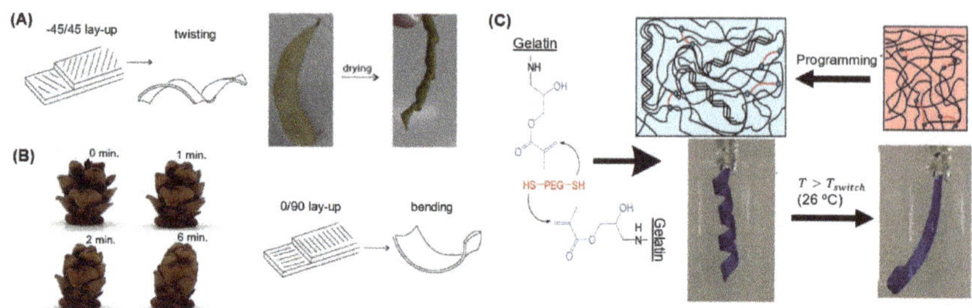

Figure 5. Twisting gelatin actuators and plant-inspired actuation. (**A**) Solvent actuated twisting actuation in edamame seedpods with clay composite twisting actuation scheme [86]. (**B**) Conifer pinecone bending motion when dried out to when fully hydrated (original images) with clay composite bending actuation scheme [86]. (**C**) Thermally stimulated shape memory gelatin hydrogels twisting actuators (Adapted) [87]. (Images (**A–C**) are licensed under CC BY 4.0 (http://creativecommons.org/licenses/by/4.0/ (accessed on 30 October 2022))).

Many of the plant-inspired twisting gelatin actuators are stimulated by swelling in water. One study took inspiration from the fact that many of these shape-changing mechanisms in nature are due to cellulose microfibrils (CMFs) arranged in specific orientations that limit externally actuated swelling or shrinking in certain directions [84]. They simulate the shape-changing effect of the CMFs with anisotropic reinforcement microparticles, such as aluminum oxide platelets coated with superparamagnetic iron oxide nanoparticles electrostatically in a gelatin matrix. A weak external magnetic field is used to orient the particles in a similar way to natural systems. Hydrating and swelling of the gels in water create bending and twisting motions that can be brought back to their original flat geometry by drying. They simulated the twisting mechanisms of orchid tree chiral seed pods, the bending mechanisms of the opening and closing of a pinecone, and the opening and closing of a wheat awn [84]. Another study took inspiration from self-winding plant tendrils [85]. They created water-responsive gelatin actuators loaded with rigid left-handed amyloid fibrils which were then roll-dry spun into wires. The concentration and distribution of the amyloid fibrils in the matrix cause wire bending, while the orientation of the amyloid fibrils causes wire twisting.

Another study designed fluid-driven hydrogel actuators in the form of origami structures called cuboid actuator units (CAUs) to achieve diverse actuation movements such as twisting, bending, and linear contraction [88]. The origami structures have a predesigned crease pattern that guides the directional movements of the fluid-driven hydrogel actuators when actuated. Combining multiple types of CAUs achieved various actuation modes, including decoupling, superposition, and reprogramming. These actuation modes were useful for different simple applications such as a two-finger gripper and a three-finger gripper for grasping tasks, and a multi-way circuit switch.

Another winding actuator stimulated by temperature was made out of shape memory hydrogel synthesized by a reaction of glycidylmethacrylated gelatin with oligo(ethylene glycol) α,ω-dithiols, a bifunctional crosslinker, to establish a polymer network (Figure 5C) [87]. When the hydrogel was heated and subsequently cooled, the temporary winding shape of the material was activated due to the formation of triple helices acting as temporary netpoints. When the hydrogel was then heated, the temporary netpoints were dissociated and the permanent shape was re-established.

3. Expanding the Use of Gelatin for Actuation

The scope of research on gelatin in actuators is not widespread because gelatin is still considered a weak material with limitations in its mechanical properties. Despite this, gelatin is seen as a good candidate material for biomechanical actuators because of its attractive properties in biocompatibility, biodegradation, biomimicry, and cell proliferation.

Thus, there have been several studies aiming to enhance and expand the mechanical properties of gelatin for actuation.

Gelatin can form various types of actuators with various stimuli methods. To expand their applications beyond what is currently available, the material properties of gelatin can be enhanced or developed with the use of reinforcing fibers or by foaming the gelatin material. Another way of expanding the use of gelatin in actuation is to use gelatin as coatings to enhance or support the functioning and efficiency of other working actuators and active substrates.

3.1. Reinforcing Gelatin with Fibers to Improve Material Characteristics

Gelatin can be reinforced with a variety of materials to improve material characteristics, improve mechanical properties, and constrain movement and actuation while maintaining biocompatibility and biodegradability. Gelatin has been reinforced with powders such as bio-ceramic powders that enable it to be used as bone implants or repair cartilage tissue [89–91]. Gelatin has also been reinforced by crystalline materials such as crystalline cellulose and cellulose nanocrystals [53,92–94], as they are found to have excellent biocompatibility and mechanical properties, along with water-resistive properties. Gelatin has been reinforced with nanosheet materials such as GO sheets that increase Young's modulus by over 50% and increase fracture stress by over 60% [95]. These additives work well for reinforcement and expand the functionality and applications of gelatin. Gelatin has also been reinforced with a variety of fibers and nanofibers that are covered in the following section (Figure 6).

Natural fibers commonly used in fabrics such as silk, jute, and cotton have been used as reinforcement in gelatin composites. Shubhra et al., fabricated silk fiber-gelatin composites using compression molding [96,97]. They found that the composite had improved mechanical properties with the tensile strength, tensile modulus, bending strength, bending modulus, and impact strength of the composite having increased by 250 to 450%. Studies conducted on jute-gelatin composites [54] found that the tensile strength and bending strength of the jute composites increased by 212% and 241% when the fiber weight percent was increased from 0% to 50% due to a higher load transfer capacity. However, both tensile strength and bending strength decreased when the fiber content was >50% because of the presence of fiber ends that may initiate cracks. These fibers can also be used to guide the deformation shape on actuation like the use of crocheted cotton to constrain the pneumatic gelatin actuator when inflated to a particular shape [3].

Other natural fibers have also been used to reinforce gelatin. Liu et al., used vegetable-tanned collagen fibers (VCF) for gelatin reinforcement [98]. They found that the composite films had significantly improved mechanical properties compared to pure gelatin films in a wet state and had higher water and thermal resistance. Wei et al., used a combination of collagen fibers and mesoporous bioactive glass particles to reinforce gelatin so that they could cure the material in situ with UV light (Figure 6A) [99]. They found that the composite's shear strength increased by 62% compared to pure gelatin. In addition, the composite's stability was enhanced in an artificial saliva solution.

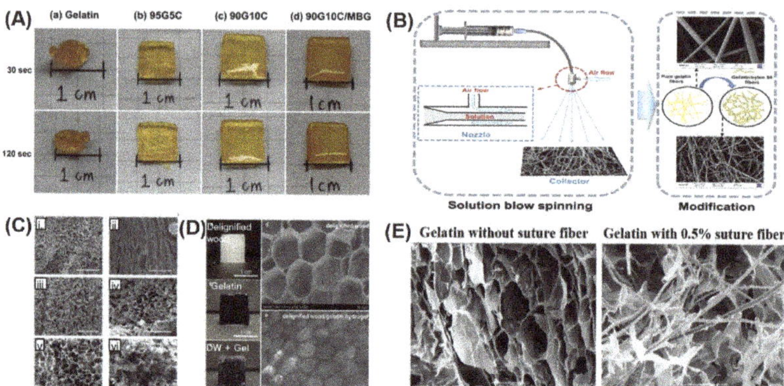

Figure 6. Fiber-reinforced gelatin enhancements. (**A**) Various sealant gelatin-collagen fibril composi-tions containing 20% mesoporous bioactive glass exposed to an LED light for different durations [99]. (**B**) Fabricating gelatin-nylon 66 nanofibers through solution blow spinning with improved mechan-ical properties [100]. (**C**) Cross- and sagittal-sectioned SEM images of (**i,ii**) gelatin; (**iii,iv**) CNFs; (**v,vi**) (Gel:CNFs)-75:25 [101]. (**D**) Delignified wood fiber reinforcement, gelatin, and delignified wood/gelatin hydrogel. SEM image of freeze-dried delignified wood with a well-preserved cellular structure, and optical microscopy image of a delignified wood/gelatin composite hydrogel [102]. (**E**) The SEM images of a gelatin biomaterial ink without suture fibers and with suture fibers [103]. (Images (**A**–**E**) are licensed under CC BY 4.0 (http://creativecommons.org/licenses/by/4.0/ (ac-cessed on 30 October 2022))).

Synthetic fabric fibers that are biocompatible but not biodegradable, such as nylon and polyester, have been used as reinforcement in gelatin composites. Yang et al., used a mixture of nylon 66 and gelatin to fabricate nanofibers through a solution blow spinning technique to create composite films with improved mechanical properties (Figure 6B) [100]. The elongation at the break of the composited film increased from 7.98% to 30.36%, and the tensile strength increased from 0.03 MPa to 1.42 MPa. The composite produced an improved water barrier. Dacron fibers are woven or knitted polyester fibers currently used in biomedical practices as grafts that have also been gelatin reinforced to improve hydrophilicity and cell adhesion [104–106]. Dacron-gelatin composites that have been used in prosthetic heart valves can release lysozyme anti-bacterial proteins to reduce infections since impregnating the Dacron with gelatin increased the lysozyme loading capacity with a sustained release 30 h after implantation [106].

In recent years, nanocelluloses such as cellulose nanofibrils (CNFs) and cellulose nanocrystals (CNCs) have been evaluated for biomedical applications due to low toxicity, biocompatibility, and excellent mechanical properties [101,107]. Due to the synergistic interaction between CNF and gelatin, their bio-composites had higher cytocompatibility compared to pure gelatin and CNFs (Figure 6C) [101]. Reinforcing the gelatin with CNFs improved the maximum compressive breaking strength 5.75-fold more than that of pure gelatin, and significantly increased the strain [108]. The Young's Modulus also increased linearly with increasing amounts of CNFs added to the gelatin [101]. Wang et al., developed a nanofiber cellulose (NFC) reinforced gelatin structure and found that it increased material hardness while reducing elasticity [109]. Wang et al., extracted lignin from wood to form delignified wood-gelatin composites crosslinked with genipin (Figure 6D) [102]. This approach takes advantage of the highly aligned cellulose nanofibril bundles in the cell walls of the delignified wood to increase mechanical support, mechanical strength, and stiffness, and also to improve liquid conduction through the naturally aligned micro and nanochannels.

Gelatin has also been reinforced with carbon nanofibers due to their excellent bio-compatibility and successful use in several clinical applications [110]. It was found that for long carbon fibers in a gelatin composite, the increase in carbon fiber volume fraction increased the tensile strength, modulus, and shear strength. For short carbon fibers in a gelatin composite, the increase in carbon fiber volume fraction initially increased tensile strength and modulus, which then later decreased, while shear strength improved and then reached a constant value. Long carbon fiber composites consistently demonstrated higher mechanical properties than short fiber composites.

Since gelatin itself can be spun into fibers, Ravishankar et al., jet-spun gelatin into fibers when blended with polycaprolactone to mimic diameters in the range found in a heart valve extracellular matrix [111]. These fibers were embedded in a methacrylated hydrogel mixture of gelatin, sodium hyaluronate, and chondroitin sulfate to create reinforced composites. The reinforced composites were able to swell higher than the hydrogel alone and were able to mimic heart valve mechanical behavior and have a high cell viability. Gelatin fibers were also hand-spun and electrospun [112]. They found that gelatin fiber acts as a foaming agent when placed in Linear Low-density Polyethylene (LLDPE) and that mechanical properties such as tensile strength, bending strength, elongation at break, tensile modulus, bending modulus, and hardness decreased with the increase in gelatin fiber in the LLDPE composite.

Gelatin actuators and structures can also be used as biomaterial ink and 3D-printed fibers could be used to reinforce the ink to enhance mechanical properties [103,108]. Jiang et al., added cellulose microfibrils to their 3D printing gelatin inks and found that the resulting parts had improved mechanical strength [108]. Choi et al., added biodegradable suture fibers to the gelatin biomaterial ink which improved their printing accuracy to 97%, their mechanical strength 6-fold, and their dimensional stability (Figure 6E) [103].

3.2. Gelatin Cellular Solids to Improve Material Properties

Cellular structure solids are porous materials that offer increased benefits due to their low density large surface-to-volume ratio, energy absorption properties, high thermal and acoustic insulation, and low power loss factor. Gelatin cellular solids are a type of bio-foam that have increased benefits due to them being non-toxic, biocompatible, and biodegradable while maintaining the desirable properties of synthetic cellular solids, including improved thermal and acoustic insulation properties, increased biochemical activities, and a high specific surface area [113]. Gelatin foams have typically been used in the biomechanical field for tissue regeneration, the release of bioactive substances, and as biodegradable packaging materials.

There are a variety of material properties that affect foam behavior such as density, pore connectivity, pore size distribution, or mechanical properties. [114,115]. Pore connectivity defines whether the solid foam is a closed-cell foam (neighboring gas pores are separated by thin films) or an open-cell foam, also known as sponges (the gas phase is continuous) [115]. Closed-cell foams are useful for structural, protective, and thermal insulation applications as mechanical stiffness is ensured and air motion is avoided [115]. Open-cell foams are useful for the propagation and growth of 3D-cell networks, and acoustic insulation applications. These benefits of gelatin foams can be exploited and expanded for actuation by using them as coatings or alongside other actuators.

Whether a foam is open-cell or closed-cell is heavily influenced by its fabrication process [113]. Open-cell gelatin foams can be fabricated through conventional foaming methods such as mechanical foaming (whipping, stirring, shaking, static mixing, and ultrasonic cavitation) and foaming through physical or chemical blowing agents [113]. Mechanical foaming is non-toxic and safe compared to using blowing agents but is difficult to control and quantify [113]. Closed-cell gelatin foams can be fabricated through Thermally Induced Phase Separation (TIPS) with freeze-drying, or other lesser-known techniques, such as microwave foaming (Figure 7A) [116]. Of all methods, freeze drying and TIPS are the most widely reported for fabricating gelatin-based cellular solids because of their multiple advantages [113]. Freeze drying results in superior cellular structure stability and batch-to-batch consistency and allows for the creation of tailored morphologies (e.g., open/closed, fibrous/porous/membrane-like) by adjusting the processing parameters [113]. There are several other techniques to fabricate gelatin foams including 3D printing, electrospinning, gas foaming, and particle leaching methods. However, most techniques including freeze-drying, gas foaming, or salt leaching result in a combination of open- and closed-cell pores [116].

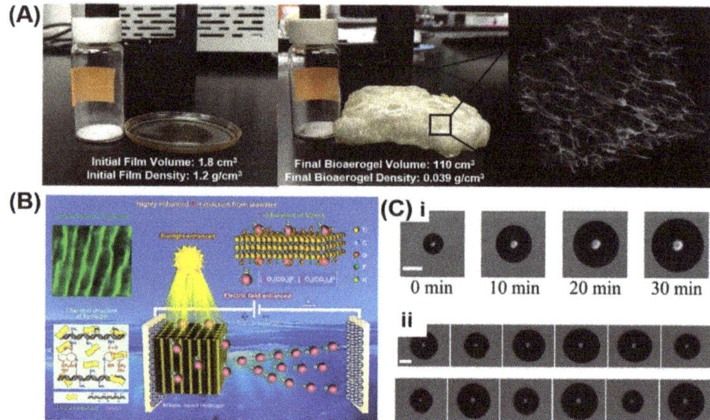

Figure 7. Gelatin foam enhancement. (**A**) Microwave-assisted fabrication of gelatin foams before and after microwaving at 700 W for 30 secs, with a view of the foam structure [116]. (**B**) The chemical structure and electric field/sunlight highly enhanced the uranium-adsorbing mechanism of the wood-mimetic directional macro-porous MXene-based hydrogel [117]. (**C**) (**i**) Bubbles generated by focusing a laser pulse into a 6 wt% gelatin gel supersaturated with dissolved air and (**ii**) finite-amplitude spherical oscillations of a bubble in the 6 wt% gelatin gel under 28 kHz ultrasound irradiation; the scale bar represents 100 μm [118]. (Images (**A–C**) are licensed under CC BY 4.0 (http://creativecommons.org/licenses/by/4.0/ (accessed on 30 October 2022))).

One of the common uses of fabricating gelatin foams is to fabricate scaffolds for tissue engineering [119,120], wound healing [14,15], artificial skin, sealants, bone repairing matrices [21,121], and blood plasma expanders [113]. In most of these applications, the gelatin foam is more of a passive foam rather than an actuated one.

However, gelatin foams can play more active roles as well. Fu et al., developed a muscle-like magnetorheological actuator with a magneto-restrictive component made of an alginate-gelatin sponge embedded with micron carbonyl iron particles and Fe_3O_4 nanoparticles wrapped with MWCNTs [122]. A polymer sponge matrix was used compared to a general elastomer because of its better flexibility and deformability, useful for soft actuation, and better ability to support magnetic particles than fluids and gels under a free boundary without encapsulation. The actuation performance was significantly influenced by the mass ratio of AL-GE and the mass fraction of the uniformly dispersed magnetic particles. The sponge exhibited superior flexibility and enhanced magneto-induced performance compared with other magnetic AL-GE sponge matrices. Chen et al., developed a wood-biomimetic macro-porous metal carbide/carbonitride (MXene) based hydrogel that can be enhanced by electric fields and sunlight irradiation to adsorb uranium from seawater (Figure 7B) [117]. The hydrogel has oriented micropores that allow for a high specific surface area for adsorption and has a good conductivity enhanced by the electric field/sunlight which accelerates the immigration of uranyl ions and the high efficiency of photothermal conversion. The uranium adsorption of the porous hydrogel increases by 79.95% when under both a 0.4 V electric field and sun irradiation.

Another active role played by gelatin foams as actuated devices are when the bubbles can be manipulated for drug delivery [123]. Ultrasound contrast agents (UCAs) are a special bubble type that were moved to tumor sites where an ultrasound or shock wave would excite it for drug delivery [124–126]. Acoustics could also be used to vaporize droplets to deliver drugs to cancer cells [127]. Similarly, gelatin was seen to liquefy when hit with a shock wave, and the shockwaves induced the collapse of gelatin bubbles with accompanying high-speed jets [128–131]. Another study used a shock tube to apply a planar shock front which provides an instantaneous pressure jump to a constant high pressure to induce the collapse of gas bubbles in a gelatinous mixture and could be used for targeted drug delivery and cancer research. [132]. Similar studies and results were found by Murakami et al., using ultrasonic irradiation to collapse bubbles in a gelatin gel where the bubbles were generated by focusing a laser pulse into a 6 wt% gelatin gel supersaturated with dissolved air and examined the role of viscoelasticity since human tissues are viscoelastic (Figure 7C) [118]. A better understanding of the role of viscoelasticity in acoustic bubble dynamics is desirable since collapsing bubbles and the cavitation of bubble activities do cause damage or heating to nearby tissues [125,131,133] and studies must be conducted on gelatin porous structures to estimate and control this damage.

Gelatin foams are also capable of enhancing actuator feedback and control by behaving as biodegradable sensors, such as capacitors [3,134]. Baumgartner et al., created a deformable gelatin foam capacitor that added tactile sense to their gelatin actuator [3]. They used food additives, mono- and di-glycerides of fatty acids (E471), to create the stable air bubbles of the foam sandwiched between two zinc electrodes such that an applied load could compress the soft foam, leading to impedance change. They were also able to use this foam to create a pressure-sensitive e-skin that was able to detect objects with complex shapes. Fan and Shen also developed a supercapacitor that is made of a microporous gelatin structure [135] since micropores can enhance electric double-layer capacitance. Since it is difficult for electrolyte ions to penetrate micropores, they improved ion transferring at high current densities with graphene oxide dispersed in the gelatin mixture. The resulting porous carbon nanosheet had excellent electrical conductivity and supercapacitor performance. Wang et al., created a gelatin-based microporous thermoelectric generator that was regulated and optimized by silica nanoparticles that improved ionic conductivity [136]. They created wearable devices such as wristbands and sleeves that could light up an LED or power a calculator by harvesting and storing low-grade body heat energy. Wearable energy harvesters such as these aid the problem of power supply for wearable devices.

3.3. Using Gelatin as Coatings on Other Actuators

Hydrogel coatings are commonly used for a variety of applications due to favorable traits such as biocompatibility, lubricity, and flexibility (Figure 8A) [137]. Gelatin coatings are widely used due to their ability to form uniform, strong, clear, and moderately flexible coatings that can readily swell and absorb water, making them ideal for the manufacture of capsules and photographic films [54]. Gelatin is currently widely used as a coating in the food industry due to its water-binding ability, gel-formation ability, water vapor barrier, film-forming ability, foam-forming ability, emulsification tendency, color-maintaining ability, and antioxidant activities [4,5]. Gelatin coatings and packaging films are also used to preserve foods and extend their shelf life [2]. In addition, gelatin coatings are used in the pharmaceutical industry because they are nonirritating, relatively low antigenic, inert, toxicity-reducing, and inexpensive, and they enhance the physical stability, targeting ability, and biocompatibility of nanoparticles [6,7]. Gelatin coatings were also used to encapsulate a probiotic yeast on chemically crosslinked gelatin hydrogels to protect the bioactive agents in different environments (Figure 8B) [138] and increase the efficiency of drug delivery into cancer cells by coating drug-encapsulating liposomes (Figure 8C) [7]. Gelatin can also be used on nanomaterials such as carbon nanotubes for superb affinity and uniformity wrapping on the surface of the carbon fiber while improving even distribution in a resin matrix to produce an advanced CNT-reinforced CFRP composite (Figure 8D) [139]. Since gelatin coatings are so versatile, they can be extended to support or expand actuating function on other actuators.

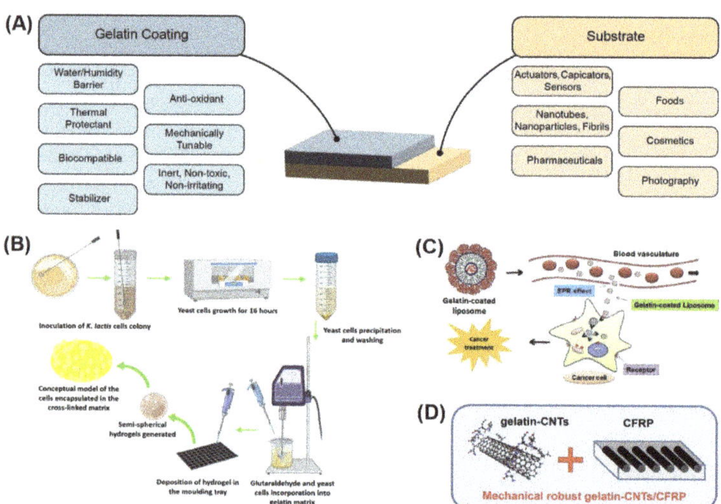

Figure 8. Gelatin used for actuator coatings. (**A**) Schematic depicting the benefits of gelatin-based coatings and their applications. (**B**) The protocol followed for the encapsulation of probiotic cells in the gelatin-glutaraldehyde (GTA) matrix [138]. (**C**) Summary of the cellular uptake of gelatin-coated liposomes [7]. (**D**) Mechanical robust gelatin-CNTs/CFRP composite [139]. (Images (**B–D**) are licensed under CC BY 4.0 (http://creativecommons.org/licenses/by/4.0/ (accessed on 30 October 2022))).

Wei et al., used gelatin methacryloyl (GelMA) to develop a biodegradable coating with mechanically tunable, anti-freezing, and long-term storage properties through facilely soaking strategies [137]. The coating is tunable due to the hydrophobic aggregation and hydrogen bonding of the hydrogel. They found that the coating was stable without disintegration on a variety of substrates of different geometrical shapes such as glass, ceramic, iron, wood, PTFE, and glass. They also found the coating to have superior interfacial adhesion and flexibility. These properties make the coating a good option for biomedical devices and actuators as this material coating can reduce friction between soft tissues and medical tools/rigid implantables.

Gelatin coatings also have thermal protective properties. Nickel-titanium shape memory alloys (NiTi SMAs) have been used for biomedical applications such as stents, orthodontic arc wires, and orthopedic staples due to their favorable features such as stable shape memory and small cross sections [140–142]. However, since they are thermally actuated, their surface temperatures limit biomedical applications. NiTi implants are also prone to nickel ion dissolution which could lead to medical complications. Simsek et al., developed a NiTi SMA with a coating made of a blend of gelatin and polyvinyl alcohol which provides a hydrophilic surface that forms a cushion at soft tissue-implant interfaces [140].

Gelatin coatings can also protect against humidity and swelling. Tan et al., used gelatin as a coating for long-period grating sensors used to measure relative humidity since gelatin is highly sensitive and can vary its index of refraction with relative humidity while protecting the sensor from water droplet condensation, short-circuiting, and corrosion [143]. Gelatin by itself exhibits poor barrier properties against water vapor due to its hydrophilic characteristics and being a hygroscopic material [94,144]. Thus, gelatin is a disadvantage in high moisture environments because the films may disintegrate in contact with water. A uniform dispersion of cellulose nanocrystals in nanocomposite films could block the permeating path of small molecules and lead to a good barrier performance [145]. The CNC-gelatin matrix was used in seed coatings and was found to reduce moisture absorption by a maximum of 39% due to CNC-gelatin hydrogen bonding interactions which reduced the chance of water molecules bonding to sorption sites and thus reduced the water uptake [94].

Baumgartner et al., developed an autonomous e-skin sensor patch made of a mixture of gelatin, glycerol, water, and citric acid and enhanced it with sensors to provide the multimodal sensing of temperature, humidity, and the state of deformation [3]. The temperature sensor was made of graphite powder and carnauba wax and could monitor temperature variations in proximity to a hot object. The humidity sensor was a designed interdigital electrode that could detect humidity changes induced by aspiration. Equipping soft actuators with such skins could bring them closer to autonomy.

4. Discussion

Soft robot actuators have several advantages over conventional robots such as safer human-machine interactions, adaptability to wearable devices, adept at navigating uneven terrains, resilience to perturbations, high conformity, and adaptability [146,147]. They have a wide range of functions, including wearables, grippers, and movement and locomotion, as well as biomechanical applications, such as rehabilitation devices, soft tools for surgery, drug delivery, artificial organs, and active simulators for training [148], and can be made in small scale [149,150]. In addition, the foreign body response to synthetic materials can dictate the long-term applicability of devices. Thus, biocompatibility and biomimicry are key considerations for soft robotics in biomedical engineering [148].

Gelatin is both biocompatible and biodegradable. The mechanical properties of gelatin are also tunable and can mimic human tissue due to its versatility. Therefore, it is a good candidate material to fabricate soft actuators for biomechanical applications. Gelatin also has many unique properties that promote cell proliferation and fluid absorbance that are beneficial for biomedical applications. The scope of research on gelatin actuators is increasing largely because their properties can be improved, and their applications in actuation can be vastly expanded with the use of plasticizers, crosslinkers, nanoparticles, fiber reinforcements, and foam structures. These additives leverage gelatin as a foundation and mold its properties to enhance and even transform them. Efforts to tune the mechanical properties of gelatin have yielded composites of a variety of stiffnesses to match a variety of biological tissues (Figure 9). The stiffness and porosity of gelatin-based hydrogels, such as GelMA hydrogel, can be controlled by tuning the hydrogel concentration, degree of functionalization, UV intensity, temperature, and additive supplementation [151–153].

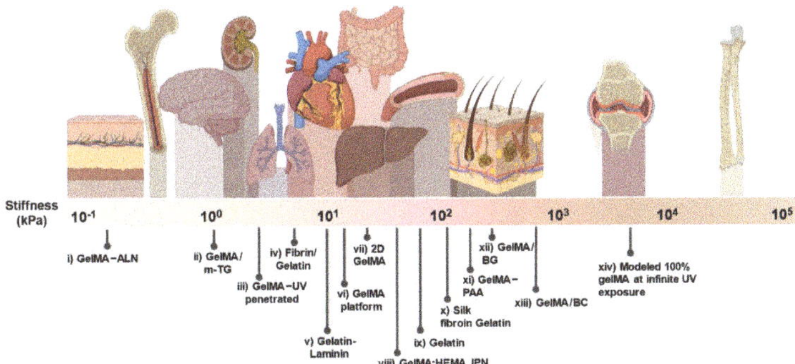

Figure 9. Depiction of gelatin tunability by comparing its stiffnesses against different tissue stiffness. (**i**) Alendronate-functionalized GelMA [154], (**ii**) GelMA/m-TG [121], (**iii**) GelMA–UV penetrated [155], (**iv**) Fibrin/gelatin [156], (**v**) Enzyme crosslinked gelatin-laminin hybrid [33], (**vi**) GelMA based platforms [33], (**vii**) 2D Photo-crosslinked GelMA [157], (**viii**) GelMA and poly(2-hydrox-ethyl methacrylate) (pHEMA) interpenetrating network (IPN) hydrogels [158], (**ix**) Gelatin of different concentrations [159], (**x**) Silk fibroin-Gelatin Bio-ink [160], (**xi**) GELMA-polyacrylamide (PAA) [161], (**xii**) GelMA/doped bioactive glass (BG) [162], (**xiii**) GelMA/bacterial cellulose (BC) [163], and (**xiv**) Theoretical value modeled for 100% GelMA gel at infinite exposure time [164]. Created with BioRender.com (accessed on 30 October 2022).

As such, gelatin has been explored for actuation with different modes, i.e., linear, bending, folding, and twisting. It also has different methods of stimulation or actuation mediums, such as pneumatic, electric, magnetic, pH, light, and temperature, due to different additives and crosslinkers. We can also find the maximum actuation deflection to size percent of the different types of actuators. The swelling-stimulated linear actuators are typically small but can swell up to 500% of their initial size when solvent-stimulated [54]. Bending gelatin actuators have a deflection to size percent of 30 to 50% when stimulated thermally or with solvent [22,41,63], and larger ratios of 100 to 130% when stimulated pneumatically, electrically, or mechanically [3,23,44,64]. The linear deflection to size percent for twisting gelatin actuators is typically around 75% for any stimulant, including solvent, pneumatic, or thermal [84,85].

In addition to actuators, gelatin is also extendable as sensors, capacitors, generators, and coatings for other non-gelatin actuators and active materials that we have discussed throughout this paper. Furthermore, using gelatin as a coating may provide enhanced biocompatibility for implantable devices. The soft robotics field is rapidly growing with innovations in biomechanical actuators. As there are many benefits to gelatin's properties and there are relevant ways that gelatin can be enhanced, there is the potential for typical actuators to be built and enhanced with gelatin. Thus, gelatin can increase the opportunities to more safely interface soft robotics in implantable systems and broaden the scope for innovation in biomechanics.

Author Contributions: Conceptualization, S.E. and H.M.G.; writing—original draft preparation, S.E.; writing—review and editing, H.M.G.; visualization, S.E. and H.M.G.; supervision, H.M.G.; All authors have read and agreed to the published version of the manuscript.

Funding: This research was funded by the Department of Bioengineering at the University of Illinois at Urbana Champaign.

Data Availability Statement: Not applicable.

Acknowledgments: The authors thank Lucy Brizzolara for her help with schematics.

Conflicts of Interest: The authors declare no conflict of interest.

References

1. Rahman, S.; Islam, M.; Islam, S.; Zaman, A.; Ahmed, T.; Biswas, S.; Sharmeen, S.; Rashid, T.U.; Rahman, M.M. Morphological Characterization of Hydrogels. In *Cellulose-Based Superabsorbent Hydrogels*; Mondal, M.I.H., Ed.; Springer: Cham, Switzerland, 2019; pp. 819–863.
2. Liu, D.; Nikoo, M.; Boran, G.; Zhou, P.; Regenstein, J.M. Collagen and Gelatin. *Annu. Rev. Food Sci. Technol.* **2015**, *6*, 527–557. [CrossRef] [PubMed]
3. Baumgartner, M.; Hartmann, F.; Drack, M.; Preninger, D.; Wirthl, D.; Gerstmayr, R.; Lehner, L.; Mao, G.; Pruckner, R.; Demchyshyn, S.; et al. Resilient yet entirely degradable gelatin-based biogels for soft robots and electronics. *Nat. Mater.* **2020**, *19*, 1102–1109. [CrossRef] [PubMed]
4. Abdallah, M.R.; Mohamed, M.A.; Mohamed, H.; Emara, M.T. Application of alginate and gelatin-based edible coating materials as alternatives to traditional coating for improving the quality of pastirma. *Food Sci. Biotechnol.* **2018**, *27*, 1589–1597. [CrossRef] [PubMed]
5. Lu, Y.; Luo, Q.; Chu, Y.; Tao, N.; Deng, S.; Wang, L.; Li, L. Application of Gelatin in Food Packaging: A Review. *Polymers* **2022**, *14*, 436. [CrossRef]
6. Babu, R.J.; Annaji, M.; Alsaqr, A.; Arnold, R.D. Animal-Based Materials in the Formulation of Nanocarriers for Anticancer Therapeutics. In *Polymeric Nanoparticles as a Promising Tool for Anti-Cancer Therapeutics*; Elsevier: Amsterdam, The Netherlands, 2019; pp. 319–341.
7. Battogtokh, G.; Joo, Y.; Abuzar, S.M.; Park, H.; Hwang, S.-J. Gelatin Coating for the Improvement of Stability and Cell Uptake of Hydrophobic Drug-Containing Liposomes. *Molecules* **2022**, *27*, 1041. [CrossRef]
8. Al-Nimry, S.; Dayah, A.; Hasan, I.; Daghmash, R. Cosmetic, Biomedical and Pharmaceutical Applications of Fish Gelatin/Hydrolysates. *Mar. Drugs* **2021**, *19*, 145. [CrossRef]
9. Calixto, S.; Ganzherli, N.; Gulyaev, S.; Figueroa-Gerstenmaier, S. Gelatin as a Photosensitive Material. *Molecules* **2018**, *23*, 2064. [CrossRef]
10. Liu, J.; Li, Y.; Hu, D.; Chao, X.; Zhou, Y.; Wang, J. An Essential Role of Gelatin in the Formation Process of Curling in Long Historical Photos. *Polymers* **2021**, *13*, 3894. [CrossRef]

11. Chen, Z.; Shi, X.; Xu, J.; Du, Y.; Yao, M.; Guo, S. Gel properties of SPI modified by enzymatic cross-linking during frozen storage. *Food Hydrocoll.* **2016**, *56*, 445–452. [CrossRef]

12. Saxena, A.; Sachin, K.; Bohidar, H.; Verma, A.K. Effect of molecular weight heterogeneity on drug encapsulation efficiency of gelatin nano-particles. *Colloids Surf. B Biointerfaces* **2005**, *45*, 42–48. [CrossRef]

13. Santoro, M.; Tatara, A.M.; Mikos, A.G. Gelatin Carriers for Drug and Cell Delivery in Tissue Engineering. *J. Control. Release* **2014**, *190*, 210–218. [CrossRef]

14. Ndlovu, S.P.; Ngece, K.; Alven, S.; Aderibigbe, B.A. Gelatin-Based Hybrid Scaffolds: Promising Wound Dressings. *Polymers* **2021**, *13*, 2959. [CrossRef]

15. Sandrasegaran, K.; Lall, C.; Rajesh, A.; Maglinte, D.T. Distinguishing Gelatin Bioabsorbable Sponge and Postoperative Abdominal Abscess on CT. *Am. J. Roentgenol.* **2005**, *184*, 475–480. [CrossRef]

16. Saw, M.M.; Chandler, B.; Ho, K.M. Benefits and Risks of Using Gelatin Solution as a Plasma Expander for Perioperative and Critically Ill Patients: A Meta-Analysis. *Anaesth. Intensive Care* **2012**, *40*, 17–32. [CrossRef]

17. Ichwan, A.M.; Karimi, M.; Dash, A.K. Use of gelatin–acacia coacervate containing benzocaine in topical formulations. *J. Pharm. Sci.* **1999**, *88*, 763–766. [CrossRef]

18. Wang, T.; Zhu, X.-K.; Xue, X.-T.; Wu, D.-Y. Hydrogel sheets of chitosan, honey and gelatin as burn wound dressings. *Carbohydr. Polym.* **2012**, *88*, 75–83. [CrossRef]

19. Luo, Z.; Sun, W.; Fang, J.; Lee, K.; Li, S.; Gu, Z.; Dokmeci, M.R.; Khademhosseini, A. Biodegradable Gelatin Methacryloyl Microneedles for Transdermal Drug Delivery. *Adv. Health Mater.* **2018**, *8*, e1801054. [CrossRef]

20. Echave, M.C.; Pimenta-Lopes, C.; Pedraz, J.L.; Mehrali, M.; Dolatshahi-Pirouz, A.; Ventura, F.; Orive, G. Enzymatic crosslinked gelatin 3D scaffolds for bone tissue engineering. *Int. J. Pharm.* **2019**, *562*, 151–161. [CrossRef]

21. Singh, D.; Tripathi, A.; Zo, S.; Singh, D.; Han, S.S. Synthesis of composite gelatin-hyaluronic acid-alginate porous scaffold and evaluation for in vitro stem cell growth and in vivo tissue integration. *Colloids Surf. B Biointerfaces* **2014**, *116*, 502–509. [CrossRef]

22. Hanzly, L.E.; Kristofferson, K.A.; Chauhan, N.; Barone, J.R. Biologically Controlled Gelatin Actuators. *Green Mater.* **2021**, *9*, 157–166. [CrossRef]

23. Shintake, J.; Sonar, H.; Piskarev, E.; Paik, J.; Floreano, D. Soft Pneumatic Gelatin Actuator for Edible Robotics. In Proceedings of the 2017 IEEE/RSJ International Conference on Intelligent Robots and Systems (IROS), Vancouver, BC, Canada, 24–28 September 2017; pp. 6221–6226.

24. Wang, H.; Xiang, J.; Wen, X.; Du, X.; Wang, Y.; Du, Z.; Cheng, X.; Wang, S. Multifunctional skin-inspired resilient MXene-embedded nanocomposite hydrogels for wireless wearable electronics. *Compos. Part A Appl. Sci. Manuf.* **2022**, *155*, 106835. [CrossRef]

25. Wang, X.; Bai, Z.; Zheng, M.; Yue, O.; Hou, M.; Cui, B.; Su, R.; Wei, C.; Liu, X. Engineered gelatin-based conductive hydrogels for flexible wearable electronic devices: Fundamentals and recent advances. *J. Sci. Adv. Mater. Devices* **2022**, *7*, 100451. [CrossRef]

26. Ebara, M.; Kotsuchibashi, Y.; Uto, K.; Aoyagi, T.; Kim, Y.-J.; Narain, R.; Idota, N.; Hoffman, J.M. Smart Hydrogels. In *Smart Biomaterials*; Springer: Tokyo, Japan, 2014; pp. 9–65.

27. Nakayama, A.; Kakugo, A.; Gong, J.P.; Osada, Y.; Takai, M.; Erata, T.; Kawano, S. High Mechanical Strength Double-Network Hydrogel with Bacterial Cellulose. *Adv. Funct. Mater.* **2004**, *14*, 1124–1128. [CrossRef]

28. Hardman, D.; Thuruthel, T.G.; Iida, F. Self-healing ionic gelatin/glycerol hydrogels for strain sensing applications. *NPG Asia Mater.* **2022**, *14*, 11. [CrossRef]

29. Lai, J.-Y. Biocompatibility of chemically cross-linked gelatin hydrogels for ophthalmic use. *J. Mater. Sci. Mater. Med.* **2010**, *21*, 1899–1911. [CrossRef]

30. Stevens, K.R.; Einerson, N.J.; Burmania, J.A.; Kao, W.J. In vivo biocompatibility of gelatin-based hydrogels and interpenetrating networks. *J. Biomater. Sci. Polym. Ed.* **2002**, *13*, 1353–1366. [CrossRef]

31. Wu, S.-C.; Chang, W.-H.; Dong, G.-C.; Chen, K.-Y.; Chen, Y.-S.; Yao, C.-H. Cell Adhesion and Pro-liferation Enhancement by Gelatin Nanofiber Scaffolds. *J. Bioact. Compat. Polym.* **2011**, *26*, 565–577. [CrossRef]

32. Gattazzo, F.; De Maria, C.; Rimessi, A.; Donà, S.; Braghetta, P.; Pinton, P.; Vozzi, G.; Bonaldo, P. Gelatin-Genipin-Based Biomaterials for Skeletal Muscle Tissue Engineering: Gelatin-Genipin-Based Biomaterials. *J. Biomed. Mater. Res.* **2018**, *106*, 2763–2777. [CrossRef]

33. Sadeghi, A.H.; Shin, S.R.; Deddens, J.C.; Fratta, G.; Mandla, S.; Yazdi, I.K.; Prakash, G.; Antona, S.; Demarchi, D.; Buijsrogge, M.P.; et al. Engineered 3D Cardiac Fibrotic Tissue to Study Fibrotic Remodeling. *Adv. Health Mater.* **2017**, *6*, 1601434. [CrossRef]

34. Sundaram, C.P.; Keenan, A.C. Evolution of hemostatic agents in surgical practice. *Indian J. Urol.* **2010**, *26*, 374–378. [CrossRef]

35. Zolfagharian, A.; Mahmud, M.A.P.; Gharaie, S.; Bodaghi, M.; Kouzani, A.Z.; Kaynak, A. 3D/4D-Printed Bending-Type Soft Pneumatic Actuators: Fabrication, Modelling, and Control. *Virtual Phys. Prototyp.* **2020**, *15*, 373–402. [CrossRef]

36. Xavier, M.S.; Tawk, C.D.; Zolfagharian, A.; Pinskier, J.; Howard, D.; Young, T.; Lai, J.; Harrison, S.M.; Yong, Y.K.; Bodaghi, M.; et al. Soft Pneumatic Actuators: A Review of Design, Fabrication, Modeling, Sensing, Control and Applications. *IEEE Access* **2022**, *10*, 59442–59485. [CrossRef]

37. Xavier, M.S.; Fleming, A.J.; Yong, Y.K. Finite Element Modeling of Soft Fluidic Actuators: Overview and Recent Developments. *Adv. Intell. Syst.* **2021**, *3*, 2000187. [CrossRef]

38. Czerner, M.; Fellay, L.S.; Suárez, M.P.; Frontini, P.M.; Fasce, L.A. Determination of Elastic Modulus of Gelatin Gels by Indentation Experiments. *Procedia Mater. Sci.* **2015**, *8*, 287–296. [CrossRef]

39. Bracq, A.; Haugou, G.; Bourel, B.; Maréchal, C.; Lauro, F.; Roth, S.; Mauzac, O. On the modeling of a visco-hyperelastic polymer gel under blunt ballistic impacts. *Int. J. Impact Eng.* **2018**, *118*, 78–90. [CrossRef]

40. Bello, J.; Bello, H.R.; Vinograd, J.R. The mechanism of gelation of gelatin the influence of pH, concentration, time and dilute electrolyte on the gelation of gelatin and modified gelatins. *Biochim. Biophys. Acta* **1962**, *57*, 214–221. [CrossRef]

41. Riedel, S.; Heyart, B.; Apel, K.S.; Mayr, S.G. Programing stimuli-responsiveness of gelatin with electron beams: Basic effects and development of a hydration-controlled biocompatible demonstrator. *Sci. Rep.* **2017**, *7*, 17436. [CrossRef]

42. Sha, X.-M.; Hu, Z.-Z.; Ye, Y.-H.; Xu, H.; Tu, Z.-C. Effect of extraction temperature on the gelling properties and identification of porcine gelatin. *Food Hydrocoll.* **2019**, *92*, 163–172. [CrossRef]

43. Liu, Y.; Sun, X.-C.; Lv, C.; Xia, H. Green nanoarchitectonics with PEDOT:PSS–gelatin composite for moisture-responsive actuator and generator. *Smart Mater. Struct.* **2021**, *30*, 125014. [CrossRef]

44. Chungyampin, S.; Niamlang, S. The Soft and High Actuation Response of Graphene Oxide/Gelatin Soft Gel. *Materials* **2021**, *14*, 7553. [CrossRef]

45. Helminger, M.; Wu, B.; Kollmann, T.; Benke, D.; Schwahn, D.; Pipich, V.; Faivre, D.; Zahn, D.; Cölfen, H. Synthesis and Characterization of Gelatin-Based Magnetic Hydrogels. *Adv. Funct. Mater.* **2014**, *24*, 3187–3196. [CrossRef] [PubMed]

46. Wu, Z.; Lin, X.; Zou, X.; Sun, J.; He, Q. Biodegradable Protein-Based Rockets for Drug Transportation and Light-Triggered Release. *ACS Appl. Mater. Interfaces* **2015**, *7*, 250–255. [CrossRef] [PubMed]

47. Chen, Z.; Huang, G.; Trase, I.; Han, X.; Mei, Y. Mechanical Self-Assembly of a Strain-Engineered Flexible Layer: Wrinkling, Rolling, and Twisting. *Phys. Rev. Appl.* **2016**, *5*, 017001. [CrossRef]

48. Hughes, J.; Rus, D. Mechanically Programmable, Degradable & Ingestible Soft Actuators. In Proceedings of the 3rd IEEE International Conference on Soft Robotics (RoboSoft), New Haven, CT, USA, 15 May–15 July 2020; pp. 836–843.

49. Heiden, A.; Preninger, D.; Lehner, L.; Baumgartner, M.; Drack, M.; Woritzka, E.; Schiller, D.; Gerstmayr, R.; Hartmann, F.; Kaltenbrunner, M. 3D Printing of Resilient Biogels for Omnidirectional and Exteroceptive Soft Actuators. *Sci. Robot.* **2022**, *7*, eabk2119. [CrossRef] [PubMed]

50. Tawk, C.; Alici, G. A Review of 3D-Printable Soft Pneumatic Actuators and Sensors: Research Challenges and Opportunities. *Adv. Intell. Syst.* **2021**, *3*, 2000223. [CrossRef]

51. Ehrenhofer, A.; Elstner, M.; Wallmersperger, T. Normalization of hydrogel swelling behavior for sensoric and actuatoric applications. *Sens. Actuators B Chem.* **2018**, *255*, 1343–1353. [CrossRef]

52. Tattanon, T.; Pongprayoon, T.; Arpornmaeklong, P.; Ummartyotin, S. Development of Hydroxy-apatite from Cuttlebone and Gelatin-Based Hydrogel Composite for Medical Materials. *J. Polym. Res.* **2022**, *29*, 364. [CrossRef]

53. Ooi, S.Y.; Ahmad, I.; Amin, M.C.I.M. Cellulose nanocrystals extracted from rice husks as a reinforcing material in gelatin hydrogels for use in controlled drug delivery systems. *Ind. Crop. Prod.* **2016**, *93*, 227–234. [CrossRef]

54. Shubhra, Q.T.H. *Gelatin Film and Fiber Reinforced Gelatin Composites. Gelatin: Production, Applications and Health Implications*; Boran, G., Ed.; Nova Publishers: New York, NY, USA, 2013.

55. Qiao, C.; Cao, X.; Wang, F. Swelling Behavior Study of Physically Crosslinked Gelatin Hydrogels. *Polym. Polym. Compos.* **2012**, *20*, 53–58. [CrossRef]

56. Hajikarimi, A.; Sadeghi, M. Free radical synthesis of cross-linking gelatin base poly NVP/acrylic acid hydrogel and nanoclay hydrogel as cephalexin drug deliver. *J. Polym. Res.* **2020**, *27*, 57. [CrossRef]

57. Barthélémy, F.; Santoso, J.W.; Rabichow, L.; Jin, R.; Little, I.; Nelson, S.F.; McCain, M.L.; Miceli, M.C. Modeling Patient-Specific Muscular Dystrophy Phenotypes and Therapeutic Responses in Reprogrammed Myotubes Engineered on Micromolded Gelatin Hydrogels. *Front. Cell Dev. Biol.* **2022**, *10*, 830415. [CrossRef]

58. Gupta, D.; Santoso, J.; McCain, M. Characterization of Gelatin Hydrogels Cross-Linked with Microbial Transglutaminase as Engineered Skeletal Muscle Substrates. *Bioengineering* **2021**, *8*, 6. [CrossRef]

59. Besser, R.R.; Bowles, A.C.; Alassaf, A.; Carbonero, D.; Claure, I.; Jones, E.; Reda, J.; Wubker, L.; Batchelor, W.; Ziebarth, N.; et al. Enzymatically Cross-linked Gelatin–Laminin Hydrogels for Applications in Neuromuscular Tissue Engineering. *Biomater. Sci.* **2020**, *8*, 591–606. [CrossRef]

60. Li, J.; Chee, H.L.; Chong, Y.T.; Chan, B.Q.Y.; Xue, K.; Lim, P.C.; Loh, X.J.; Wang, F. Hofmeister Effect Mediated Strong PHEMA-Gelatin Hydrogel Actuator. *ACS Appl. Mater. Interfaces* **2022**, *14*, 23826–23838. [CrossRef]

61. Wang, Y.; Bamdad, F.; Song, Y.; Chen, L. *Hydrogel Particles and Other Novel Protein-Based Methods for Food Ingredient and Nutraceutical Delivery Systems: Encapsulation Technologies and Delivery Systems for Food Ingredients and Nutraceuticals*; Elsevier: Amsterdam, The Netherlands, 2012; pp. 412–450.

62. Gomes, C.M.; Liu, C.; Paten, J.A.; Felton, S.M.; Deravi, L.F. Protein-Based Hydrogels That Actu-ate Self-Folding Systems. *Adv. Funct. Mater.* **2019**, *29*, 1805777. [CrossRef]

63. Stroganov, V.; Al-Hussein, M.; Sommer, J.-U.; Janke, A.; Zakharchenko, S.; Ionov, L. Reversible Thermosensitive Biodegradable Polymeric Actuators Based on Confined Crystallization. *Nano Lett.* **2015**, *15*, 1786–1790. [CrossRef]

64. Harris, H.; Radecka, A.; Malik, R.; Pineda Guzman, R.A.; Santoso, J.; Bradshaw, A.; McCain, M.; Kersh, M.; Golecki, H. Development and Characterization of Biostable Hydrogel Robotic Actuators for Implantable Devices: Tendon Actuated Gelatin. In Proceedings of the 2022 Design of Medical Devices Conference, American Society of Mechanical Engineers, Minneapolis, MN, USA, 11–14 April 2022; p. V001T05A004.

65. Sardesai, A.N.; Segel, X.M.; Baumholtz, M.N.; Chen, Y.; Sun, R.; Schork, B.W.; Buonocore, R.; Wagner, K.O.; Golecki, H.M. Design and Characterization of Edible Soft Robotic Candy Actuators. *MRS Adv.* **2018**, *3*, 3003–3009. [CrossRef]

66. Nagai, T.; Kurita, A.; Shintake, J. Characterization of Sustainable Robotic Materials and Finite Element Analysis of Soft Actuators Under Biodegradation. *Front. Robot. AI* **2021**, *8*, 760485. [CrossRef]

67. Tungkavet, T.; Seetapan, N.; Pattavarakorn, D.; Sirivat, A. Graphene/Gelatin Hydrogel Compo-sites with High Storage Modulus Sensitivity for Using as Electroactive Actuator: Effects of Surface Area and Electric Field Strength. *Polymer* **2015**, *70*, 242–251. [CrossRef]

68. Wang, W.; Wang, Z.; Liu, Y.; Li, N.; Wang, W.; Gao, J. Preparation of Reduced Graphene Ox-ide/Gelatin Composite Films with Reinforced Mechanical Strength. *Mater. Res. Bull.* **2012**, *47*, 2245–2251. [CrossRef]

69. Chambers, L.D.; Winfield, J.; Ieropoulos, I.; Rossiter, J. Biodegradable and Edible Gelatine Actuators for Use as Artificial Muscles. In *Electroactive Polymer Actuators and Devices (EAPAD)*; Bar-Cohen, Y., Ed.; SPIE: San Diego, CA, USA, 2014; p. 90560B.

70. Shin, S.R.; Bolagh, B.A.G.; Dang, T.; Topkaya, S.N.; Gao, X.; Yang, S.Y.; Jung, S.M.; Oh, J.H.; Dokmeci, M.R.; Tang, X.; et al. Cell-laden Microengineered and Mechanically Tunable Hybrid Hydrogels of Gelatin and Graphene Oxide. *Adv. Mater.* **2013**, *25*, 6385–6391. [CrossRef] [PubMed]

71. Haider, S.; Park, S.-Y.; Saeed, K.; Farmer, B. Swelling and electroresponsive characteristics of gelatin immobilized onto multi-walled carbon nanotubes. *Sens. Actuators B Chem.* **2007**, *124*, 517–528. [CrossRef]

72. Bigi, A. Relationship between Triple-Helix Content and Mechanical Properties of Gelatin Films. *Biomaterials* **2004**, *25*, 5675–5680. [CrossRef] [PubMed]

73. Firouzeh, A.; Salerno, M.; Paik, J. Soft Pneumatic Actuator with Adjustable Stiffness Layers for Multi-DoF Actuation. In Proceedings of the IEEE/RSJ International Conference on Intelligent Robots and Systems (IROS), Hamburg, Germany, 28 September 2015; pp. 1117–1124.

74. Su, H.; Hou, X.; Zhang, X.; Qi, W.; Cai, S.; Xiong, X.; Guo, J. Pneumatic Soft Robots: Challenges and Benefits. *Actuators* **2022**, *11*, 92. [CrossRef]

75. Na, H.; Kang, Y.-W.; Park, C.S.; Jung, S.; Kim, H.-Y.; Sun, J.-Y. Hydrogel-based strong and fast actuators by electroosmotic turgor pressure. *Science* **2022**, *376*, 301–307. [CrossRef]

76. Dick, W.B. Cements and Uniting Bodies, Part 3. In *Encyclopedia of Practical Receipts and Processes*; Dick and Fitzgerald Publishers: New York, NY, USA, 1872.

77. Mosadegh, B.; Polygerinos, P.; Keplinger, C.; Wennstedt, S.; Shepherd, R.; Gupta, U.; Shim, J.; Bertoldi, K.; Walsh, C.J.; Whitesides, G.M. Pneumatic Networks for Soft Robotics that Actuate Rapidly. *Adv. Funct. Mater.* **2014**, *24*, 2163–2170. [CrossRef]

78. Greer, A.H.; King, E.; Lee, E.H.; Sardesai, A.N.; Chen, Y.; Obuz, S.E.; Graf, Y.; Ma, T.; Chow, D.Y.; Fu, T.; et al. Soluble Polymer Pneumatic Networks and a Single-Pour System for Improved Accessibility and Durability of Soft Robotic Actuators. *Soft Robot.* **2021**, *8*, 144–151. [CrossRef]

79. Elhi, F.; Karu, K.; Rinne, P.; Nadel, K.-A.; Järvekülg, M.; Aabloo, A.; Tamm, T.; Ivaništšev, V.; Põhako-Esko, K. Understanding the Behavior of Fully Non-Toxic Polypyrrole-Gelatin and Polypyrrole-PVdF Soft Actuators with Choline Ionic Liquids. *Actuators* **2020**, *9*, 40. [CrossRef]

80. Wan, C.; Frydrych, M.; Chen, B. Strong and Bioactive Gelatin–Graphene Oxide Nanocomposites. *Soft Matter* **2011**, *7*, 6159. [CrossRef]

81. Choe, G.; Tang, X.; Wang, R.; Wu, K.; Jin Jeong, Y.; Kyu An, T.; Hyun Kim, S.; Mi, L. Printing of Self-Healable Gelatin Conductors Engineered for Improving Physical and Electrical Functions: Exploring Potential Application in Soft Actuators and Sensors. *J. Ind. Eng. Chem.* **2022**, *116*, 171–179. [CrossRef]

82. Li, D.; Fan, D.; Zhu, R.; Lei, Q.; Liao, Y.; Yang, X.; Pan, Y.; Wang, Z.; Wu, Y.; Liu, S.; et al. Origami-Inspired Soft Twisting Actuator. *Soft Robot.* **2021**. [CrossRef]

83. Roche, E.T.; Wohlfarth, R.; Overvelde, J.T.B.; Vasilyev, N.V.; Pigula, F.A.; Mooney, D.J.; Bertoldi, K.; Walsh, C.J. A Bioinspired Soft Actuated Material. *Adv. Mater.* **2014**, *26*, 1200–1206. [CrossRef]

84. Erb, R.M.; Sander, J.S.; Grisch, R.; Studart, A.R. Self-shaping composites with programmable bioinspired microstructures. *Nat. Commun.* **2013**, *4*, 1712. [CrossRef]

85. Lutz-Bueno, V.; Bolisetty, S.; Azzari, P.; Handschin, S.; Mezzenga, R. Self-Winding Gelatin–Amyloid Wires for Soft Actuators and Sensors. *Adv. Mater.* **2020**, *32*, 2004941. [CrossRef]

86. Zhang, Y.; Le Ferrand, H. Bioinspired Self-Shaping Clay Composites for Sustainable Development. *Biomimetics* **2022**, *7*, 13. [CrossRef]

87. Neffe, A.; Löwenberg, C.; Julich-Gruner, K.; Behl, M.; Lendlein, A. Thermally-Induced Shape-Memory Behavior of Degradable Gelatin-Based Networks. *Int. J. Mol. Sci.* **2021**, *22*, 15892. [CrossRef]

88. Huang, Z.; Wei, C.; Dong, L.; Wang, A.; Yao, H.; Guo, Z.; Mi, S. Fluid-driven hydrogel actuators with an origami structure. *Iscience* **2022**, *25*, 104674. [CrossRef]

89. Sasaki, N.; Umeda, H.; Okada, S.; Kojima, R.; Fukuda, A. Mechanical Properties of Hydroxyapatite-Reinforced Gelatin as a Model System of Bone. *Biomaterials* **1989**, *10*, 129–132. [CrossRef]

90. Lin, F.-H.; Yao, C.-H.; Sun, J.-S.; Liu, H.-C.; Huang, C.-W. Biological effects and cytotoxicity of the composite composed by tricalcium phosphate and glutaraldehyde cross-linked gelatin. *Biomaterials* **1998**, *19*, 905–917. [CrossRef]

91. Avila-Ramirez, A.; Catzim-Ríos, K.; Guerrero-Beltrán, C.E.; Ramírez-Cedillo, E.; Ortega-Lara, W. Reinforcement of Alginate-Gelatin Hydrogels with Bioceramics for Biomedical Applications: A Comparative Study. *Gels* **2021**, *7*, 184. [CrossRef]
92. Bhowmik, S.; Islam, J.; Debnath, T.; Miah, M.; Bhattacharjee, S.; Khan, M. Reinforcement of Gelatin-Based Nanofilled Polymer Biocomposite by Crystalline Cellulose from Cotton for Advanced Wound Dressing Applications. *Polymers* **2017**, *9*, 222. [CrossRef] [PubMed]
93. Alves, J.S.; dos Reis, K.C.; Menezes EG, T.; Pereira, F.V.; Pereira, J. Effect of Cellulose Nano-crystals and Gelatin in Corn Starch Plasticized Films. *Carbohydr. Polym.* **2015**, *115*, 215–222. [CrossRef] [PubMed]
94. Dogaru, B.-I.; Stoleru, V.; Mihalache, G.; Yonsel, S.; Popescu, M.-C. Gelatin Reinforced with CNCs as Nanocomposite Matrix for *Trichoderma harzianum* KUEN 1585 Spores in Seed Coatings. *Molecules* **2021**, *26*, 5755. [CrossRef] [PubMed]
95. Panzavolta, S.; Bracci, B.; Gualandi, C.; Focarete, M.L.; Treossi, E.; Kouroupis-Agalou, K.; Rubini, K.; Bosia, F.; Brely, L.; Pugno, N.M.; et al. Structural reinforcement and failure analysis in composite nanofibers of graphene oxide and gelatin. *Carbon* **2014**, *78*, 566–577. [CrossRef]
96. Shubhra, Q.T.H.; Alam, A.M.; Khan, M.A.; Saha, M.; Saha, D.; Khan, J.A.; Quaiyyum, M.A. The Preparation and Characterization of Silk/Gelatin Biocomposites. *Polym. Technol. Eng.* **2010**, *49*, 983–990. [CrossRef]
97. Shubhra, Q.T.; Alam, A.; Beg, M.D.H. Mechanical and degradation characteristics of natural silk fiber reinforced gelatin composites. *Mater. Lett.* **2011**, *65*, 333–336. [CrossRef]
98. Liu, J.; Liu, C.-K.; Brown, E. Development and Characterization of Genipin Cross-Linked Gelatin Based Composites Incorporated with Vegetable-Tanned Collagen Fiber (VCF). *J. Am. Leather Chem. Assoc.* **2017**, *112*, 410–419.
99. Wei, S.-M.; Pei, M.-Y.; Pan, W.-L.; Thissen, H.; Tsai, S.-W. Gelatin Hydrogels Reinforced by Absorbable Nanoparticles and Fibrils Cured In Situ by Visible Light for Tissue Adhesive Applications. *Polymers* **2020**, *12*, 1113. [CrossRef]
100. Yang, Z.; Shen, C.; Zou, Y.; Wu, D.; Zhang, H.; Chen, K. Application of Solution Blow Spinning for Rapid Fabrication of Gelatin/Nylon 66 Nanofibrous Film. *Foods* **2021**, *10*, 2339. [CrossRef]
101. Campodoni, E.; Montanari, M.; Dozio, S.M.; Heggset, E.B.; Panseri, S.; Montesi, M.; Tampieri, A.; Syverud, K.; Sandri, M. Blending Gelatin and Cellulose Nanofibrils: Biocomposites with Tunable Degradabil-ity and Mechanical Behavior. *Nanomaterials* **2020**, *10*, 1219. [CrossRef]
102. Wang, S.; Li, K.; Zhou, Q. High strength and low swelling composite hydrogels from gelatin and delignified wood. *Sci. Rep.* **2020**, *10*, 17842. [CrossRef]
103. Choi, D.J.; Choi, K.; Park, S.J.; Kim, Y.-J.; Chung, S.; Kim, C.-H. Suture Fiber Reinforcement of a 3D Printed Gelatin Scaffold for Its Potential Application in Soft Tissue Engineering. *Int. J. Mol. Sci.* **2021**, *22*, 11600. [CrossRef]
104. Singh, C.; Wong, C.S.; Wang, X. Medical Textiles as Vascular Implants and Their Success to Mimic Natural Arteries. *J. Funct. Biomater.* **2015**, *6*, 500–525. [CrossRef]
105. Hiob, M.A.; She, S.; Muiznieks, L.D.; Weiss, A.S. Biomaterials and Modifications in the Development of Small-Diameter Vascular Grafts. *ACS Biomater. Sci. Eng.* **2017**, *3*, 712–723. [CrossRef]
106. Kuijpers, A.; van Wachem, P.; van Luyn, M.; Engbers, G.; Krijgsveld, J.; Zaat, S.; Dankert, J.; Feijen, J. In vivo and in vitro release of lysozyme from cross-linked gelatin hydrogels: A model system for the delivery of antibacterial proteins from prosthetic heart valves. *J. Control. Release* **2000**, *67*, 323–336. [CrossRef]
107. Hassan, M.L.; Fadel, S.M.; El-Wakil, N.A.; Oksman, K. Chitosan/Rice Straw Nanofibers Nano-composites: Preparation, Mechanical, and Dynamic Thermomechanical Properties. *J. Appl. Polym. Sci.* **2012**, *125*, E216–E222. [CrossRef]
108. Jiang, Y.; Xv, X.; Liu, D.; Yang, Z.; Zhang, Q.; Shi, H.; Zhao, G.; Zhou, J. Preparation of cellulose nanofiber-reinforced gelatin hydrogel and optimization for 3D printing applications. *Bioresources* **2018**, *13*, 5909–5924. [CrossRef]
109. Wang, W.; Zhang, X.; Teng, A.; Liu, A. Mechanical Reinforcement of Gelatin Hydrogel with Nanofiber Cellulose as a Function of Percolation Concentration. *Int. J. Biol. Macromol.* **2017**, *103*, 226–233. [CrossRef]
110. Wan, Y.Z.; Wang, Y.L.; Luo, H.L.; Cheng, G.X.; Yao, K.D. Carbon Fiber-Reinforced Gelatin Composites. I. Preparation and Mechanical Properties. *J. Appl. Polym. Sci.* **2000**, *75*, 987–993. [CrossRef]
111. Ravishankar, P.; Ozkizilcik, A.; Husain, A.; Balachandran, K. Anisotropic Fiber-Reinforced Gly-cosaminoglycan Hydrogels for Heart Valve Tissue Engineering. *Tissue Eng. Part A* **2021**, *27*, 513–525. [CrossRef]
112. Khan, R.A.; Khan, M.A.; Sarker, B.; Saha, S.; Das, A.K.; Noor, N.; Huq, T.; Khan, A.; Dey, K.; Saha, M. Fabrication and Characterization of Gelatin Fiber-based Linear Low-density Polyethylene Foamed Composite. *J. Reinf. Plast. Compos.* **2010**, *29*, 2438–2449. [CrossRef]
113. Torrejon, V.M.; Song, J.; Yu, Z.; Hang, S. Gelatin-Based Cellular Solids: Fabrication, Structure and Properties. *J. Cell. Plast.* **2022**, *58*, 797–858. [CrossRef]
114. Gibson, L.J.; Ashby, M.F. *Cellular Solids: Structure and Properties*; Cambridge University Press: Cambridge, UK, 1997.
115. Andrieux, S.; Quell, A.; Stubenrauch, C.; Drenckhan, W. Liquid foam templating—A route to tailor-made polymer foams. *Adv. Colloid Interface Sci.* **2018**, *256*, 276–290. [CrossRef]
116. Frazier, S.D.; Aday, A.N.; Srubar, W. On-Demand Microwave-Assisted Fabrication of Gelatin Foams. *Molecules* **2018**, *23*, 1121. [CrossRef] [PubMed]
117. Chen, L.; Sun, Y.; Wang, J.; Ma, C.; Peng, S.; Cao, X.; Yang, L.; Ma, C.; Duan, G.; Liu, Z.; et al. A wood-mimetic porous MXene/gelatin hydrogel for electric field/sunlight bi-enhanced uranium adsorption. *E-Polymers* **2022**, *22*, 468–477. [CrossRef]

118. Murakami, K.; Yamakawa, Y.; Zhao, J.; Johnsen, E.; Ando, K. Ultrasound-Induced Nonlinear Oscillations of a Spherical Bubble in a Gelatin Gel. *J. Fluid Mech.* **2021**, *924*, A38. [CrossRef]

119. Pezeshki Modaress, M.; Mirzadeh, H.; Zandi, M. Fabrication of a Porous Wall and Higher Inter-connectivity Scaffold Comprising Gelatin/Chitosan via Combination of Salt-Leaching and Lyophilization Methods. *Iran. Polym. J.* **2012**, *21*, 191–200. [CrossRef]

120. Lu, B.; Wang, T.; Li, Z.; Dai, F.; Lv, L.; Tang, F.; Yu, K.; Liu, J.; Lan, G. Healing of skin wounds with a chitosan–gelatin sponge loaded with tannins and platelet-rich plasma. *Int. J. Biol. Macromol.* **2016**, *82*, 884–891. [CrossRef]

121. La Gatta, A.; Tirino, V.; Cammarota, M.; La Noce, M.; Stellavato, A.; Pirozzi, A.V.A.; Portaccio, M.; Diano, N.; Laino, L.; Papaccio, G.; et al. Gelatin-Biofermentative Unsulfated Glycosaminoglycans Semi-Interpenetrating Hydrogels via Microbial-Transglutaminase Crosslinking Enhance Osteogenic Poten-tial of Dental Pulp Stem Cells. *Regen. Biomater.* **2021**, *8*, rbaa052. [CrossRef]

122. Fu, Y.; Yao, J.; Zhao, H.; Zhao, G.; Wan, Z.; Guo, R. A muscle-like magnetorheological actuator based on bidisperse magnetic particles enhanced flexible alginate-gelatin sponges. *Smart Mater. Struct.* **2020**, *29*, 015019. [CrossRef]

123. Ohl, S.-W.; Klaseboer, E.; Khoo, B.C. Bubbles with shock waves and ultrasound: A review. *Interface Focus* **2015**, *5*, 20150019. [CrossRef]

124. Ohl, C.D.; Ikink, R. Shock-Wave-Induced Jetting of Micron-Size Bubbles. *Phys. Rev. Lett.* **2003**, *90*, 214502. [CrossRef]

125. Ter Haar, G. Ultrasonic imaging: Safety considerations. *Interface Focus* **2011**, *1*, 686–697. [CrossRef]

126. Liang, H.-D.; Noble, J.A.; Wells, P.N.T. Recent advances in biomedical ultrasonic imaging techniques. *Interface Focus* **2011**, *1*, 475–476. [CrossRef]

127. Wong, Z.Z.; Kripfgans, O.D.; Qamar, A.; Fowlkes, J.B.; Bull, J.L. Bubble evolution in acoustic droplet vaporization at physiological temperature via ultra-high speed imaging. *Soft Matter* **2011**, *7*, 4009–4016. [CrossRef]

128. Dear, J.P.; Field, J.E. A study of the collapse of arrays of cavities. *J. Fluid Mech.* **1988**, *190*, 409–425. [CrossRef]

129. Dear, J.P.; Field, J.E.; Walton, A.J. Gas compression and jet formation in cavities collapsed by a shock wave. *Nature* **1988**, *332*, 505–508. [CrossRef]

130. Bourne, N.K.; Field, J.E. Shock-induced collapse of single cavities in liquids. *J. Fluid Mech.* **1992**, *244*, 225–240. [CrossRef]

131. Kodama, T.; Takayama, K. Dynamic behavior of bubbles during extracorporeal shock-wave lithotripsy. *Ultrasound Med. Biol.* **1998**, *24*, 723–738. [CrossRef]

132. Hopfes, T.; Wang, Z.; Giglmaier, M.; Adams, N.A. Collapse Dynamics of Bubble Pairs in Gelati-nous Fluids. *Exp. Therm. Fluid Sci.* **2019**, *108*, 104–114. [CrossRef]

133. Gillitzer, R.; Neisius, A.; Wöllner, J.; Hampel, C.; Brenner, W.; Bonilla, A.A.; Thüroff, J. Low-Frequency Extracorporeal Shock Wave Lithotripsy Improves Renal Pelvic Stone Disintegration in a Pig Model. *BJU Int.* **2009**, *103*, 1284–1288. [CrossRef] [PubMed]

134. Saliu, O.D.; Mamo, M.; Ndungu, P.; Ramontja, J. The Making of a High Performance Supercapac-itor Active at Negative Potential Using Sulphonic Acid Activated Starch-Gelatin-TiO$_2$ Nano-Hybrids. *Arab. J. Chem.* **2021**, *14*, 103242. [CrossRef]

135. Fan, H.; Shen, W. Gelatin-Based Microporous Carbon Nanosheets as High Performance Superca-pacitor Electrodes. *ACS Sustain. Chem. Eng.* **2016**, *4*, 1328–1337. [CrossRef]

136. Wang, S.; Han, L.; Liu, H.; Dong, Y.; Wang, X. Ionic Gelatin-Based Flexible Thermoelectric Gener-ator with Scalability for Human Body Heat Harvesting. *Energies* **2022**, *15*, 3441. [CrossRef]

137. Wei, L.; Huang, J.; Yan, Y.; Cui, J.; Zhao, Y.; Bai, F.; Liu, J.; Wu, X.; Zhang, X.; Du, M. Substrate-Independent, Mechanically Tunable, and Scalable Gelatin Methacryloyl Hydrogel Coating with Drag-Reducing and Anti-Freezing Properties. *ACS Appl. Polym. Mater.* **2022**, *4*, 4876–4885. [CrossRef]

138. Patarroyo, J.L.; Florez-Rojas, J.S.; Pradilla, D.; Valderrama-Rincón, J.D.; Cruz, J.C.; Reyes, L.H. Formulation and Characterization of Gelatin-Based Hydrogels for the Encapsulation of Kluyveromyces lactis—Applications in Packed-Bed Reactors and Probiotics Delivery in Humans. *Polymers* **2020**, *12*, 1287. [CrossRef]

139. Zeng, L.; Tao, W.; Zhao, J.; Li, Y.; Li, R. Mechanical performance of a CFRP composite reinforced via gelatin-CNTs: A study on fiber interfacial enhancement and matrix enhancement. *Nanotechnol. Rev.* **2022**, *11*, 625–636. [CrossRef]

140. Simsek, G.M.; Barthes, J.; Muller, C.; McGuinness, G.B.; Vrana, N.E.; Yapici, G.G. PVA/gelatin-based hydrogel coating of nickel-titanium alloy for improved tissue-implant interface. *Appl. Phys. A* **2021**, *127*, 387. [CrossRef]

141. Tung, A.T.; Park, B.-H.; Liang, D.H.; Niemeyer, G. Laser-machined shape memory alloy sensors for position feedback in active catheters. *Sens. Actuators A Phys.* **2008**, *147*, 83–92. [CrossRef]

142. Muhammad, N.; Whitehead, D.; Boor, A.; Oppenlander, W.; Liu, Z.; Li, L. Picosecond Laser Micromachining of Nitinol and Platinum–Iridium Alloy for Coronary Stent Applications. *Appl. Phys. A* **2012**, *106*, 607–617. [CrossRef]

143. Tan, K.M.; Tay, C.M.; Tjin, S.C.; Chan, C.C.; Rahardjo, H. High Relative Humidity Measurements Using Gelatin Coated Long-Period Grating Sensors. *Sens. Actuators B Chem.* **2005**, *110*, 335–341. [CrossRef]

144. Nur Hanani, Z.A.; Roos, Y.H.; Kerry, J.P. Use and Application of Gelatin as Potential Biodegradable Packaging Materials for Food Products. *Int. J. Biol. Macromol.* **2014**, *71*, 94–102. [CrossRef]

145. Miao, C.; Hamad, W.Y. In-situ polymerized cellulose nanocrystals (CNC)—Poly(l-lactide) (PLLA) nanomaterials and applications in nanocomposite processing. *Carbohydr. Polym.* **2016**, *153*, 549–558. [CrossRef]

146. Lee, C.; Kim, M.; Kim, Y.J.; Hong, N.; Ryu, S.; Kim, H.J.; Kim, S. Soft Robot Review. *Int. J. Control Autom. Syst.* **2017**, *15*, 3–15. [CrossRef]

147. Zhou, Y.; Li, H. A Scientometric Review of Soft Robotics: Intellectual Structures and Emerging Trends Analysis (2010–2021). *Front. Robot. AI* **2022**, *9*, 868682. [CrossRef]

148. Cianchetti, M.; Laschi, C.; Menciassi, A.; Dario, P. Biomedical applications of soft robotics. *Nat. Rev. Mater.* **2018**, *3*, 143–153. [CrossRef]

149. Huang, W.; Shang, W.; Huang, Y.; Long, H.; Wu, X. Insect-Scale SMAW-Based Soft Robot With Crawling, Jumping, and Loading Locomotion. *IEEE Robot. Autom. Lett.* **2022**, *7*, 9287–9293. [CrossRef]

150. Yang, X.; Shang, W.; Lu, H.; Liu, Y.; Yang, L.; Tan, R.; Wu, X.; Shen, Y. An agglutinate magnetic spray transforms inanimate objects into millirobots for biomedical applications. *Sci. Robot.* **2020**, *5*, eabc8191. [CrossRef]

151. Bupphathong, S.; Quiroz, C.; Huang, W.; Chung, P.-F.; Tao, H.-Y.; Lin, C.-H. Gelatin Methacrylate Hydrogel for Tissue Engineering Applications—A Review on Material Modifications. *Pharmaceuticals* **2022**, *15*, 171. [CrossRef]

152. Burdick, J.A.; Prestwich, G.D. Hyaluronic acid hydrogels for biomedical applications. *Adv. Mater.* **2011**, *23*, H41–H56. [CrossRef]

153. Fernandez, P.; Bausch, A.R. The compaction of gels by cells: A case of collective mechanical activity. *Integr. Biol.* **2009**, *1*, 252–259. [CrossRef] [PubMed]

154. Liu, L.; Li, X.; Shi, X.; Wang, Y. Injectable Alendronate-Functionalized GelMA Hydrogels for Mineralization and Osteogenesis. *RSC Adv.* **2018**, *8*, 22764–22776. [CrossRef] [PubMed]

155. Kim, C.; Young, J.L.; Holle, A.W.; Jeong, K.; Major, L.G.; Jeong, J.H.; Aman, Z.; Han, D.-W.; Hwang, Y.; Spatz, J.P.; et al. Stem Cell Mechanosensation on Gelatin Methacryloyl (GelMA) Stiffness Gradient Hydrogels. *Ann. Biomed. Eng.* **2020**, *48*, 893–902. [CrossRef] [PubMed]

156. Wachendörfer, M.; Schräder, P.; Buhl, E.M.; Palkowitz, A.L.; Ben Messaoud, G.; Richtering, W.; Fischer, H. A Defined Heat Pretreatment of Gelatin Enables Control of Hydrolytic Stability, Stiffness, and Microstructural Architecture of Fibrin–Gelatin Hydrogel Blends. *Biomater. Sci.* **2022**, *10*, 5552–5565. [CrossRef]

157. Sun, Y.; Deng, R.; Ren, X.; Zhang, K.; Li, J. 2D Gelatin Methacrylate Hydrogels with Tunable Stiffness for Investigating Cell Behaviors. *ACS Appl. Bio Mater.* **2019**, *2*, 570–576. [CrossRef]

158. Kilic Bektas, C.; Hasirci, V. Cell Loaded GelMA: HEMA IPN Hydrogels for Corneal Stroma Engineering. *J. Mater. Sci. Mater. Med.* **2020**, *31*, 2. [CrossRef]

159. Karimi, A.; Navidbakhsh, M. Material Properties in Unconfined Compression of Gelatin Hydro-gel for Skin Tissue Engineering Applications. *Biomed. Eng. Biomed. Tech.* **2014**, *59*, 6.

160. Singh, Y.P.; Bandyopadhyay, A.; Mandal, B.B. 3D Bioprinting Using Cross-Linker-Free Silk–Gelatin Bioink for Cartilage Tissue Engineering. *ACS Appl. Mater. Interfaces* **2019**, *11*, 33684–33696. [CrossRef]

161. Serafim, A.; Tucureanu, C.; Petre, D.-G.; Dragusin, D.-M.; Salageanu, A.; Van Vlierberghe, S.; Dubruel, P.; Stancu, I.-C. One-Pot Synthesis of Superabsorbent Hybrid Hydrogels Based on Methacrylamide Gela-tin and Polyacrylamide. Effortless Control of Hydrogel Properties through Composition Design. *New J. Chem.* **2014**, *38*, 3112–3126. [CrossRef]

162. Moghanian, A.; Portillo-Lara, R.; Sani, E.S.; Konisky, H.; Bassir, S.H.; Annabi, N. Synthesis and characterization of osteoinductive visible light-activated adhesive composites with antimicrobial properties. *J. Tissue Eng. Regen. Med.* **2020**, *14*, 66–81. [CrossRef]

163. Gu, L.; Li, T.; Song, X.; Yang, X.; Li, S.; Chen, L.; Liu, P.; Gong, X.; Chen, C.; Sun, L. Preparation and Characterization of Methacrylated Gelatin/Bacterial Cellulose Composite Hydrogels for Cartilage Tissue En-gineering. *Regen. Biomater.* **2020**, *7*, 195–202. [CrossRef]

164. Schuurman, W.; Levett, P.A.; Pot, M.W.; van Weeren, P.R.; Dhert, W.J.A.; Hutmacher, D.W.; Melchels, F.P.W.; Klein, T.J.; Malda, J. Gelatin-Methacrylamide Hydrogels as Potential Biomaterials for Fabrication of Tissue-Engineered Cartilage Constructs: Gelatin-Methacrylamide Hydrogels as Potential Bio-materials for Fabrication. *Macromol. Biosci.* **2013**, *13*, 551–561. [CrossRef]

Article

Performance Assessment of a Low-Cost Miniature Electrohydrostatic Actuator

Brendan Deibert and Travis Wiens *

Department of Mechanical Engineering, University of Saskatchewan, Saskatoon, SK S7N 5A9, Canada
* Correspondence: t.wiens@usask.ca

Abstract: Low-cost small-scale (<100 W) electrohydrostatic actuators (EHAs) are not available on the market, largely due to a lack of suitable components. Utilizing plastic 3D printing, a novel inverse shuttle valve has been produced which, when assembled with emerging small-scale hydraulic pumps and cylinders from the radio-controlled hobby industry, forms a low-cost and high-performance miniature EHA. This paper presents experimental test results that characterize such a system and highlight its steady, dynamic, and thermal performance capabilities. The results indicate that the constructed EHA has good hydraulic efficiency downstream of the pump and good dynamic response but is limited by the efficiency of the pump and the associated heat generated from the pump's losses. The findings presented in this paper validate the use of a 3D printed plastic inverse shuttle valve in the construction of a low-cost miniature EHA system.

Keywords: fluid power; hydraulics; electrohydrostatic actuator; EHA; 3D printing; additive manufacturing; plastic; step response; efficiency; heat generation

Citation: Deibert, B.; Wiens, T. Performance Assessment of a Low-Cost Miniature Electrohydrostatic Actuator. *Actuators* 2022, 11, 334. https://doi.org/10.3390/act11110334

Academic Editor: Tatiana Minav

Received: 25 October 2022
Accepted: 11 November 2022
Published: 18 November 2022

Publisher's Note: MDPI stays neutral with regard to jurisdictional claims in published maps and institutional affiliations.

1. Introduction

Fluid power (hydraulic) systems are popular for their excellent power and force capabilities relative to their size and weight. An electrohydrostatic actuator (EHA) is a particular hydraulic circuit configuration which directly couples the hydraulic pump to the actuator creating a pump-controlled circuit and eliminates losses from directional and relief valves used in more traditional valve-controlled circuits [1]. An example application of an larger-scale EHA is in the actuation of aerospace flight control surfaces, where EHAs have been chosen over previous centralized hydraulic systems due to their improved efficiency and lower weight [2,3].

While hydraulic systems typically dominate in many large-scale applications, a lack of appropriate small-scale hydraulic components limits the use of hydraulics in smaller applications (<100 W). This lack of components has been noted by others looking to incorporate small-scale hydraulic systems into their prosthetic, orthotic, and exoskeleton designs [4–6]. Actuation on this smaller scale is typically achieved by electromechanical screw type actuators which tend to suffer from low power density compared to the small scale EHA introduced by Wiens and Deibert [7] in addition to limited reliability, difficult overload protection, and low force capabilities [3,8]. Furthermore, EHAs can recover energy under assistive loads using the electric motor as a generator to create and store power or share power between other actuators reducing the overall power supply required [9]. Further work is required to create low-power hydraulic cylinder drives in order to be competitive with electro-mechanical drives [1].

Recently there has been a selection of small-scale hydraulic components introduced into the radio-controlled hobby industry intended for the use in model construction equipment. While these small-scale pieces of machinery use traditional valve-controlled circuit designs, some of their components such as pumps and cylinders may be repurposed for the use in more demanding applications. Wiens and Deibert have introduced a very low cost inverse shuttle valve design that can be used to combine these newly available small-scale pumps and cylinders in an EHA configuration [7,10]. The novel shuttle-valve design is entirely 3D printed from polyethylene terephthalate glycol-modified (PETG) and functions to handle the unbalanced cylinder flows occurring with a typical asymmetric hydraulic cylinder in an EHA circuit.

The EHA system can operate in each of four quadrants defined by cylinder force and velocity as shown schematically in Figure 1. Quadrants I and III represent pumping modes where the pump is supplying power to the cylinder. Quadrants II and IV represent motoring modes where the cylinder is supplying power to the pump which acts as a motor. The function of the inverse shuttle valve is also illustrated in Figure 1; the fluid required to balance the cylinder flow can be seen entering or leaving the reservoir through the inverse shuttle valve, which connects the lower pressure pump port to the reservoir. Due to equipment and time limitations, the analysis of the EHA system in the motoring modes (Quadrants II and IV) were left outside the scope of this work.

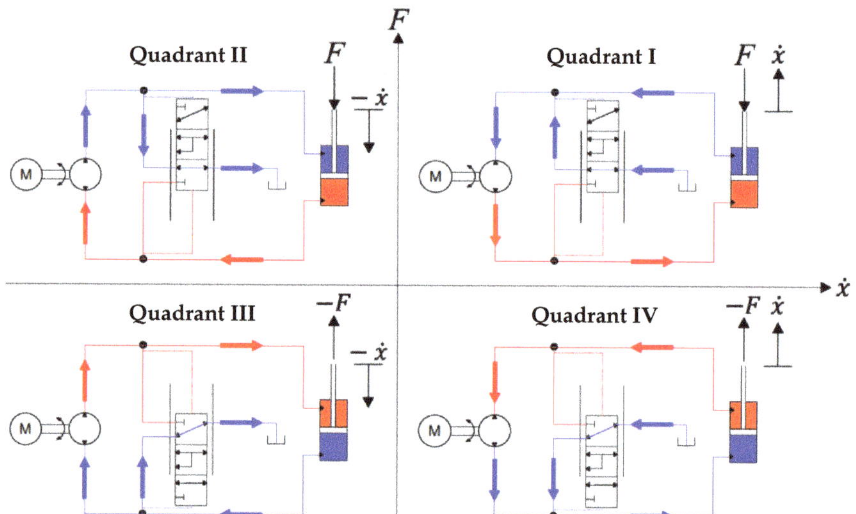

Figure 1. The EHA may operate in any of four quadrants based on the cylinder force F and velocity $\dot{x}$. Each quadrant has a unique valve position and flow direction combination. Red indicates high pressure and blue indicates low pressure.

While Wiens and Deibert have investigated the steady-state performance [7] and simulated dynamic responses [10] of a prototype miniature EHA system utilizing their 3D printed plastic inverse shuttle valve, a wider range of performance capabilities of an improved system are studied here. The preliminary valve design has been improved with less restrictive flow paths (eliminating the need for a charge pump) and better sealing poppet geometry (to reduce power losses). An investigation into 3D printed plastic poppet valve sealing performance and 3D printed plastic pressure vessel strength was performed by Deibert et al. [11]. This paper explores some of the performance capabilities of the EHA system with the improved valve design including steady state pump characteristics, actuator force and speed limitations, system step response, and system thermal performance. These tests highlight the impressive performance that may be obtained for such a low-cost small-scale actuator.

2. Materials and Methods

2.1. Testing Methodology

The objectives of this work include:

- characterizing the performance of the pump and motor combination
- identifying the EHA speed and force limits
- measuring the hydraulic efficiency of the EHA system excluding the pump
- measuring the step response of the EHA system, and
- assessing the thermal limits of the EHA system.

2.1.1. Pump Characterization Methods

Characterizing the pump and motor combination was done to establish a map of the pump's output fluid power respective to the motor's input electrical power at various points across its operating range and to provide an assessment of the pump's performance. By fitting the pump's pressure and flow output to the motor's speed and current inputs, a performance map was created. This map facilitated an accurate prediction of pump pressure and flow rate in the EHA system from only motor speed and current removing the requirement of intrusive flow meters and pressure transducers.

The pump was tested by running its flow over a relief valve, varying the pump speed at a number of set relief valve settings. The commanded pump flows were 0 to 0.7 L/min, increasing in 0.1 L/min increments, which was repeated for each of eight relief valve settings. The same ideal flow rates (pump speeds) were commanded for each run, so the decrease in measured flow rates observed at higher pressures is evidence of internal leakage in the pump increasing with pressure.

The data collected from the tests was compiled and processed in MATLAB where the pump characterization maps were created. The pump's pressure and flow outputs were characterized by fitting several typical pump and motor performance parameter coefficients to the experimental data. The fitted coefficients include the motor speed constant K_v, the pump pressure friction coefficient C_f, the pump viscous damping coefficient C_d, the fluid viscosity μ, the pump breakaway torque T_C, the pump displacement D, and the pump slip leakage coefficient C_S. The equations fitted to the data are derived in Appendix A.

2.1.2. Steady State EHA System Performance Methods

Identifying the speed and force limits of the EHA system was done to define the allowable operating range for the EHA. The actuator speed limits were established by identifying the points at which the relationship between pump speed and cylinder speed became non-linear, indicating the onset of fluid cavitation at the pump inlet. The actuator force limits were determined by extrapolating the relationship between fluid pressure and actuator load to the working pressure rating of the tubing.

Testing the steady state actuator performance involved commanding a series of cylinder velocities with a set load. The process was repeated at various cylinder loads. The pump speed N was set using a basic control algorithm relating the commanded cylinder velocity $\dot{x}_{cmd}$ to the cylinder head end and rod end areas A_A or A_B and pump displacement D based on the sign of the pressure differential across the inverse shuttle valve P_A and P_B as illustrated by Equations (1) and (2).

$$if\ P_A > P_B: \qquad N = \frac{\dot{x}_{cmd} A_A}{D} \qquad (1)$$

$$if\ P_B > P_A: \qquad N = \frac{\dot{x}_{cmd} A_B}{D} \qquad (2)$$

This assumes no internal or external leakage. A more advanced control algorithm such as a closed loop control scheme could also be implemented to adjust the pump speed based on the cylinder speed to compensate for leakage. This was not performed due to time constraints and in order to simplify testing. High performance control that minimizes error between commanded and actual cylinder velocity was not required as it was not within the focus of this work. The sign convention for the cylinder speeds and forces follows that used in other areas of this work, with positive cylinder forces resisting cylinder extension and positive cylinder speeds during cylinder extension.

The maximum cylinder speed was first found in order to establish an upper cylinder speed limit on the following tests. The maximum cylinder speed is governed by fluid cavitation at the fluid inlet. This occurs when the absolute fluid pressure at the pump inlet drops too low and vapor bubbles form. The pressure at the pump inlet is directly related to the pressure drop across the pump's supply path between the tank and the pump inlet, which is a function of the flow rate and flow restriction of the flow path. When the cylinder speed, and thus pump flow, increases beyond the allowable threshold, the pressure drop along the pump supply path becomes too large to sustain a pump inlet pressure sufficient to prevent fluid cavitation.

To establish the maximum cylinder speed, a series of increasing commanded cylinder speeds were tested until a plateau in the actual cylinder speed was observed. This was performed without a load applied to the cylinder since the loading mechanism became unstable at very high cylinder speeds. Since the valve connects the lower pump port pressure to the reservoir, it is largely insensitive to loading in pumping quadrants.

The efficiency of the EHA system, including the valve, cylinder, and flow paths, and excluding the pump, was measured to gain an understanding of the efficiency of the pump relative to the remainder of the EHA system. The pump's electrical and hydraulic efficiencies were of secondary importance since the pump's design was excluded from the scope of this work. The circuit's input fluid power from the pump was estimated using the motor's speed and current in combination with the pump performance map constructed in the first experimental objective. The system's output power was measured by using a known weight to apply a mechanical load to the cylinder and recording the cylinder's velocity.

2.1.3. Dynamic Step Response Methods

The EHA system's dynamic response to a step increase in commanded cylinder velocity (by a commanded pump speed) from a steady initial condition was measured. The output response was assessed by recording the cylinder velocity measured by a linear potentiometer. A known weight functioned as the mechanical load and mass.

Testing of the EHA's dynamic response was conducted by commanding a step increase in cylinder velocity from $-100\,\mathrm{mm/s}$ to $-150\,\mathrm{mm/s}$. The procedure was repeated multiple times to gain statistical confidence in the results. Three different loads were tested to assess the system's response over a range of operating points in Quadrant III. The metric used to assess the step response of the system was the time constant of a first order system model fitted to the cylinder velocity response as represented by

$$\dot{x} = \begin{cases} \dot{x}_0 & if\ t < t_0 \\ \dot{x}_0 + \Delta\dot{x}\left(1 - e^{\frac{-t-t_0}{\tau_{vel}}}\right) & otherwise \end{cases} \tag{3}$$

where $\dot{x}$ is the cylinder velocity [m/s], $\dot{x}_0$ is the initial cylinder velocity [m/s], t is the elapsed time [s], t_0 is the time at which the step occurs [s], $\Delta\dot{x}$ is the cylinder velocity step size [m/s], and τ_{vel} is the system time constant [s]. The first order model fitted to the system response approximates the EHA system and neglects compliant components such as fluid compressibility, tubing compliance, and printed plastic compliance. This simplification was deemed to be acceptable for the small load masses and pressures experimentally tested but may not be an appropriate model for the system response under all operating conditions.

2.1.4. Thermal Performance Methods

Lastly, the thermal limit of the EHA system was examined. An assessment of the transient temperature of critical areas was made to estimate how long the system may operate at a given power level before overheating occurs. Excessive heat was predicted to cause failure in the printed plastic components and plastic tubing, which have a much lower temperature rating than typical large-scale hydraulic components. Polyamide tubing, such as that used in the construction of this EHA system, has a working pressure that is sensitive to operating temperature and is expected to be the limiting factor in the system [12].

The system was programmed to extend and retract the cylinder just short of the cylinder stroke limits (approximately 45 mm of travel) in a continuous cycle while the temperatures of the critical areas were monitored over time. This was repeated at several power levels to extrapolate the steady state temperature and effective thermal time constant of the system. The commanded extension speed was 100 mm/s, and the commanded retraction speed was 50 mm/s. The mismatch of commanded extension and retraction speeds was used to increase the testing apparatus stability. The system electrical input power was recorded by the motor controller. System output power was calculated by multiplying the cylinder velocity (derived from the recorded cylinder potentiometer signal) with the force of the loading weight applied to the apparatus. The difference between the input electrical power and the output mechanical power yields the heat generated in the system. The time averaged calculated heat generation was used as an equivalent continuous heat generation.

A thermal imaging camera was used during preliminary tests to reveal the areas of the EHA system that may be prone to failure due to thermal effects. The images indicated that the pump and motor were substantially the hottest components, and that the cylinder's temperature was not likely to be a concern, as shown in Figure 2a,b, respectively. The tubing at the compression fitting on the high-pressure side of the circuit was deemed to be the most likely failure point due to material weakening.

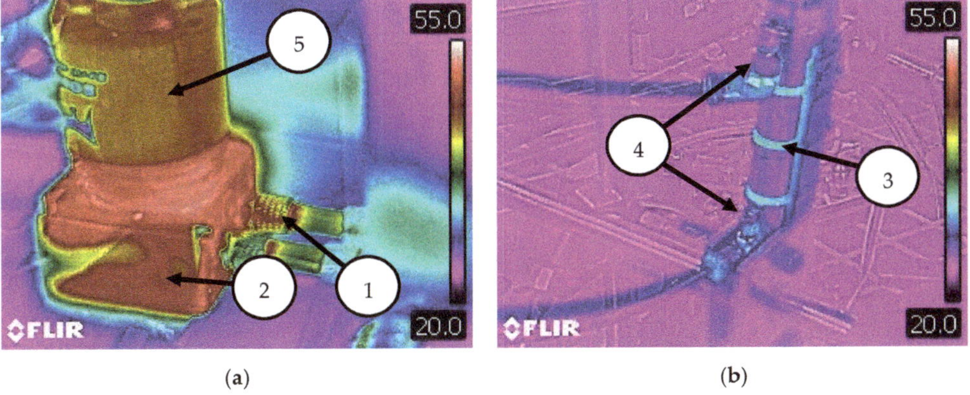

Figure 2. (**a**) The temperature of the (1) compression fitting on the high-pressure side of the circuit was close to that of the (2) pump housing. (**b**) The temperature of the (3) cylinder and (4) fittings was low relative to that of the pump and (5) motor.

With the point most susceptible to thermal failure identified, the temperatures of the components of concern were instrumented and monitored during testing of the system. A thermocouple probe was attached to the pump housing as opposed to the compression fitting of interest due to practicalities of mounting the probe to the small fitting. This was justified by Figure 2a, as the temperature of the fitting was near that of the pump. Another thermocouple probe was submerged in the reservoir fluid near the reservoir ports of the inverse shuttle valve.

A first-order thermal system model was fit to each of the temperature responses to extrapolate an effective thermal time constant for the system. The model is expressed as

$$T = T_0 + (T_\infty - T_0)\left(1 - e^{-\frac{t}{\tau_{therm}}}\right) \tag{4}$$

where T is the temperature of the component [°C], T_0 is the steady temperature of the component before testing (near ambient) [°C], T_∞ is the steady state temperature of the component during operation [°C], t is the elapsed time from the start of the test [s], and τ_{therm} is the thermal time constant [s].

2.2. Apparatus and Instrumentation

The apparatus used to characterize the pump and motor combination is shown in Figure 3. The pump [13] supplied Nuto 32 hydraulic fluid from the reservoir through a Vickers relief valve, followed by a Flomec EGM004S511-821 flow meter, and back to the reservoir. The relief valve was used to set the pump pressure and an XIDIBEI XDB303 pressure transducer was installed in line close to the pump outlet to measure the pump pressure. An ODrive motor controller [14] controlled by an Arduino microcontroller powered the pump and an encoder on the motor shaft supplied closed-loop velocity feedback to the motor controller. Motor speed and current data was retrieved and logged from the motor controller.

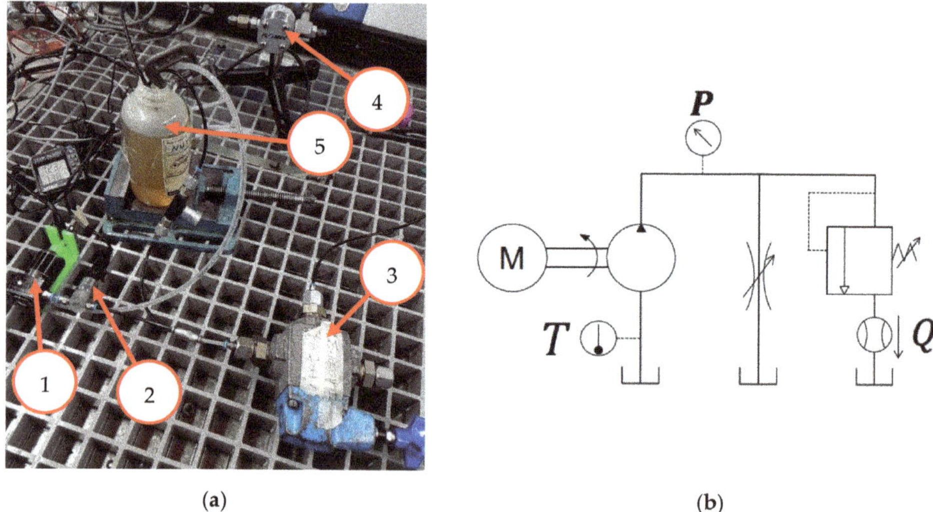

(a)   (b)

Figure 3. (**a**) Experimental apparatus and (**b**) schematic used for characterizing the pump and motor combination. Notable components include the (1) pump and motor, (2) pressure transducer, (3) relief valve, (4) flow meter, (5) fluid reservoir.

The apparatus used to test the performance of the EHA system is shown in Figure 4. The same pump and motor combination was used, as well as the same motor control system. The pump was connected to a fully 3D printed hydraulic power unit which enclosed the inverse shuttle valve within a fluid reservoir. Two XIDIBEI XDB303 pressure transducers were used to measure the fluid pressure on either side of the inverse shuttle valve. The power unit was connected to a hobby-grade single-rod hydraulic cylinder with a 10 mm bore diameter, 4 mm rod diameter, and 50 mm stroke. The cylinder was loaded using steel weights and a 2:1 lever arm which could be configured to apply a tensile or compressive load to the cylinder. The potentiometer from an Actuonix L16-P linear actuator was used to measure cylinder position which was differentiated to find cylinder velocity. An Omega

HH306A datalogging thermometer (not shown) was used to monitor the temperature of the system and a Flir E60 thermal imaging camera was used to supplement findings.

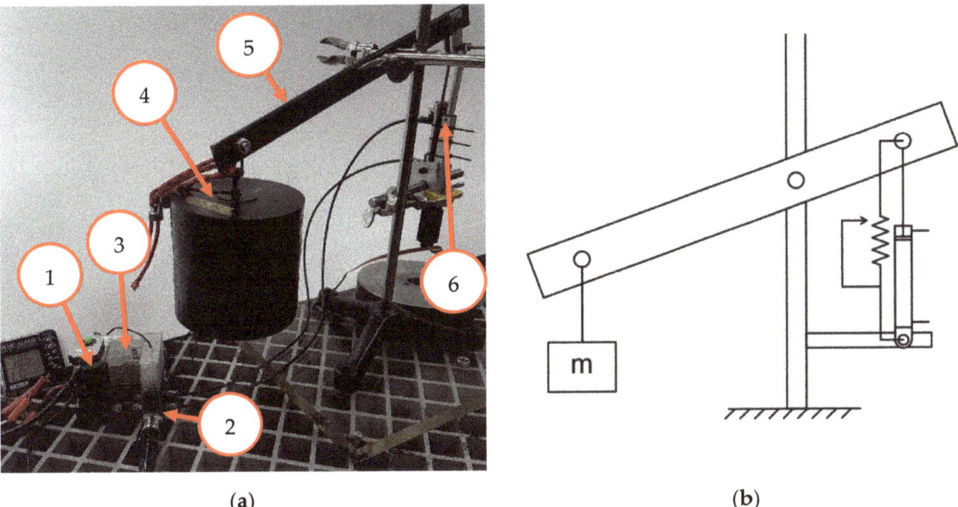

(a)   (b)

Figure 4. (**a**) Experimental apparatus and (**b**) schematic used testing the EHA system. Notable components include the (1) pump and motor, (2) pressure transducer, (3) 3D printed hydraulic power unit, (4) loading weights, (5) 2:1 lever arm, and (6) hydraulic cylinder.

3. Results and Discussion

3.1. Pump Characterization Results

The range of pump pressures and flows achieved for each run is shown in Figure 5. The nonlinear trend of the data is due to the typical nonlinear flow-pressure curve of the relief valve. Note that multiple runs were taken at the highest and lowest relief valve settings at different points throughout the experiment to ensure consistency. The error bars represent a 95% confidence interval in the measurements.

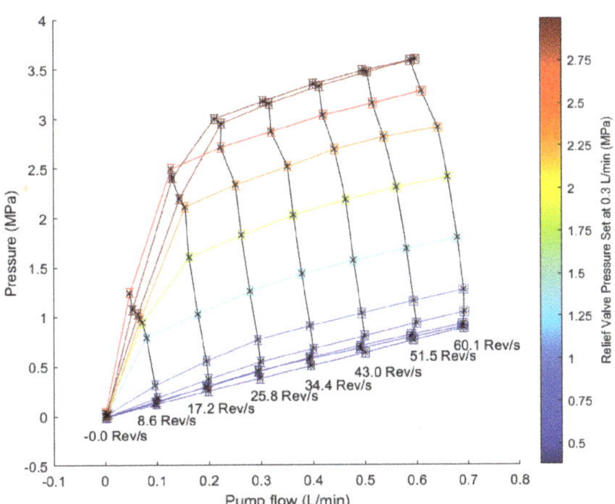

Figure 5. Pump flows and pressures measured for each run during the pump characterization tests. Line color represents the relief set pressure at 0.3 L/min. Black lines indicate constant pump speed.

The characterization maps for the pump's flow, pressure, and overall efficiency are shown in Figure 6. The black crosses indicate the points at which data was taken, and the color gradient of the plot indicates the flow, pressure, or efficiency quantity. Note that the trend in motor speed and current observations very closely match the trend in measured flow and current shown in Figure 5.

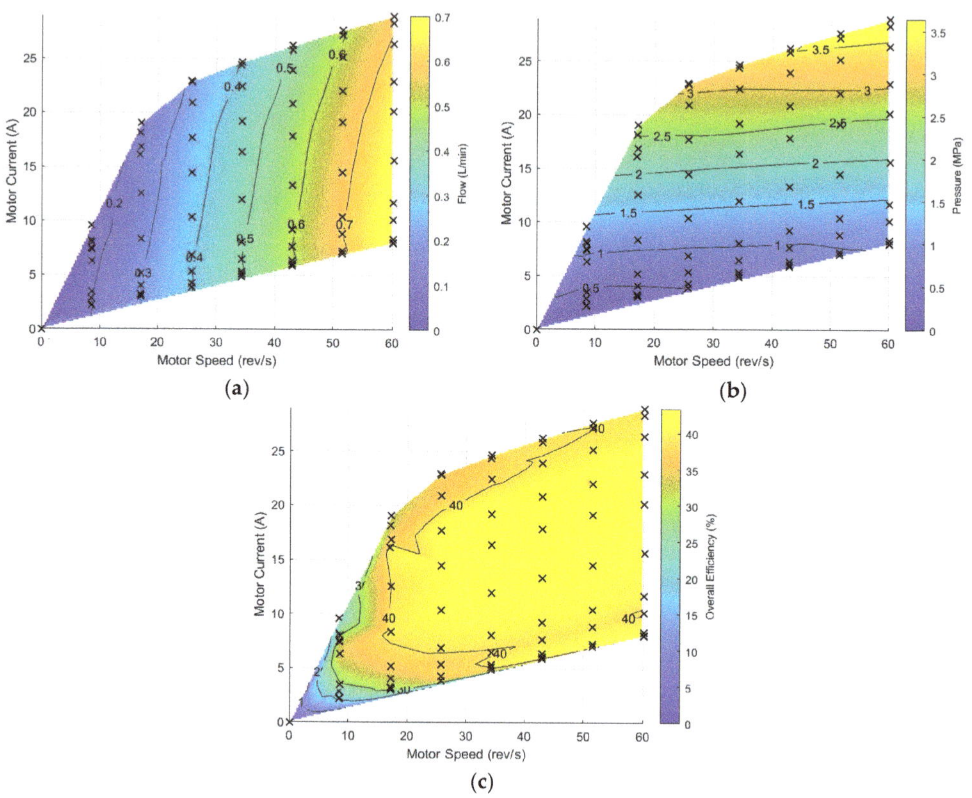

Figure 6. Pump characterization map relating pump (**a**) flow, (**b**) pressure, and (**c**) combined pump and motor efficiency to motor current and speed.

The pump characterization maps presented in Figure 6 appear to be reasonable. Figure 6a shows the flow rate is primarily dependent on motor speed and Figure 6b shows pressure is primarily dependent on current. This is expected since motor speed, and thus flow rate, is ideally linear to motor voltage, and motor torque, and thus pressure, is ideally linear with motor current. Some slight nonlinearities are evident in each of the maps which can be attributed to the real performance of the pump and motor. Figure 6a shows a decrease in flow rate with an increase in motor current at a given motor speed. This is attributed to the increase in internal pump leakage at high pressures. Figure 6b shows a decrease in fluid pressure with an increase in motor speed at a given motor current. This is attributed to increased torque losses due to viscous friction in the pump. Finally, Figure 6c shows an overall energy efficiency of the motor and pump combination over the range of operating points tested. This efficiency represents the combined electrical and mechanical efficiencies of the motor and the mechanical and volumetric efficiencies of the pump. A relatively low efficiency of 35% to 40% is observed for most operating points. Though no datasheets are available for these small hobby-grade pumps, industrial manufacturers are producing external gear pumps 5–10 times the displacement which generally achieve efficiencies of 60% to 80% [15–18]. The low efficiencies measured for the small pump in this work may be due to poor pump and/or motor design and/or manufacturing tolerances. These small hobby-grade pumps may suffer from increased leakage due to large gear tooth clearances relative to the pump size. Additionally, larger commercial offerings typically employ wear- and pressure-compensating designs to increase efficiency which are not used in these hobby-grade offerings.

The values for each fitted coefficient are presented in Table 1 along with their 95% confidence intervals. The slip leakage coefficient and viscous damping coefficient are presented in terms of fluid viscosity since the fluid viscosity was not directly measured. Utilizing the fitted equations yields smoother pump performance maps compared to the maps generated from the experimental data. The fitted maps agree closely with the experimental data taken, typically with less than 10% error for most operating points.

Table 1. Summary of pump performance characteristics fitted to the experimental data.

Parameter	Fitted Value
K_v [RPM/V]	1108 (1088–1128)
C_f [-]	0.98 (0.94–1.03)
$C_d\mu$ [Pa·s]	885 (678–1093)
T_c [Nm]	1.91×10^{-3} (0.66×10^{-3}–3.2×10^{-3})
D [m³/rev]	2.02×10^{-7} (2.01×10^{-7}–2.04×10^{-7})
C_s/μ [1/(Pa·s)]	2.65×10^{-6} (2.49×10^{-6}–2.80×10^{-6})

3.2. Steady State System Performance Results

Figure 7 shows the relationship between commanded cylinder speed and pump speed to the actual steady cylinder speed achieved. The plateau in relationship of cylinder speed to pump speed indicates the onset of pump inlet cavitation. The maximum positive and negative cylinder speeds achieved were approximately 150 mm/s and −187 mm/s, respectively. The maximum cylinder retraction speed is greater than the maximum cylinder extension speed due to the differences in cylinder end areas, and both speeds correspond to a pump speed of approximately 60 rev/s. Although maximum cylinder speed under various cylinder loads was not assessed, those maximum speeds should be close to that of the unloaded cylinder as the systems pressures on the low-pressure pump inlet side of the EHA circuit are largely dependent on cylinder speed and independent of load. This assumption remains a limitation of this work that could be explored in later work.

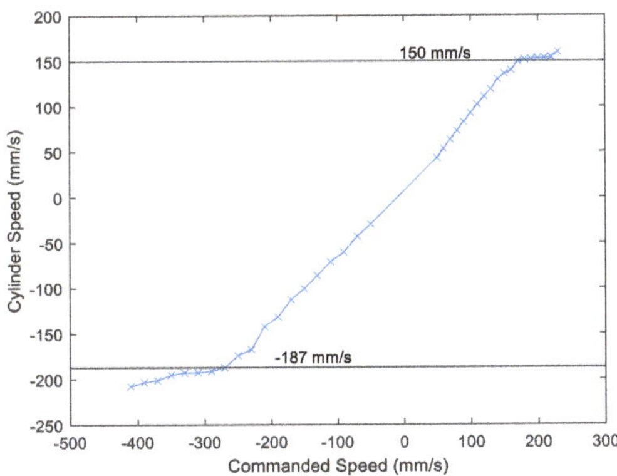

Figure 7. Actual cylinder speed versus the commanded cylinder speed or pump speed revealing the pump speed at the onset of cavitation.

With the maximum cylinder speed established, attention turned to investigating the maximum cylinder forces in the pumping operating Quadrants I and III. The limiting factor of the maximum cylinder forces is the 4 MPa safe working pressure rating of the polyamide tubing used to connect the pump and cylinder to the power unit. In the pumping quadrants the maximum fluid pressure occurs at the pump outlet, therefore system pressures were evaluated there for a range of cylinder speeds and loads. The cylinder loads tested yielded system pressures close to but below the maximum allowable as a conservative measure to avoid unsafe testing conditions. Loaded cylinder test speeds were below the maximum cylinder speeds due to instability in the lever loading mechanism at high speeds. Results were extrapolated to the remaining range between the testing points and up to the tubing pressure limit and maximum cylinder speeds.

Figure 8 shows the fluid pressures for the high-pressure side of the circuit as a function of cylinder speed and force. The system pressure increases with cylinder load due to the balances of forces at the cylinder's piston and with increasing cylinder speed due to friction in the fluid flow paths and cylinder seals. The black crosses indicate the operating points at which data was taken. The horizontal shift in the cylinder speed operating points with increasing cylinder load magnitude is attributed to the increase in internal leakage in the pump, poppets, and cylinder with increasing fluid pressure. The bold black lines indicate the maximum allowable cylinder force due to the tubing pressure rating and the bold red lines indicate the maximum cylinder speed to avoid pump cavitation.

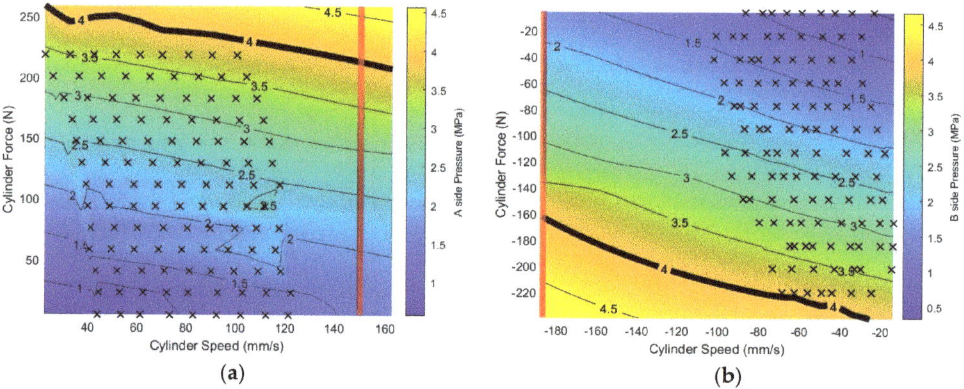

Figure 8. Fluid pressure on the (**a**) "A" and (**b**) "B" side of the circuit over a range of cylinder speeds and loads in (**a**) operating Quadrant I and (**b**) operating Quadrant III.

The hydraulic efficiency of the system was measured for both pumping quadrants. These efficiencies are defined as the of the ratio of load power to pump output power and represent the efficiency of the flow paths, inverse shuttle valve, and cylinder. The pump's efficiency is excluded here since it was outside the scope of the system design. Figure 9 shows the hydraulic efficiency of the EHA system excluding the pump over a range of tested operating points in operating Quadrants I and III. Again, the black crosses indicate the operating points tested and the bold black and red lines represent the maximum cylinder force and speed, respectively. Peak efficiencies of approximately 70% occur within a region near cylinder speeds of 50 mm/s and for both quadrants. Most operating points in the high-pressure regions of either quadrant achieve a 60% to 70% hydraulic efficiency.

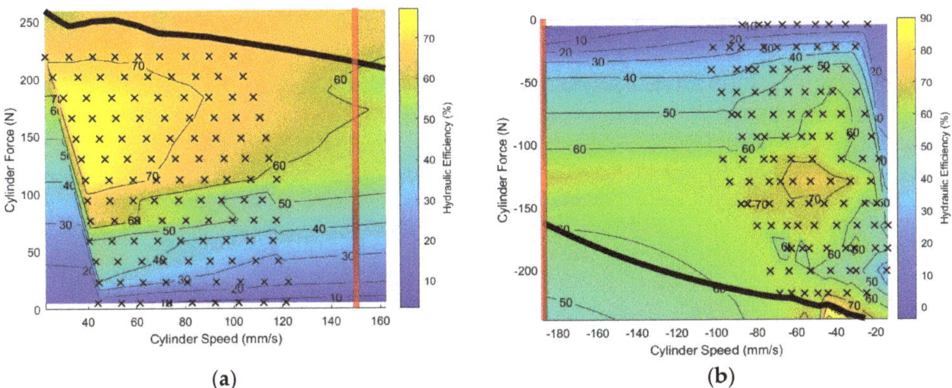

Figure 9. Hydraulic efficiency of the system (excluding the pump) for (**a**) operating Quadrant I and (**b**) operating Quadrant III.

Although the hydraulic power losses due to each component were not specifically measured, they were estimated to give a better understanding of where improvements to system efficiency may be made. The power loss due to the tubing connecting the HPU to the cylinder was calculated using the cylinder flows and measured tubing and fitting flow resistances. The power loss due to internal leakage was calculated from the difference in pump flow and cylinder flow on the high-pressure side of the circuit multiplied by the pressure of the high-pressure side of the circuit. The power loss due to cylinder friction was calculated by finding the effective friction force using a summation of forces on the piston and multiplying that friction force by the cylinder speed. Plots of the calculated power loss contributions are shown in Figure 10 for operating Quadrants I and III.

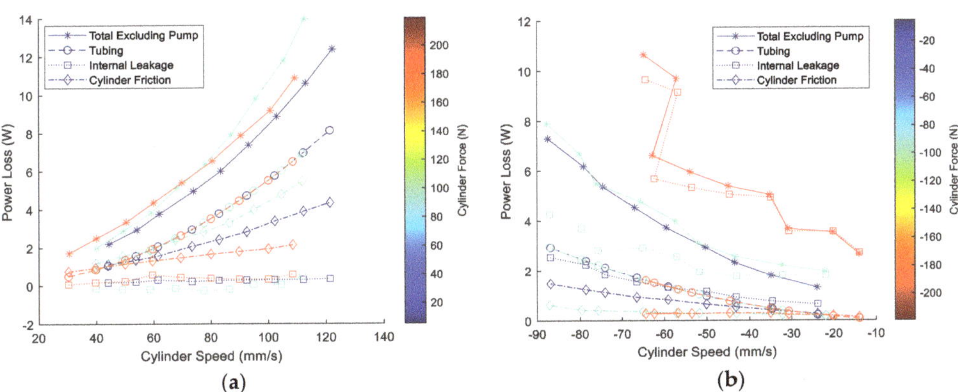

Figure 10. Estimations of the power losses due to the tubing restrictions, poppet leakage, and cylinder friction for (**a**) operating Quadrant I and (**b**) operating Quadrant III.

Several observations may be made from the power loss contributions shown in Figure 10. The sum of the losses due to tubing resistance, internal leakage, and cylinder friction approximately equals that of the difference in total input and output hydraulic power, suggesting the significant power losses are accounted for. The internal leakage resistances were close to the poppet leakage resistances reported in the poppet leakage investigation performed by Deibert et al. [11], indicating that the poppet leakage is likely the dominating internal leakage. The losses due to tubing resistance were independent of fluid pressure (cylinder load) whereas the losses due to internal leakage were largely independent of flow rate (cylinder speed) and the losses due to cylinder friction were dependent on both cylinder force and speed. For most operating points the losses due to the tubing resistance are the largest, closely followed by cylinder friction. This is a reasonable outcome because 600 mm of 4.0 mm (outer diameter) × 2.5 mm (inner diameter) tubing was used to connect each side of the HPU circuit to the cylinder. Applications allowing mounting of the cylinder closer to the HPU or the use of larger internal diameter tubing would significantly reduce the power lost due the tubing restrictions. A couple of cases in Figure 10 indicate that the internal leakage is the dominating power loss, likely due to poor poppet sealing. More careful lapping of the poppets could be used to reduce the power loss in those cases.

Finally, the EHA output power capabilities for both pumping quadrants can be established. This power output is limited by the combination of limits on tubing pressure and cylinder speed before pump cavitation. Figure 11 shows the EHA's power output capabilities in operating Quadrants I and III. The maximum power output in Quadrant I is 32 W and the maximum power output in Quadrant III is 30 W. The output power capability in the two quadrants are similar but occur at significantly different operating points. The highest power operating point in Quadrant I is slower but at a higher force than that in Quadrant III. This is attributed to the difference in cylinder end areas, where operating in Quadrant I pressurizes the head end of the cylinder, requiring a higher flow rate but less pressure for a given cylinder speed and load than an equivalent operating point in Quadrant III pressurizing the rod end of the cylinder.

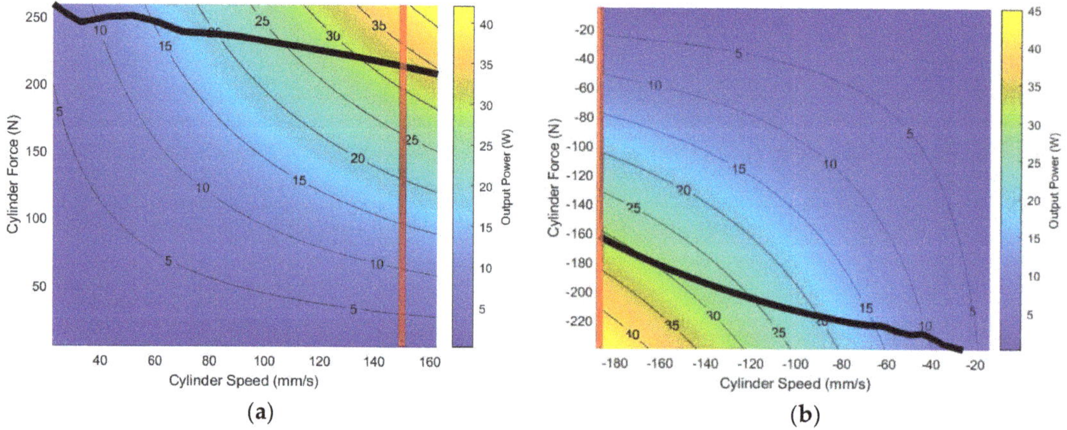

Figure 11. Output power in (**a**) operating Quadrant I and (**b**) operating Quadrant III is limited by maximum system pressure (black line) and pump cavitation (red line).

3.3. Dynamic Step Response Results

Samples of the system's response to step changes in commanded velocity are shown in Figure 12 for three different load masses. As previously observed in the steady state EHA performance testing, the steady state cylinder velocity achieved by the system was less than the commanded speed due to internal leakage, mainly in the pump and shuttle valve poppets. Ten step responses were measured for each load mass, and the average time constants and standard deviations are presented in Table 2.

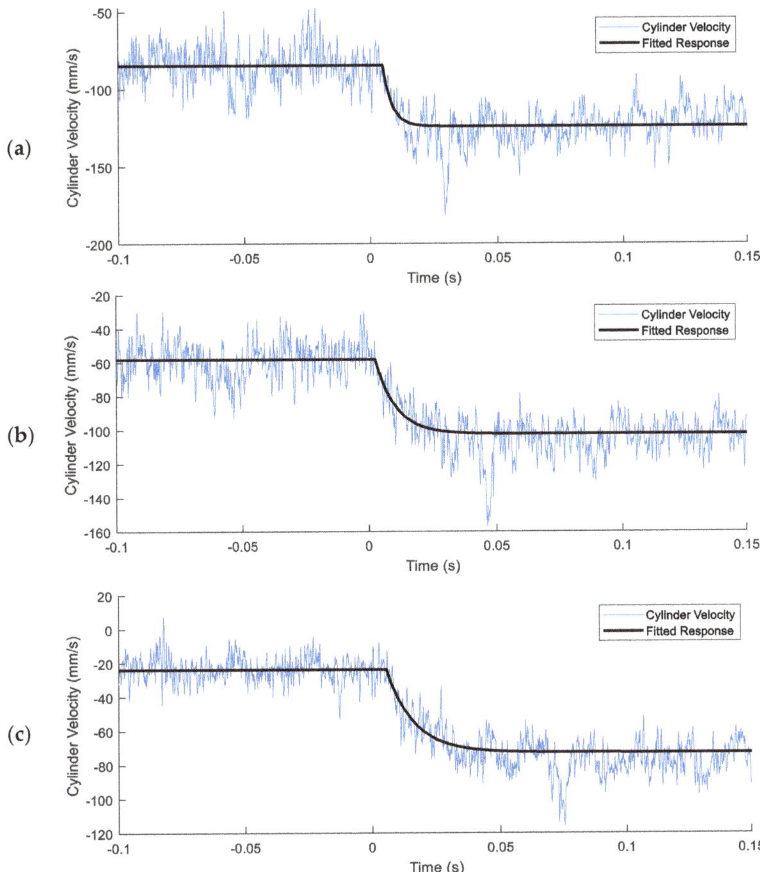

Figure 12. Sample experimental EHA system response to a step increase in commanded speed from −100 mm/s to −150 mm/s with load masses of (**a**) 0 kg, (**b**) 4.8 kg, and (**c**) 9.4 kg.

Table 2. First-order time constants fitted to the system response for various load masses tested, averaged over ten trials.

Load Mass (kg)	Fitted Time Constant τ_{vel} (ms)	
	Average	Standard Deviation
0	3.89	1.72
4.8	7.65	1.34
9.4	12.5	1.01

The results shown in Figure 12 and Table 2 indicate that the response of the system is fast. The effect of the load mass on the system's time constant is as expected, where doubling the mass approximately doubles the time constant. This is a reasonable result as the mass attached to the cylinder is likely the dominant inertia of the system when compared to the relatively lightweight electric motor rotor and small fluid inertance. The approximation of a first order system for the velocity response is a good fit, especially with small load masses. With larger masses the cylinder velocity begins to have a small amount of overshoot, indicating the compliance in the compressible volumes, tubing, and printed plastics may be becoming significant, which could be modelled as a second order system if desired.

3.4. Thermal Performance Results

Figure 13 shows a sample of the recorded and calculated powers of the EHA system running at steady state during the thermal performance testing. The solid black line indicates the time-averaged heat generation level for the system.

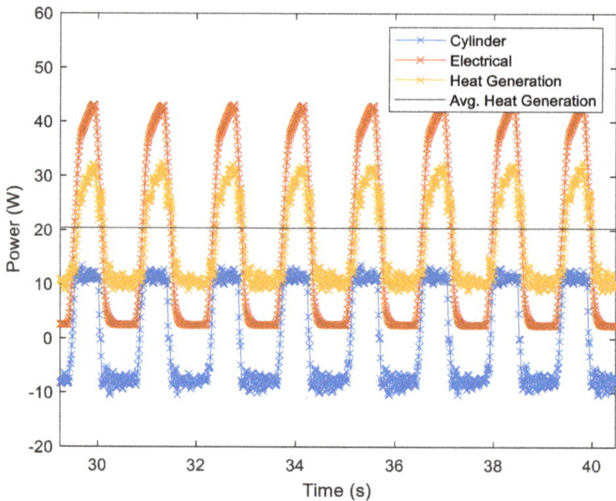

Figure 13. Sample of cylinder and electrical power measured which was used to calculate system heat generation.

The system was run with several different loads applied creating different levels of heat input to the system, and the thermocouple temperatures were logged over the duration of the tests and the results are shown in Figure 14. It was observed that the temperatures of the two thermocouples were very similar to each other for every power level, indicating a high rate of heat transfer from the pump housing to the fluid being pumped. Ketelsen et al. [19] used a similar approach which lumped the thermal capacitances of multiple components including the pump, motor, and oil to predict the temperature of a larger EHA system with accuracy suitable for design purposes. The metal compression fitting conducts heat from the pump housing very well and appears to reach temperatures near that of the pump housing, as shown in Figure 2a. Parker Hannifin Corp. Ref. [12] shows that the working pressure of their 4×2.5 mm polyamide tubing reduces from 5.2 MPa at 20 °C to just 3.0 MPa at 60 °C. Overheating of the compression fitting attaching the tubing to the pump would cause excessive softening of the plastic fluid tubing which would likely lead to failure of either the mechanical connection between the fitting and the tubing or in the tubing itself. Furthermore, the temperature of the fluid within the shuttle valve and reservoir is of concern, as excessive fluid temperatures in those locations may cause failure of the plastic printed parts. The heat deflection temperature of the Prusa PETG material used for these printed parts is 68 °C [20]. The investigation in this work neglects any thermal related failure that may occur in the cylinder, pump, or electric motor since the design of these components were not in the scope of this work, and their temperature limits are expected to be higher than the plastic components.

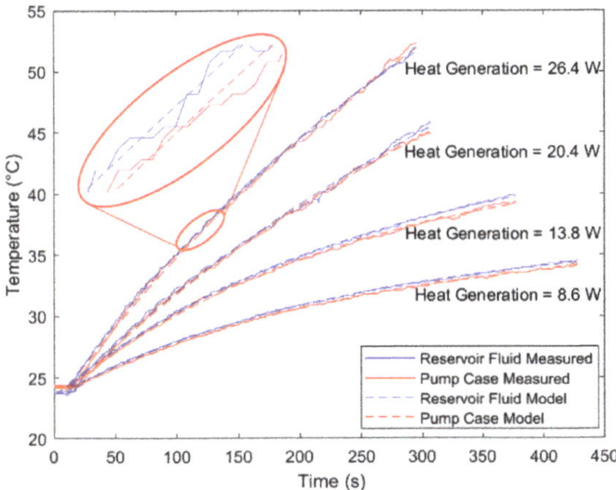

Figure 14. Pump case and reservoir fluid temperature over time at four different heat generation levels. Solid lines indicate measured data and dashed lines indicate the fitted first order model.

As shown in Figure 14, the thermal models follow the measured temperatures closely. The results of the fitted coefficients for the pump housing temperature are provided in Table 3. There are some discrepancies between the thermal time constants of the tests at the lower two and higher two power levels. These differences may be attributed to uncertainty in the system thermal response much past one time constant which was not measured since the tests at high power levels were stopped at a conservative temperature to avoid potentially destructive failure of the system. In general, the fitted parameters provide a good estimate of the system temperatures of concern.

Table 3. Fitted thermal model parameters for the temperature of the pump housing at four different heat generation levels.

Cylinder Load, F (N)	Heat Generated, Q (W)	Time Constant, τ_{therm} (s)	Steady Temperature, T_∞ (°C)
50	8.6	242	36.3
94	13.8	246	43.6
139	20.4	440	67.1
184	26.4	459	84.1

Video captured by the FLIR thermal imaging camera provided further insight to the heat transfers occurring within the EHA system. Still frames from the thermal video were taken at one-minute intervals during the 184 N test and are shown in Figure 15. The measurement callouts in Figure 15 show the temperature of the pump, reservoir, and cylinder as measured by the thermal camera. The pump temperature measured by the thermal camera closely agreed with that measured by the thermocouple shown in Figure 14. The reservoir temperature measured by the thermal camera was typically lower than that measured by the thermocouples since the thermal imaging camera was measuring the plastic reservoir surface temperature whereas the thermocouple was measuring the actual fluid temperature. Observing the still frames in Figure 15, the majority of the heat appears to build first in the pump housing which then transfers to the fluid and finally to the other components including the tubing, reservoir, and cylinder. Since the cylinder load was applied only in one direction, the pressure differential between the two sides of the circuit would not have been changing sign, thus the inverse shuttle valve would not have been switching the sides of the circuit connected to reservoir. This explains the temperature of one of the tubes supplying the cylinder being higher than the other. The fluid in the high-pressure side of the circuit would have been isolated from the reservoir fluid (neglecting leakage), whereas the low-pressure side was able to exchange fluid, and thus heat, with the reservoir.

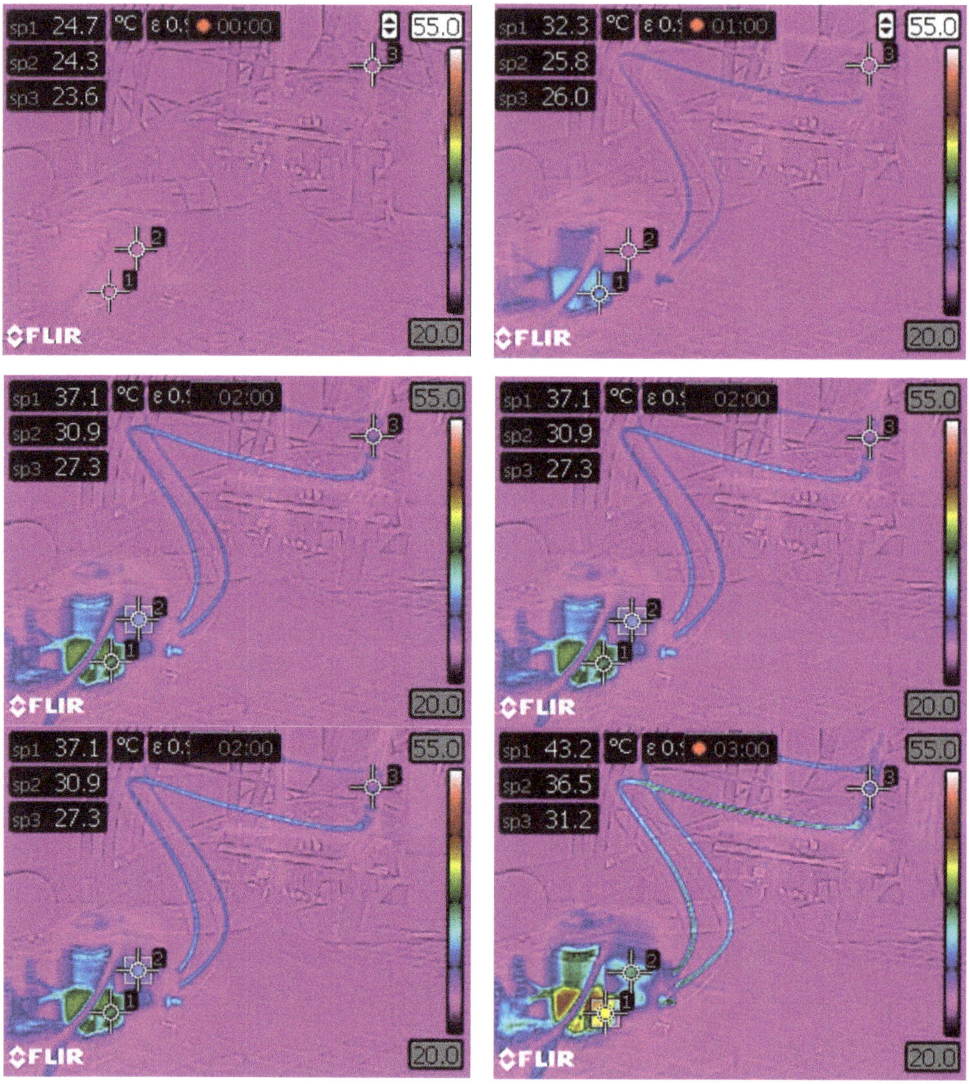

Figure 15. Thermal images of the EHA system at one-minute intervals during the first five minutes of the 184 N load test. The three measurement locations in each frame show the pump housing, reservoir surface, and cylinder temperatures.

The results shown in Figure 14 and Table 3 above are expressed in terms of heat generated, or useful power lost by the system. This is a simplification made to exploit the fact that a given amount of heat may be generated across a range of different operating points. Estimations of the power losses of the EHA system including the pump and motor were created from data obtained during the steady state EHA testing performed and are shown in Figure 16 for operating Quadrants I and III. Though not experimentally verified in this work, the system temperature responses for any heat generation level should be similar regardless at which cylinder speed and load combination it occurs at. This is an estimation and simplification of other factors that would affect the steady and transient heat transfers within the system but provides an estimate of the thermal limitations of the EHA system developed thus far. Future work is required to determine how much of the generated heat is absorbed by the fluid relative to the amount of heat dissipated directly to atmosphere from the motor and pump housing.

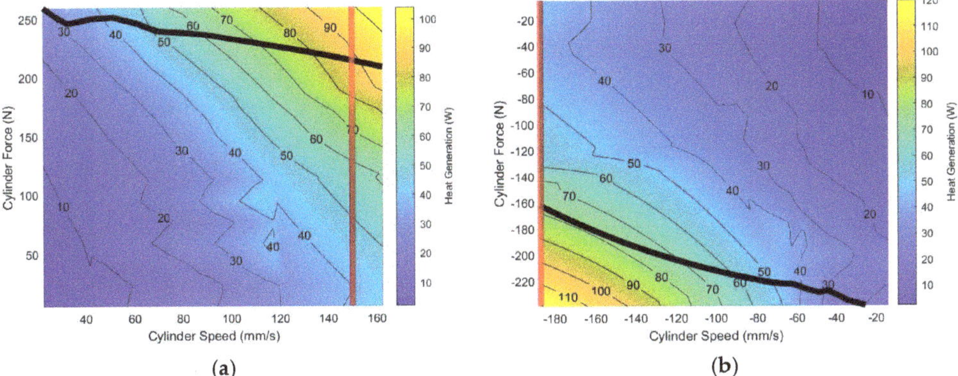

Figure 16. EHA system power losses (heat generation) in (**a**) operating Quadrant I and (**b**) operating Quadrant III. Operating limits exist as maximum system pressure (black line) and pump cavitation (red line).

The data shown in Table 3, and Figure 16 may be used to interpolate a safe equivalent continuous heat generation level and permissible EHA operating points. For example, if a reduction in tubing working pressure to 3.0 MPa at 60 °C was permissible, the maximum equivalent continuous heat generation would be approximately 18.4 W. Note that this implies that at a 50% actuator duty cycle the permissible heat generation would be approximately 36.8 W. With the maximum equivalent continuous heat generation established, Figure 16 may be used to identify the permissible EHA operating points that would not exceed that heat generation level at that duty cycle. Concluding the example given here, the actuator would be able to run with a 50% duty cycle at either 90 mm/s and 150 N or 130 mm/s and 50 N. Alternatively, the data in Table 3 may be used with Equation (4) to predict the maximum time the actuator may operate at a higher heat generation level continuously before exceeding temperature limits.

Several opportunities for improvement of the thermal performance of the system were recognized. Perhaps the largest improvement in thermal performance may be increasing the energy efficiency of the pump and motor package, which was identified as the dominant source of heat during testing of the EHA system. Utilizing the design freedom of 3D printing, many creative elements may be employed in the reservoir design, including cooling fins or internal baffles, which may improve heat dissipation from the reservoir fluid. External methods of cooling such as a liquid-cooled heat exchanger or air fans may be added to the high temperature components like the pump and motor to allow continuous high-power operation. Finally, the strength of the polyamide tubing is again identified as a major limiting factor in the design of the EHA system. A tubing that can withstand higher pressures at higher temperatures would allow the system to operate at a higher power level for a longer duration. These considerations are left for future development of the small-scale EHA due to time constraints.

4. Conclusions

This paper summarizes the performance capabilities of a low-cost small-scale EHA system constructed using a hobby-grade pump and cylinder and a 3D-printed plastic inverse shuttle valve. Combined, these components form an actuator which has been shown in preliminary work to rival comparable electromechanical actuators in terms of cost, force density, and power density. This work extends that preliminary work and further explores the steady and dynamic limitations of the EHA system and identifies areas which may require further development.

The pump's performance was characterized and the EHA's force and speed limitations, hydraulic efficiency, step response, and thermal performance were assessed. The pump had a relatively low efficiency of 35% to 40% for most operating points, resulting in considerable power loss and corresponding heat generation within the EHA system. Maximum actuator extension and retraction speeds were found to be 150 mm/s and -187 mm/s, respectively. Maximum actuator extension and retraction forces were determined to be approximately 250 N and -220 N, respectively. The hydraulic efficiency of the 3D printed inverse shuttle valve, tubing, and cylinder (excluding pump) was satisfactory at 60% to 70%, with potential for improvement with the selection of less restrictive tubing to the cylinder. The step response of the EHA system was fast with time constants between 4 ms and 13 ms for the loads tested, confirming the applicability of the EHA in high bandwidth applications. Limitations of the current tubing's working pressure at elevated temperatures combined with the low pump efficiency restrict the system's capabilities to handle high loads (>10 W) for extended periods (> 5 min), though the actuator is capable of approximately 30 W output power for short duty cycles. These results indicate that the low-cost EHA system presented has the potential for impressive performance especially with continued development of other small-scale components such as pumps and tubing.

Author Contributions: Conceptualization, B.D. and T.W.; methodology, B.D. and T.W.; validation, B.D. and T.W.; formal analysis, B.D.; investigation, B.D.; writing—original draft preparation, B.D.; writing—review and editing, B.D. and T.W.; visualization, B.D.; supervision, T.W.; project administration, T.W.; funding acquisition, T.W. All authors have read and agreed to the published version of the manuscript.

Funding: This research and APC were funded by Natural Sciences and Engineering Research Council of Canada, grant number 2017-05906.

Data Availability Statement: Not applicable.

Acknowledgments: The authors wish to thank Doug Bitner for his support and suggestions during the experiments.

Conflicts of Interest: The authors declare no conflict of interest.

Appendix A

This appendix develops the equations used to characterize the pump and motor combination. The pump's pressure and flow outputs were characterized by fitting several typical pump performance parameter coefficients to the experimental data. The torque produced by the pump T_{pump} [Nm] is expressed as

$$T_{pump} = \frac{D\left(1 + C_f\right)\Delta P}{2\pi} + C_d\mu DN + T_c \tag{A1}$$

where D is the pump displacement [m³/rev], C_f is the pressure friction coefficient [-], ΔP is the fluid pressure increase across the pump [Pa], C_d is the viscous damping coefficient [-], μ is the fluid viscosity [Pa·s], N is the pump's rotational speed [rev/s] (assumed to be positive), and T_c is the breakaway torque [Nm].

The flow produced by the pump Q [m³/s] is expressed as

$$Q = DN - \frac{C_s D\Delta P}{\mu} \tag{A2}$$

where C_s is the slip leakage coefficient [-]. The brushless electric motor was assumed to follow the standard equation relating motor torque T to current I [amps] given as

$$T = \frac{30I}{\pi K_v} \tag{A3}$$

where K_v is the motor speed constant [RPM/V].

Combining Equations (A1) through (A3) and rearranging gives the expressions relating pump flow and pressure to motor speed and current as

$$\Delta P = \frac{60I}{K_v D\left(1 + C_f\right)} - \frac{2\pi C_d \mu N}{(1 + Cf)} - \frac{2\pi T_c}{D\left(1 + C_f\right)} \tag{A4}$$

and

$$Q = DN - \frac{C_s D \Delta P}{\mu}. \tag{A5}$$

Equations (A4) and (A5) were fit to the experimental data to characterize the pump's flow and pressure in terms of motor speed and current.

References

1. Ketelsen, S.; Padovani, D.; Andersen, T.O.; Ebbesen, M.K.; Schmidt, L. Classification and Review of Pump-Controlled Differential Cylinder Drives. *Energies* **2019**, *12*, 1293. [CrossRef]
2. Alle, N.; Hiremath, S.S.; Makaram, S.; Subramaniam, K.; Talukdar, A. Review on Electro Hydrostatic Actuator for Flight Control. *Int. J. Fluid Power* **2016**, *17*, 125–145. [CrossRef]
3. Van den Bossche, D. The A380 flight control electrohydrostatic actuators, achievements and lessons learnt. In Proceedings of the ICAS 25TH International Congress of the Aeronautical Sciences, Hamburg, Germany, 3 September 2006.
4. Xia, J.; Durfee, W.K. Analysis of Small-Scale Hydraulic Actuation Systems. *J. Mech. Des.* **2013**, *135*, 091001. [CrossRef]
5. Houle, K.L. A Power Transmission Design for an Untethered Hydraulic Ankle Orthosis. Master's Thesis, University of Minnesota, Minnesota, MN, USA, 2012.
6. Neubauer, B.C. Principles of Small-Scale Hydraulic Systems for Human Assistive Machines. Ph.D. Thesis, University of Minnesota, Minnesota, MN, USA, 2017.
7. Wiens, T.; Deibert, B. A Low-Cost Miniature Electrohydrostatic Actuator System. *Actuators* **2020**, *9*, 130. [CrossRef]
8. Hagen, D.; Pawlus, W.; Ebbesen, M.K.; Andersen, T.O. Feasibility Study of Electromechanical Cylinder Drivetrain for Offshore Mechatronic Systems. *Model. Identif. Control. A Nor. Res. Bull.* **2017**, *38*, 59–77. [CrossRef]
9. Ristic, M.; Whaler, M. Electrification of hydraulics opens new ways for intelligent energy-optimized systems. In Proceedings of the 11th International Fluid Power Conference, Aachen, Germany, 19 March 2018.
10. Wiens, T.; Deibert, B. A Low-cost miniature electrohydrostatic actuator. In Proceedings of the 1st International Electronic Conference on Actuator Technology: Materials, Devices and Applications MDPI, Online, 20 November 2020.
11. Deibert, B.; Scott, S.; Dolovich, A.; Wiens, T. The use of additive manufactured plastic in small-scale poppet valves and pressure vessels (Accepted). In Proceedings of the BATH/ASME 2022 Symposium on Fluid Power and Motion Control, American Society of Mechanical Engineers, Bath, UK, 14–16 September 2020.
12. Parker Hannifin Corp. Parker Legris Technical Tubing & Hose. 2014. Available online: https://www.farnell.com/datasheets/1905436.pdf (accessed on 25 July 2022).
13. AliExpress Brushless Hydraulic Lift Oil Pump for 1/14 Tamiya RC Truck Trailer Tipper Scania Actros Volvo MAN LESU JDM Excavator DIY Parts | Parts & Accessories |. Available online: //www.aliexpress.com/item/1005003462607936.html?src=ibdm_d03p0558e02r02&sk=&aff_platform=&aff_trace_key=&af=&cv=&cn=&dp= (accessed on 25 May 2022).
14. ODrive. Available online: https://odriverobotics.com (accessed on 3 February 2022).
15. Parker Hannifin Corp. Fixed Displacement Gear Pumps D/H/HD Series. 2002. Available online: https://www.parker.com/Literature/Pump%20&%20Motor%20Division/Catalogs/PDFs/DHHD_0910.pdf (accessed on 5 October 2022).
16. Parker Hannifin Corp. Gear Pumps/Motors Series PGP/PGM. 2017. Available online: https://www.parker.com/literature/PMDE/Catalogs/Gear_Units/PGP_PGM/HY30-3300-UK.pdf (accessed on 4 October 2022).
17. Duplomatic MS S.p.A GP External Gear Pumps Series 20. 2020. Available online: https://duplomaticmotionsolutions.com/docs/2020/11100-ed_4d72e112c3.pdf (accessed on 4 October 2022).
18. Bucher Hydraulics S.p.A Gear Pumps AP100. 2015. Available online: https://www.bucherhydraulics.com/datacat/files/Katalog/Pumpen/Aussenzahnradpumpen/Aussenzahnradpumpen%20AP100/AP100_200-P-991218-en.pdf (accessed on 5 October 2022).

19. Ketelsen, S.; Michel, S.; Andersen, T.O.; Ebbesen, M.K.; Weber, J.; Schmidt, L. Thermo-Hydraulic Modelling and Experimental Validation of an Electro-Hydraulic Compact Drive. *Energies* **2021**, *14*, 2375. [CrossRef]
20. Technical Data Sheet Prusament PETG by Prusa Polymers. Available online: https://prusament.com/media/2020/01/PETG_TechSheet_ENG.pdf (accessed on 1 March 2022).

 *actuators*

Article

Evaluation of a Soft Sensor Concept for Indirect Flow Rate Estimation in Solenoid-Operated Spool Valves

Simon Hucko *, Hendrik Krampe and Katharina Schmitz 

Institute for Fluid Power Drives and Systems, RWTH Aachen University, 52074 Aachen, Germany
* Correspondence: simon.hucko@ifas.rwth-aachen.de

Abstract: Concepts, such as power and condition monitoring or smart systems, are becoming increasingly important but require extensive insight into the process or machine, which is mostly gained by additional sensors. However, this development is contrasted by growing global competition and cost pressure. The measurement concept of a soft flow rate sensor presented here, addresses this discrepancy by means of a cost-effective software-based sensor, utilizing the dependence between flow rate and flow force in hydraulic spool valves. In the presented work, the feasibility of the introduced approach and the anticipated challenges, such as the modeling of disturbances in the mechanical, fluidic and electromagnetic subdomain, are assessed. For this purpose, important phenomena influencing measurement accuracy are identified on the basis of previous work. With the help of these findings, a greatly simplified version of the soft sensor is built and evaluated with a commercial valve to perform an initial test of the concept. Regardless of future implementations, the soft sensor concept presented here may have limitations in terms of dynamics and accuracy. However, the aim is not to replace a classic flow rate sensor, such as a gear sensor, but rather to create a cost-effective way to determine the flow rate without any additional integration effort.

Keywords: components; condition monitoring; flow rate sensor; predictive maintenance; soft sensor; solenoid; valve

Citation: Hucko, S.; Krampe, H.; Schmitz, K. Evaluation of a Soft Sensor Concept for Indirect Flow Rate Estimation in Solenoid-Operated Spool Valves. *Actuators* **2023**, *12*, 148. https://doi.org/10.3390/act12040148

Academic Editor: Andrea Vacca

Received: 1 February 2023
Revised: 13 March 2023
Accepted: 16 March 2023
Published: 30 March 2023

1. Introduction

Numerous efforts are being made to integrate more functionalities into hydraulic drive components. During this development, power and condition monitoring, smart systems and the creation of redundancies to increase safety are becoming more and more important but require extensive insight into the process or the machine. This information is mostly obtained using measurement data and additional sensors. However, the development is contrasted by cost pressure as well as the aim to reduce system complexity and integration effort, fueled by growing global competition.

The measurement concept presented here addresses this predicament with cost-effective software-based flow rate sensors, taking advantage of the dependence between flow rate and flow force.

A spool valve is a device that regulates the flow within a hydraulic system. It typically consists of a spool, moving within a valve body. An actuator, such as a solenoid, is used to move the spool. A position sensor, if present, detects the position of the spool and provides feedback to the control system.

The fluid flowing through the valve causes a flow force acting on the spool, which particularly depends on the spool position and flow rate. For most industrially relevant valves, this relationship is unambiguous, meaning, if the spool position and flow force are known, the flow rate can be inferred. The flow force commonly acts in the closing direction of the valve, counteracting the solenoid or spring (depending on the configuration). As shown in Figure 1, this results in an equilibrium of forces, consisting of the axial force F_{ax} of the actuator, the flow force F_{ff}, the inertial force $F_{iner,hydr}$, the friction force $F_{fric,hydr}$ and a spring force $F_{spr,hydr}$, holding the spool in a predefined position. If the inertial, frictional and spring forces are known, the flow force can be deduced from the actuator force. The works [1,2] refer to this possibility. The axial force can be estimated from the state variables of the solenoid (current, voltage and position). In this study, spool valves with proportional solenoids and position feedback are considered.

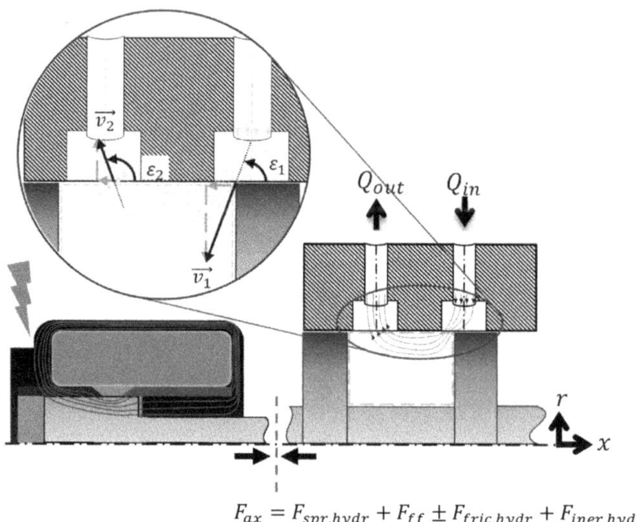

$$F_{ax} = F_{spr,hydr} + F_{ff} \pm F_{fric,hydr} + F_{iner,hydr}$$

Figure 1. Valve with Solenoid and Axial Forces.

In summary, as illustrated in Figure 2, with the measurement concept, the state variables of the solenoid current i, voltage U and the armature position x are used to infer the solenoid force F_{ax}, which, in turn, is used to determine the flow force F_{ff} and, thus, the flow rate Q. Depending on the valve configuration, the variables required to determine the flow rate are already available in the control electronics. Thus, additional functionality can be provided cost effectively by extending the control software.

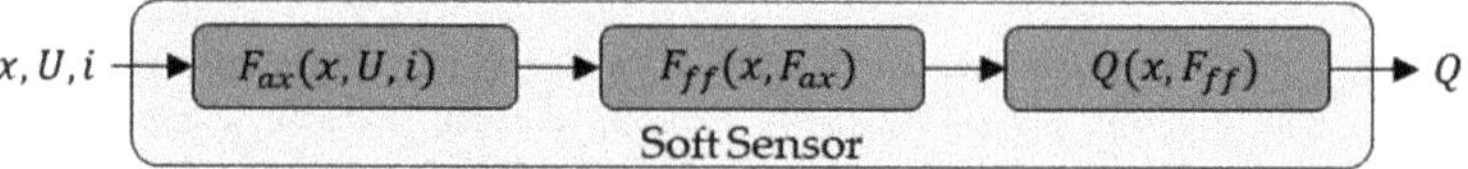

Figure 2. Measurement Concept of the Soft Sensor.

The measurement concept presented might be limited in terms of dynamics and accuracy. However, the aim is not to replace a classic flow rate sensor, such as a gear sensor, but rather to create a cost-effective way to determine the flow rate without an additional integration effort. Especially in machines and plants where no flow rate sensors can be installed for economic reasons, this concept offers additional information. The information can be used for power and condition monitoring, for monitoring tasks or for application-specific purposes.

Due to the wide use of magnetically actuated spool valves, there are various preliminary works within individual subdomains [1–5]. However, none of these works combine these findings into a sensor concept comparable to the one proposed in this paper.

The basis of the presented soft sensor is a model of the valve system, which allows for the deduction of the flow rate from given state variables.

A model, especially a real-time model, is always a trade-off between model accuracy and computation time. As such, important mechanisms significantly affecting the system behavior must be identified. Hence, in the following section, based on relevant preliminary work, system properties are identified that could influence the measurement concept and must, therefore, be taken into account in the modeling. Subsequently, a simplified model is created on the basis of experimental data, which is intended to represent a static/low dynamic system. Finally, an initial metrological investigation using a commercially available valve is carried out, and the concept is tested in practice.

2. Literature Review, Conception of the Soft Sensor and Identification of Relevant System Properties

In this section, with the help of relevant literature, the concept of soft sensors is briefly explained, a model structure is developed and the main mechanisms of the system are identified.

A soft sensor is a software-based system using mathematical algorithms to estimate process variables or properties that are not directly measured by physical sensors from known variables. It can provide real-time monitoring and control of industrial processes by predicting key process variables. While soft sensors are also models, they are more specialized and are focused on estimating the value of a specific variable in a particular process or system [6].

According to [6,7], it is possible to distinguish soft sensors into model- or physically driven sensors and data-driven sensors. The former models are also called white box models and have complete phenomenological knowledge about the process. Data-driven models are also called black box models because they are based on empirical observations and have no knowledge about the process. White and black box models represent the two extremes. Between these two main types, numerous gradations/combinations are possible. For example, parts of a model-driven soft sensor can be realized by data-driven sub models that cannot be easily represented by physical models. These models are referred to as hybrid or grey box models.

Soft sensors have been used for several decades in monitoring and control, especially in the process industry.

Partly, this is because process industries typically involve complex processes and systems that include a wide range of process variables. Many of these variables are difficult or expensive to measure and require the use of costly sensors or specialized instrumentation. Soft sensors provide a cost-effective way to estimate these variables. Due to the complexity of the processes, analytical descriptions are often difficult. Given the historical data that are often available, data-driven approaches are commonly used. In [6–8], detailed examples and an overview of the applications and the design of soft sensors are presented.

In fluid technology, soft sensors have not yet been widely used. Examples can be found for the determination of the flow rate of pumps, for example, by [9–11]. In the works of [12,13], several soft sensors for hydraulic components based on simple physical models are presented and combined to sensor networks. In contrast to the approaches presented in [12,13], no additional sensors are required for the proposed measurement concept. Furthermore, the focus lies on a more detailed modeling of the component.

Therefore, it is aimed at a physically motivated model with physical- and data-driven sub models, i.e., a grey box model.

Based on the descriptions in Section 1, a conceptual structure of the soft sensor and its sub models is derived, as shown in Figure 3. The soft sensor can be divided into two main subsystems: The actuator and the valve.

These sub models are coupled by the axial force F_{ax}, as shown in Equation (1). The axial force F_{ax} exerted by the actuator is obtained by offsetting the estimated magnetic force F_{mag} with the force of the actuator spring $F_{spr,act}$, the frictional force acting in the actuator $F_{fric,act}$ and the acceleration force on the armature $F_{iner,act}$. In the valve model, the axial force F_{ax} is offset with the force of the valve spring $F_{spr,hydr}$, the friction force in the valve

$F_{fric,hydr}$ and the acceleration force at the spool $F_{iner,hydr}$ to obtain the flow force. The flow force is then used to determine the flow rate Q.

$$F_{ax,act} = F_{ax,valve}$$
$$F_{ax,act} = F_{mag} + F_{spr,act} \pm F_{fric,act} - F_{iner,act} \tag{1}$$
$$F_{ax,valve} = F_{spr,hydr} + F_{ff} \pm F_{fric,hydr} + F_{iner,hydr}$$

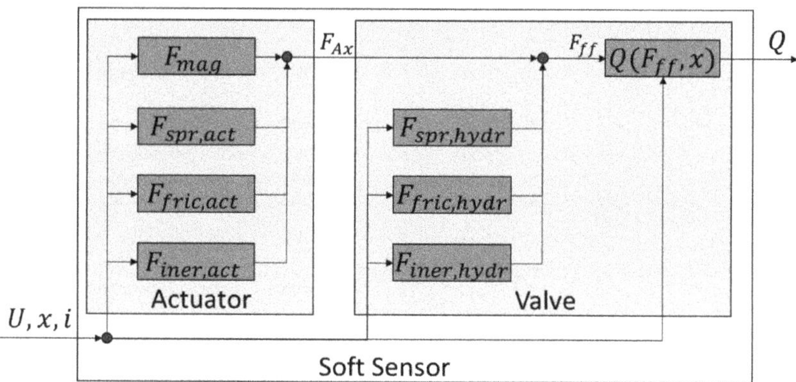

Figure 3. Conceptual Structure of the Soft Sensor.

The described forces, their underlying mechanisms and the relationship between flow force and flow rate are considered in the following sections. Following the measurement procedure shown in Figure 2, the analysis begins with the relationship between the flow rate and the flow force and, thus, with the valve. Subsequently, the actuator and the influences on it are considered.

2.1. Flow Force–Flow Rate Relationship

Flow forces act on the spool and the sleeve/housing due to changes in momentum of the fluid in the valve chamber. These forces are usually considered as a disturbance variable in the position control of the spool. They can be divided into static and dynamic flow forces. Static flow forces are those continuously exerted by the fluid, even when the spool is at rest and the flow rate is constant. Static flow forces usually act in a closing manner. They have been investigated in numerous works, such as [3,14–16], often with the aim of reducing the flow forces in order to decrease the required actuation force. Dynamic flow forces usually result from the acceleration or deceleration of the fluid in the valve chamber, for instance, when the spool moves, or pressure or flow rate pulsation occurs. Dynamic flow forces can act in both opening and closing directions.

There are different approaches to determine the flow forces, including analytical calculation, determination by CFD simulation, as well as by measurement. For the analytical calculation of the flow forces, the approach formulated in Equation (2) proposed by Schmitz and Murrenhoff [17] can be used. It has been applied in several works [3,14,18,19] and has been extensively studied by Bordowski [3].

$$F_{ff} = \rho Q^2 \frac{1}{A_1} \frac{\cos \varepsilon_1}{\sin \varepsilon_1} - \rho Q^2 \frac{1}{A_2} \frac{\cos \varepsilon_2}{\sin \varepsilon_2} - \rho l \frac{\partial Q}{\partial t} \tag{2}$$

The first two expressions in Equation (2) describe the steady-state flow force; the third describes the dynamic flow force. The areas A_1 and A_2 denote the in- and outlet metering areas, the angles ε_1 and ε_2 the in- and outflow angles, ρ the density of the fluid, Q the flow rate and F_{ff} the flow force acting on the spool in axial direction. When examining the analytical description, it becomes evident that influences which change the fluid density, the flow angles or geometric parameters, have a direct effect on the relationship between the flow force and flow rate. These influences are considered in more detail subsequently.

2.1.1. Influence of the Oil Properties

Considering Equation (2), it can be seen that the density and, thus, the fluid temperature influence the magnitude of the flow forces. For a common hydraulic oil, for example, mineral oil of type HLP46, the density changes up to 3% with a change in oil temperature of 50 °C.

The results of Schuster et al. [20], Yuan et. al [1], Bordowski et al. [21] and Hucko et al. [19] suggest that changes in viscosity due to temperature variation also affect the flow forces. In the latter investigation, a temperature change of 50 °C (30→80 °C) at a constant flow rate caused a maximum decrease in flow force of 23%.

Relating these findings to the presented measurement concept, such a temperature-induced change in the flow force–flow rate relationship, if not considered, would lead to a significant estimation error. In addition to investigating the temperature influence on the flow force–flow rate relationship, Hucko et al. [19] presented a method to compensate for this influence at a known temperature.

In many systems, the oil temperature is roughly known. In addition, some valves already have temperature sensors or have the capability to receive temperature values from other modules within a modern automation architecture [22]. In summary, the influence of the fluid temperature can and should be considered in the soft sensor.

2.1.2. Geometric Factors

In addition to the oil properties, geometric factors, such as the distance between in- and outflow ports [23], metering edges, radial clearance [24], geometry of the spool chamber and metering area, have a considerable influence on the flow forces.

Changes in valve geometry, for example, in the metering edges, can occur due to wear, which may alter the in- and outflow angles and, thus, the flow force–flow rate characteristics. At present, the author is not aware of any studies on the extent to which wear of the metering edges affects the flow force–flow rate relationship.

In addition to wear, geometric parameters can vary due to temperature changes and the associated material expansion. For the valves considered, the clearance between the spool and sleeve/housing as well as the metering areas are of particular interest. The change in clearance can be compensated for by matching the expansion coefficients of the spool and sleeve to another. To ensure safe operation in the operating range, this is usually accounted for in the design. According to Hucko et al. [19], the change in opening areas has a negligible effect on the flow force–flow rate relationship for the nominal size 6 valves considered there. However, a change in temperature can have an effect on more than just the geometric variables. The sensors installed in valves to detect the spool or armature position can be influenced by temperature changes. This drift is usually specified in the data sheets of the sensors and can be adjusted for with little effort, provided the temperature is known. For commercially available linear variable differential transformers (LVDTs), often used in valves, the drift is 0.01–0.02% F.S./K.

In summary, to date, little research has been devoted to wear-related influences on the flow force. Additionally, the temperature-related influence of geometric variables has a minor effect on the flow force–flow rate relationship, according to current research. The influence of position sensors, however, should be considered, depending on the mounting position and operating temperature span.

2.1.3. Influence of Frictional Forces in the Valve

The central idea of the presented soft sensor is to infer the flow rate from the flow forces. Flow forces cannot be measured directly. However, they can be calculated from the sum of the axial forces acting on the spool. For this purpose, the frictional forces acting on the spool must be taken into account.

Friction can be differentiated according to the type of friction, into solid, boundary, mixed and fluid friction, as well as according to the kinematics in static, sliding and rolling friction [25]. The annular seal usually used in spool valves causes a permanent leakage flow, resulting in a liquid film in the clearance between the spool and the valve sleeve/housing. In most cases, solid sliding friction and liquid friction are superimposed. The frictional force depends on the surface condition, the clearance, the viscosity of the oil, the spool velocity and the acting normal force [16].

Lu and Tiainen [26] investigated friction in spool valves, both simulatively and experimentally, and were able to identify radial flow and pressure forces as the main contributors. These occur when the flow around the metering edge is not symmetrical and radial forces, therefore, do not completely negate each other. It was also shown that the radial forces caused by static pressure increase linearly with the pressure difference.

The friction occurring in spool valves was investigated for valves of different sizes in [18,19]. Here, an assumption was made that the amount of friction during opening and closing of the valve is the same. By averaging the axial force recorded during the opening and closing process, the flow forces can be calculated. The measurements showed a high repeatability of the measured axial forces and, thus, good reproducibility of the friction. It was shown that the frictional forces account for a non-negligible proportion of the axial force. The investigations mentioned were limited to the quasi-static movement of the spool. If the Stribeck curve described in [25] is taken into consideration, a decrease in the frictional force can be expected as the spool velocity increases.

In conclusion, it can be stated that friction forces should be considered in the sensor concept, in order to improve the determination of the flow forces. The described dependence on operating parameters, such as speed, should also be accounted for. In case the dependence of the friction forces on pressure difference or flow rate is observed for the investigated valve, the flow rate calculated by the soft sensor could be fed back and used for a better estimation of the friction.

2.1.4. Dynamic Influences

As previously mentioned, movement of the spool as well as pressure pulsations lead to an acceleration of the fluid in the valve chamber and cause dynamic flow forces. Del Vescovo and Lippolis [27] analyzed these, using the law of the conservation of momentum and CFD simulation. They found that the influence of the spool movement up to a velocity of 1 m/s, which corresponds to an opening time of 20 ms (for a common stroke of 2 mm), as well as pressure pulses up to 1 kHz are negligible and the consideration of static flow forces is sufficient. Manring and Zhang [28] investigated the influence of the dynamic flow forces experimentally and argued that the pressure transient term needs to be considered but did not elaborate further.

Nakada and Ikebe [29] also studied the influence of dynamic flow forces experimentally and found a correlation between the damping length and the magnitude of the dynamic flow forces. In the valves investigated, the static flow forces were dominant. An influence of the dynamic flow forces was found from 20 Hz.

Tanaka and Kamata [2] examined the flow forces occurring in a spool valve using three-dimensional unsteady CFD simulations and analytical models. The spool-sleeve geometry was described as complex, due to radial cylindrical notches, which were explicitly taken into account. Validation was performed mathematically using the law of momentum as well as experimentally by comparing measured static flow forces to the calculated ones. It was found that in the analytical calculation, both the inflow and outflow angles must be considered. Furthermore, it was shown that an assumption of constant flow angles leads to increased deviation. The influence of dynamic flow forces was examined for a spool movement of 100 Hz at full stroke. It was concluded that the velocity and acceleration terms are negligible. Only the damping term contributes 7% to the flow forces at the given excitation.

In summary, the preliminary simulative and experimental work of [2,27–29], amongst others, dealt with the influence of the spool movement and the pressure pulsation on the flow forces. The studies show that, for the valves considered, the influence is most significant in case of increased dynamics.

2.2. Disturbances Occurring at the Solenoid

In the presented measurement principle, the electromagnet used for positioning the spool simultaneously serves to measure the axial force. For this purpose, the unknown actuator force, generated by the solenoid, needs to be inferred from the known state variables, voltage, current and armature position.

The current force characteristic of proportional solenoids is usually manipulated by constructive measures, so that the force generated is proportional to the current, regardless of the valve position. Despite intensive efforts, this proportionality is limited, even when restricted to certain stroke ranges. In addition to a nonlinear current and position dependence, a pronounced force hysteresis can usually be observed, especially in proportional solenoids. The hysteresis is caused by various overlapping effects. For a precise and computationally efficient force estimation, the effects with the most significant influence have to be modeled.

According to [30], mechanical friction between the armature and the pole tube, eddy currents and magnetic hysteresis contributes significantly to the force hysteresis. These effects and common approaches to modeling them are described in [30]. In static conditions, experience shows that the force hysteresis using dithers is between 2 and 10% of the total force of the actuator. As the current rate increases, so does the effect of eddy currents. Neglecting this effect would lead to a significant deviation between the model and system and, thus, must be accounted for.

2.3. Summary

The described system properties and mechanisms influence the accuracy of the presented soft sensor. By measuring and modeling these influences, the estimation error can be reduced.

For the valve sub model, the influence of the fluid temperature as well as that of the state-dependent friction and, at higher spool speeds, dynamic influences should be considered. The latter ought to be modeled as part of the flow force–flow rate relationship.

For the actuator sub model, the friction occurring between the armature and the pole tube, the magnetic hysteresis and eddy currents must be taken into account. The magnetic hysteresis and eddy currents should be considered within the magnetic force estimation.

3. Simplified Model and Parameterization

To verify the feasibility of the described sensor concept, it was implemented in simplified form, using a commercially available valve, as shown in Figure 4. The modeled valve is a proportional 4/3-way valve of nominal size 6. It is equipped with position feedback and built-in electronics.

As discussed, friction in both the solenoid and the valve is dependent on system variables that change during operation, thus requiring a complex model to sufficiently describe the friction within the system. To provide an initial assessment, the friction is, therefore, not modeled. Instead, the measured force and the flow rate hysteresis are averaged. Since the validation is limited to a quasi-static case, the acceleration forces are omitted. This simplifies the model shown in Figure 3 to that in Figure 4.

As depicted in this figure, only the two most important relationships, the estimation of the solenoid force and of the flow rate, are considered. In this way, the model offers the possibility to validate the basic function of the concept. Since all other phenomena described in Section 2 are neglected, an upper estimation error can be determined. The magnetic force is estimated using the electrical quantities and position. The estimation of the flow rate is made using the flow force and the position. The spring forces are also taken into account.

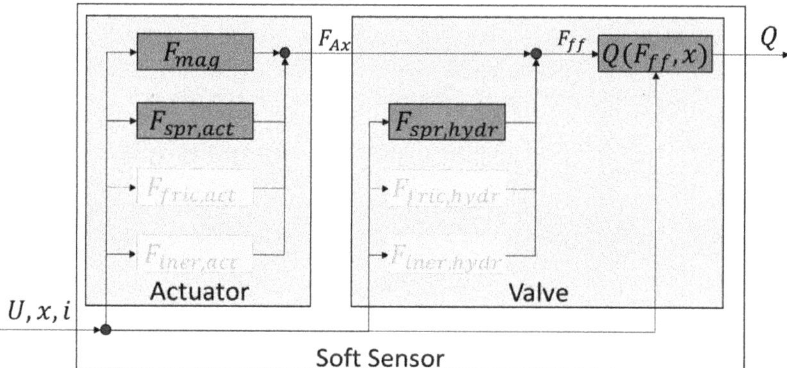

Figure 4. Simplified Soft Sensor Model.

To model the generation of the magnetic force and the flow force–flow rate characteristic, lookup tables determined by measurements were used. Their determination is explained in the following.

Using the setup described in [19], the axial force was determined for different flow rates and piston positions. Pressure and temperature, upstream and downstream of the considered metering edge, as well as the flow rate, were also recorded. The axial force was measured with a load cell. The purpose of these measurements was to determine the relationship between flow force, position and flow rate. Intuitively, it would, therefore, make sense to adjust the flow rate across the considered metering edge and to open and close the valve. However, since pressure sensors have significantly higher dynamics than flow rate sensors, better results could be achieved by controlling the pressure difference.

To record the characteristic diagram, the valve was opened and closed, while a constant pressure difference across the metering edge was maintained. This was repeated for the different pressure differences shown in Figure 5.

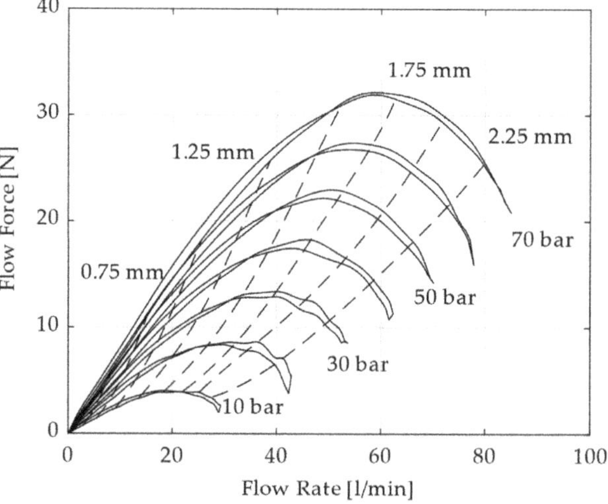

Figure 5. Relationship Between Spool Position, Axial Force and Flow Rate.

Since the valve was opened and closed at each set pressure difference, the hysteresis of the axial force was also recorded. To determine the flow force, the assumption is made that the friction in both directions of movement has the same value but a different sign, as described in [19]. Hence, the flow force was obtained by averaging the two measured curves of axial force of each pressure difference. The curves were then interpolated, so that the flow rate can be determined for any spool position and flow force. In Figure 5, it can be seen that if the hysteresis is neglected, the relationship between flow rate, flow force and position is always unique for the measured valve. Thus, for a given flow force and position, the flow rate can be calculated. A further examination of the measured curves shows that the slope of the dashed position curves is a good indicator for the sensitivity of the used correlation. As the slope of these lines increases, so does the accuracy with which the flow rate can be inferred from the flow force and position.

For the determination of the magnetic force, a metrologically obtained lookup table was used. It was generated by measuring the magnetic force at different currents and positions. Due to the simplified approach, the applied voltage was not used for modeling. The setup shown in Figures 6 and 7 was used to measure the magnetic force. The test setup was designed to allow for precise measurement of a wide variety of valve solenoids. Figure 6 shows the mechanical part of the test rig.

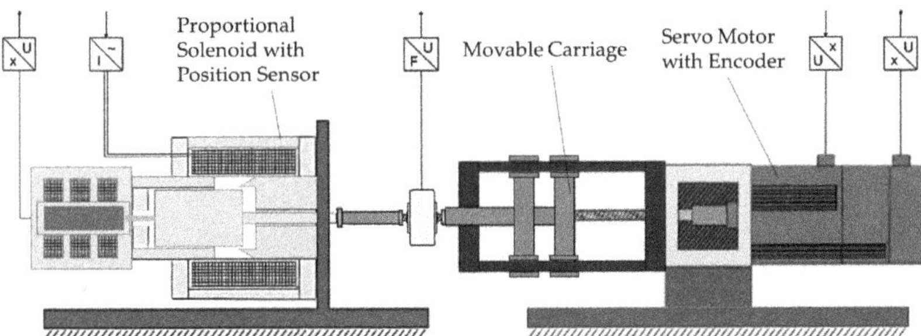

Figure 6. Schematic Depiction of the Mechanical Setup for Measuring Solenoids. On the left, the Solenoid to be Measured.

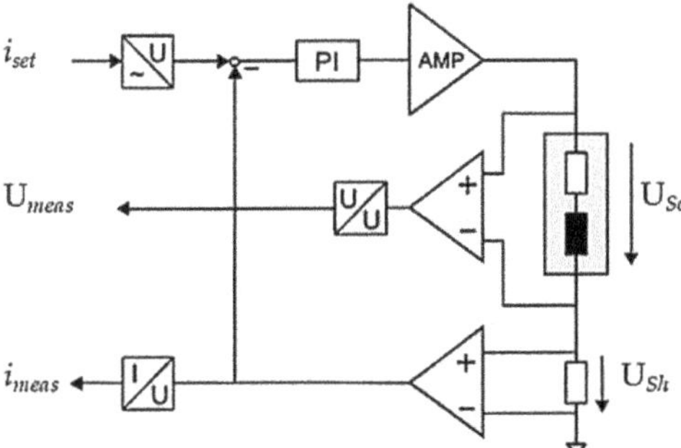

Figure 7. Electrical Schematic for Measuring Solenoids.

On the left side, the solenoid to be measured is mounted in a rigid frame. A movable carriage driven by a servo motor allows for precise, dynamic positioning of a mechanical stop against which the armature presses. The force generated by the armature is recorded by a force sensor (MES KM40). Both the position of the armature and the position of the carriage are measured to precisely control the armature position and eliminate possible position deviations, e.g., due to backlash or the limited stiffness of the force sensor.

Figure 7 schematically shows the electrical part of the experimental setup. The circuit consists of the solenoid itself, an amplifier and a low-side measurement shunt ($3\ \Omega \pm 0.02\%$) for current measurement. The amplifier (Kikusuiu PBZ80-5) allows for, in push/pull configuration, the application of arbitrary currents and voltages. The solenoid is highlighted by a gray rectangle and is represented by an inductor and an ohmic resistor. The temperature change in the measurement shunt is small due to sufficient cooling. The resulting change in resistance is, therefore, negligible. The voltage is measured across the solenoid. The synchronized acquisition of voltage, current and force is carried out using a measuring amplifier (MC USB 404-60).

The force characteristics determined with this measurement setup are shown in Figure 8. In order to avoid static friction at a constant current, the armature was moved quasi-statically between the stops using the servo motor shown in Figure 6. The position was detected using the actuator's own position sensor.

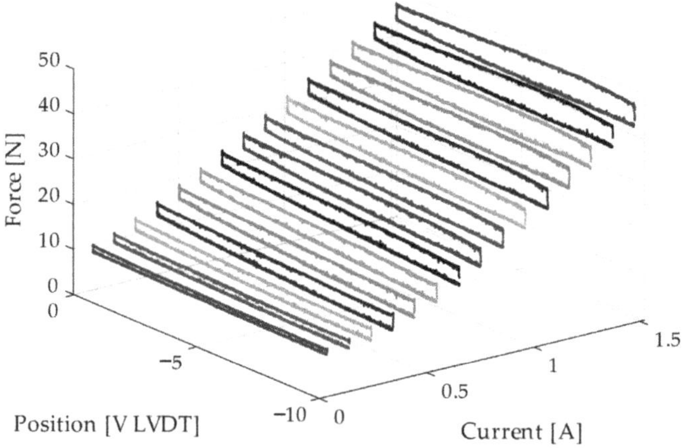

Figure 8. Measured Force Characteristics of a Solenoid.

The equally shaded lines on top of each other represent the force progression with ascending and descending directions of movement for each current step. The force hysteresis is clearly visible. The solenoid's air gap is reduced with decreasing voltage of the LVDT. The force increasing with the growing air gap can be observed, especially with proportional magnets, and is due to the design influence on the force characteristics. The hysteresis width increases with the current. This can be seen from the growing distance between the force curves measured at the same current. Without using a dither signal, the force hysteresis for this solenoid is 1–5 N. According to [30], the static force hysteresis of the magnet depends mainly on the radial magnetic flux and the material hysteresis and, thus, on the momentary state of excitation. However, modeling this relationship would exceed the scope of this paper. Therefore, a simplified approach was chosen to validate the sensor concept. The individually recorded hysteresis curves were averaged and then interpolated, resulting in a lookup table, which provides the magnetic force for a given current and position.

4. Validation of the Sensor/Measurement Concept

In order to validate the soft sensor concept, the determined characteristic diagrams and the measured spring constants were combined to form a simplified model, as shown in Figure 4. This model was then tested with measured data. The test data were generated on a test rig using the methods described in Section 3. A differential pressure of 60 bar was applied across the control edge PB of the valve, and the spool was opened and closed at a rate of 5%/s. During the measurement, the current through the coil and the flow rate were recorded. The measured flow rate is shown by a solid line in Figure 9. Simultaneously, the flow rate was estimated using the soft sensor and is represented by a dashed line.

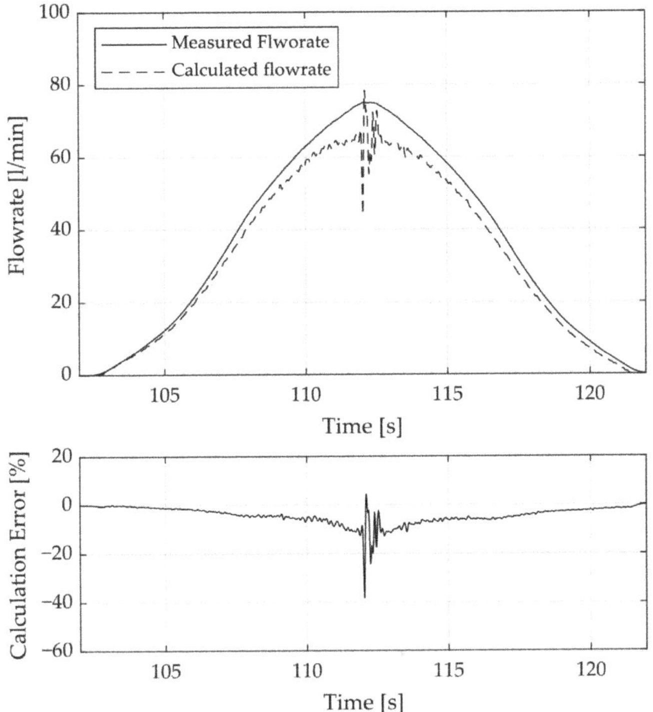

Figure 9. Measured and Estimated Flow Rate while Opening and Closing the Valve.

Examining Figure 9, it can be seen that a qualitative estimation of the flow rate is possible even with the use of a condensed model. The error between measured and estimated flow rate grows with increasing flow rates to up to 9 L/min or 11% FS and is maximum at the reversal point of the signal, with up to 30 L/min or 38% FS. The maximum flow rate of the valve is 80 L/min and defines the full-scale value. Other measurements also showed an increased estimation error with growing valve openings and flow rate.

It is assumed that with increasing flow rate, the flow and pressure distribution around the metering edge is no longer symmetrical. Therefore, radial flow forces and static pressures no longer fully compensate. This leads to a force, which acts perpendicular to the direction of movement. It creates friction and depends on the flow rate and the position of the spool. The frictional force depends on the perpendicular acting force and, therefore, on the operating point. This was confirmed by further measurements.

The oscillation clearly visible at the reversal point, when the spool is at rest, increases with the flow rate or flow force. In a more detailed examination, the current imposed by the valve control and the resulting forces were examined. The current profile shows that the position controller exhibits pronounced oscillations when the spool is at rest, increasing with the flow rate.

It was observed that if the spool is stationary with only a slight control error, the current and, therefore, the axial force is increased by the controller until static friction is overcome. This causes an overshoot of the spool position and the force is reduced until the spool slips again below the target position. For a conventional application, this behavior is not a problem, as the change in spool position and flow rate is negligible. In this concept, however, the mechanism is relevant.

The solenoid current serves as the input of the model and is used to determine the actuator force and, subsequently, the flow force. Without modelling static friction forces, the determination of the flow force from the axial force exhibits high errors.

In terms of frictional forces, increasing dynamics could have a positive effect, as static friction forces would be reduced. However, when increasing the dynamics, the acceleration forces, the dynamic flow forces described in Equation (2) and the eddy currents must be taken into account.

As already shown in Figure 5, various combinations of spool positions and flow rates are possible during operation. For this reason, the validation was also carried out for other pressure differences, as depicted in Figure 10. Like before, over the control edge PB of the valve, five various differential pressures of 30, 40, 50, 60 and 70 bar were applied. At each pressure level, the spool was opened and closed at a rate of 5%/s. This leads to the measured triangular increasing and decreasing flow rate signal represented by a solid line. The estimated flow rate is indicated by a dashed line. The error between the estimated and measured flow rate is given as before. Figure 10 shows that for the other pressure levels considered, the error also grows with increasing valve opening or flow rate. In all measurements, the error is greatest when the spool is at rest, while the direction of movement is changed. This observation can be explained by the occurrence of frictional forces described above. Both at the second pressure level of 40 bar and at the last level of 70 bar, sharp increases in the measurement error can be observed. After further examination, these could be traced back to acquisition errors of the position signal. In order to make the soft sensor more robust against such errors, future models will aim at filtering the position signal.

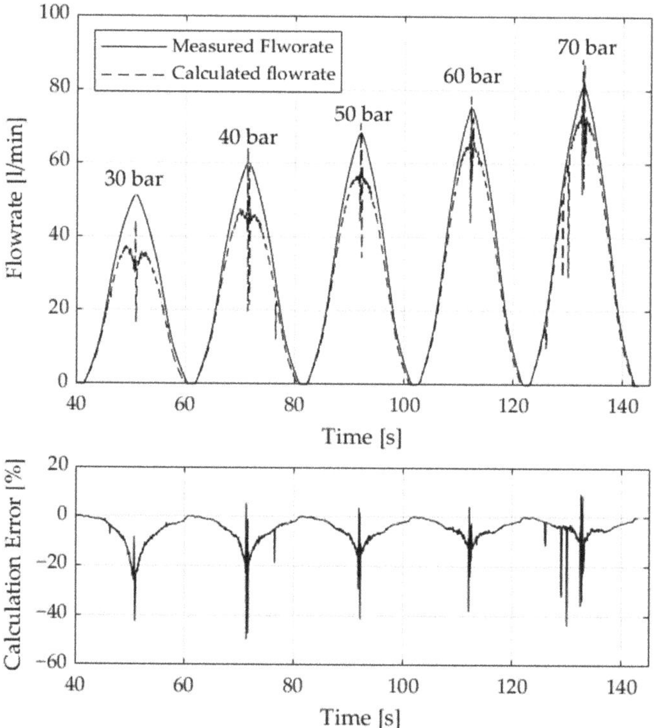

Figure 10. Measured and Estimated Flow Rate while Opening and Closing the Valve at differential Pressure Levels of 30–70 bar.

5. Conclusions and Results

In this paper, a new concept for measuring the flow rate with a soft sensor is presented. The concept uses the relationship between the flow rate and the flow forces in valves. To measure the flow forces, the solenoid is not only used to position the spool but also as a force sensor. With the electrical parameters and the valve position, which are both already recorded by the valve controller, the flow forces and from these the flow rate can be derived with a suitable model. In the first section of the paper, the basic concept was explained, and relevant preliminary work was briefly summarized. Subsequently, important mechanisms and phenomena which affect the measurement concept were identified. To examine whether the presented concept can be practically implemented, a condensed version of the soft sensor was realized with a commercially available valve. In this model, only the flow and spring forces were considered to determine the flow rate. Other influences, such as friction, inertial and dynamic flow forces, were not taken into account. In the actuator sub model, phenomena, such as magnetic hysteresis, eddy currents and friction, causing hysteretic behavior of the actuator force, were not considered. During validation of the soft sensor, the flow rate could be measured at low dynamics with an error of less than 9 L/min or 11% FS (FS: 80 L/min). The maximum measured flow rate was 76 L/min. The accuracy of estimation decreases significantly with decreasing flow rate or smaller valve opening. When the spool is at rest, frictional forces cause oscillations of the position controller, increasing the error up to 30 L/min or 38% FS. It is expected that by taking the frictional forces into account, the estimation error can be significantly reduced.

Overall, it is shown that the presented soft sensor concept applied to a commercially available valve can be used to estimate the flow rate. The estimation of the flow rate can be further improved in future works by more detailed modeling of the valve and the actuator to include identified relevant phenomena. After further enhancements of the soft sensor, a sensitivity analysis for the valve and actuator variables as well as an uncertainty analysis are planned for future work.

However, the aim of this soft sensor is not to replace conventional flow rate sensors. With the achievable accuracy, this concept is more suitable for cost-effective applications to estimate the flow rate from data already known in the valve control. Possible applications include power or condition monitoring as well as smart systems.

Author Contributions: Conceptualization, S.H. and H.K.; Methodology, S.H.; Software, S.H. and H.K.; Validation, S.H. and H.K.; Formal analysis, S.H.; Investigation, S.H.; Data curation, H.K.; Writing—original draft, S.H.; Writing—review & editing, S.H., H.K. and K.S.; Visualization, S.H. and H.K.; Supervision, S.H. and K.S.; Project administration, S.H.; Funding acquisition, S.H. and K.S. All authors have read and agreed to the published version of the manuscript.

Funding: This research was funded by Forschungskuratorium Maschinenbau e.V.–FKM, grant number 7052200.

Data Availability Statement: The data are not publicly available. The data presented in this study are available upon request from the corresponding author if the institution funding the project agrees to their disclosure.

Conflicts of Interest: The authors declare no conflict of interest.

Nomenclature

ρ	Oil density [kg/m^3]
Q	Flow Rate [
A_1, A_2	Metering Areas [m^3]
ϵ_1, ϵ_2	Flow angle [°]
x	Spool Position [mm]
I	Momentum [kg m/s]

References

1. Yuan, Q.; Singh, S.; Kulkarni, A. Flow Forces Investigation through Computational Fluid Dynamics and Experimental Study. In Proceedings of the Symposium of 9th International Fluid Power Conference, Aachen, Germany, 24–26 March 2014; pp. 101–109.
2. Tanaka, K.; Kamata, K. Steady and Unsteady Flow Force acting on a Spool Valve. In Proceedings of the Fluid Power and Motion Control, FPMC, Bath, Great Britain, 12–14 September 2012; pp. 451–465.
3. Bordovský, P. *Evaluation of Steady-State Flow Forces in Spool Valves*; Shaker Verlag: Düren, Germany, 2019; ISBN 9783844066999.
4. Kallenbach, E.; Eick, R.; Ströhla, T.; Feindt, K.; Kallenbach, M.; Radler, O. *Elektromagnete: Grundlagen, Berechnung, Entwurf und Anwendung*, 5th ed.; Vieweg: Wiesbaden, Germany, 2018; ISBN 9783658147884.
5. Vaughan, N.D.; Gamble, J.B. The Modeling and Simulation of a Proportional Solenoid Valve. *J. Dyn. Syst. Meas. Control* **1996**, *118*, 120–125. [CrossRef]
6. Fortuna, L.; Graziani, S.; Rizzo, A.; Xibilia, M. *Soft Sensors for Monitoring and Control of Industrial Processes*; Springer: London, UK, 2007; ISBN 9781846284809.
7. Kadlec, P.; Gabrys, B.; Strandt, S. Data-driven Soft Sensors in the process industry. *Comput. Chem. Eng.* **2009**, *33*, 795–814. [CrossRef]
8. Desai, K.; Badhe, Y.; Tambe, S.S.; Kulkarni, B.D. Soft-sensor development for fed-batch bioreactors using support vector regression. *Biochem. Eng. J.* **2006**, *27*, 225–239. [CrossRef]
9. Ahonen, T. *Monitoring of Centrifugal Pump Operation by a Frequency Converter*; Lappeenranta University of Technology: Lappeenranta, Finland; Lappeenrannan Teknillinen Yliopisto: Lappeenranta, Finland, 2011; ISBN 9789522650757.
10. Leonow, S.; Monnigmann, M. Soft sensor based dynamic flow rate estimation in low speed radial pumps. In Proceedings of the European Control Conference, Zurich, Switzerland, 17–19 July 2013; IEEE: Piscataway, NJ, USA, 2013; pp. 778–783, ISBN 978-3-033-03962-9.
11. Jia, Y.F.; Gu, L.C.; Tian, Q.Q. Soft-Sensing Method for Flow of the Variable Speed Drive Constant Pump. *AMM* **2013**, *318*, 55–58. [CrossRef]
12. Pelz, P.F.; Dietrich, I.; Schänzle, C.; Preuß, N. Towards digitalization of hydraulic systems using soft sensor networks. In Proceedings of the 11 International Fluid Power Conference, Aachen, Germany, 21–23 March 2022.
13. Hartig, J.; Schänzle, C.; Pelz, P.F. Validation of a soft sensor network for condition monitoring in hydraulic systems. In Proceedings of the 12th International Fluid Power Conference, Dresden, Germany, 19–21 March 2020; Volume 2, pp. 167–173.
14. Schrank, K.; Murrenhoff, H. Beschreibung der Strömungskraft in Längsschieberventilen mittels Impulserhaltung. *O+P-Journal* **2013**, *4*, 4–15.
15. Opitz, H. *Über Die Dynamische Stabilität Hydraulischer Steuerungen Unter Berücksichtigung der Strömungskräfte*; VS Verlag fur Sozialwissenschaften GmbH: Wiesbaden, Germany, 1964; ISBN 9783663073451.
16. Merrit, H.E. *Hydraulic Control Systems*; Wiley: New York, NY, USA, 1967; ISBN 0471596175.
17. Schmitz, K.; Murrenhoff, H. *Umdruck zur Vorlesung Grundlagen der Fluidtechnik/Hubertus Murrenhoff*; Vollständig neu Bearbeitete Shaker Verlag: Düren, Germany, 2018; ISBN 9783844062465.
18. Bordovsky, P.; Schmitz, K.; Murrenhoff, H. CFD Simulation and Measurement of Flow Forces Acting on a Spool Valve. In Proceedings of the 10th International Fluid Power Conference, Dresden, Germany, 8–10 March 2016.
19. Hucko, S.; Vonderbank, V.; Krampe, H.; Schmitz, K. Investigation of the influence of Fluid Temperature on Flow Forces in Spool Valves. In Proceedings of the 18th Scandinavian International Conference on Fluid Power, SICFP'23, Linköping, Sweden, 30 May–1 June 2023.
20. Schuster, G. *CFD-Gestützte Maßnahmen zur Reduktion von Strömungskraft und Kavitation am Beispiel eines hydraulischen Schaltventils*; Shaker: Aachen, Germany, 2005; ISBN 3832244662.
21. Bordovsky, P.; Murrenhoff, H. Investigation of Steady-State Flow Forces in Spool Valves of Different Geometries and at Different Oil Temperatures With the Help of Measurements and CFD Simulations. In Proceedings of the Symposium on Fluid Power and Motion Control, FPMC2016. Bath, UK, 7–9 September 2016.
22. Alt, R.; Murrenhoff, H.; Schmitz, K. A survey of "Industrie 4.0" in the field of Fluid Power–challenges and opportunities by the example of field device integration. In Proceedings of the 11th International Fluid Power Conference, Aachen, Germany, 19–21 March 2018.
23. Tatar, H. Störkräfte bei Elektromagnetisch Betätigten Wegeventilen. Ph.D. Thesis, RWTH, Aachen, Germany, 1974.
24. Kipping, M. Experimentelle Untersuchungen und numerische Berechnungen zur Innenströmung in Schieberventilen der Ölhydraulik. *O+P-Journal* **1997**, *42*, 12.
25. Czichos, H.; Habig, K.-H. (Eds.) *Tribologie-Handbuch: Tribometrie, Tribomaterialien, Tribotechnik*; Springer: Wiesbaden, Germany, 2015; ISBN 9783834822369.
26. Lu, Q.; Tiainen, J.; Kiani-Oshtorjani, M.; Wu, Y. Lateral Force Acting on the Sliding Spool of Control Valve Due to Radial Flow Force and Static Pressure. *IEEE Access* **2021**, *9*, 126658–126669. [CrossRef]
27. Del Vescovo, G.; Lippolis, A. A Review Analysis of Unsteady Forces in Hydraulic Valves. *Int. J. Fluid Power* **2006**, *7*, 29–39. [CrossRef]
28. Manring, N.D.; Zhang, S. Pressure Transient Flow Forces for Hydraulic Spool Valves. *J. Dyn. Syst. Meas. Control* **2012**, *134*, 034501. [CrossRef]

29. Nakada, T.; Ikebe, Y. Measurement of the Unsteady Axial Flow Force on a Spool Valve. *IFAC Proc. Vol.* **1980**, *13*, 193–198. [CrossRef]

30. Hucko, S.; Matthiesen, G.; Reinertz, O.; Schmitz, K. Investigation of Magnetic Force Hysteresis in Electromechanical Actuators. *Chem. Eng. Technol* **2023**, *46*, 158–166. [CrossRef]

Article

Design of a Non-Back-Drivable Screw Jack Mechanism for the Hitch Lifting Arms of Electric-Powered Tractors

Marco Claudio De Simone [1,*], **Salvio Veneziano** [2] **and Domenico Guida** [1]

1 Department of Industrial Engineering, University of Salerno, Via Giovanni Paolo II, 84084 Fisciano, SA, Italy
2 MEID4 Academic Spin-Off of the University of Salerno, Via Giovanni Paolo II, 132, 84084 Fisciano, SA, Italy
* Correspondence: mdesimone@unisa.it

Abstract: The agricultural sector is constantly evolving. The rise in the world's population generates an increasingly growing demand for food, resulting in the need for the agroindustry to meet this demand. Tractors are the vehicles that have made a real difference in agriculture's development throughout history, lowering costs in soil tillage and facilitating activities and operations for workers. This study aims to successfully design and build an autonomous, electric agricultural tractor that can autonomously perform recurring tasks in open-field and greenhouse applications. This project is fully part of the new industrial and agronomic revolution, known as Factory 4.0 and Agriculture 4.0. The predetermined functional requirements for the vehicle are its lightweight, accessible price, the easy availability of its spare parts, and its simple, ordinary maintenance. In this first study, the preliminary phases of sizing and conceptual design of the rover are reported before subsequently proceeding to the dynamical analysis. To optimize the design of the various versions of the automated vehicle, it is decided that a standard chassis would be built based on a robot operating inside a greenhouse on soft and flat terrains. The SimScape multi-body environment is used to model the kinematics of the non-back-drivable screw jack mechanism for the hitch-lifting arms. The control unit for the force exerted is designed and analyzed by means of an inverse dynamics simulation to evaluate the force and electric power consumed by the actuators. The results obtained from the analysis are essential for the final design of the autonomous electric tractor.

Keywords: automated self-driving tractor; agriculture; robot; multibody; control systems

Citation: De Simone, M.C.; Veneziano, S.; Guida, D. Design of a Non-Back-Drivable Screw Jack Mechanism for the Hitch Lifting Arms of Electric-Powered Tractors. *Actuators* **2022**, *11*, 358. https://doi.org/10.3390/act11120358

Academic Editor: Ioan Ursu

Received: 31 October 2022
Accepted: 28 November 2022
Published: 2 December 2022

Publisher's Note: MDPI stays neutral with regard to jurisdictional claims in published maps and institutional affiliations.

1. Introduction

In the course of human history, there has been an evolution in agricultural production concerning yield, endurance, and requirements [1,2]. Throughout the centuries, humankind always has looked for various types of alternative energy sources for farming activities; suffice it to say that the sector only started relying on human and animal power a century ago [3]. The first tractors and farming machines were introduced only in vast areas due to the high costs of the new technology. Nevertheless, the mass production of tractors and machines allowed for such tools to be accessible to small land owners [4]. Moreover, advancements in electronics and automatism resulted in a milestone that fostered the development and production of automated self-driving vehicles [5]. This study aims to design and develop an autonomous electric tractor capable of performing activities both in the open field and in greenhouses [6–8]. Agriculture 4.0 is the systematic use of innovative technologies, and the utilization of electric vehicles is an integral part of it [9–11]. Currently, many companies are developing technologies and actual models that fit the perspectives of Agriculture 4.0. For this reason, a preliminary market investigation has been conducted to understand the state-of-the-art technology in propulsion and automation [12,13]. Such analysis showed that the propulsion used in such applications is either electric or hybrid for easier vehicle management. For these valid reasons, this type of propulsion has been adopted for our model [14].

2. Materials and Methods

The authors have been involved for some time in the dissemination and application of Agriculture 4.0 methods and in the research and development activities of medium-small companies, which are typical in the agricultural sector of southern Italy [15–20]. This work reports a study concerning the design of a new electric tractor for precision agriculture applications, one that is capable of working in both open fields and greenhouses. For this purpose, it was decided that a modular machine would be designed based on the same frame, but with the possibility of being equipped with ad hoc devices according to the needs encountered. Given the high versatility of the robot, three configurations have been identified [21]. The first substantial difference between an electric tractor and a traditional machine lies in the lifting and in the towed machine's driving systems. In a traditional machine, the operating machine is lifted by a hydraulic lift, while a Cardan joint is operated by the power take-off. In our vehicle, both lifting and driving are guaranteed by electro-mechanical drives [19]. The electric tractor must be equipped with a lift mechanism that is capable of operating all the operating machines that are already on the market. With such consideration, an operational width of 130 cm was identified, which would allow many activities to be carried out both in the open field and in greenhouses [22]. In this way, based on the catalogues of agricultural machinery manufacturers, it was possible to identify the power and width of the tractor as a function of the working width [23]. In Figure 1a,b, it can be seen that for an operating machine of 130 cm, a tractor with a power of 15 kW is required, and this must be as wide as the connected tool [24]. For the missing parameters, an analysis of the most widespread agricultural machine on the market was performed. With the typical characteristics and through appropriate scaling, the final model was then defined [25].

MOD	MOVEMENT MACHINE			WORKING WIDTH	WORKING DEPTH	Standard tools
mod.	cm		cm	cm	cm	nr (4×)
ZLC 100	62	50	--	100	16	20
ZLC 130	77	65	--	130	16	24
ZLC 150	87	75	--	150	16	28

(**a**) Working width of the rotary tillers of operating machines

MOD							
mod.	cod.	HP	rpm/min	type	cm	kg	
ZLC 100	1.002.656	15	540	T-40	112×60×92	130	
ZLC 130	1.002.657	20	540	T-40	142×60×92	145	
ZLC 150	1.002.658	25	540	T-40	162×60×92	160	

(**b**) Tractor power

Figure 1. Data from zanon.it.

As for the locomotion, it was decided that a pair of steering axle shafts without a braking system would be installed for the front transmission, with a square section of 40 mm and a four-hole Balilla junction [26], and that the rover would be rear-wheel drive. By placing the entire propulsion system on the rear axle, as is often the case in traditional tractors, the batteries could then be used to balance the vehicle by positioning them on the front chassis. For the rear actuators, Benevelli TX2 series motors were used (see Figure 2).

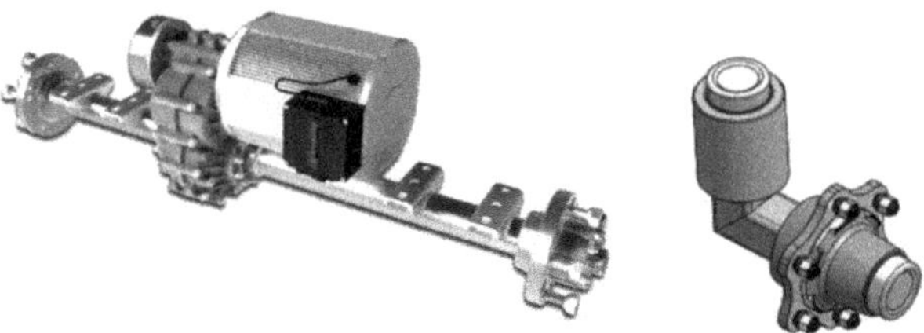

Figure 2. Rear axle and front axle shaft.

In this phase, the sizing relative to the estimation of the position of the vehicle's centre of gravity concerns the known masses and their position with respect to a reference system [27]. The position of the centre of gravity will help to calculate the weight distribution on the front and rear axles. As for the isodiamctric vehicle, the weight must be distributed evenly, with 50% on the front axle and 50% on the rear axle [28]. In our calculations, the allocation of the centre of gravity resulted in the vehicle's longitudinal axis of symmetry being 89.2 mm higher than the vertical axis and 725.4 mm further away from the longitudinal axis. Once the position of the centre of gravity was determined, the focus shifted to the weight dispersion on the axes [29]. By placing the battery pack on the front axle without any tool connected to the vehicle, the front axle would then bear 89.6 % of the weight. However, it should be borne in mind that the three-point hitch and the second engine for the power take-off (PTO) in the rear part of the vehicle still need to be evaluated [30].

Another aspect to consider is the choice of rims and tires to use for the vehicle. The European Tire and Rim Organization, a Belgian-based entity established in order to address the needs of tire manufacturers, provides specifications and harmonises sizes and nominal dimensions to avoid any ambiguity (see Table 1). However, to evaluate all the forces that are discharged on the tires, it is impossible to neglect the dynamic effects of the lifter's handling of the operating machines [31]. For this reason, in the following section, all the activities carried out to evaluate the actions on the frame caused by the presence of the operating machine are reported.

Table 1. Tyre measure and matching rim.

TIRE SIZE		SERVICE DESCRIPTION	ADDITIONAL CHARACTERISTICS			
Standard code designation						
11.5/70 -		16 135 A6 14 PR	-			
-Metric designation-						
Tire size	PR	Pattern	Diameter	Width	Max.Load (kg)	Rim size
6.50/80-12	4	KT801	604	165	500	5J12

3. Numerical Activity

For the sizing of the chassis, it is necessary not to neglect the effect of the dynamic actions of the operating machine manoeuvred during the various processes [17]. For this reason, the SimScape environment of Mathworks was used to model a generic operating machine and the lifting mechanism. These geometries, shown in Figure 3, were modelled in Solidworks and imported into the multibody modelling environment.

An MBD (model base design) approach was adopted for the development of the multibody model. In this way, starting from a simplified model of the system and the drives, it was possible to build a particularly efficient and effective model in stages in order to determine the requirements that the drive system must possess. For the numerical activity reported, an electromechanical model of the drive was developed, the parameters of which are given in Table 2. The kinematic model of the lead screw mechanism and a friction model were also added to the model, to evaluate the irreversible behaviour. The lifting system had been designed by taking into account the ISO730 directives and category 1, which include vehicles with a PTO power of up to 48 kW.

Table 2. Electric motor data.

Nominal electric engine speed	5200 rpm
Electric engine mechanical power	1200 W
Nominal CC supply voltage	48 V
Transmission Ratio	4

Furthermore, this choice was justified by the desire not to use a hydraulic system on the vehicle, a system that is generally used on traditional machines to operate the hydraulic pistons for lifting operating machines.

Figure 3. A 3D Model of the equipment attached to the three-point hitch, designed in the Solidworks environment.

Furthermore, the choice of using a worm screw drive allowed for the exploitation of the irreversibility of the mechanism itself, in order to keep the lifting mechanism in position without having to supply torque and, in this way, to save energy [32].

The mechanism's representation and its equivalent mechanical model are reported in Figure 4. The load is pushed up or down, depending on the direction of the screw's rotation. Due to the presence of the load, frictional stresses are expected to arise between the thread and the slider, which the actuator will have to overcome [33,34]. The screw is essentially characterized by the pitch and the nominal radius that define the slope of the thread according to the following relation:

$$\tan \phi = \frac{p}{\pi \cdot d} \tag{1}$$

From the equilibrium at the translation of the system schematized in Figure 4, it is possible to write the final expression as follows:

$$P = W \tan(\lambda - \phi) \tag{2}$$

where W is the load, P is the peripheral active force ($P = T/r$), and λ is the friction cone ($\mu = \tan\lambda$).

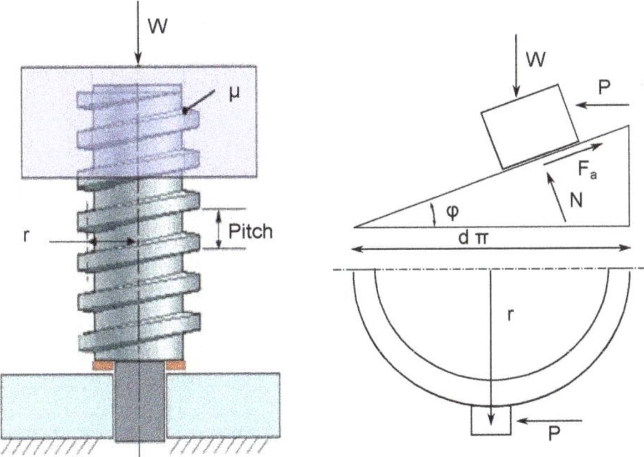

Figure 4. Representation of self-locking screw and equivalent mechanical model during down motion.

With (2), it is possible to identify the condition in which the mechanism becomes irreversible. In the case of $\phi < \lambda$, the active force P must be applied to move the load.

In order to analyse the behaviour of a more detailed model, a multibody model was developed by importing a three-dimensional CAD into the SimScape multidomain simulation environment [18,35].

The study of multibody systems consists of the analysis of the dynamic behaviour of rigid or flexible bodies interconnected by constraints, each of which can undergo large translational and rotational displacements, depending on the force fields to which they are subjected.

Figure 5 shows the 3D rendering of the electromechanical drive and its sub-model in the SimScape environment.

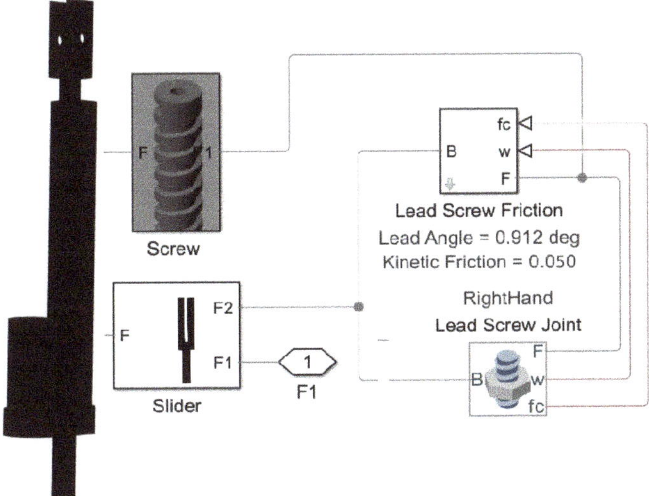

Figure 5. Multibody model of the lead screw mechanism.

The performance of this mechanism can be assessed by the following relations:

$$\eta = \frac{L_u}{L_m} = \frac{Ws\sin\phi}{Ps\cos\phi} = \frac{\tan\phi}{\tan(\phi+\lambda)} \tag{3}$$

As shown in Figure 5, the helix angle ϕ is equal to 0.912 deg, while the friction cone angle λ is equal to 2.86 deg. For these values, the mechanical efficiency in the retrograde mechanism η is equal to 0.245. Inside the lead screw friction block, visible in Figure 5, a continuous stick-slip friction model is implemented, which is used to determine the friction force and torque coefficients according to the rotation speed of the screw–nut system connected to the joint [36,37]. An irreversible drive will allow the electric motors to be used only while manoeuvring the operating machine [38].

Smaller values for the velocity threshold and the low-pass filter constant make the model more realistic but result in a numerically stiffer system. The hypothesized agricultural machine must be able to work mainly autonomously and, only in particular cases, be managed remotely [39].

Several numerical simulations were carried out to evaluate the dynamic behaviour of the electromechanical system and the effects attributable to the inertial actions on the frame [40,41].

In this paper, three different case studies are presented to evaluate the correlation between the robustness of the control system, the mechanism's coefficient of friction, and the related constraint reactions that are transferred to the frame.

The first scenario led back to an autonomous mission. The operating machine from the rest position initially had to be brought into contact with the ground to be worked, and only then could it be sunk by the amount indicated in Figure 1a.

In the second case, the activation of the stress control system was emulated during a machining operation due to an excessive resistance to advancement by the operating machine. In such cases, the lifter management system needs to raise the machine in order to free the vehicle.

Finally, the third case study aimed to test the realization of the non-back-drivable condition, which depends on the friction coefficient value in the case of a sudden change in the vehicle altitude (for example, a difference in the height of the ground). Figure 6a,b show the reference positions and the real positions occupied by the operating machine with two different PID controllers, for two of the scenarios mentioned above [42]. However, Figure 6c reports the difference in height of the ground used for testing the non-back-drivable behaviour of the mechanism. The goal for the first two simulations was to analyse the dynamic effects that would be transmitted onto the frame for two particular operating conditions, considering both the effects due to the robustness of the control system and the friction coefficient of the mechanism. The last analysis, however, had the goal of testing the effectiveness of the irreversible mechanism in the presence of variable inertial loads [43,44]. Having chosen the size of the electric motor by means of a preliminary inverse dynamic analysis, the aim of the authors was to investigate the correlation between the robustness of the PID controller and the constraint reactions that would be discharged on the frame. A control system with lower performance was indicated for the controllers with PID I, while a much more robust PID controller was considered for those with PID II.

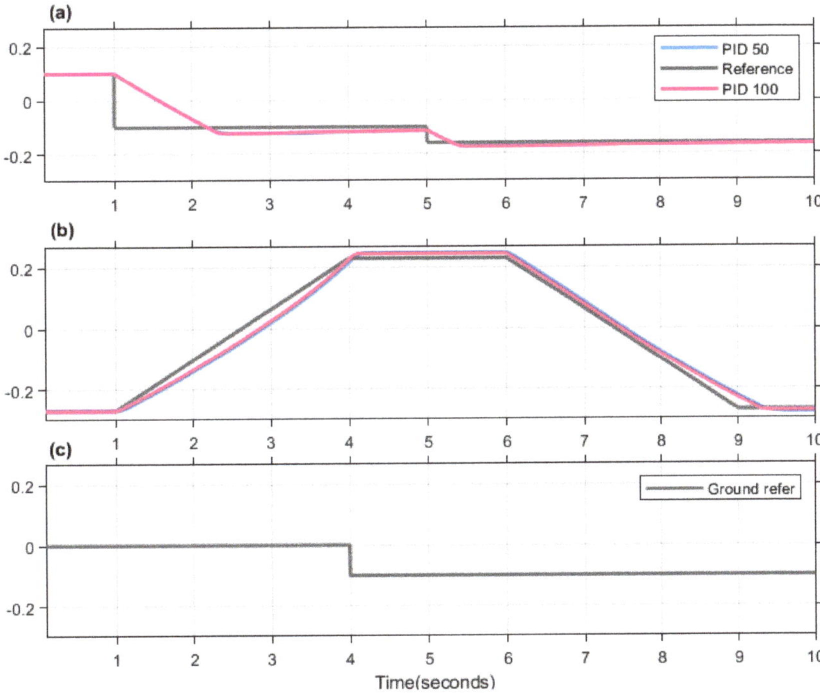

Figure 6. Reference and actual motion of the multibody model of the operating machine for: (**a**) sinking of the operating machine; (**b**) handling for effort control; (**c**) Abrupt change in the terrain profile.

4. Results

As can be seen in the first two graphs of Figure 6, the two different controllers qualitatively return the same positioning for the operating machine, both during the lowering and the beginning of the tillage (case a) and as regards the activation of the effort management system to overcome a possible obstacle present in the ground (case b).

On the other hand, this argument cannot be made when commenting on Figure 7. This image shows the constraint reactions on the longitudinal plane for the first four points indicated in Figure 3. In particular, it is possible to appreciate how the use of a more robust PID controller results in a doubling of the reaction values. This behaviour is also recorded in a decidedly significant way in the second case study; the results are shown in Figure 8. In both cases, the particularly stressed joints are the drive attachment point (see point 2 of Figure 3) and the lifting lever (see point 3 of Figure 3). Such results suggest a possible future improvement of the kinematic chain used for the lifter.

Figure 9 shows the correlation between friction and the behaviour of the mechanism in the event of a sudden change in vehicle altitude. In particular, this analysis aimed to capture the reversibility and irreversibility conditions for the chosen mechanism, depending on the friction value. In Figure 9a, a red line indicates the irreversible behaviour of the lift chosen for the electric vehicle. For a friction value of 0.017, a self-braking condition by the mechanism was identified, while for a value of 0.024, a self-locking condition was found. Furthermore, in Figure 9b, for a value of 0.015, it is possible to see the reversible behaviour of the mechanism with the lowering of the load, even before the irregularity (t > 4 s). The subsequent behaviour by the load is due to the geometric limit switches of the drive.

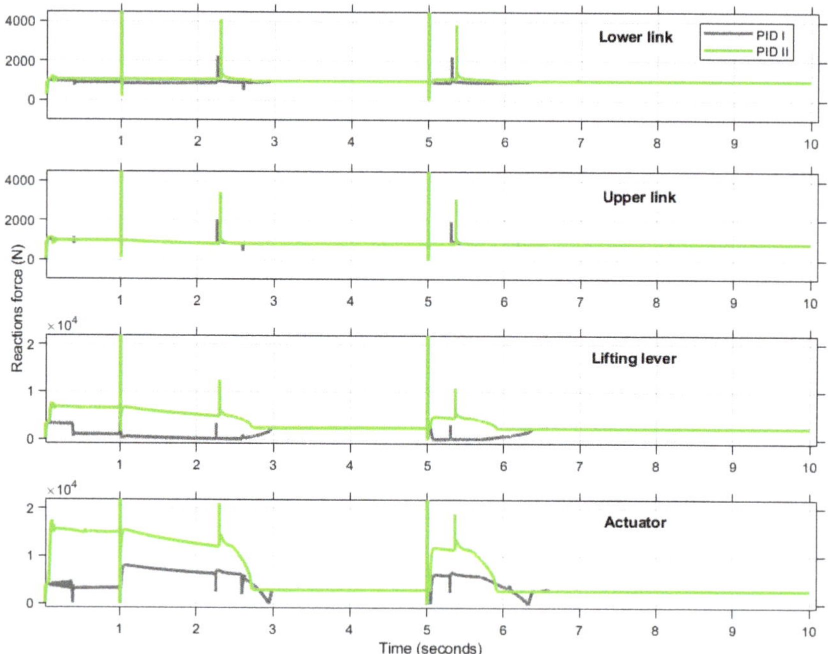

Figure 7. Correlation of reaction forces calculated with various PID.

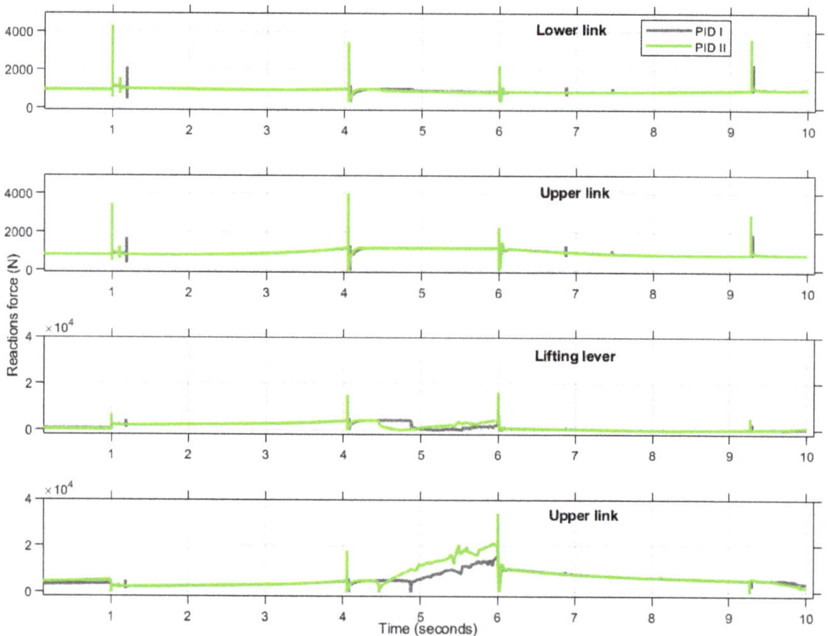

Figure 8. Correlation of the reaction forces determined with increasing dynamic friction coefficients.

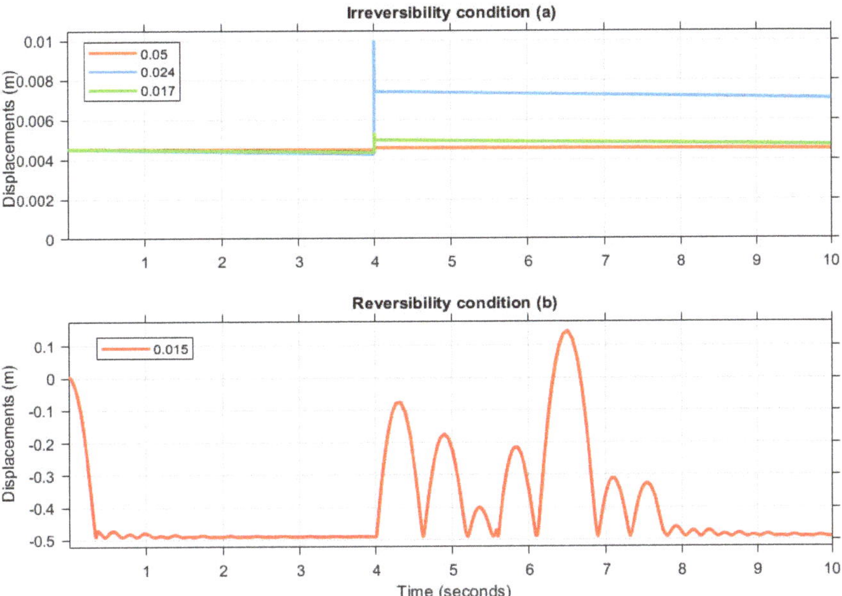

Figure 9. Correlation of the reaction forces determined with increasing dynamic friction coefficients: (**a**) Irreversibility condition; (**b**) Reversibility condition.

The self-locking condition is obtained when the angle of the friction cone is greater than the helix angle. The following self-locking factor r can be defined in the following way:

$$r = \frac{\tan(\lambda)}{\tan(\phi)} \tag{4}$$

If the self-locking factor is greater than unity, there will be no movement in static conditions. In our case, by applying (4) and considering the helix angle of the mechanism, it was possible to estimate the friction coefficient necessary for self-locking with a factor of 1.1.

$$\tan(\lambda) = 1.1 \tan(0.912) \rightarrow \mu = 0.017 \tag{5}$$

On the other hand, the automatic braking condition is verified when the friction energy that is transmitted from the nut is greater than the energy adduce from the screw. In this case, the braking factor, indicated with b, is evaluated in the following way:

$$b = 1 - \frac{\tan(\phi - \lambda)}{\tan(\phi)} \tag{6}$$

In this case study, b must be equal to 1.5 for a low feed rate [45].

$$1.5 = 1 - \frac{\tan(0.912 - \lambda)}{\tan(0.912)} \rightarrow \lambda = 1.37 \rightarrow \mu = 0.024 \tag{7}$$

It is important to always keep in mind that the self-locking action (static condition) does not necessarily include the self-braking action (dynamic condition). In Figure 9, the cases of irreversible condition and reversible condition are illustrated.

To summarize what has been highlighted, the areas of reversibility and irreversibility as a function of the thread angle and the friction coefficient of the mechanism were graphically represented. Figure 10 illustrates the so-defined stability map of the mechanism, which shows both the helix angle values and the friction value used (metal contact on lubricated metal) for this application as well as the conditions shown in the previous graph. On the stability map, it is possible to see how the mechanical performance changes as a function of the two parameters mentioned above. Furthermore, this map is useful in the eventual redesign of the actuation device, which will modify the dynamic behaviour of the lifter [46].

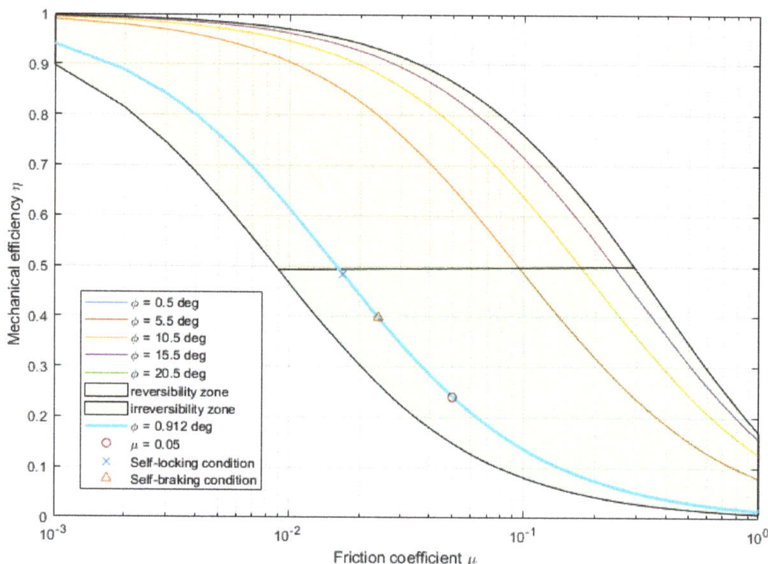

Figure 10. Correlation of the reaction forces determined with increasing dynamic friction coefficients.

The preliminary concept of the electric tractor is shown in Figure 11. However, the results exhibited in this work and a subsequent phase of geometry optimization will significantly influence the final shape of the vehicle.

Figure 11. A 3D rendering of the concept.

5. Conclusions

This paper reports the preliminary phase of a much larger project, which is focused on the development of a new autonomous electric tractor for greenhouse and open-field applications.

The use of virtual modelling and prototyping techniques together with the low cost of electronic components makes it possible to significantly lower both the cost and the development time of a new machine. It also completely breaks down the technology gaps that have precluded less industrialized countries from competing with industrialized countries on equal footing.

After identifying the basic components for the construction of the vehicle, such as the chassis, tires, transmission, engine, battery, and lift, the authors focused on the reference standard for the sizing and design of the vehicle in order to estimate the centre of gravity of the whole system. The rover's centre of gravity remained the most sensitive variable needed to proceed with the dynamic analyses. To estimate the centre of gravity, it was necessary to evaluate the constraint reactions transmitted onto the frame by the lifter and the power take-off because the centre of gravity was positioned too far forward, and therefore the weight distribution between the two axes was unbalanced (front and rear). For this reason, the study focused on the rear part of the tractor, and in particular, on the lifting mechanism, which had been designed and sized. To evaluate the reactions transmitted to the frame by the lifter, the system was modelled and simulated in Mathworks' SimScape multidomain simulation environment. Inverse dynamic analyses made it possible to identify the characteristics required of the electric motors. The direct dynamic analysis, on the other hand, made it possible to test the reversible and irreversible behaviour of the lifter for an optimal component design. The results were essential for the subsequent chassis design phase and for the final vehicle construction phase. The irreversible electromechanical drive proposed by the authors was designed for a newly developed all-electric or hybrid tractor. However, nothing detracts from the fact that such a solution could also be thought of for the retrofitting of machines that are already on the market. Such a scenario, although unlikely given that hydraulic circuits are also used to operate the steering and power external users in conventional tractors, would require a retrofit of the machine itself. In such a case, it would be necessary to equip the vehicle with the appropriate energy accumulators to ensure lifter manoeuvrability. It would also be necessary to install a microcontroller that is capable of handling the drive. Such a solution could be part of a retrofitting process for obsolete machines, which would extend the useful life of machines destined for decommissioning. On the other hand, using an irreversible mechanism on tractors that use hydraulic systems would make the machine fail-safe for the circuit itself.

Author Contributions: Conceptualization, M.C.D.S. and S.V.; validation, M.C.D.S. and D.G. All authors have read and agreed to the published version of the manuscript.

Funding: This research received no external funding.

Institutional Review Board Statement: Not applicable.

Informed Consent Statement: Not applicable.

Conflicts of Interest: The authors declare no conflict of interest.

References

1. Mehla, A.; Deora, S. Use of Machine Learning and IoT in Agriculture. In *EAI/Springer Innovations in Communication and Computing*; Springer: Berlin/Heidelberg, Germany, 2022; pp. 277–293. [CrossRef]
2. De Simone, M.; Celenta, G.; Rivera, Z.; Guida, D. Mechanism Design for a Low-Cost Automatic Breathing Applications for Developing Countries. *Lect. Notes Netw. Syst.* **2022**, *472*, 345–352. [CrossRef]
3. Vitlox, O. Technological Developments in Agricultural Machinery. *Eur. J. Mech. Environ. Eng.* **1983**, *30*, 19–25.
4. In Proceedings of the 11th International Symposium on Farm Machinery and Processes Management in Sustainable Agriculture, FMPMSA 2022, Bari, Italy, June 2023; Lecture Notes in Civil Engineering; 2022; Volume 289.
5. De Simone, M.; Rivera, Z.; Guida, D. Obstacle avoidance system for unmanned ground vehicles by using ultrasonic sensors. *Machines* **2018**, *6*, 18. [CrossRef]
6. Olalla, E.; Cadena-Lema, H.; Domínguez Limaico, H.; Nogales-Romero, J.; Zambrano, M.; Vásquez Ayala, C. Irrigation Control System Using Machine Learning Techniques Applied to Precision Agriculture (Internet of Farm Things IoFT). *Lect. Notes Netw. Syst.* **2022**, *512*, 329–342. [CrossRef]
7. Celenta, G.; De Simone, M. Retrofitting Techniques for Agricultural Machines. *Lect. Notes Netw. Syst.* **2020**, *128*, 388–396. [CrossRef]
8. Casillo, M.; Colace, F.; Lorusso, A.; Marongiu, F.; Santaniello, D. An IoT-Based System for Expert User Supporting to Monitor, Manage and Protect Cultural Heritage Buildings. *Stud. Comput. Intell.* **2022**, *1030*, 143–154. [CrossRef]

9. Di Filippo, A.; Lombardi, M.; Marongiu, F.; Lorusso, A.; Santaniello, D. Generative design for project optimization. In *DMSVIVA*; KSI Research: Pittsburgh, PA, USA, 2021; pp. 110–115.

10. Casillo, M.; Gupta, B.; Lombardi, M.; Lorusso, A.; Santaniello, D.; Valentino, C. Context Aware Recommender Systems: A Novel Approach Based on Matrix Factorization and Contextual Bias. *Electronics* **2022**, *11*, 1003. . [CrossRef]

11. La Regina, R.; Pappalardo, C.; Guida, D. Dynamic Analysis and Attitude Control of a Minisatellite. *Lect. Notes Netw. Syst.* **2022**, *472*, 244–251. [CrossRef]

12. Sato, S.; Jiang, Y.; Russell, R.; Miller, J.; Karavalakis, G.; Durbin, T.; Johnson, K. Experimental driving performance evaluation of battery-powered medium and heavy duty all-electric vehicles. *Int. J. Electr. Power Energy Syst.* **2022**, *141*, 108100. [CrossRef]

13. Ghobadpour, A.; Monsalve, G.; Cardenas, A.; Mousazadeh, H. Off-Road Electric Vehicles and Autonomous Robots in Agricultural Sector: Trends, Challenges, and Opportunities. *Vehicles* **2022**, *4*, 843–864. [CrossRef]

14. Manrique-Escobar, C.; Pappalardo, C.; Guida, D. A multibody system approach for the systematic development of a closed-chain kinematic model for two-wheeled vehicles. *Machines* **2021**, *9*, 245. [CrossRef]

15. De Simone, M.; Laiola, V.; Rivera, Z.; Guida, D. Dynamic Analysis of a Hybrid Heavy-Vehicle. *Lect. Notes Netw. Syst.* **2022**, *472*, 236–243. [CrossRef]

16. Pappalardo, C.; Vece, A.; Galdi, D.; Guida, D. Developing a Reciprocating Mechanism for the Emergency Implementation of a Mechanical Pulmonary Ventilator using an Integrated CAD-MBD Procedure. *FME Trans.* **2022**, *50*, 238–247. [CrossRef]

17. De Simone, M.; Ventura, G.; Lorusso, A.; Guida, D. Attitude Controller Design for Micro-satellites. *Lect. Notes Netw. Syst.* **2021**, *233*, 21–31. [CrossRef]

18. Curcio, M.; Pappalardo, C.; Guida, D. Multibody Modeling and Dynamical Analysis of a Fixed-Wing Aircraft. *Lect. Notes Netw. Syst.* **2022**, *472*, 77–84. [CrossRef]

19. Sicilia, M.; De Simone, M. Development of an Energy Recovery Device Based on the Dynamics of a Semi-trailer. In *Lecture Notes in Mechanical Engineering*; Springer: Cham, Switzerland, 2020; pp. 74–84. [CrossRef]

20. Pappalardo, C.; Lettieri, A.; Guida, D. Identification of a Dynamical Model of the Latching Mechanism of an Aircraft Hatch Door using the Numerical Algorithms for Subspace State-Space System Identification. *IAENG Int. J. Appl. Math.* **2021**, *51*, 1–14.

21. Shanmugasundar, G.; Sivaramakrishnan, R.; Venugopal, S. Modeling, design and static analysis of seven degree of freedom articulated inspection robot. *Adv. Mater. Res.* **2013**, *655–657*, 1053–1056. [CrossRef]

22. Menon, A.; Prabhakar, M. Intelligent IoT-Based Monitoring Rover for Smart Agriculture Farming in Rural Areas. *Lect. Notes Netw. Syst.* **2023**, *401*, 619–630. [CrossRef]

23. Botta, A.; Moreno, E.; Baglieri, L.; Colucci, G.; Tagliavini, L.; Quaglia, G. Autonomous Driving System for Reversing an Articulated Rover for Precision Agriculture. *Mech. Mach. Sci.* **2022**, *120 MMS*, 412–419. [CrossRef]

24. Henry, D.; Aubert, H.; Galaup, P.; Veronese, T. Dynamic Estimation of the Yield in Precision Viticulture from Mobile Millimeter-Wave Radar Systems. *IEEE Trans. Geosci. Remote Sens.* **2022**, *60*, 4704915. [CrossRef]

25. Ribeiro, J.; Gaspar, P.; Soares, V.; Caldeira, J. Computational Simulation of an Agricultural Robotic Rover for Weed Control and Fallen Fruit Collection-Algorithms for Image Detection and Recognition and Systems Control, Regulation, and Command. *Electronics* **2022**, *11*, 790. [CrossRef]

26. Kuska, M.; Heim, R.; Geedicke, I.; Gold, K.; Brugger, A.; Paulus, S. Digital plant pathology: A foundation and guide to modern agriculture. *J. Plant Dis. Prot.* **2022**, *129*, 457–468. [CrossRef] [PubMed]

27. Netthonglang, C.; Jongrukchob, T.; Thongtan, T.; Bairaksa, J. Real-time and online post-processing kinematic positioning services. In Proceedings of the 2022 19th International Conference on Electrical Engineering/Electronics, Computer, Telecommunications and Information Technology, Prachuap Khiri Khan, Thailand, 24–27 May 2022. [CrossRef]

28. Dobretsov, R.; Dobretsova, S.; Voinash, S.; Shcherbakov, A.; Dolmatov, S.; Sokolova, V.; Taraban, V.; Alekseeva, S.; Taraban, M. Elements of the mathematical support for the design of an autonomous tractor. *IOP Conf. Ser. Earth Environ. Sci.* **2021**, *723*, 032039. [CrossRef]

29. Nunez-Quispe, J.; Lleren-Sernaque, J.; Lara-Chavez, E. Mechanical Design of a ROVER prototype for Exploration tasks on Mars: Structural and Transient Dynamics simulation analysis. In Proceedings of the 2021 IEEE MIT Undergraduate Research Technology Conference (URTC), Cambridge, MA, USA, 8–10 October 2021. [CrossRef]

30. Manrique-Escobar, C.; Pappalardo, C.; Guida, D. On the Analytical and Computational Methodologies for Modelling Two-wheeled Vehicles within the Multibody Dynamics Framework: A Systematic Literature Review. *J. Appl. Comput. Mech.* **2022**, *8*, 153–181. [CrossRef]

31. Renius, K.T. *Fundamentals of Tractor Design*. Springer: Cham, Switzerland, 2020.

32. Cho, M.S.; Hwang, H.S.; Lee, M.H.; Kim, B.; Zinn, M.R. A screwjack mechanism based separation device driven by a piezo actuator. *Int. J. Precis. Eng. Manuf.* **2012**, *13*, 2079–2082. [CrossRef]

33. Gallina, P. Vibration in screw jack mechanisms: Experimental results. *J. Sound Vib.* **2005**, *282*, 1025–1041. [CrossRef]

34. Liguori, A.; Armentani, E.; Bertocco, A.; Formato, A.; Pellegrino, A.; Villecco, F. Noise reduction in spur gear systems. *Entropy* **2020**, *22*, 1306. [CrossRef]

35. Cammarata, A.; Maddío, P.D. A system-based reduction method for spatial deformable multibody systems using global flexible modes. *J. Sound Vib.* **2021**, *504*, 116118. [CrossRef]

36. De Simone, M.; Guida, D. Modal coupling in presence of dry friction. *Machines* **2018**, *6*, 8. [CrossRef]

37. Formato, A.; Ianniello, D.; Pellegrino, A.; Villecco, F. Vibration-based experimental identification of the elastic moduli using plate specimens of the olive tree. *Machines* **2019**, *7*, 46. [CrossRef]
38. Noormohamed, A.; Mercan, O.; Ashasi-Sorkhabi, A. Optimal active control of structures using a screw jack device and open-loop linear quadratic gaussian controller. *Front. Built Environ.* **2019**, *5*, 43. [CrossRef]
39. Formato, A.; Romano, R.; Villecco, F. A Novel Device for the Soil Sterilizing in Sustainable Agriculture. *Lect. Notes Netw. Syst.* **2021**, *233*, 858–865. [CrossRef]
40. Shanmugasundar, G.; Sivaramakrishnan, R. Design and analysis of a newly developed seven degree of freedom robot for inspection. *Int. J. Control Theory Appl.* **2016**, *9*, 393–402.
41. Cammarata, A.; Sinatra, R.; Maddìo, P.D. Static condensation method for the reduced dynamic modeling of mechanisms and structures. *Arch. Appl. Mech.* **2019**, *89*, 2033–2051. [CrossRef]
42. Araki, Y.; Asai, T.; Kimura, K.; Maezawa, K.; Masui, T. Nonlinear vibration isolator with adjustable restoring force. *J. Sound Vib.* **2013**, *332*, 6063–6077. [CrossRef]
43. Thesiya, D.; Srinivas, A.; Shukla, P. A novel axial foldable mechanism for a segmented primary mirror of space telescope. *J. Astron. Space Sci.* **2015**, *32*, 269–279. [CrossRef]
44. Li, T.; Kou, Z.; Wu, J.; Yahya, W.; Villecco, F. Multipoint Optimal Minimum Entropy Deconvolution Adjusted for Automatic Fault Diagnosis of Hoist Bearing. *Shock Vib.* **2021**, *2021*, 6614633. [CrossRef]
45. *Manuale Degli Organi Delle Macchine*; Tecnologie Industriali: Padova, Italy, 2006.
46. Sun, X.; Liu, H.; Song, W.; Villecco, F. Modeling of eddy current welding of rail: Three-dimensional simulation. *Entropy* **2020**, *22*, 947. [CrossRef]

 *actuators*

Article

Parameter Optimization of Large-Size High-Speed Cam-Linkage Mechanism for Kinematic Performance

Guodong Zhu [1], Yong Wang [1,*], Guo-Niu Zhu [2,*], Minghao Weng [1], Jianhui Liu [3], Ji Zhou [3] and Bing Lu [3]

[1] School of Mechanical Engineering, Hefei University of Technology, Hefei 230009, China
[2] School of Mechanical and Aerospace Engineering, Nanyang Technological University, Singapore 639798, Singapore
[3] Zhejiang Xinyuhong Intelligent Equipment Co., Ltd., Hangzhou 311115, China
* Correspondence: ywang9868@163.com (Y.W.); guoniu.zhu@gmail.com (G.-N.Z.)

Abstract: The cam-linkage mechanism is a typical transmission mechanism in mechanical science and is widely used in various automated production equipment. However, conventional modeling methods mainly focus on the design and dimensional synthesis of the cam-linkage mechanism in the slow-speed scenario. The influence of component dimensions is not taken into consideration. As a result, the model accuracy dramatically falls when analyzing large-size cam-linkage mechanisms, especially in high-speed environments. The kinematic aspects of cam design have been investigated, but there are few studies discussing the motion characteristic and accuracy analysis models of the large-size cam-linkage mechanism under high-speed scenarios. To handle such issues, this paper proposes a parameter optimization methodology for the design analysis of the large-size high-speed cam-linkage mechanism considering kinematic performance. Firstly, the mathematical model of the cam five-bar mechanism is presented. The cam curve and motion parameters are solved forward with linkage length and output speed. Then, a particle swarm-based multi-objective optimization method is developed to find the optimal structure parameters and output speed curve to minimize cam pressure angle and roller acceleration and maximize linkage mechanism drive angle. A Monte Carlo-based framework is put forward for the reliability and sensitivity analysis of kinematic accuracy. Finally, a transverse device of a sanitary product production line is provided to demonstrate the applicability of the proposed method. With the parameter optimization, the productivity of the transverse device is doubled, from 600 pieces per minute (PPM) to 1200 PPM.

Keywords: cam-linkage mechanism; parameter optimization; reliability analysis; kinematics; transverse device

Citation: Zhu, G.; Wang, Y.; Zhu, G.-N.; Weng, M.; Liu, J.; Zhou, J.; Lu, B. Parameter Optimization of Large-Size High-Speed Cam-Linkage Mechanism for Kinematic Performance. *Actuators* **2023**, *12*, 2. https://doi.org/10.3390/act12010002

Academic Editor: Zhuming Bi

Received: 24 October 2022
Revised: 16 December 2022
Accepted: 18 December 2022
Published: 21 December 2022

1. Introduction

As a combination of the cam and linkage mechanisms, the cam-linkage mechanism is one of the most popular transmission mechanisms in mechanical science. By integrating the merits of both kinds of mechanisms, the cam-linkage mechanism can achieve superior kinematic performance while maintaining high reliability and compact structures. Due to such prominent abilities in realizing complex motion laws, the cam-linkage mechanism has been widely used in various machinery and automatic production equipment, such as textile machinery [1], packaging machines [2], rehabilitation devices [3], bionic horse robots [4], and parallel manipulators [5].

In view of the significant role of the cam-linkage mechanism in mechanical transmission, lots of studies have been proposed to investigate the design optimization of the cam and connecting rod. For example, Rybansky et al. [6] studied the topological optimization of the internal shape of the biaxial spring cam mechanism. The weight/stiffness trade-off in the cam design was investigated during the topology optimization. Abderazek et al. [7] discussed the motion law of the disk cam mechanism with a roller follower. Li et al. [8] generated the design of a mold substructure with cams in the form of an assembly. In addition to the cam optimization, Zhang et al. [9] designed a 1-DOF (degree of freedom) cam-linked bi-parallelogram mechanism. This mechanism was applied to a double-deck parking system. Wu et al. [10] presented a new robot with a five-bar spatial linkage design form. The robot has the advantage of a larger working space. For the cam profile that people are generally concerned about, Arabaci et al. [11] proposed a dimensionless design method for a double-arc cam mechanism. The motion equation of the cam profile was obtained during the design process. Moreover, Xia et al. [12] and Ouyang et al. [13] constructed the optimization design model of the new cam profile. Chen et al. [14] investigated the X- and Y-shaped cam profiles. It was found that the cam linkage polishing device has a bicircular polishing trajectory with zero velocity deviation.

Chang et al. [15] developed a design method and an optimization model for the rotational balance of disc cams. Li et al. [16] constructed a rehabilitation device based on the six-bar linkage mechanism. A novel optimization algorithm was proposed to obtain the optimal structural design parameters. Furthermore, the compactness and stability of the mechanism are currently key concerns in academic circles. Yang et al. [17,18] proposed a new coaxial cam mechanism, which consists of conjugate cams and parallelogram linkage. Its structure is more compact and can be used in high-speed working conditions. Radaelli et al. [19] applied the compliant revolute joint to the linkage mechanism. The mechanical stability and performance were significantly improved. Wang et al. [20] developed a cam angular velocity model for a high-pressure oil pump system. The high stability of the mechanism was determined.

Although these studies have provided a number of impressive techniques for contour design, motion analysis, pressure angle calculation, and overall dimensional optimization of the traditional cam-linkage mechanism, most of them focus mainly on design analysis in the slow-speed environment. The influence of component dimensions is not taken into account. However, when oriented to high-speed operation, the kinematics performance of the cam-linkage mechanism differs significantly from low- or normal-speed scenarios. Moreover, the high-speed conditions bring significant challenges to the stable operation and fatigue life of the mechanism. The parameter optimization of the large-size high-speed cam-linkage mechanism remains to be resolved. There is scant research discussing the multi-objective optimization of the cam-linkage mechanism under such challenging environments.

To bridge this gap, this paper proposes a parameter optimization method for a large-size high-speed cam-linkage mechanism considering kinematic performance. Specifically, the cam five-bar mechanism is introduced as an example. First of all, the modeling analysis of the cam five-bar mechanism is presented. Then, the multi-objective optimization of the cam five-bar mechanism is investigated under high-speed scenarios. Finally, the reliability and sensitivity analysis is conducted to investigate the kinematic performance of the optimized structure. The main contributions of this paper are as follows.

(1) A mathematical model is constructed to determine the performance parameters of the cam five-bar mechanism. The motion characteristics of the mechanism are obtained by resolving the mathematical model.

(2) A multi-objective optimization method for a large-size cam-linkage mechanism is proposed. The optimal kinematic parameters are identified by solving the optimization problems.

(3) A computer-aided platform is developed for the design analysis of the cam five-bar mechanism. The parameter calculation, optimization, and reliability analysis are well-handled with the aid of the software package.

(4) A real-world case study of the transverse device is put forward to demonstrate the effectiveness of the proposed method. The productivity of the transverse device is substantially improved.

The rest of this paper is structured as follows. Section 2 outlines the modeling analysis of the cam five-bar mechanism. Section 3 proposes the multi-objective optimization method. Section 4 presents the reliability analysis of kinematic accuracy. Section 5 provides the model validation and discussion. Finally, the paper is concluded in Section 6.

2. Cam Five-Bar Mechanism

2.1. Mechanism Principle

Compared with traditional cam-linkage mechanisms, the compactness of the large-size cam-linkage mechanism is poor. The fatigue damage and wear also differ significantly. In this study, we take the cam five-bar mechanism as an example to investigate the design optimization of the large-size cam-linkage mechanism, which is shown in Figure 1.

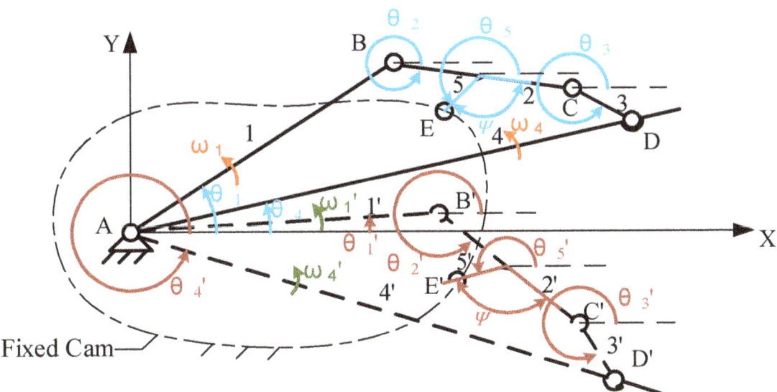

Figure 1. Schematic diagram of the cam five-bar mechanism.

In general, the objective of the cam five-bar mechanism is to achieve a controlled change of the rotational angular speed with the rotation angle. As illustrated in Figure 1, the cam five-bar mechanism is composed of one cam and five bars. Rod 1 is the prime mover, whose angular velocity is constant. Rod 4 is the output member, whose angular velocity is supposed to meet working requirements. In addition, rods 2 and 5 are fixedly connected. Rods 2 and 3 generate the triangle BCD. During the system operation, the angle BCD would be changed if the motion state of rod 2 is adjusted through the cam. Correspondingly, the distance between B and D is also changed. As a result, the movement of the output member (i.e., rod 4) can be determined by adjusting the distance between B and D.

Theoretically, the output angular velocity and cam shape can be uniquely determined if the angular velocity of rod 1 and the length of each rod are known. This process of deriving the output velocity and cam shape from the given input angular velocity and rod structure is known as a positive solution. By contrast, the input and output angular velocities are often given in real applications, whereas the linkage mechanism characteristics and cam shape need to be resolved. This process is known as inverse solving. Compared with the positive solution, reverse solving may produce multiple solutions in which the motion characteristics of different structures can vary significantly. For example, some cam shapes may generate certain points where the pressure angle is too large, resulting in uneven motion and poor forces. Similarly, some linkage structures may produce points where the transmission angle is too small or even dead in motion. Thus, it is quite challenging to identify excellent motion characteristics while performing inverse solving.

2.2. Mathematical Model

As shown in Figure 1, a right-angle coordinate system Oxy is established to facilitate the theoretical analysis. The center of rotation of the prime mover (point A) is set as the origin (i.e., O). Suppose l_1, l_2, l_3, l_4, and l_5 are the lengths of the corresponding rods. ω_1 is the angular velocity of the prime mover. θ_1 is the angle of rotation. ω_4 and θ_4 are the angular velocity and angle of rotation to be satisfied by the follower, respectively. The vector equation is obtained as:

$$l_1 + l_2 = l_3 + l_4 \tag{1}$$

where l_1, l_2, l_3, and l_4 are vectors of magnitude l_1, l_2, l_3, and l_4, respectively.

By expressing Equation (1) in complex form and expanding it according to Euler's formula, it can be written as:

$$\begin{bmatrix} cos\theta_1 & cos\theta_2 \\ sin\theta_1 & sin\theta_2 \end{bmatrix} \begin{bmatrix} l_1 \\ l_2 \end{bmatrix} = \begin{bmatrix} cos\theta_3 & cos\theta_4 \\ sin\theta_3 & sin\theta_4 \end{bmatrix} \begin{bmatrix} l_3 \\ l_4 \end{bmatrix} \tag{2}$$

Then, the rotation angle of each rod is obtained as:

$$\theta_3 = 2arctan\frac{B_1 \pm \sqrt{D_1}}{A_1 - C_1} + \theta_4 \tag{3}$$

$$\theta_2 = arctan\frac{B_1 + l_3 \sin(\theta_3 - \theta_4)}{A_1 + l_3 \cos(\theta_3 - \theta_4)} + \theta_4 \tag{4}$$

$$\theta_5 = 360 + \theta_2 - \psi \tag{5}$$

where $A_1 = l_4 - l_1 cos(\theta_1 - \theta_4)$, $B_1 = -l_1 sin(\theta_1 - \theta_4)$, $C_1 = \left(A_1^2 + B_1^2 + l_3^2 - l_2^2\right)/(2l_3)$, and $D_1 = A_1^2 + B_1^2 - C_1^2$.

To find the final solution, i.e., the trajectories of points A, B, C, D, and E, the values of l_1, l_2, l_3, l_4, l_5, Ψ, ω_4, and ω_1 and the initial value of θ_{14} (the angle between rod 1 and rod 4) need to be determined. In real practice, ω_4 is usually given as ω_{41} and ω_{42}, which is shown in Figure 2. The connection curve between ω_{41} and ω_{42} can have many choices. The angular velocity ω_4 is completely determined only after selecting the connection curve concerning the actual working condition requirement.

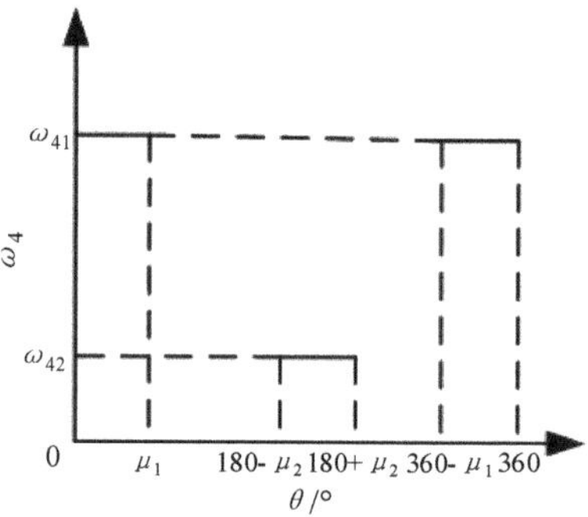

Figure 2. The desired motion law of the output member.

Thus, the coordinates of each point in the motion of the cam five-bar mechanism are determined as:

$$\begin{bmatrix} x_b \\ y_b \\ x_c \\ y_c \\ x_d \\ y_d \\ x_e \\ y_e \end{bmatrix} = \begin{bmatrix} cos\theta_1 & 0 & 0 & 0 \\ sin\theta_1 & 0 & 0 & 0 \\ cos\theta_1 & cos\theta_2 & 0 & 0 \\ sin\theta_1 & sin\theta_2 & 0 & 0 \\ 0 & 0 & cos\theta_4 & 0 \\ 0 & 0 & sin\theta_4 & 0 \\ cos\theta_1 & 0 & 0 & cos\theta_5 \\ sin\theta_1 & 0 & 0 & sin\theta_5 \end{bmatrix} \begin{bmatrix} l_1 \\ l_2 \\ l_3 \\ l_4 \end{bmatrix} \tag{6}$$

where x_i and y_i ($i = b,c,d,e$) are the horizontal and vertical coordinates of points B, C, D, and E, respectively.

By solving Equation (6), the following mechanism motion characteristics can be obtained.

(1) Cam theoretical profile curve:

From the above analysis, it can be concluded that the trajectory of point E is the theoretical contour curve of the cam.

(2) Acceleration at point E during the motion of the mechanism:

$$a_e = \sqrt{\left(x_e''\right)^2 + \left(y_e''\right)^2} \tag{7}$$

where x_e'' and y_e'' are the horizontal and vertical accelerations of point E, respectively. a_e is the acceleration at point E.

(3) Transmission angle:

$$\gamma = \begin{cases} \varphi_1, & \varphi_1 \leq 90^\circ \\ 180 - \varphi_1, & \varphi_1 > 90^\circ \end{cases} \tag{8}$$

where φ_1 is the angle between rods 2 and 3.

(4) Cam pressure angle:

$$\alpha = \begin{cases} |\beta_1 - \beta_2|, & |\beta_1 - \beta_2| \leq 90^\circ \\ 180 - |\beta_1 - \beta_2|, & |\beta_1 - \beta_2| > 90^\circ \end{cases} \tag{9}$$

where $\beta_1 = arctan\frac{y_e - y_d}{x_e - x_d}$ and $\beta_2 = arctan\frac{dy_e}{dx_e}$.

3. Multi-Objective Optimization Method

After determining the motion characteristics and the cam profile curve, a particle swarm-based multi-objective optimization method is proposed to determine the optimal solution for the inverse solving, which is conducted as follows.

3.1. Selection of Optimization Variables

In light of the mathematical analysis of the cam five-bar mechanism, the initial values of θ_{14}, ω_4, and rod lengths l_1, l_2, l_3, and l_5 are selected as optimization variables to enhance the motion characteristics of the mechanism.

First of all, a combination of five straight lines and four curves is introduced for the representation of ω_4, which is depicted in Figure 3. Specifically, the angular velocity of the $0{\sim}T_1$, $T_4{\sim}T_6$, and $T_9{\sim}T_{10}$ periods are given in advance. Therefore, the remaining curve is divided into two parts. The first part is the $T_2{\sim}T_3$ and $T_7{\sim}T_8$ periods, while the second is the period of $T_1{\sim}T_2$, $T_3{\sim}T_4$, $T_6{\sim}T_7$, and $T_8{\sim}T_9$. In particular, the angular velocity keeps constant in the first part, whereas the second part (i.e., the buffer section) introduces a motion law curve as the corresponding angular velocity curve. The integration of ω_4 in the 0 to T_{10} period is 360°.

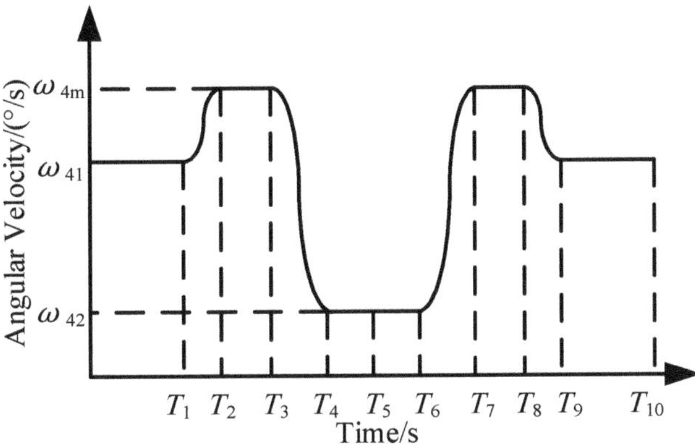

Figure 3. Angular velocity curve of the output component.

As illustrated in Figure 3, the first part of the curve can be considered symmetric about $t = T_5$. Then, the time parameters are determined as:

$$\begin{bmatrix} T_1 \\ T_4 \\ T_6 \\ T_7 \\ T_8 \\ T_9 \end{bmatrix} = \begin{bmatrix} \frac{1}{\omega_{41}} & 0 & 0 & 0 \\ 0 & -\frac{1}{\omega_{42}} & 0 & 0 \\ 0 & \frac{1}{\omega_{42}} & 0 & 0 \\ 0 & \frac{1}{\omega_{42}} & 0 & -1 \\ 0 & 0 & -1 & 0 \\ -\frac{1}{\omega_{41}} & 0 & 0 & 0 \end{bmatrix} \begin{bmatrix} \mu_1 \\ \mu_2 \\ T_2 \\ T_3 \end{bmatrix} + \begin{bmatrix} 0 \\ \frac{180}{\omega_{41}} \\ \frac{180}{\omega_1} \\ \frac{180}{\omega_1} \\ \frac{360}{\omega_1} \\ \frac{360}{\omega_1} \end{bmatrix} \tag{10}$$

In addition, the curve of the buffer section is chosen as a centrosymmetric curve to simplify the calculation. The rotation angle of the output member 4 can be determined by Equation (11).

$$\omega_{41} T_2 + \frac{1}{2}(\omega_{4m} - \omega_{41})(T_2 - T_1) + \omega_{4m}(T_3 - T_2) + \frac{1}{2}(\omega_{4m} - \omega_{42})(T_4 - T_3) + \omega_{42}(T_5 - T_3) = 180^\circ \tag{11}$$

By solving Equation (11), ω_{4m} is obtained as:

$$\omega_{4m} = \frac{360 - \omega_{41}(T_1 + T_2) + \omega_{42}(T_3 + T_4 - 2T_5)}{T_3 + T_4 - T_2 - T_1} \tag{12}$$

For the second part of the curve, the commonly-used follower motion law curves are modified sine, modified iso-velocity, and quintuple polynomials [21]. As the first-order derivative of the dwell-free modified iso-velocity curve is too small, the dwell-free revised iso-velocity curves are used for the buffer section.

In sum, the curve of ω_4 can be linearly represented by T_2 and T_3. Therefore, the final optimization variable x can be determined as $x = (l_1, l_2, l_3, l_5, \theta_{14}, T_2, T_3)^{\mathrm{T}}$.

3.2. Constraint Establishment

When rods 2 and 3 are in a straight line, the mechanism is in the limit state. At this time, the angle between rods 1 and 4 is expressed as $\theta_{\max}$. The following constraint should be satisfied:

$$\sqrt{l_1^2 + l_4^2 - 2l_1 l_4 \cos\theta_{max}} \leq l_2 + l_3 \tag{13}$$

$$T_2 - T_3 < 0 \tag{14}$$

Moreover, the following constraint should be satisfied for the large-size cam-linkage mechanism:

$$\begin{cases} 100 \leq l_1 \leq 400 \\ 150 \leq l_2 + l_3 \leq 400 \\ 0^\circ \leq \theta_{14i} \leq 10^\circ \end{cases} \tag{15}$$

where θ_{14i} is the initial value of the angle between rods 1 and 4.

3.3. Model for the Optimal Design

Based on the above analysis, the objective of the optimization is to maximize the transmission angle of the five-bar mechanism, minimize the cam pressure angle, and minimize the acceleration at point E, which is represented as:

$$\min[a_{Emax}, \alpha_{max}, -\gamma_{min}] \tag{16}$$

Then, the optimization model is obtained as:

$$\begin{cases} \min\left[a_{Emax}, \alpha_{max}, -\gamma_{min}\right] \\ s.t. \ \sqrt{l_a^2 + l_d^2 - 2l_a l_d \cos\theta_{max}} \le l_b + l_c \\ T_3 - T_2 \ge 0 \\ 100 \le l_1 \le 400 \\ 150 \le l_2 + l_3 \le 400 \\ 0° \le \theta_{14i} \le 10° \end{cases} \tag{17}$$

Finally, the particle swarm optimization (PSO) algorithm is introduced to resolve the optimization problem, which is described as:

$$v_i^{k+1} = wv_i^k + c_1 rand_1\left(p_i^k - x_i^k\right) + c_2 rand_2\left(p_g^k - x_i^k\right) \tag{18}$$

$$x_i^{k+1} = x_i^k + v_i^{k+1} \tag{19}$$

where w is the inertia weight, c_1 and c_2 are positive constants, and $rand_1$ and $rand_2$ are two random numbers in the [0, 1] interval. x_i^k and v_i^k are the current position and velocity of particle i, respectively. p_i^k is the optimal position of particle i. p_g^k is the best position among all the particles in the population. x_i^{k+1} and v_i^{k+1} are the updated position and velocity of particle i, respectively.

As a population-based optimization method, the optimal solution can be determined by updating the position and velocity of the particles. Once the optimization model is constructed, the solving process can be handled by the POS module in MATLAB.

4. Reliability Analysis of Kinematic Accuracy

In the process of reliability analysis, the mathematical model is generally established according to the reliability design principle and practical problems. Finally, a suitable algorithm is used to solve the problem. The general solution process is shown in Figure 4.

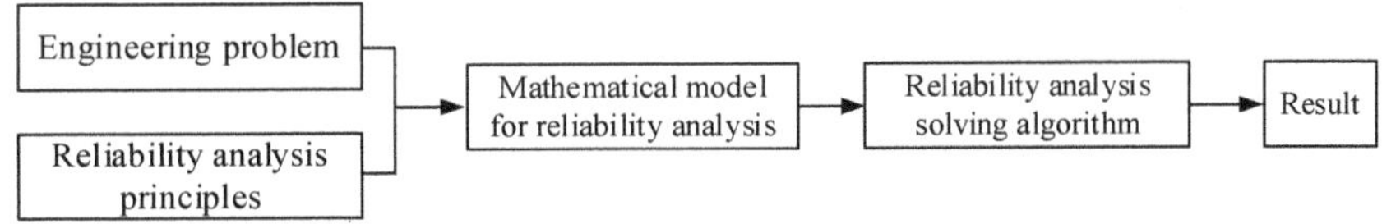

Figure 4. The general process of reliability analysis.

4.1. Mathematical Model of Reliability Analysis

Compared with the traditional cam-linkage mechanism, the geometric shape error of the cam five-bar mechanism with a large size has a more significant impact on motion accuracy, especially in the high-speed scenario. To investigate the kinematic accuracy of the optimized structure, a reliability analysis of motion accuracy is presented based on the Monte Carlo methodology.

Firstly, motion analysis is conducted before the reliability analysis. Specifically, the errors in the machining process of the rod are investigated in the motion analysis. Assuming l_1', l_2', l_3', l_4', and l_5' are the actual lengths of the corresponding rod shown in Figure 1, the intersection point with the theoretical contour line of the cam is $E'(x_E', y_E')$. Then, the vector equation of the linkage mechanism considering the errors is built as:

$$\begin{cases} l_1' + l_2' = l_3' + l_4' \\ l_1' + l_5' = l_{AE'} \end{cases} \tag{20}$$

By the transformation operation similar to Equation (1), Equation (20) is also written as:

$$\begin{bmatrix} cos\theta_1 & cos\theta_2 & cos\theta_3 & cos\theta_4 \\ sin\theta_1 & sin\theta_2 & sin\theta_3 & sin\theta_4 \end{bmatrix} \begin{bmatrix} l'_1 \\ l'_2 \\ l'_3 \\ l'_4 \end{bmatrix} = \begin{bmatrix} 0 \\ 0 \end{bmatrix} \tag{21}$$

$$\begin{bmatrix} cos\theta_1 & cos\theta_5 \\ sin\theta_1 & sin\theta_5 \end{bmatrix} \begin{bmatrix} l'_1 \\ l'_5 \end{bmatrix} = \begin{bmatrix} x_{E'} \\ x_{E'} \end{bmatrix} \tag{22}$$

Accordingly, θ_3 and θ_4 are obtained as:

$$\theta_3 = 2arctan\left[\frac{B_0 \pm \sqrt{B_0^2 + A_0^2 - C_0^2}}{A_0 - C_0}\right] \tag{23}$$

$$\theta_4 = arctan\frac{l'_3 sin\theta_3 + P_0 - R_0}{l'_3 cos\theta_3 + O_0 - Q_0} \tag{24}$$

where $A_0 = 2l'_3(O_0 - Q_0)$, $B_0 = 2l'_3(P_0 - R_0)$, $C_0 = l'^2_3 + (O_0 - Q_0)^2 + (P_0 - R_0)^2 - l'^2_4$, and

$$\begin{bmatrix} O_0 \\ P_0 \\ Q_0 \\ R_0 \end{bmatrix} = \begin{bmatrix} -cos\theta_1 & 0 \\ -sin\theta_1 & 0 \\ 0 & cos\theta_2 \\ 0 & sin\theta_2 \end{bmatrix} \begin{bmatrix} l'_1 \\ l'_2 \end{bmatrix}.$$

In addition to the linkage mechanism, the cam and the roller errors are also investigated. As shown in Figure 5, Δl is the cam's shape error. r_g is the roller radius error.

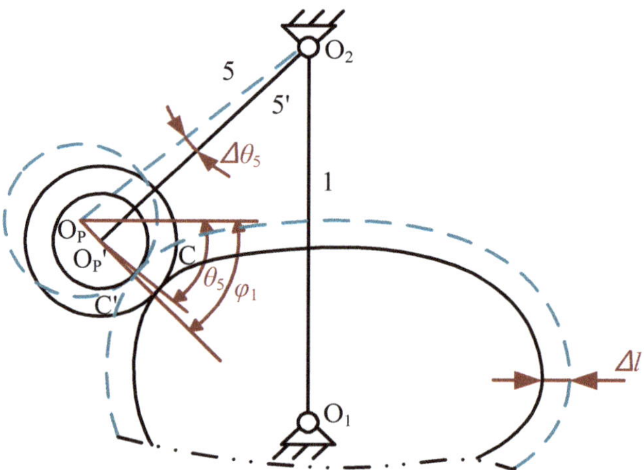

Figure 5. Error diagram of the swing cam mechanism.

Ideally, the cam and the roller are in tangential contact at point C. However, the point would change from C to C', when Δl and r_g are taken into account. Then, the angular error $\Delta\theta_5$ of rod 5 is calculated as:

$$\Delta\theta_5 = \frac{\delta}{l'_5 \cos(\varnothing_1 - \theta_5)} \tag{25}$$

where δ indicates the combined error of roller radius and cam geometry, $\delta = \Delta l + r_g$.

The rotational angle θ_5 of rod 5 is obtained as:

$$\theta_5 = arctan\frac{l'_1 sin\theta_1 - y_{E'}}{l'_1 cos\theta_1 - y_{E'}} + \frac{\delta}{l'_5 cos(\alpha - \theta_5)} \tag{26}$$

Therefore, the actual output angle of the mechanism at any given moment can be obtained from the above motion analysis. In addition, the dimensional errors of the components can be considered mutually independent and normally distributed random variables [13]. The probability distribution function $N(\mu,\sigma)$ of each rod size can be obtained according to the "3σ principle".

4.2. Reliability Analysis

After the mathematical model is established, the Monte Carlo strategy is introduced to conduct the reliability analysis, which is illustrated in Figure 6.

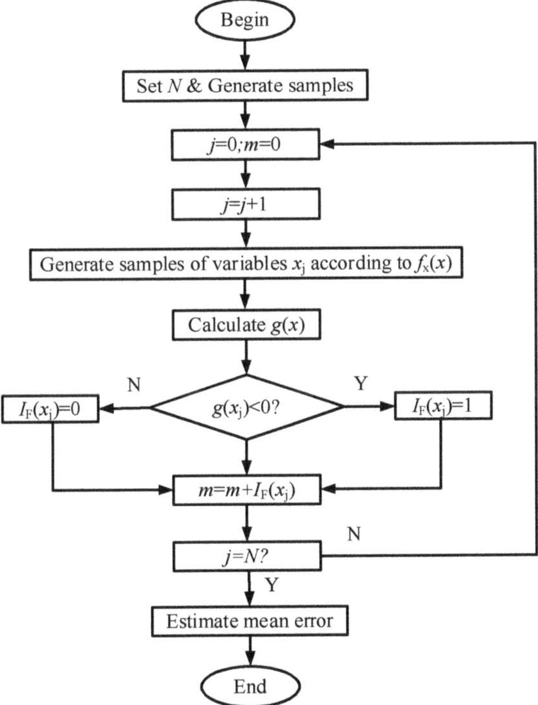

Figure 6. Flow chart of Monte Carlo reliability analysis.

In Figure 6, $I(X)$ indicates the number of samples that meet the requirements. $g(X)$ is the limit state function, by which the product is judged to be failing or not.

In the process of reliability analysis by Monte Carlo strategy, first set the sampling number, input the probability distribution function of the variables to generate N samples, and then substitute each sample into the limit state function for calculation. The Monte Carlo method is a numerical calculation technique guided by probabilistic statistical theory. It estimates the overall probability of failure by the failure frequency of the sample, which is described as:

$$\hat{P}_f = \frac{N_F}{N} \tag{27}$$

where $\hat{P}_f$ is the probability of failure estimate. N_F indicates the number of $g(X_j) \in F$, where F is the failure domain.

In Equation (27), the number of samples N should be chosen with a balance of computational speed and accuracy. The appropriate number of samples is determined based on the failure probability of the mechanism. Thus, the calculated failure probability will not be seriously distorted and large errors can be avoided.

Based on the probability of failure, the evaluation parameters (e.g., the mean value of errors) can be calculated.

4.3. Software Development

In addition to the theoretical analysis, this paper proposes a computer-aided design platform to assist the design analysis of the cam five-bar mechanism, which is developed based on MATLAB. As depicted in Figure 7, the software package is composed of three modules: optimization, parameter calculation, and reliability analysis.

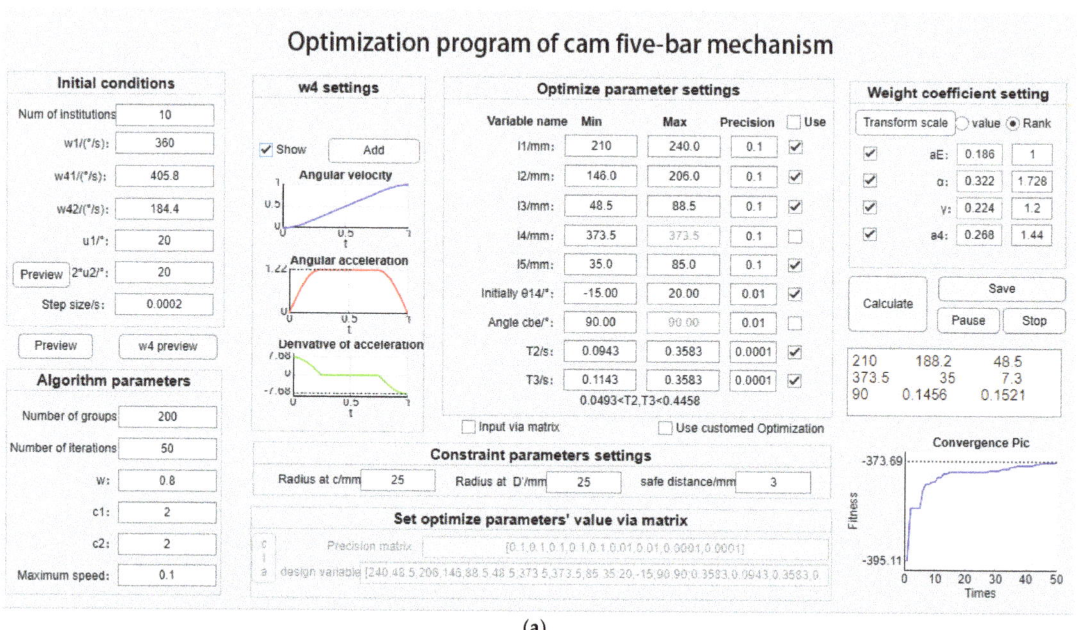

(**a**)

Figure 7. *Cont.*

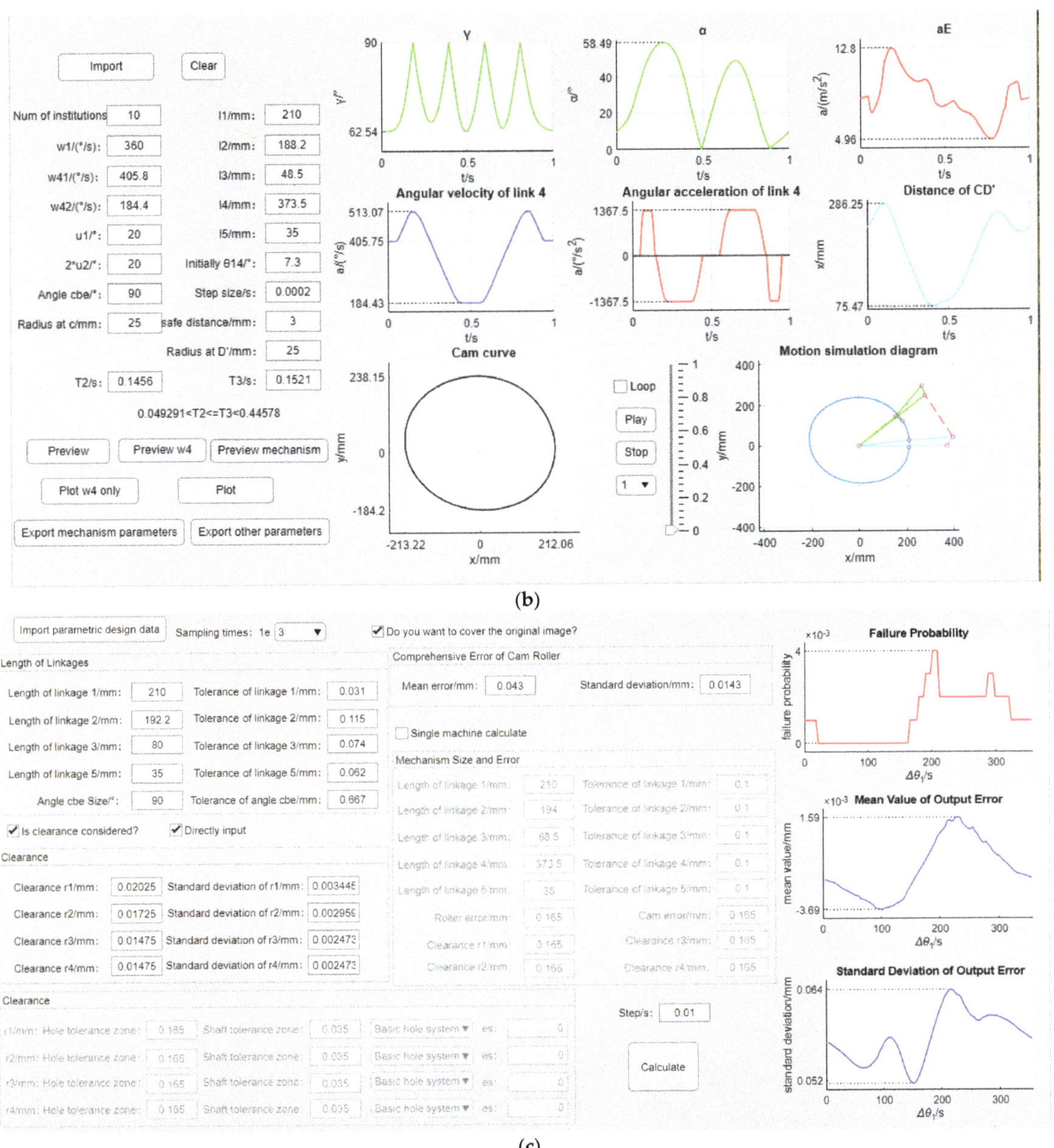

Figure 7. Software interface. (**a**) Optimization module (**b**) Parameter calculation module (**c**) Reliability analysis module.

5. Model Validation and Discussion

To demonstrate the effectiveness of the proposed method, a real-world case study of the design analysis for a transverse device is presented in this section. As illustrated in Figure 8, the transverse device is core equipment in a sanitary product production line. It is mainly used to shift the product from a horizontal to a vertical placement.

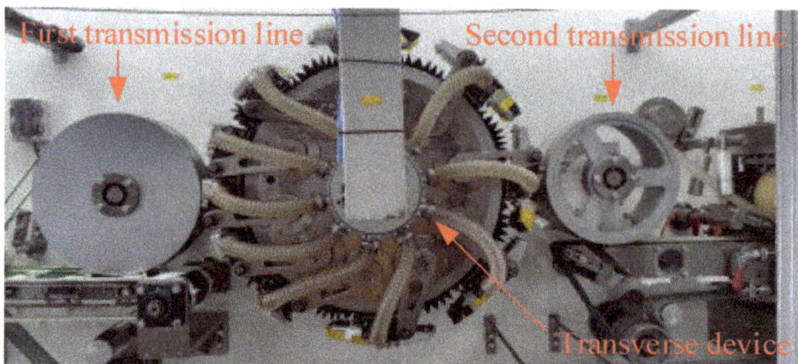

Figure 8. Physical diagram of the transverse device.

Corresponding to the physical structure, the working diagram of the production line is shown in Figure 9. Initially, the products are placed horizontally on the first conveyor, moving with speed v_1. x_1 is the spacing between products on the first conveyor. During the system operation, the intermediate mechanism (i.e., the transverse device, width: d) would pass the products from the first conveyor to the second conveyor, with the placement changing from horizontal to vertical. Accordingly, the spacing and moving speed are changed to x_2 and v_2, respectively.

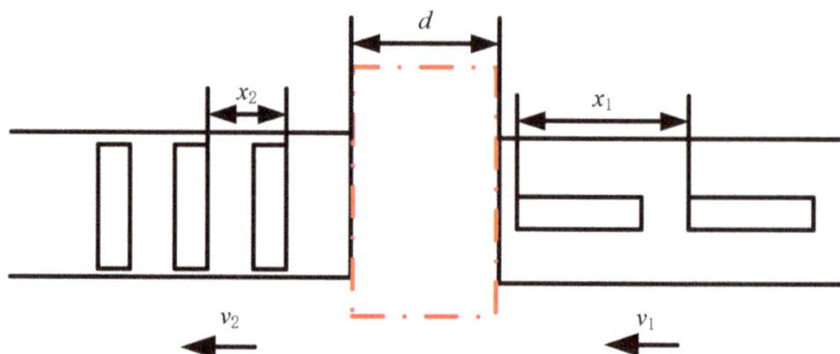

Figure 9. Production line working diagram.

According to the working requirement of the production line, a cam five-bar mechanism is developed for the transverse device. As shown in Figure 10, the cam five-bar mechanism consists of a linkage mechanism 1-2-3-4 and a cam mechanism 5-6, where member 1 is a rotating disc, member 2 is a rocker arm, and member 3 is a linkage. Output member 4 is composed of a wind box assembly and a circular guideway.

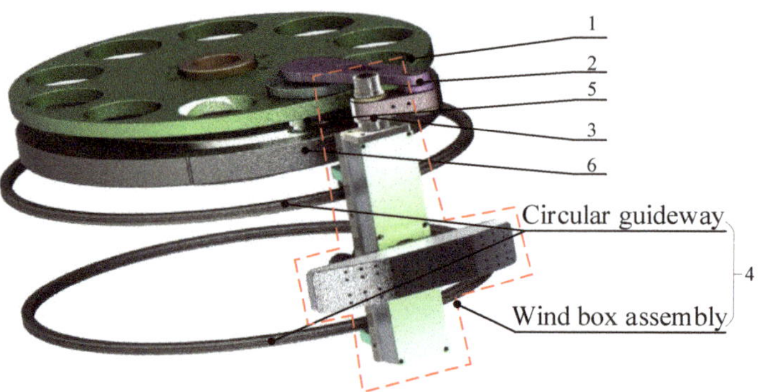

Figure 10. Cam five-bar mechanism in the transverse device.

During the system operation, the cam is fixed. Rotating disc 1 acts as the prime mover to rotate the whole mechanism. Rocker arm 2 is connected to the cam. As the cam constrains one free degree of the structure, the system can be considered a cam five-bar mechanism.

5.1. Original Mechanism Motion Characteristics

Figure 11 shows the principle illustration of the cam five-bar mechanism used in the transverse device. The values of l_1, l_2, l_3, l_4, and l_5 are 210 mm, 176 mm, 58.5 mm, 373.5 mm, and 35 mm, respectively. $\omega_1 = 360°/s$, $\Psi = 90°$, and the initial value of θ_{14} is $2°$.

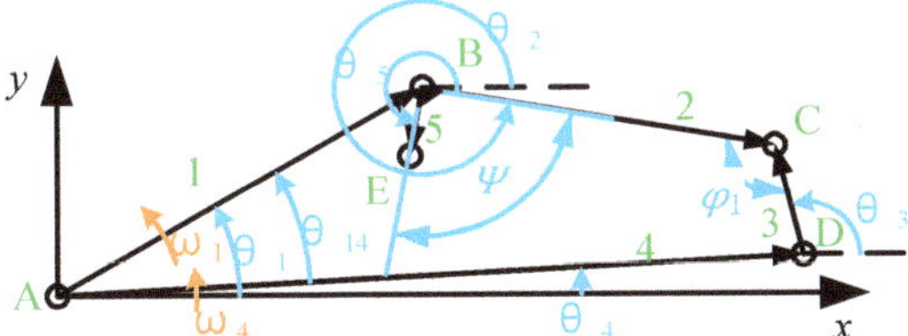

Figure 11. Principle illustration of the cam five-bar mechanism.

According to the working requirements, the ω_{41} and ω_{42} are determined as $405.75°/s$ and $184.43°/s$, respectively. A dwell-free modified iso-velocity curve [21] is adopted in the buffer section. Then, the motion law of the output member is obtained, as shown in Figure 12.

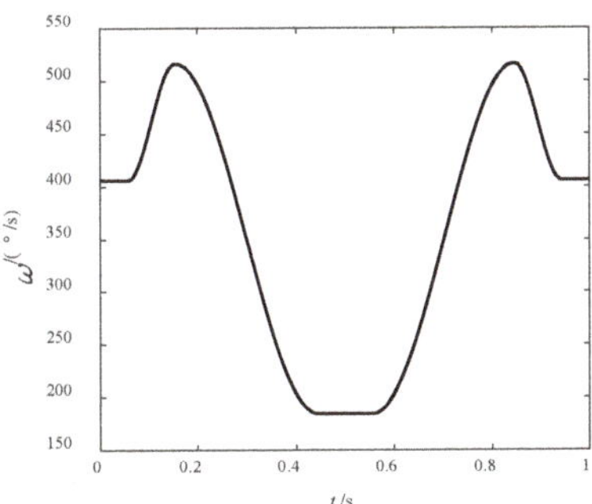

Figure 12. ω_4 curve before optimization.

Finally, the motion characteristics of the mechanism and the cam theory curve are determined by resolving the model for the forward solution, which are shown in Figures 13 and 14 respectively. In Figure 13, γ_{min} is $37.90°$, α_{max} is $66.97°$ and a_{Emax} is 12.29 m/s^2.

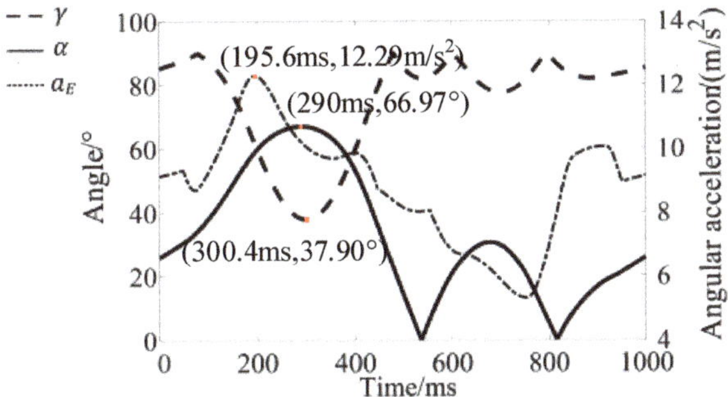

Figure 13. Motion characteristics before optimization.

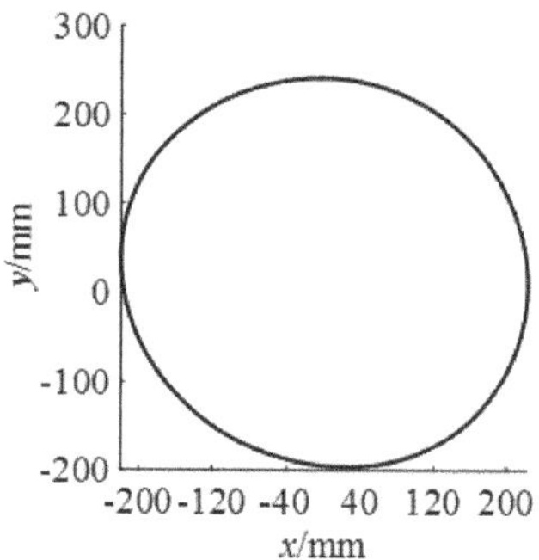

Figure 14. Cam theory curve before optimization.

5.2. Optimization Results Analysis

In line with the multi-objective optimization method proposed in Section 3, the optimal solution is obtained as l_1 = 210 mm, l_2 = 192.2 mm, l_3 = 80 mm, l_5 = 35 mm, θ_{14} = 15.38°, T_2 = 0.0943 s, and T_3 = 0.2534 s. Correspondingly, the optimized motion characteristics and cam theory curve are depicted in Figures 15 and 16, respectively.

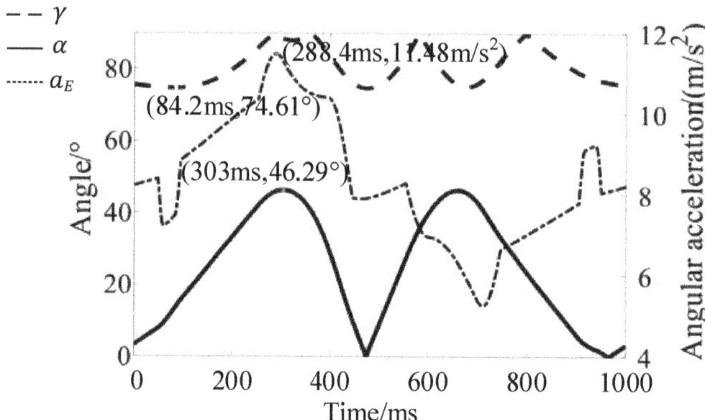

Figure 15. Motion characteristics after optimization.

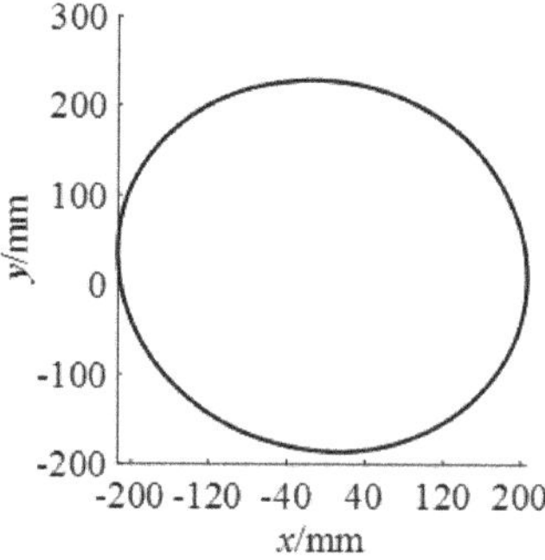

Figure 16. Cam theory curve after optimization.

In Figure 15, $\gamma_{\min}$ is 74.61°, $\alpha_{\max}$ is 46.29° and $a_{E\max}$ is 11.48 m/s². Compared with Figure 13, the relevant performance parameters are optimized. The new motion characteristics meet the design requirements (i.e., $\gamma_{\min} \geq 40°$ and $\alpha_{\max} \leq 50°$).

As well as the motion characteristics, the angular velocity ω_4 and angular acceleration a_4 curves are obtained, which are shown in Figure 17. Compared with the ω_4 curve before optimization (Figure 12), the angular acceleration of the output member becomes smaller.

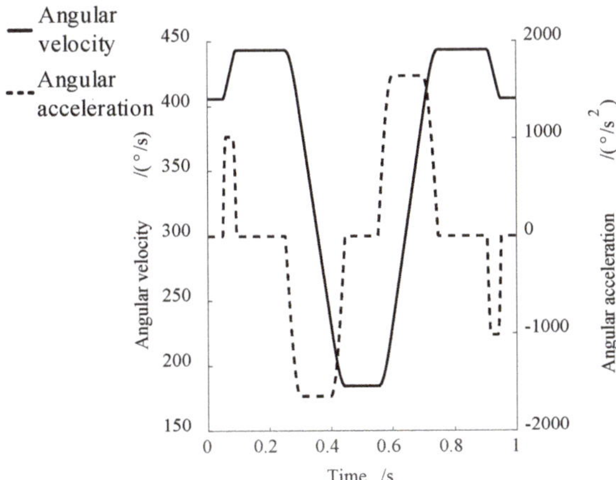

Figure 17. ω_4 curve after optimization.

5.3. Reliability Analysis of Kinematic Accuracy

Corresponding to the framework proposed in Section 4, the reliability analysis of kinematic accuracy is conducted for the optimized mechanism. The parameters of each rod are listed in Table 1.

Table 1. Parameters related to each rod.

Parameter Name	Mean Value	Tolerance	Standard Deviation
l_1/mm	210	0.03	0.0052
l_2/mm	192.2	0.12	0.0192
l_3/mm	80	0.07	0.0123
l_4/mm	35	0.06	0.0104
$\Psi/°$	90	0.67	0.111
δ/mm	0.043	—	0.0143

Based on the above reliability analysis model, the average value of component size error is shown in Figure 18.

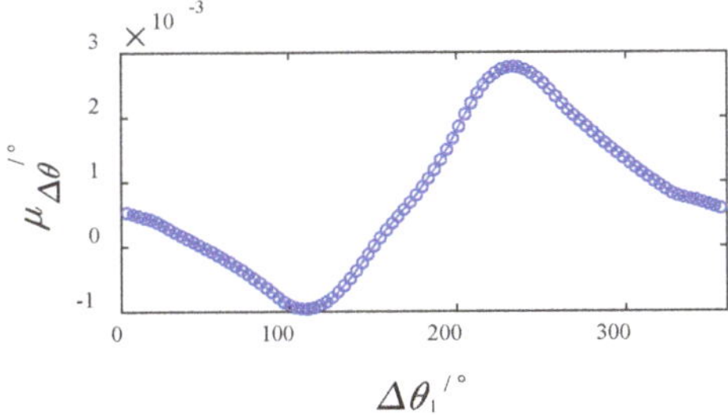

Figure 18. Mean value of rod 4 angle error.

From Figure 18, it can be concluded that the absolute value of the average angle error of rod 4 is less than 0.003° in a movement cycle. It meets the production requirements.

5.4. Comparison of Structure and Motion Characteristics

Tables 2 and 3 present the specific parameters before and after optimization. The comparison shows that the maximum acceleration at point E is slightly reduced after optimization. The transmission angle and pressure angle are optimized significantly, with the minimum value of the transmission angle increased by 96.8% and the maximum value of the pressure angle reduced by 30.9%. The new motion characteristics meet the design requirements (i.e., $\gamma_{min} \geq 40°$ and $\alpha_{max} \leq 50°$).

Table 2. Comparison of motion characteristics parameters.

Parameter Name	$\alpha_{max}/(°)$	$\gamma_{min}/(°)$	a_{Emax} (m/s^2)
Before optimization	66.97	37.90	12.90
After optimization	46.29	74.61	11.48

Table 3. Comparison of connecting rod parameters.

Status	l_1/mm	l_2/mm	l_3/mm	l_5/mm	Initially $\theta_{14}/(°)$
Before optimization	210	176	58.5	35	2
After optimization	210	192.2	80	35	15.38

Figure 19 illustrates the comparison of the cam curves before and after optimization. As shown in Figure 19, the optimized cam profile size is smaller. It improves the compactness of the overall mechanism and ensures excellent motion characteristics.

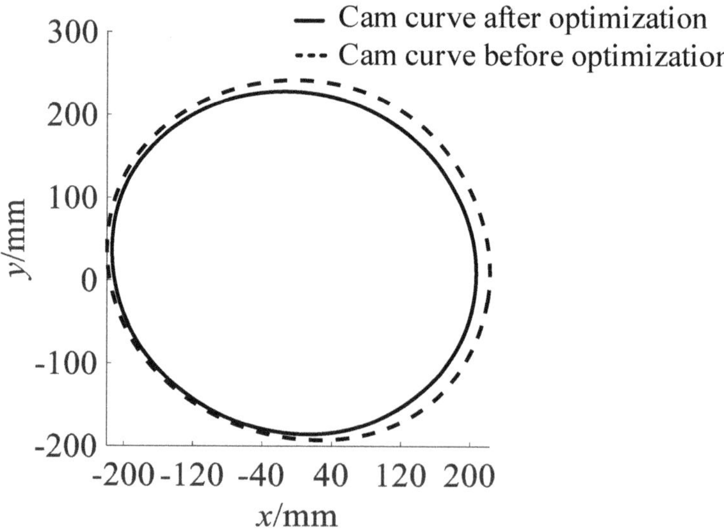

Figure 19. Comparison before and after cam optimization.

5.5. Engineering Applications

In light of the optimized rod parameters and cam curve, the transverse device is redesigned, which is shown in Figure 20. Figure 21 illustrates the working state of the new transverse device.

As depicted in Figure 21, the optimized transverse device can reach a maximum speed of 1200 pieces per minute (PPM). Compared with the original productivity of 600 PPM, the productivity is significantly improved.

Figure 20. Optimized transverse device (side).

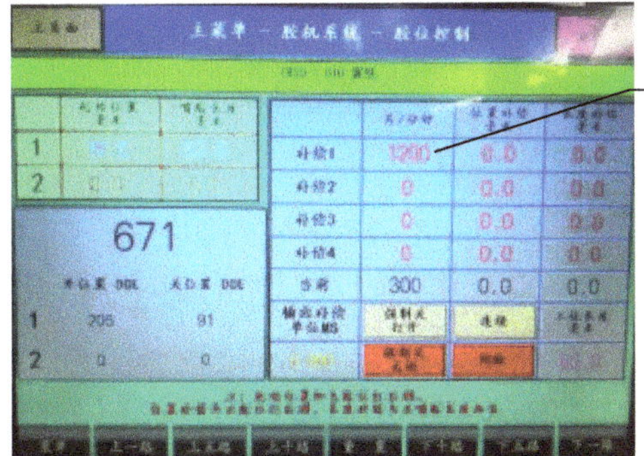

Figure 21. The maximum running speed of the transverse device.

6. Conclusions

To investigate the motion characteristics and kinematic accuracy of the cam-linkage mechanism under high-speed scenarios, this paper proposes a series of methods for the kinematic modeling, optimization, and reliability analysis of the large-size cam-linkage mechanism considering kinematic performance. The conclusions are as follows.

(1) A mathematical model is constructed to determine the performance parameters of the cam-linkage mechanism, such as E-point acceleration, transmission angle, and cam pressure angle.

(2) A multi-objective optimization methodology is proposed for the parameter optimization of the large-size cam-linkage mechanism. The co-linear of rods 2 and 3 is set as constraint, while the maximum drive angle, minimum cam pressure angle, and minimum E-point acceleration are taken as the objective function.

(3) The kinematic model of the cam-linkage mechanism is presented. Reliability analysis is conducted to evaluate the kinematic accuracy of the optimized mechanisms.

(4) A computer-aided platform is developed to assist the parameter calculation, optimization, and reliability analysis of the cam-linkage mechanism.

(5) The effectiveness of the proposed method is validated by a real-world case study. The productivity of the transverse device is increased from 600 PPM to 1200 PPM.

Author Contributions: G.Z.: conceptualization, data collection, data analysis, visualization, computer-aided design, computer-aided manufacturing, software, and original draft writing; Y.W.: conceptualization, data analysis, project administration, computer-aided manufacturing, supervision, and manuscript review and editing; G.-N.Z.: conceptualization, investigation, supervision, and manuscript review and editing; M.W.: data analysis, visualization, software, and manuscript review and editing; J.L.: data collection and visualization; J.Z.: data collection, data analysis, software and computer-aided design; B.L.: data collection and visualization. All authors have read and agreed to the published version of the manuscript.

Funding: This work is supported by the Department of Science and Technology of Anhui Province, China, under the Foundation of Anhui Province Science and Technology Major Project (grant number: 202003a0502041).

Data Availability Statement: All data generated during this work are available in the manuscript.

Conflicts of Interest: The authors declare no conflict of interest.

References

1. Yu, H.; Yin, H.; Peng, J.; Wang, L. Quasi static analysis of heald frame driven by rotary Dobby with cam-linkage modulator. *J. Text. Eng.* **2020**, *66*, 37–45. [CrossRef]
2. Li, M.; Zhang, X.H.; Yan, J.Q. Mathematical modeling and optimization of transverse heat sealing and cutting mechanism. *J. Chin. Inst. Eng.* **2022**, *45*, 385–390. [CrossRef]
3. Zhao, P.; Zhu, L.; Zi, B.; Li, X. Design of Planar 1-DOF cam-linkages for lower-limb rehabilitation via kinematic-mapping motion synthesis framework. *J. Mech. Robot.* **2019**, *11*, 041006. [CrossRef]
4. Liu, Y.; Zhang, J.; Wang, L.; Li, L.; Shi, Y.; Xie, G.; Wang, X. Optimization design of a bionic horse's leg system driven by a cam-linkage mechanism. *Proc. Inst. Mech. Eng. Part C J. Mech. Eng. Sci.* **2022**. [CrossRef]
5. Wen, H.; Cong, M.; Xu, W.; Zhang, Z.; Dai, M. Optimal design of a linkage-cam mechanism-based redundantly actuated parallel manipulator. *Front. Mech. Eng.* **2021**, *16*, 451–467. [CrossRef]
6. Rybansky, D.; Sotola, M.; Marsalek, P.; Poruba, Z.; Fusek, M. Study of Optimal Cam Design of Dual-Axle Spring-Loaded Camming Device. *Materials* **2021**, *14*, 1940. [CrossRef] [PubMed]
7. Abderazek, H.; Yildiz, A.R.; Mirjalili, S. Comparison of recent optimization algorithms for design optimization of a cam-follower mechanism. *Knowl.-Based Syst.* **2020**, *191*, 105237. [CrossRef]
8. Li, G.; Long, X.; Zhou, M. A new design method based on feature reusing of the non-standard cam structure for automotive panels stamping dies. *J. Intell. Manuf.* **2019**, *30*, 2085–2100. [CrossRef]
9. Zhang, C.; Zhang, X.; Ye, H.; Wei, M.; Ning, X. An efficient parking solution: A cam-linkage double-parallelogram mechanism based 1-degrees of freedom stack parking system. *J. Mech. Robot.* **2019**, *11*, 045001. [CrossRef]
10. Wu, X.; Wang, K.; Wang, Y.; Bai, S. Kinematic design and analysis of a 6-DOF spatial five-bar linkage. *Mech. Mach. Theory* **2021**, *158*, 104227. [CrossRef]
11. Arabaci, E. Dimensionless design approach to translating flat faced follower mechanism with two-circular-arc cam. *Mech. Sci.* **2019**, *10*, 497–503. [CrossRef]
12. Xia, B.Z.; Liu, X.C.; Shang, X.; Ren, S.Y. Improving cam profile design optimization based on classical splines and dynamic model. *J. Cent. South Univ.* **2017**, *24*, 1817–1825. [CrossRef]
13. Ouyang, T.; Wang, P.; Huang, H.; Zhang, N.; Chen, N. Mathematical modeling and optimization of cam mechanism in delivery system of an offset press. *Mech. Mach. Theory* **2017**, *110*, 100–114. [CrossRef]
14. Chen, T.H. The systematic design and prototype verification of cam-linkage polishing devices with double-circle polishing traces. *Adv. Mech. Eng.* **2018**, *10*, 1687814018818974. [CrossRef]
15. Chang, W.T.; Hu, Y.E. An integrally formed design for the rotational balancing of disk cams. *Mech. Mach. Theory* **2021**, *161*, 104282. [CrossRef]
16. Li, M.; Yan, J.; Zhao, H.; Ma, G.; Li, Y. Mechanically Assisted Neurorehabilitation: A Novel Six-Bar Linkage Mechanism for Gait Rehabilitation. *IEEE Trans. Neural Syst. Rehabil. Eng.* **2021**, *29*, 985–992. [CrossRef] [PubMed]
17. Yang, Y.; Xie, R.; Wang, J.; Tao, S. Design of a novel coaxial cam-linkage indexing mechanism. *Mech. Mach. Theory* **2022**, *169*, 104681. [CrossRef]
18. Yang, Y.; Wang, J.; Zhou, S.; Huang, T. Design of a novel coaxial eccentric indexing cam mechanism. *Mech. Mach. Theory* **2019**, *132*, 1–12. [CrossRef]
19. Radaelli, G. Reverse-twisting of helicoidal shells to obtain neutrally stable linkage mechanisms. *Int. J. Mech. Sci.* **2021**, *202*, 106532. [CrossRef]
20. Wang, X.; Wang, T.; He, X.; Ma, Z. Effect of Cam Angular Velocity on Working Efficiency of Fuel Engine. *J. Phys. Conf. Ser.* **2020**, *1533*, 042064.
21. Li, J.; Zhao, H.; Zhang, L. Transmission and tooth profile equation of swing output movable teeth cam mechanism. *J. Mech. Eng.* **2018**, *54*, 23–31. (In Chinese) [CrossRef]

 *actuators*

Article

Effect of the Magnetorheological Damper Dynamic Behaviour on the Rail Vehicle Comfort: Hardware-in-the-Loop Simulation

Filip Jeniš [1],*, Michal Kubík [1], Tomáš Michálek [2], Zbyněk Strecker [1], Jiří Žáček [1] and Ivan Mazůrek [1]

[1] Institute of Machine and Industrial Design, Faculty of Mechanical Engineering, Brno University of Technology, Technicka 2, 616 69 Brno, Czech Republic

[2] Department of Transport Means and Diagnostics, Faculty of Transport Engineering, University of Pardubice, Studentska 95, 532 10 Pardubice, Czech Republic

* Correspondence: filip.jenis@vutbr.cz; Tel.: +420-541-143-216

Abstract: Many publications show that the ride comfort of a railway vehicle can be significantly improved using a semi-active damping control of the lateral secondary dampers. However, the control efficiency depends on the selection of the control algorithm and the damper dynamic behaviour, i.e., its force rise response time, force drop response time and force dynamic range. This paper examines the influence of these parameters of a magnetorheological (MR) damper on the efficiency of S/A control for several control algorithms. One new algorithm has been designed. Hardware-in-the-loop simulation with a real magnetorheological damper has been used to get close to reality. A key finding of this paper is that the highest efficiency of algorithms is not achieved with a minimal damper response time. Furthermore, the force drop response time has been more important than the force rise response time. The Acceleration Driven Damper Linear (ADD-L) algorithm achieves the highest efficiency. A reduction in vibration of 34% was achieved.

Keywords: hardware-in-the-loop; Acceleration Driven Damper; response time; dynamic range; semi-active; magnetorheological; damper; railway vehicle; lateral vibration

Citation: Jeniš, F.; Kubík, M.; Michálek, T.; Strecker, Z.; Žáček, J.; Mazůrek, I. Effect of the Magnetorheological Damper Dynamic Behaviour on the Rail Vehicle Comfort: Hardware-in-the-Loop Simulation. *Actuators* **2023**, *12*, 47. https://doi.org/10.3390/act12020047

Academic Editor: Norman M. Wereley

Received: 19 December 2022
Revised: 12 January 2023
Accepted: 17 January 2023
Published: 19 January 2023

1. Introduction

Rail transport has recently become increasingly important around the world. One of the most important parameters of a railway vehicle is comfort (carbody vibration). The damping system is primarily responsible for carbody vibration mitigation. Today, commonly-used passive dampers have limits and can no longer solve new problems. The carbody vibration can be reduced using a semi-active or active system. The active system contains actuators instead of springs and dampers. The actuator can precisely create the force required to dampen the sprung mass. The significant disadvantages of active systems are the relative complexity, high cost, energy consumption and difficulties in implementing a fail-safe system [1,2]. An alternative to an active system is a semi-active system. The semi-active damper can change the damping characteristic using data from various sensors [3,4]. Using magnetorheological (MR) dampers for semi-active (S/A) control is advisable because MR dampers have very good transient behaviour [5]. One of the advantages of the MR damper is the possibility of fail-safe behaviour using a permanent magnet [6].

In the case of railway vehicles, several publications have verified the potential benefits of semi-active control of lateral secondary MR dampers in reducing carbody vibrations. Codeca et al. [7] tested the semi-actively controlled lateral secondary dampers and their effect on the carbody vibrations on a laboratory stand. The experimental stand was excited by a signal obtained from a real measurement on the track when running at high speed. A combination of Skyhook and Acceleration Driven Damping algorithms achieved the highest efficiency, reducing vibrations by 34%, compared to passive damping. Lau and Liao [8] examined the effect of S/A control of the lateral secondary dampers on ride comfort. The semi-active algorithm is simple and switches between Low and High damping states based on the magnitude of carbody lateral velocity. If the carbody velocity exceeds the critical speed, the damper switches to a High state. The simulations were performed on a complex railway vehicle dynamic model. The carbody lateral vibration was reduced by 39% compared to the passive damper. Shin et al. [9] dealt with the simulation and experimental verification of the secondary lateral damper semi-active control effect. A half-vehicle model with nine degrees of freedom was used for the simulation. For experimental validation, a 1:5 scale real simplified model was used. The Skyhook algorithm decreased vibration by 77% in the simulation compared to passive damping and 67% in the experiment. Hudha et al. [10] used two types of Skyhook algorithm: body-based and bogie-based for the lateral secondary MR dampers. The vehicle's body lateral deviation, yawing, and rolling motion were evaluated in a dynamic model of 17 degrees of freedom. The algorithms worked well, both improving all three criteria.

Papers [11,12] show that damper force response time is essential for the performance of semi-actively controlled damping systems. The shorter the response time, the better the S/A control performance. Paper [13] shows that the time response is further dependent on the piston velocity, the magnitude of the electric current change, the stiffness of the damper mounting and the control electronics. The response time is negatively affected by the formation of eddy currents in the piston core, but this problem can be solved by using a suitable core shape [5,14]. Response time also depends on the dwell time of Fe particles in the active-magnetic zone and the concentration of Fe particles in the carrier fluid [13]. Papers [5,7] show that force response time differs for force rise and drop. The damper dynamic range (the ratio of the damping force in the activated and non-activated state) is also essential for the efficiency of the S/A control [15].

The MR dampers exhibit highly non-linear behaviour, which is very difficult to model properly. The MR damper models used in simulations are always simplified and often do not correspond to the force generated by real MR damper for given working conditions, degrading the results from simulations of semi-active suspension. To avoid the problem, many teams used the Hardware-in-the-loop method to evaluate the semi-active algorithms. Choi et al. [16] used HILS to design a semi-active control seat for a truck vehicle, and Lee et al. [17] used it to study the S/A control of a car suspension. Misselhorn et al. [18] successfully tested HILS usability in suspension design, comparing HILS with a virtual simulation on a single corner model and a simplified physical model of motorcycle suspension. Kwak et al. [19] used HILS to verify the benefit of the Skyhook algorithm for the lateral movement of the railway vehicle carbody. Oh et al. [20] studied the active control of tramcars independently rotating wheels in HILS. This approach enables to use of force generated directly by the real MR damper excited by the pulsator. The pulsator position in the time is controlled by the output of a virtual model running on a suitable platform.

Problem Formulation

It is known that force response time and dynamic range have an essential effect on the effectiveness of S/A control but the impact of this effect on different S/A algorithms has not been directly investigated. It is also known that the response time is different for the force rise and force drop when the electric current changes but in the case of simulation with the implemented response time of the damper, the response time is considered to be the same for both force rise and force drop. It is not known whether this simplification is permissible. It can be assumed the longer response time and the lower dynamic range lead to the lower effectiveness of the strategy used, but it is not known what maximum response time is acceptable for different algorithms.

The paper's main aim is to investigate the effect of selected parameters of the MR damper on rail suspension performance in detail. The response time of damping force rise, damping force drop, and dynamic force range are studied separately. The hardware-in-the-loop simulation with a real MR damper will be used to accommodate several non-linearity in damper dynamics. Simulations are evaluated for several known S/A damper control algorithms and one newly-designed algorithm (Acceleration Driven Damper Linear).

2. Materials and Methods

2.1. Vehicle Model

A lateral motion model of a rail vehicle with two degrees of freedom has been developed. The model represents one wheelset, half of the bogie frame and a quarter of a carbody. The model layout can be seen in Figure 1, and the following equations describe the model:

$$m_2\ddot{y}_2 = -k_2\left(y_2 - y_1\right) + F_{mr} \tag{1}$$

$$m_1\ddot{y}_1 = k_2\left(y_2 - y_1\right) - F_{mr} - k_1\left(y_1 - y_0\right) - c_1\left(\dot{y}_1 - \dot{y}_0\right) \tag{2}$$

where F_{mr} is s force of MR damper and y_2, y_1 and y_0 are lateral displacements of carbody, bogie frame and wheelset. Other model parameters are listed in Table 1. The model parameters are derived from a four-axle locomotive weighing 90 tons. The lateral damper used is reduced in scale. Therefore the parameters of the vehicle model are also reduced by a ratio of 1:5.

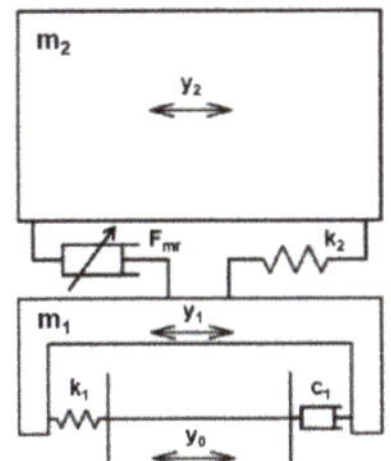

Figure 1. Railway vehicle simplified model schema.

Table 1. Model parameters.

Parameter	Symbol	Original	1:5 Scale
half bogie frame weight	m1	5000 kg	1000 kg
quarter carbody weight	m2	13,750 kg	2750 kg
wheelset-bogie frame bond stiffness	k1	10 kN/mm	2 kN/mm
bogie frame-carbody bond stiffness	k2	1 kN/mm	0.2 kN/mm
wheelset-bogie frame bond damping	c1	10 kNs/m	2 kNs/m

The model was excited by defined lateral wheelset motion y_0, which was obtained from a simulation of a complex multi-body model of a railway vehicle (electric locomotive) running on a straight track with irregularities at 160 km/h, which is the speed limit on Czech railways. This complex model with 58 degrees of freedom was created in the multi-body simulation tool "SJKV" [21]. The track irregularities were generated using PSD of track irregularities in Europe according to ORE B 176. The wheelset-track coupled dynamic is neglected, for a more objective comparison of the results of S/A algorithms, so excitation y_0 is still the same for every case. Figure 2 shows the wheelset lateral displacement course and the FFT analysis of this course. The duration of the excitation signal was 10 s. The excitation magnitude was at the original scale, so even the scale of the damper stroke corresponds to reality.

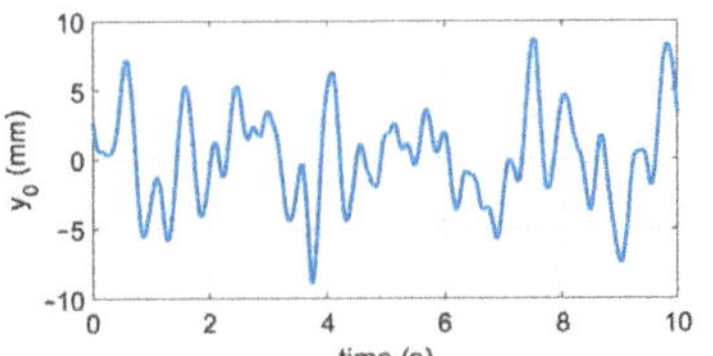
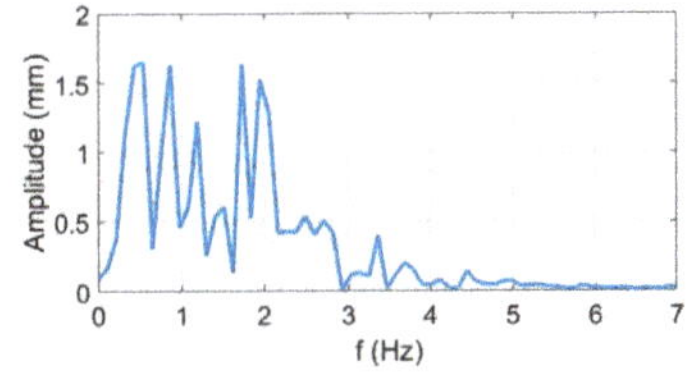

Figure 2. Excitation signal y_0 (**left**) and its FFT analysis (**right**).

2.2. Hardware-in-the-Loop Simulation

Mathematically describing the behaviour of a real damper is very difficult because it is necessary to include hysteresis (magnetic, hydraulic, etc.), temperature dependence, etc., in the model. For simplified models, it is difficult to say whether they neglect some essential property. Therefore, it is necessary to use HIL simulation to bring it closer to reality. This simulation allows the damper response time and the dynamic range effect on the S/A control to be investigated on a real MR damper without needing a real vehicle or physical model.

The damper is mounted in a hydraulic pulsator, controlled by a control system (dSpace RTI1104 in this case). A virtual railway vehicle model was built in Matlab/Simulink and transferred to the ControlDesk program, which controls the dSpace system. Based on the damper-measured force, the simulation calculates the virtual position of sprung and unsprung masses (y_2 and y_1) in real time. The calculated control signal 1 (see Figure 3 left) with the damper stroke $y_2 - y_1$ (see Figure 1) is sent to the hydraulic pulsator, which excites the MR damper. The control signal 2 with the desired current (see Figure 3 left) is also sent in real-time to the current controller, which excites the MR damper.

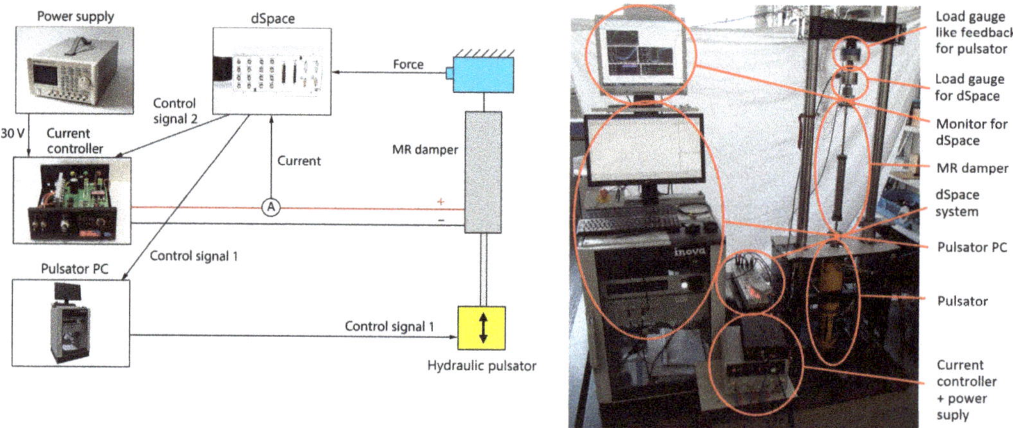

Figure 3. Experimental setup for HILS (**left**), HILS assembly (**right**).

The current controller is a self-made device that works in analogue voltage input-analogue current output mode. The highest achievable current is $I = 3$ A. The rise time of the electric current from 0 to 2 A is 1 ms as a response to input signal step 0–2 V, and the drop time from 2 to 0 A is 0.5 ms as a response to step 2–0 V. The pulsator assembly has a built-in load gauge HBM U2AD1/2 that measures the current damping force. Load gauge deflection at maximal damper force is 7 μm, so gauge mechanical response time is possible to neglect. The signal from the load gauge is amplified by an analogue bridge amplifier with no delay. The damping force data are input for the virtual model. The HILS schema is shown in Figure 3 left, and the HILS assembly is in Figure 3 right. It showed that the response time of the pulsator was significantly shorter than the response time of the damper, so it did not cause any problems.

2.3. Magnetorheological Damper

This study used a real MR damper with a stroke of 160 mm and a maximum damping force of 2000 N (at the piston velocity of 0.3 m/s). In the piston, the MR damper has one coil with an electric resistance of $R = 1.39$ Ω. The coil inductance in the damper electric circuit with MR fluid is $L = 50$ mH. The magnetic circuit is made of Sintex SMC material to achieve a fast transient response. The damper is described in more detail in the paper [5].

2.3.1. F-v-I Map

The F-v-I (force-velocity-electric current) map expresses the dependence of the damper force on the actual piston velocity and the electric current in the coil. The measurement of the F-v-I map was performed on the hydraulic pulsator in the configuration described in Figure 3. However, in the case of the F-v-I map measurement, the device did not work in HILS mode (pulsator stroke generated based on vehicle model simulations), but the course of the damper stroke was fixedly specified. The logarithmic sweep with a constant amplitude of 20 mm was used as an excitation signal in the frequency range of 0.05–1.6 Hz. Therefore, the velocity was increasing during the test. The maximum velocity was 0.2 m/s. The load gauge HBM U2AD1/2 was used. The electric current was measured by current clamps Fluke i30s. Signals were recorded by dSpace. The F-v-I map was calculated from measured data choosing the points with zero acceleration (centre of the stroke).

The measured F-v curves are shown in Figure 4. The damper F-v curves are symmetrical for both tension and compression, whereas the graph shows only the positive F-v curves part. The damper dynamic force range at piston velocity 0.1 ms^{-1} is $dr = 7.6$.

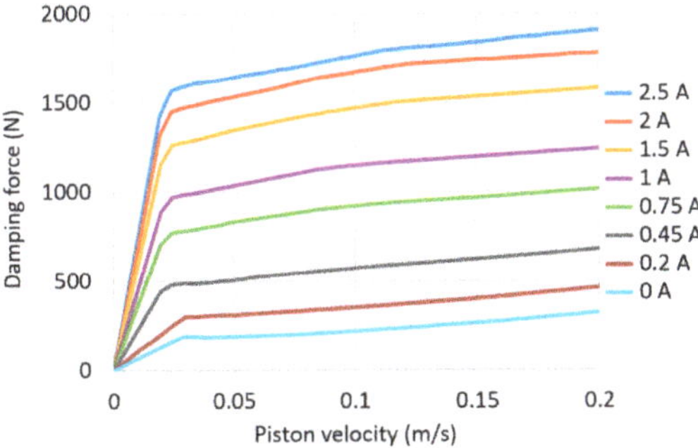

Figure 4. Measured F-v-I map of the used damper.

2.3.2. Response Time

The dynamic (transient) behaviour of the MR damper is usually assumed to be a first-order system, and the response time is usually defined as a time constant τ_{63} [22–24]. The force in time can then be described by the equation:

$$F(v,t) = F_0(v) + (F_1(v) - F_0(v)) \cdot \left(1 - e^{-\frac{t}{\tau_{63}}} \right) \tag{3}$$

where $F_0(v)$ is a force at $t = 0$, and $F_1(v)$ is force corresponding to the desired force for the given velocity and current, v is piston velocity, t is the time from the change of the control signal, and τ_{63} is the time constant.

The time constant τ_{63} means the time required to reach 63.2% of the final steady-state force (see Figure 5). Similarly, the time constant for the force drop τ_{36} expresses how long it takes for the force to drop back to 36.8% of the initial force. Response times for damper force rise τ_{63}, and for damper force drop τ_{63} on unit-step electric current rise and drop will be determined from measured data in this paper.

The measurement configuration was the same as described in the previous sections (Figure 3). The damper was tested without flexible silent blocks. The electric current was activated (to $I = 2$ A) and deactivated (to $I = 0$ A) when the damper was mid-stroke. The transient force response was measured for electric current change from 0 A to 2 A and from 2 A to 0 A, and for piston velocity of 0.1 m/s, approximately 70% of the highest piston velocity reached during simulations. The methodology is described in more detail in [25].

Damper force response time was measured at $\tau_{63} = 1.8$ ms for force rise and $\tau_{36} = 1.1$ ms for force drop (see Figure 5). Every response time value is an average of five measurements.

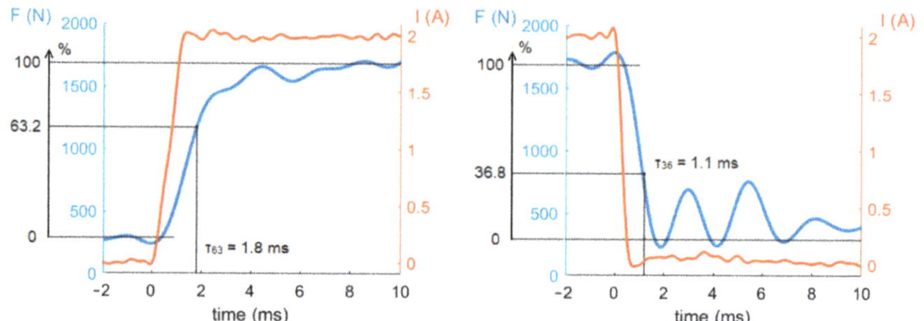

Figure 5. Example of measured response time for rise (**left**) and drop (**right**). I (A) is the measured electric current, and F (N) is the measured damping force.

2.4. Semi-Active Control

For evaluation of the response time impact on different control strategies, this research used passive mode, three well-known semi-active algorithms: Skyhook, Skyhook linear, and Acceleration Driven Damper and one newly designed: Acceleration Driven Damper Linear.

2.4.1. Skyhook

Skyhook with two states (SH-2) is a commonly used algorithm for comfort improvement. It was designed by Karnopp [26]. The control rule for this algorithm is based on two input signals: the carbody lateral velocity and the relative lateral velocity between the carbody and the bogie frame. If the carbody velocity is higher than the relative velocity between the carbody and the bogie frame, the damper is in the high damping state. Mathematically:

$$F(v) = \begin{cases} F_{max}(v), & \dot{y}_2(\dot{y}_2 - \dot{y}_1) \geq 0 \\ F_{min}(v), & \dot{y}_2(\dot{y}_2 - \dot{y}_1) < 0 \end{cases} \tag{4}$$

where $F(v)$ is a required damping force, $F_{max}(v)$ is damping force at a current of I_{max}, $F_{min}(v)$ is damping force at I_{min}, $\dot{y}_2$ is carbody lateral velocity, and $\dot{y}_1$ is bogie frame lateral velocity.

2.4.2. Skyhook Linear

The Skyhook Linear (SH-L) is an improved version of the ON/OFF Skyhook, published by Sammier [27]. This algorithm changes the damping characteristics continuously. Initially, the algorithm is designed as follows:

$$F(v) = \begin{cases} F_{min}(v), & \dot{y}_2(\dot{y}_2 - \dot{y}_1) \leq 0 \\ sat\left(\dfrac{\alpha \cdot F_{max}(v) \cdot (\dot{y}_2 - \dot{y}_1) + (1-\alpha) \cdot F_{max}(v) \cdot \dot{y}_2}{(\dot{y}_2 - \dot{y}_1)} \right), & \dot{y}_2(\dot{y}_2 - \dot{y}_1) > 0 \end{cases} \tag{5}$$

where $F(v)$ is a required damping force, $F_{max}(v)$ is damping force at a current of I_{max}, and $F_{min}(v)$ is damping force at I_{min}, $\dot{y}_2$ is carbody lateral velocity, $\dot{y}_1$ is bogie frame lateral velocity, $\alpha \in [0,1]$ is tuning, and sat denotes that $F_c \in [F_{min}(v), F_{max}(v)]$.

The algorithm has the best performance for $\alpha = 0$. For a simpler algorithm application, the damping force in the equation has been replaced by an electric current. So, the final calculation of the current looks as follows:

$$I = \begin{cases} sat\left(\dfrac{I_{max} \cdot \dot{y}_2}{(\dot{y}_2 - \dot{y}_1)} \right), & \dot{y}_2(\dot{y}_2 - \dot{y}_1) \geq 0 \\ I_{min}, & \dot{y}_2(\dot{y}_2 - \dot{y}_1) < 0 \end{cases} \tag{6}$$

where I is a required electric current, I_{max} is a maximal electric current, I_{min} is a minimal electric current, $\dot{y}_2$ is carbody lateral velocity, $\dot{y}_1$ is bogie frame lateral velocity, and *sat* denotes that $I \in [I_{min}(v), I_{max}(v)]$.

The damping force is not linearly dependent on the electric current but converting the damping force to electric current so that the force corresponds to the equation above would be unnecessarily demanding. Therefore, there is some bias in the algorithm from the original version, but the control strategy still works well.

2.4.3. Acceleration Driven Damper

Acceleration Driven Damper (ADD-2) was designed by Savaresi [28]. This algorithm works such as Skyhook but uses carbody acceleration instead of carbody velocity. So, its realisation is more accessible in practice. Mathematically:

$$F(v) = \begin{cases} F_{max}(v), & \left| \ddot{y}_2(\dot{y}_2 - \dot{y}_1) \geq 0 \right. \\ F_{min}(v), & \left| \ddot{y}_2(\dot{y}_2 - \dot{y}_1) < 0 \right. \end{cases} \tag{7}$$

where $F(v)$ is a required damping force, $F_{max}(v)$ is damping force at a current of I_{max}, $F_{min}(v)$ is damping force at I_{min}, $\ddot{y}_2$ is carbody lateral acceleration, $\dot{y}_2$ is carbody lateral velocity and $\dot{y}_1$ is bogie frame lateral velocity.

The algorithm has a problem with unwanted fast-switching behaviour. A rapid change in damping force when switching the state causes a change in the acceleration orientation, which again causes the state switch. Thus, the damper switches state very quickly at low piston velocity, reducing the damper's life. Savaresi also encountered this phenomenon of the problem of the fast-switching behaviour of ADD, but he did not deal with it [29]. However, this problem could be eliminated by switching the damper state only if the carbody lateral acceleration is larger than $\ddot{y}_2 > 0.2\ \text{ms}^{-2}$. If $\ddot{y}_2 < 0.2\ \text{ms}^{-2}$, the damper remains in the state from the previous step, regardless of the result of the equation $\ddot{y}_2(\dot{y}_2 - \dot{y}_1)$. This equation is solved only if $\ddot{y}_2 > 0.2\ \text{ms}^{-2}$. Eliminating unwanted chattering behaviour further improved the efficiency of the algorithm.

2.4.4. Acceleration Driven Damper Linear

This variant of the algorithm Acceleration Driven Damper is new. It has not been described in the literature yet. The ADD algorithm was modified to a linear form according to the Skyhook linear pattern:

$$I = \begin{cases} sat\left(\frac{I_{max} \cdot \ddot{y}_2}{(\dot{y}_2 - \dot{y}_1)} \right), & \ddot{y}_2(\dot{y}_2 - \dot{y}_1) \geq 0 \\ I_{min}, & \ddot{y}_2(\dot{y}_2 - \dot{y}_1) < 0 \end{cases} \tag{8}$$

where I is a required electric current, I_{max} is a maximal electric current, I_{min} is a minimal electric current, $\ddot{y}_2$ is carbody lateral acceleration, $\dot{y}_2$ is carbody lateral velocity, $\dot{y}_1$ is bogie frame lateral velocity, and *sat* denotes that $I \in [I_{min}(v), I_{max}(v)]$.

The new algorithm is called Acceleration Driven Damper Linear (ADD-L). This algorithm has the same problem with unwanted damping force oscillations as ADD-2. This problem has been solved similarly as in the previous case.

2.5. Plan of Experiments and Evaluation Method

The simulations of S/A control were performed in HIL mode (Section 2.2), where the damper stroke $y_2 - y_1$ is obtained based on the vehicle model simulation (Figure 1). Four case studies will be reported: (1) the influence of force rise response time, (2) the influence of force drop response time, (3) the influence of force rise and drop response time together in real proportions, and (4) the influence of dynamic force range. The variables for sensitivity analysis are in Table 2.

Table 2. The variables for sensitivity analysis.

Case	τ_{63} (ms)	τ_{36} (ms)	DR at 0.1 m/s (-)
1	1.8–56	*	7.6
2	*	1.1–56	7.6
3	1.8–56	$\tau_{63}/1.7$	7.6
4	*	*	2–7.6

* Ideal response time for the selected algorithm.

In case 3 the real damper does not have the same long response time for force rise and drop. The ratio between the force rise response time and force drop response time is about 1.7.

The influence of the response time was monitored from the shortest possible response times, determined by the damper design (τ_{63} = 1.8 ms and τ_{36} = 1.1 ms, see Section 2.3.2), to response times of 56 ms. The longer force response times, then 1.8 ms or 1.1 ms, were created using the current controller. The current rises (or drops) exponentially according to the equation:

$$I(t) = I_0 + (I_1 - I_0) \cdot \left(1 - e^{-\frac{t}{\tau}}\right) \tag{9}$$

where I_0 is an electric current at $t = 0$, I_1 is the desired current, t is the time from the step of the control signal, and τ is an artificial current response time that is adjusted as needed.

When response time was slowed down using the current ramp-up, its real value was measured by the methodology described in Section 2.3.2. The required dynamic force range (DR) was set by I_{min}. It showed that S/A algorithms have the best performance when the maximal electric current is I_{max} = 2 A, and passive damping has the best performance when the current is I = 0.5 A. Thus, these currents were used in HILS.

The lateral acceleration of the carbody testifies, above all, to ride comfort. The overall RMS of lateral acceleration in relevant track sections was used to evaluate the comfort of the railway vehicle in on-track tests. This evaluation is also used by the standard EN 14363 [30].

3. Results and Discussion

3.1. Response Time Effect

Figure 6 shows the dependence of carbody lateral acceleration overall RMS on the response time of damping force rise τ_{63} (case 1). The dependency is the same for all four algorithms. With decreasing τ_{63}, the vibrations decrease linearly to τ_{63} = 15 ms. After that, no significant improvement was observed for shorter response times. However, with a shorter time response, noise appears. The noise is caused by a large force impact during damper state switching. Enormous force impact is caused by a rapid change in damping force when a very short response time is used. We assume these force impacts cause unwanted vibrations and thus degrade the results for very short response times. The ideal force rise and drop response times for each algorithm are shown in Table 3. In case 1, the force drop response time was set statically, ideal for each algorithm, according to Table 3.

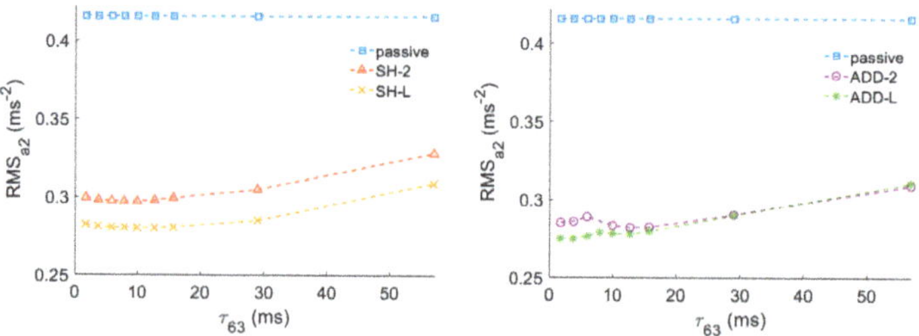

Figure 6. Damper force rise response time τ_{63} effect for passive mode, SH-2 and SH-L algorithms (left) and passive mode, ADD-2 and ADD-L algorithms (right), HILS results.

Table 3. Ideal response times for selected algorithm.

Algorithm	τ_{63} (ms)	τ_{36} (ms)
SH-2	7.9	1.1
SH-L	10	9.5
ADD-2	12.8	46
ADD-L	3.8	46

Figure 7 shows the effect of force drop response time τ_{36} on carbody lateral acceleration overall RMS (case 2). The response time trend on RMS for algorithm SH-2 decreases to $\tau_{36} = 1.1$ ms. In the case of SH-L, the vibrations decrease linearly up to 20 ms, similarly to τ_{63}, and the lowest value is at $\tau_{36} = 9.5$ ms (Figure 7 left). The performance of the ADD-2 and ADD-L algorithms is the best, surprisingly, for a force drop response time around $\tau_{36} = 46$ ms, see Figure 7 right. The force rise response time τ_{63} was set statically, ideal for each algorithm, according to Table 3.

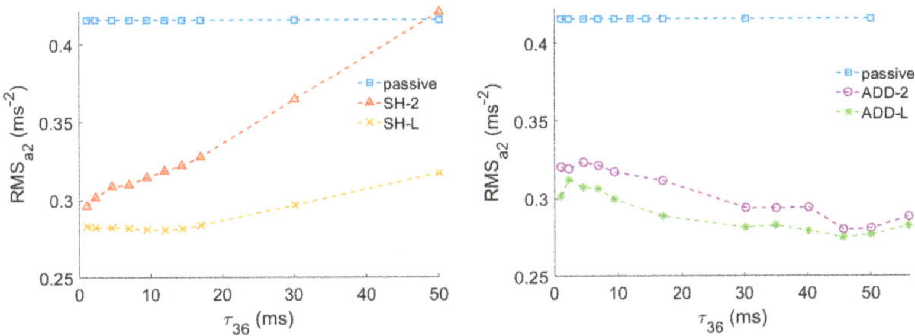

Figure 7. Damper force drop response time τ_{36} effect for passive mode, SH-2 and SH-L algorithms (left) and passive mode, ADD-2 and ADD-L algorithms (right), HILS results.

For SH-2 and SH-L algorithms, the response time for the damping force drop is significantly more critical than for the damping force rise. It is caused by the input electric current being switched to $I_{max} = 2$ A when the damper piston velocity $v_2 - v_1$ is zero (when the force for activated and non-activated states is 0 N) and switched to $I_{min} = 0$ A when the piston velocity and damping force are non-zero, see Figure 8 left. In Figure 8 right, it can be seen that the artificial long response time τ_{36} makes the function of the ADD-2 algorithm similar to the SH-L algorithm function. This similarity explains the importance of a long force drop response time for good results from the ADD-2 and ADD-L algorithms.

The effect of the response time of real damper (including response time for rise and drop) was tested (case 3), see Figure 9. It can be seen that the long response time significantly degrades the efficiency of the SH-2 algorithm. Dependency is linear. The shorter the response time, the lower RMS. However, the algorithm SH-L exhibits a significant improvement in RMS up until a response time of 30 ms. After exceeding this value, the improvement is not observed. For ADD-2 and ADD-L algorithms, it would be ideal to use a damper with a response time of around $\tau_{63} = 16$ ms and the corresponding $\tau_{36} = 9.5$ ms. When using the SH-L, ADD-2, or ADD-L algorithm, it is possible to achieve a vibration reduction of about 22% even with a response time of $\tau_{63} = 57$ ms and the corresponding $\tau_{36} = 33.5$ ms. Strecker et al. [31] wrote that a successful semi-active control requires a short response time of at least 20 ms, but they used the SH-2 algorithm, a different dynamic system for simulating, and an obstacle crossing as an excitation method. In this case, excitation with a relatively low frequency (2 Hz) is used, and more advanced algorithms for S/A control are used, so semi-active control works well even with a relatively long response time. Obviously, reducing the damper response time τ_{63} to less than 8 ms does not make sense for a lateral movement of rail vehicle carbody. Therefore, our MR damper for a railway vehicle with force rise response time of $\tau_{63} = 7.8$ ms [32] is fast enough.

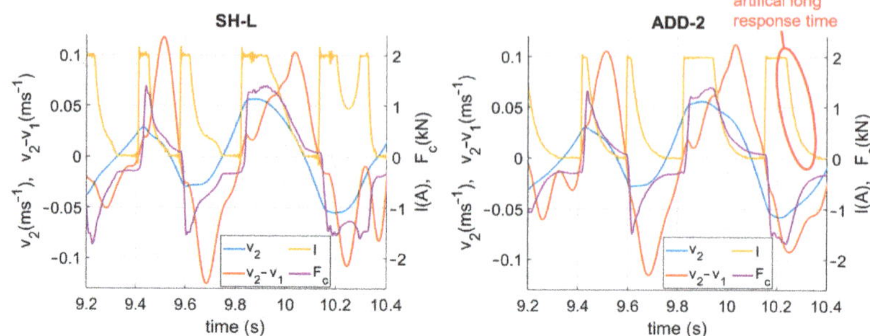

Figure 8. Course of v_2–carbody lateral velocity from the model, $v_2 - v_1$ damper piston velocity from the model, I –measured electric current in damper, F_c–measured damper force, for SH-L control (left) and ADD-2 control (right), HILS results.

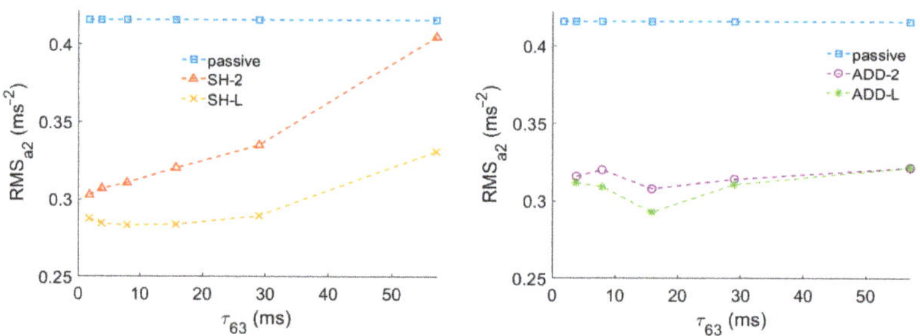

Figure 9. Damper force drop response time τ_{63} effect for passive mode, SH-2 and SH-L algorithms (**left**) and passive mode, ADD-2 and ADD-L algorithms (**right**), HILS results.

3.2. Dynamic Force Range Effect

Figure 10 shows the dependence of the S/A control efficiency on the force dynamic range of the damper (at piston velocity 0.1 ms^{-1}). The vibration maximum RMS value increases with decreasing dynamic range for all algorithms. The Skyhook algorithm is more sensitive than other algorithms. It would be appropriate to change the damper design and make the dynamic range higher than 7.6 to achieve better results. It is possible to achieve a dynamic range of around 12 [33]. From the dependence course, it seems that the ideal value of the dynamic range could be around 10. The force rise and drop response times were set statically, ideal for each algorithm, according to Table 3. All cases of S/A control, except SH-2 with the dynamic range of 2, reduced carbody vibration.

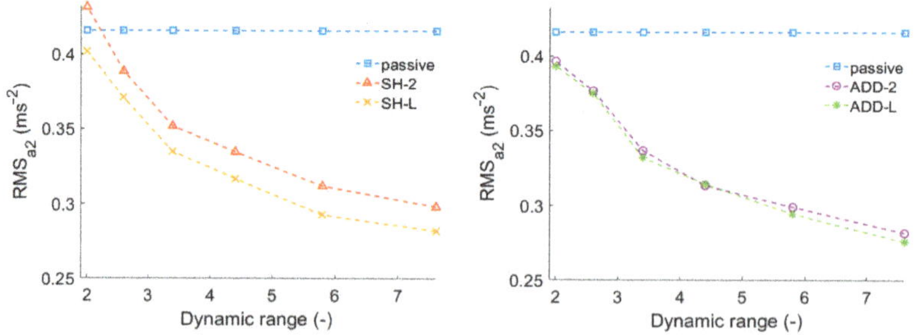

Figure 10. Damper dynamic range DR effect for passive mode, SH-2 and SH-L algorithms (left) and for passive mode, ADD-2 and ADD-L algorithms (right), HILS results.

3.3. Benefits of Each Algorithm

Table 4 shows the overall RMS of the lateral carbody vibrations determined by ideal response times. The most significant vibration reduction was performed by the ADD-L algorithm–33.6%. This value corresponds to the results of most previous research, e.g., [4,9].

Table 4. Carbody lateral acceleration RMS (ms^{-2}) for each mode and percentage reduction compared to passive mode.

Mode	RMS (ms^{-2})	Improvement (%)
passive	0.416	0
SH-2	0.298	28.3
SH-L	0.282	32.2
ADD-2	0.282	32.2
ADD-L	0.276	33.6

In relation to this, SH-L achieved the same result as ADD-2, which was 32.2%. It is an important finding because ADD-2 is easier to apply to a real control system. Only an accelerometer for measuring the sprung mass acceleration and a displacement sensor for measuring the damper piston velocity are required for the ADD-2 application. When using the SH-2 or SH-L algorithm, it is necessary to integrate the sprung mass acceleration to obtain the sprung mass velocity, which leads to problems with the integration constant. The same problem must be solved for the ADD-L algorithm.

Courses of lateral acceleration of vehicle body in the time domain for selected results are shown in Figure 11.

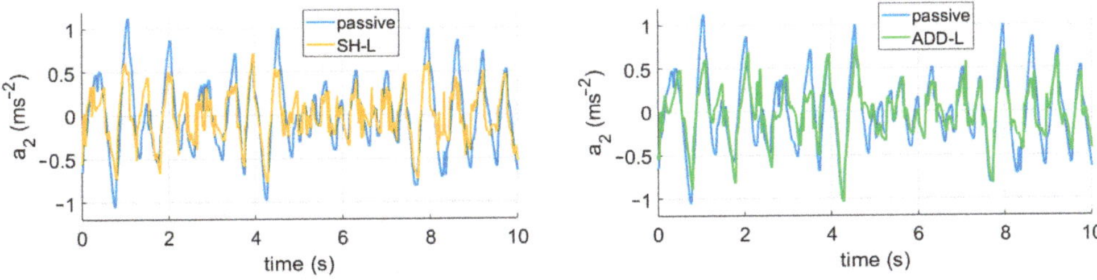

Figure 11. Courses of lateral body acceleration in the time domain for passive mode, SH-L control (**left**) and ADD-L control (**right**), HILS results.

4. Conclusions

The paper deals with the dynamic MR damper behaviour and its influence on the efficiency of four algorithms for semi-active control. A simple model of the lateral movement of a railway vehicle carbody with two degrees of freedom was used for the study. The Hardware-in-the-loop simulation was used. It has been confirmed that S/A control of dampers can significantly reduce carbody vibrations and increase crew comfort. For the selected dynamic system, the chosen excitation method and four selected control strategies, the key findings of this study can be summarised as follows:

- Force drop response time is more important than force rise response time for S/A control performance.
- In this dynamic system, there is no point in shortening the response time to less than $\tau_{63} = 8$ ms.
- The newly designed Acceleration Driven Damper Linear algorithm is best suited for damping the railway vehicle's carbody lateral movement.
- Under ideal conditions, vibrations were reduced by 34%.
- Acceleration Driven Damper (two states) achieves the same effectiveness as Skyhook Linear, but Acceleration Driven Damper is easier to implement in real vehicles.

- For better results, it would be appropriate to increase the dynamic range by at least 10.

In this research, the damper was mounted in the pulsator without silentblocks, which will probably not be possible in practice. However, the damper's soft mounting will worsen the system's hysteretic behaviour and reduce the effectiveness of the S/A control. Silentblocks of various stiffnesses are produced. It would be appropriate to investigate the dependence of the S/A control effectiveness on the damper mounting stiffness and determine what stiffness of the silentblock is acceptable for the S/A control. It will be part of follow-up research.

Author Contributions: Conceptualisation, F.J., M.K., Z.S.; methodology, F.J., Z.S.; software, F.J., T.M., J.Ž.; validation, F.J., J.Ž.; formal analysis, M.K., Z.S.; investigation, F.J., M.K., J.Ž.; data curation, F.J., T.M.; writing—original draft preparation, F.J., T.M.; writing—review and editing, F.J., M.K., T.M., Z.S., J.Ž., I.M.; visualisation, F.J., M.K.; supervision, M.K., Z.S., I.M.; project administration, M.K., I.M.; funding acquisition, M.K., I.M. All authors have read and agreed to the published version of the manuscript.

Funding: This work was supported partly by Czech Science Foundation under Grant GF21-45236L, partly by Technology Agency of the Czech Republic under Grant CK03000052 and party by Brno University of Technology under Grant FSI-S-20-6247.

Institutional Review Board Statement: Not applicable.

Informed Consent Statement: Not applicable.

Data Availability Statement: The data presented in this study are available on request from the corresponding author.

Conflicts of Interest: The authors declare no conflict of interest.

References

1. Goodall, R.M. Control Engineering Challengs for Railway Trains of the Future. *Meas. Control* **2011**, *44*, 16–24. [CrossRef]
2. Pérez, J.; Busturia, J.M.; Goodall, R.M. Control Strategies for Active Steering of Bogie-Based Railway Vehicles. *Control Eng. Pract.* **2002**, *10*, 1005–1012. [CrossRef]
3. Shin, Y.J.; You, W.H.; Hur, H.M.; Park, J.H. Semi-Active Control to Reduce Carbody Vibration of Railway Vehicle by Using Scaled Roller Rig. *J. Mech. Sci. Technol.* **2012**, *26*, 3423–3431. [CrossRef]
4. Codecà, F.; Savaresi, S.M.; Spelta, C.; Montiglio, M.; Ieluzzi, M. Semiactive Control of a Secondary Train Suspension. In Proceedings of the 2007 IEEE/ASME International Conference on Advanced Intelligent Mechatronics, Zurich, Switzerland, 4–7 September 2007. [CrossRef]
5. Strecker, Z.; Jeniš, F.; Kubík, M.; Macháček, O.; Choi, S.B. Novel Approaches to the Design of an Ultra-Fast Magnetorheological Valve for Semi-Active Control. *Materials* **2021**, *14*, 2500. [CrossRef] [PubMed]
6. Jeniš, F.; Kubík, M.; Macháček, O.; Šebesta, K.; Strecker, Z. Insight into the Response Time of Fail-Safe Magnetorheological Damper. *Smart Mater. Struct.* **2021**, *30*, 017004. [CrossRef]
7. Spelta, C.; Savaresi, S.M.; Codecà, F.; Montiglio, M.; Ieluzzi, M. Smart-Bogie: Semi-Active Lateral Control of Railway Vehicles. *Asian J. Control* **2012**, *14*, 875–890. [CrossRef]
8. Lau, Y.K.; Liao, W.H. Design and Analysis of Magnetorheological Dampers for Train Suspension. *Proc. Inst. Mech. Eng. Part F J. Rail Rapid Transit* **2005**, *219*, 261–276. [CrossRef]
9. Shin, Y.-J.; You, W.-H.; Hur, H.-M.; Park, J.-H.; Lee, G.-S. Improvement of Ride Quality of Railway Vehicle by Semiactive Secondary Suspension System on Roller Rig Using Magnetorheological Damper. *Adv. Mech. Eng.* **2014**, *6*, 298382. [CrossRef]
10. Hudha, K.; Harun, M.H.; Harun, M.H.; Jamaluddin, H. Lateral Suspension Control of Railway Vehicle Using Semi-Active Magnetorheological Damper. In Proceedings of the 2011 IEEE Intelligent Vehicles Symposium (IV), Baden-Baden, Germany, 5–9 June 2011; pp. 728–733. [CrossRef]
11. Žáček, J.; Šebesta, K.; Mohammad, H.; Jeniš, F.; Strecker, Z.; Kubík, M. Experimental Evaluation of Modified Groundhook Car Suspension with Fast Magnetorheological Damper. *Actuators* **2022**, *11*, 354. [CrossRef]
12. Jeyasenthil, R.; Yoon, D.S.; Choi, S.B. Response Time Effect of Magnetorheological Dampers in a Semi-Active Vehicle Suspension System: Performance Assessment with Quantitative Feedback Theory. *Smart Mater. Struct.* **2019**, *28*, 054001. [CrossRef]
13. Koo, J.H.; Goncalves, F.D.; Ahmadian, M. A Comprehensive Analysis of the Response Time of MR Dampers. *Smart Mater. Struct.* **2006**, *15*, 351–358. [CrossRef]
14. Yoon, D.S.; Park, Y.J.; Choi, S.B. An Eddy Current Effect on the Response Time of a Magnetorheological Damper: Analysis and Experimental Validation. *Mech. Syst. Signal Process.* **2019**, *127*, 136–158. [CrossRef]

15. Macháček, O.; Kubík, M.; Strecker, Z.; Roupec, J.; Mazůrek, I. Design of a Frictionless Magnetorheological Damper with a High Dynamic Force Range. *Adv. Mech. Eng.* **2019**, *11*, 1687814019827440. [CrossRef]
16. Choi, S.B.; Nam, M.H.; Lee, B.K. Vibration Control of a MR Seat Damper for Commercial Vehicles. *J. Intell. Mater. Syst. Struct.* **2001**, *11*, 936–944. [CrossRef]
17. Lee, H.S.; Choi, S.B. Control and Response Characteristics of a Magneto-Rheological Fluid Damper for Passenger Vehicles. *J. Intell. Mater. Syst. Struct.* **2000**, *11*, 80–87. [CrossRef]
18. Misselhorn, W.E.; Theron, N.J.; Els, P.S. Investigation of Hardware-in-the-Loop for Use in Suspension Development. *Veh. Syst. Dyn.* **2006**, *44*, 65–81. [CrossRef]
19. Kwak, M.K.; Lee, J.H.; Yang, D.H.; You, W.H. Hardware-in-the-Loop Simulation Experiment for Semi-Active Vibration Control of Lateral Vibrations of Railway Vehicle by Magneto-Rheological Fluid Damper. *Veh. Syst. Dyn.* **2014**, *52*, 891–908. [CrossRef]
20. Oh, Y.J.; Lee, J.K.; Liu, H.C.; Cho, S.; Lee, J.; Lee, H.J. Hardware-in-the-Loop Simulation for Active Control of Tramcars with Independently Rotating Wheels. *IEEE Access* **2019**, *7*, 71252–71261. [CrossRef]
21. Zelenka, J.; Michalek, T.; Kohout, M. Comparative Simulations of Guiding Behaviour of an Electric Locomotive. In Proceedings of the 20th International Conference Engineering Mechanics 2014, Svratka, Czech Republic, 12–15 May 2014; pp. 740–743.
22. Goncalves, F.D.; Koo, J.H.; Ahmadian, M. Experimental Approach for Finding the Response Time of Mr Dampers for Vehicle Applications. In *Proceedings of the ASME Design Engineering Technical Conference*; ASME: New York, NY, USA, 2003; Volume 5 A, pp. 425–430.
23. Guan, X.; Guo, P.; Ou, J. Study of the Response Time of MR Dampers. In Proceedings of the Second International Conference on Smart Materials and Nanotechnology in Engineering, Weihai, China, 8–11 July 2009; Volume 7493, p. 74930U. [CrossRef]
24. Koo, J.-H.; Goncalves, F.D.; Ahmadian, M. Investigation of the Response Time of Magnetorheological Fluid Dampers. In *Smart Structures and Materials 2004: Damping and Isolation*; SPIE: Bellingham, WA, USA, 2004; Volume 5386, p. 63. [CrossRef]
25. Strecker, Z.; Roupec, J.; Mazurek, I.; Machacek, O.; Kubik, M.; Klapka, M. Design of Magnetorheological Damper with Short Time Response. *J. Intell. Mater. Syst. Struct.* **2015**, *26*, 1951–1958. [CrossRef]
26. Karnopp, D.; Crosby, M.J.; Harwood, R.A. Vibration Control Using Semi-Active Force Generators. *J. Eng. Ind.* **1974**, *96*, 619. [CrossRef]
27. Sammier, D.; Sename, O.; Dugard, L. Skyhook and H∞ Control of Semi-Active Suspensions: Some Practical Aspects. *Veh. Syst. Dyn.* **2003**, *39*, 279–308. [CrossRef]
28. Savaresi, S.M.; Silani, E.; Bittanti, S. Acceleration-Driven-Damper (ADD): An Optimal Control Algorithm for Comfort-Oriented Semiactive Suspensions. *J. Dyn. Syst. Meas. Control. Trans. ASME* **2005**, *127*, 218–229. [CrossRef]
29. Savaresi, S.M.; Spelta, C. Mixed Sky-Hook and ADD: Approaching the Filtering Limits of a Semi-Active Suspension. *J. Dyn. Syst. Meas. Control. Trans. ASME* **2007**, *129*, 382–392. [CrossRef]
30. EN 14363:2016+A1; Railway Applications—Testing and Simulation for the Acceptance of Running Characteristics of Railway Vehicles—Running Behaviour and Stationary Tests. CEN (European Committee for Standardization): Brusel, Belgium, 2020.
31. Strecker, Z.; Mazůrek, I.; Roupec, J.; Klapka, M. Influence of MR Damper Response Time on Semiactive Suspension Control Efficiency. *Meccanica* **2015**, *50*, 1949–1959. [CrossRef]
32. Kubík, M.; Strecker, Z.; Jeniš, F.; Macháček, O.; Přikryl, M.; Špalek, P. Magnetorheological Yaw Damper with Short Response Time for Rail- Way Vehicle Bogie. In Proceedings of the International Conference and Exhibition on New Actuator Systems and Applications 2021, Online, 17–19 February 2021; Volume 36, pp. 373–376.
33. Yang, G.; Spencer, B.F.; Carlson, J.D.; Sain, M.K. Large-Scale MR Fluid Dampers: Modeling and Dynamic Performance Considerations. *Eng. Struct.* **2002**, *24*, 309–323. [CrossRef]

Communication

Fault Detection and Localisation of a Three-Phase Inverter with Permanent Magnet Synchronous Motor Load Using a Convolutional Neural Network

Dominik Łuczak *, Stefan Brock and Krzysztof Siembab

Faculty of Automatic Control, Robotics and Electrical Engineering, Poznan University of Technology, 60-965 Poznań, Poland
* Correspondence: dominik.luczak@put.poznan.pl

Abstract: Fault-tolerant control of a three-phase inverter can be achieved by performing a hardware reconfiguration of the six-switch and three-phase (6S3P) topology to the four-switch and three-phase (4S3P) topology after detection and localisation of the faulty phase. Together with hardware reconfiguration, the SVPWM algorithm must be appropriately modified to handle the new 4S3P topology. The presented study focuses on diagnosing three-phase faults in two steps: fault detection and localisation. Fault detection is needed to recognise the healthy or unhealthy state of the inverter. The binary state recognition problem can be solved by preparing a feature vector that is calculated from phase currents (i_a, i_b, and i_c) in the time and frequency domains. After the fault diagnosis system recognises the unhealthy state, it investigates the signals to localise which phase of the inverter is faulty. The multiclass classification was solved by a transformation of the three-phase currents into a single RGB image and by training a convolutional neural network. The proposed methodology for the diagnosis of three-phase inverters was tested based on a simulation model representing a laboratory test bench. After the learning process, fault detection was possible based on a 128-sample window (corresponding to a time of 0.64 ms) with an accuracy of 99 percent. In the next step, the localisation of selected individual faults was performed on the basis of a 256-sample window (corresponding to a time of 1.28 ms) with an accuracy of 100 percent.

Keywords: feature extraction; fault diagnosis; convolution neural network; deep neural networks; inverter fault; fault classification

Citation: Łuczak, D.; Brock, S.; Siembab, K. Fault Detection and Localisation of a Three-Phase Inverter with Permanent Magnet Synchronous Motor Load Using a Convolutional Neural Network. *Actuators* **2023**, *12*, 125. https://doi.org/10.3390/act12030125

Academic Editor: Giorgio Olmi

Received: 27 January 2023
Revised: 9 March 2023
Accepted: 11 March 2023
Published: 15 March 2023

1. Introduction

Households and industrial plants are equipped with many electric drives that have limited lifetimes. The proper maintenance of electric drives with mechanical loads is essential for reducing the amount of waste electrical and electronic equipment (WEEE) [1–6]. The time that a machine can satisfactorily operate before requiring repair or replacement is called the remaining useful life (RUL). The RUL can be monitored by changes in characteristics over time, which are caused by the drive and rotating machine. Mechanical vibrations will change due to bearing failure [7–12], unbalance, changes in the mechanical stiffness of the shaft, and changes in the moment of inertia [13–16]. Mechanical loading can be characterized by one mechanical resonance in a two-mass mechanical system [17–21], two or three mechanical resonances in a three- or four-mass system [22–25], or multiple mechanical resonances in a multi-mass system [13–15]. Mechanical resonance is characterized by mechanical vibrations [26]. These problems can lead to inverter overloading and faults in the three-phase inverter. According to [27–31], the inverter fault is mainly caused by an electronic switch. The problem is a fault in the control circuit of a single inverter switch. In the previous stage of the study, it was shown that it is possible to effectively control a three-phase inverter having a 6S3P (six-switch and three-phase) topology when one or two switches in the same phase of the inverter fail; such control requires reconfiguration to a 4S3P (four switches and three phases) topology [32,33]. Therefore, in this research stage, the authors focused on the feature extraction and classification of inverter faults in the cloud. This approach can limit the system deployment time compared with hard embedded programming.

2. Fault-Tolerant Control System

The considered fault-tolerant control system is a control system that can operate under constrained conditions compared with a healthy three-phase inverter. After fault detection and localisation of the faulty inverter phase, the hardware is reconfigured to a 4S3P topology. The fault-tolerant control system consists of two hardware capacitors connected to the PMSM (permanent magnet synchronous motor) phase instead of the faulty inverter phase. Figure 1 shows a healthy 6S3P topology and the reconfiguration to 4S3P in phase C of the PMSM after fault detection and localisation.

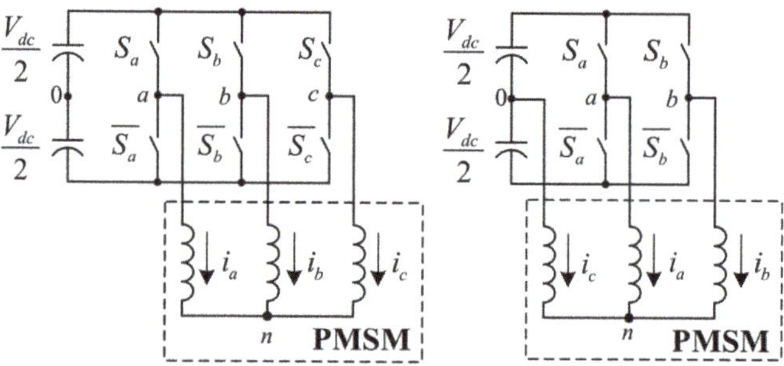

Figure 1. Inverter basic for the 6S3P topology (**left**); fault-tolerant reconfiguration of phase C in the 4S3P topology (**right**).

In a previous stage of investigation, the 6S3P and 4S3P topologies were successfully tested on a laboratory bench (Figure 2) [32,33]. The laboratory rig was equipped with the ALFINE-TIM ALS-G3-1369 controller board, which was based on using the Analog Devices SHARC® ADSP-21369 digital signal processor (DSP), ALFINE-TIM three-phase inverter, and LABINVERTER P3-5.0/550MFE. Further details of the laboratory setup, including the PMSM parameters, have been published in [34].

Figure 2. Photography of the laboratory stand.

The reconfiguration between the 6S3P and 4S3P topology is accomplished by the P_{fault} switch (see Figure 3), which is controlled by the fault diagnosis module described in the following sections. Furthermore, after hardware reconfiguration from 6S3P to 4S3P, the SVPWM (space vector pulse width modulation) switching method must be modified for proper operation [32,33] in the new topology. Other parts of the control system remain unchanged. However, the number of available voltage vectors is reduced.

The parameters of the current controllers R(i_q), R(i_d), and the speed controller R(ω) are unchanged. Empirical tests conducted in a previous research phase confirm that only a modification of the SVPWM is required when the topology is changed to 4S3P after a single-phase fault.

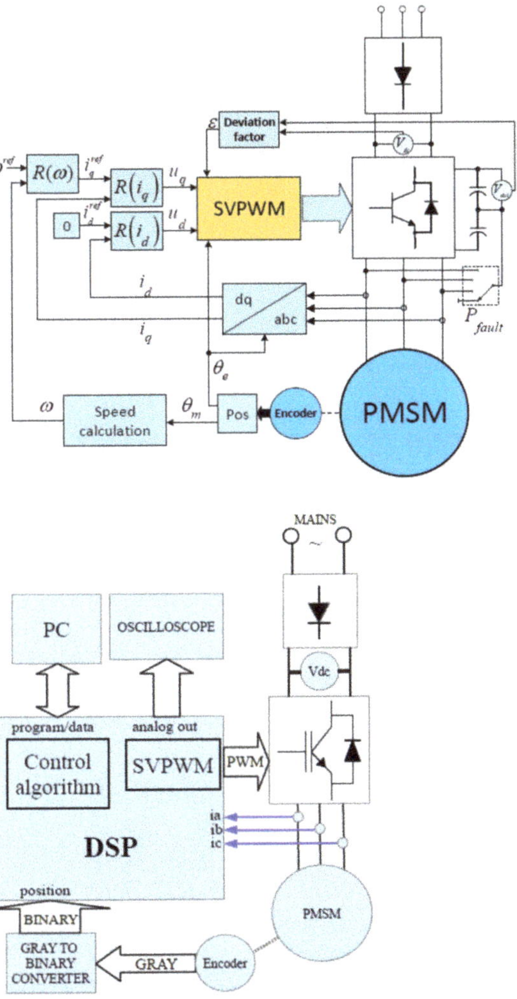

Figure 3. Block diagram of closed-loop vector control (**top**), the general structure of the laboratory stand (**bottom**).

A simulation model of the controller and hardware was created based on the laboratory setup. The inverter model and SVPWM were designed using the MathWorks Simscape Electrical™ tool. The healthy 6S3P inverter was simulated with a reference speed n_{ref} equal to 1200 rpm. The simulation time was set to 2 s with a sampling time of 5 μs. The velocity and current controllers on the q and d axes operate correctly as shown in Figure 4. The fault was simulated at a time equal to 1 s for each switch in the inverter. A total of six datasets were recorded for the fault of the upper and lower switches in each of phases A, B, and C. The data for the upper switches are shown in Figure 5. However, due to the 1 s time scale, only the long-term response of the system speed is easily visible. Therefore, the time zoom for the first 4000 samples after the fault is shown in Figure 6, where the shape of the currents can be observed.

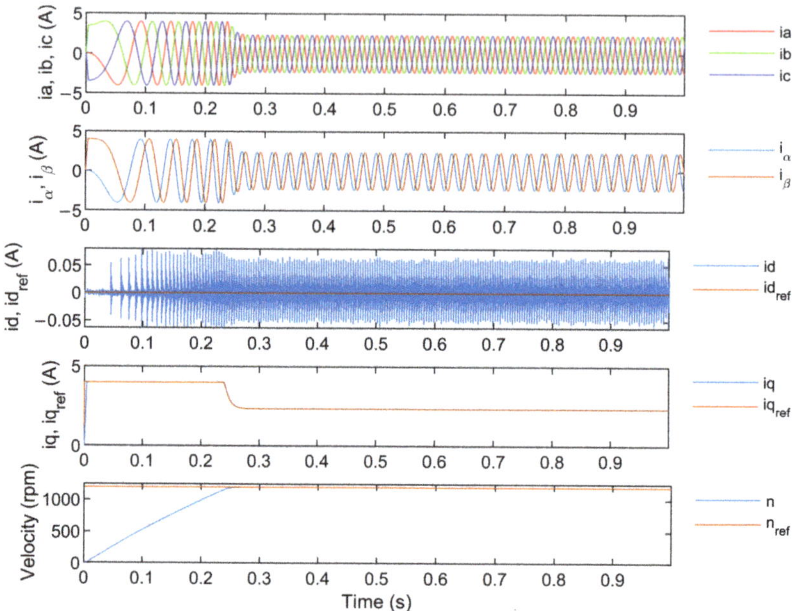

Figure 4. Simulation data collected for the healthy 6S3P topology.

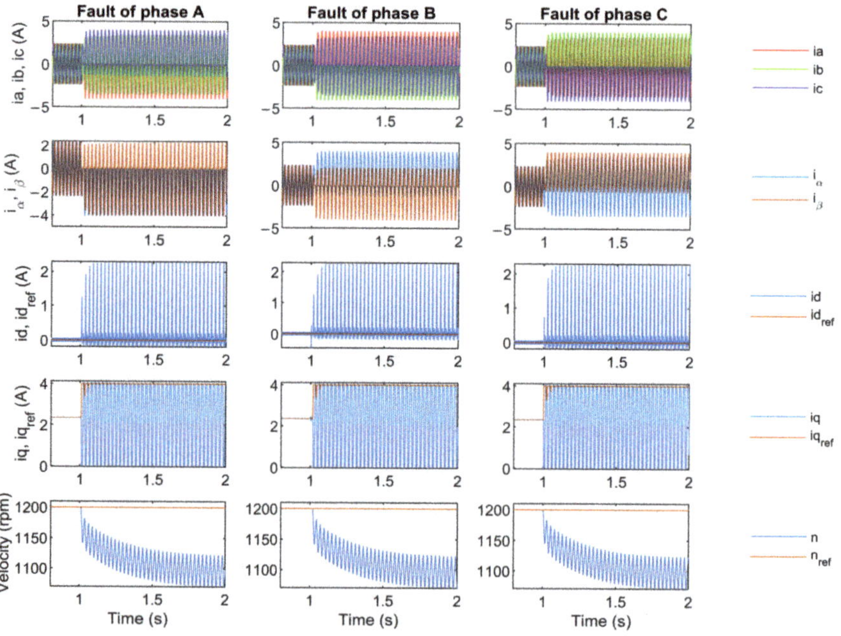

Figure 5. Simulation data collected after the fault of the upper switch in phases A, B, or C of the 6S3P topology.

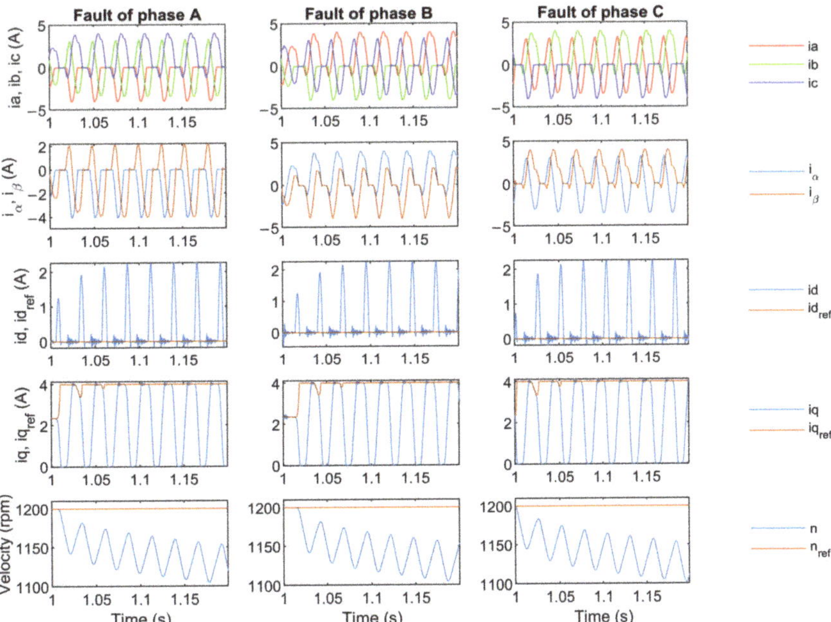

Figure 6. First 4000 samples of simulation after the fault of the upper switch at phase A, B, or C of the 6S3P topology.

3. Fault Detection

A data-driven fault diagnosis system can be developed with fault detection and fault localisation as shown in Figure 7. In this section, the focus is on fault detection, which means that it is only important to have information about the health of the system in one of two states: normal or abnormal. The normal state means that the system is working properly and that there are no symptoms to be concerned about. The abnormal state means that some symptoms of the system are outside of the range that is considered normal. The designed system needs to recognize these two states. This problem can be solved by performing a binary classification with the class labels of "normal" and "fault". The first stage of preparation for the binary classification is data collection for each class and features extraction in the desired time, frequency, or time–frequency domain.

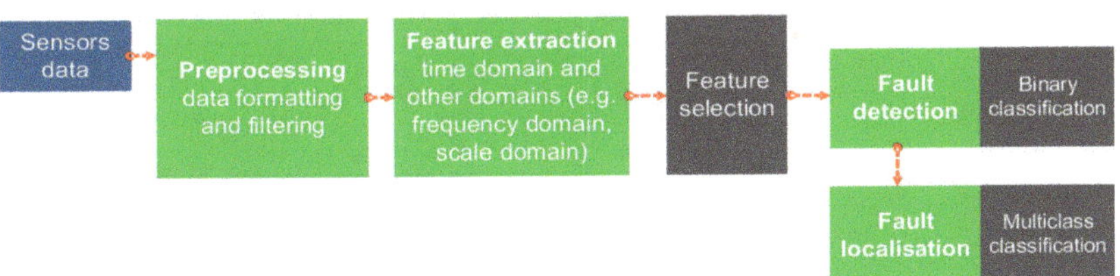

Figure 7. General structure of a data-driven fault detection and localisation system.

The sensor presented in Figure 7 can be of any kind; it can be an additional sensor only for fault diagnosis or a sensor that is already present in the system and being used by the control algorithms. The electromechanical machine or power system can be investigated by using many different sensors and signals: current [35,36] and voltage [37,38], torque [12,39], angular velocity/position [40,41], linear three-axis acceleration/speed/position [42,43], Doppler laser vibrometer [44], transmission coefficient and reflexion coefficient of an omnidirectional antenna [45], strain/tension [46–49], power consumption [50–53], internal/external temperature at selected points [11,54] or surface temperature using a thermal camera [55,56]; furthermore, depending on the frequency range, displacement [57], vibrations [58–61], sound [62–64], sound from several microphones [65] or ultrasound [66,67], vibro-acoustic [7], chemical analyses of lubrication [68,69], chemical analyses using spectral imaging [70–73], camera imaging in the human colour spectrum [74–77], and converting signals to virtual image [78–81] are also possible.

In the current research, data were collected from the currents in phases A, B, and C. These data were divided into a time domain window of 128 samples (Figure 8). Six time domain features were extracted for each current in the window: the standard deviation, variance, median, minimum, maximum, and peak-to-peak. Three frequency-domain features were also extracted: the maximum magnitude frequency component index, minimum magnitude frequency component index, and peak-to-peak frequency magnitude. In total, 27 features were used to train the classification to detect the "normal" or "fault" class. A single-switch fault was considered for each switch in the 6S3P topology. At the current stage of research, 11934 observations were used in the training process and 1325 observations were used in the test. The observation time window had 12 samples overlapping with the previous time window. Training was performed in MathWorks Matlab R2022b using the Statistics and Machine Learning Toolbox version 12.4. More than 16 classifiers with different structures (i.e., linear discriminant; SVM—support vector machine; KNN—k-nearest neighbours; narrow neural network; decision tree; bagged tree) were trained, with a test accuracy of greater than 99% for the selected features (Figure 9).

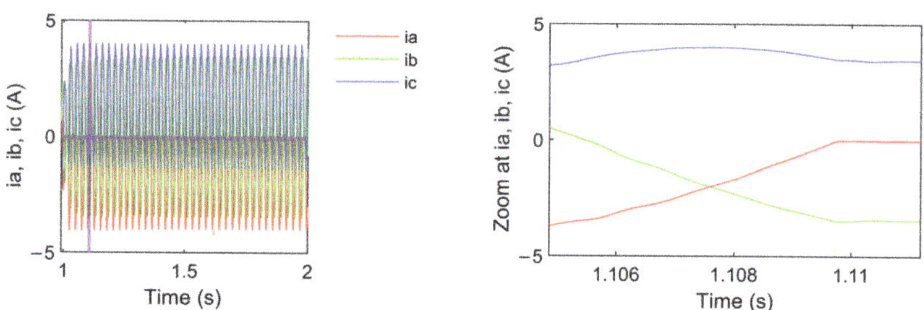

Figure 8. The fault of the upper switch in phase A of the 6S3P topology with a purple time window (**left**) and 128 samples in the selected time domain window (**right**).

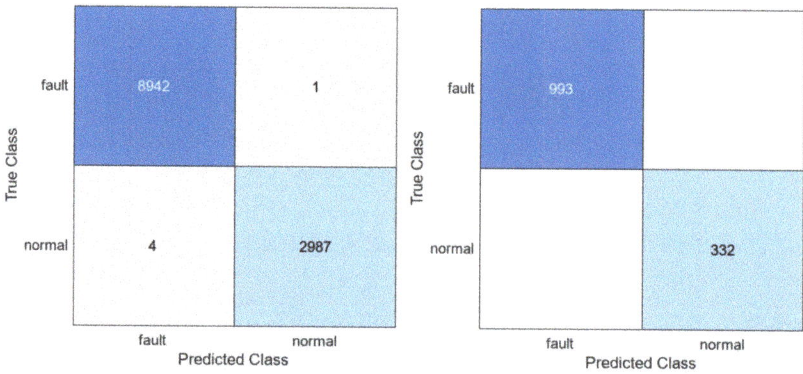

Figure 9. Confusion matrix of a single trained classificator: validation data (**left**) and test data (**right**).

4. Fault Localisation

Fault location provides more information about fault detection. Fault localisation indicates which part of the system is faulty and the extent of the fault. Therefore, the inverter fault diagnosis has been divided into separate tasks: (1) fast fault detection; and (2) the inverter fault localisation phase. Investigating the inverter fault in the 6S3P topology requires identifying the phase where the switch is broken in order to properly change the structure to a 4S3P topology; this problem can be solved by multi-label classification using the 'A', 'B', or 'C' class labels of the faulty phase.

At the current stage of the research, the time domain signal is transformed into an image. The dataset prepared for each class was used to train a CNN (convolutional neural network). The single RGB image consists of three channels: red, green, and blue. The data collection was studied in time window equal to 256 samples, which results in an image size of 16 × 16 pixels. In this approach, the time domain signal was converted into an image with a size of 16 × 16 × 3 (Figure 10), where red, green, and blue colours represent the current in phase A, current in phase B, and current in phase C, respectively. The time domain data of the 256 samples of a single phase are transformed into a 16 × 16 matrix, which is treated as an image. The image columns contain the consecutive samples of the signal. Example images at the same observation time for each phase are shown in Figure 11. There were 618 RGB images for each class, which provided a total of 1854 images. The time window of a single observation had 64 samples overlapping with the previous time window. All images were divided into training, validation, and test sets.

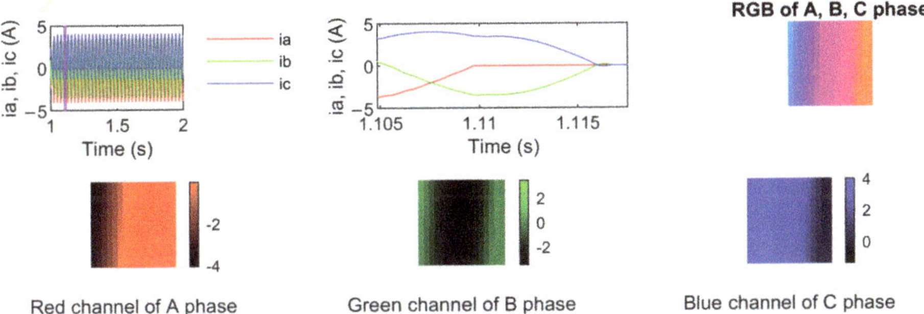

Figure 10. Transformation of the i_a, i_b, and i_c currents into an RGB image for the fault of an upper switch in phase A of the 6S3P topology. The frame of the time window and its RGB image (**top**); each channel of the image (**bottom**).

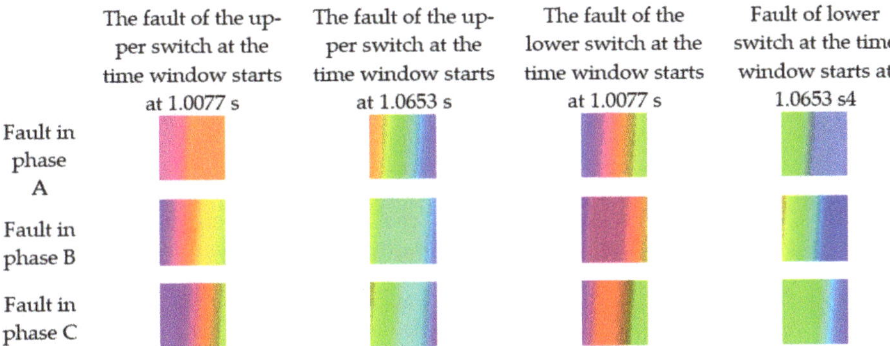

Figure 11. Example images in fault phases A, B, or C for the upper and lower switches of the 6S3P topology.

The three-label classifier ('A', 'B', and 'C') was designed as a convolutional neural network (CNN). An essential part of the design process is the selection of a CNN structure capable of RGB image recognition. At the current stage of the research, the CNN consists of seven layers (Figure 12 [left]): (1) an image input with a size of $16 \times 16 \times 3$; (2) a 2D convolution of $8 \times 8 \times 3$ convolutions; (3) a batch normalization with 20 channels; (4) ReLU; (5) three fully connected layers; (6) softmax; and (7) a classification output with class 'A', 'B', and 'C'. The CNN training was performed in MathWorks Matlab R2022b using Deep Learning Toolbox version 14.5. The training result is shown in Figure 12 (bottom right). The total number of trained parameters in the CNN is 8763. The test accuracy was 100% for the test dataset.

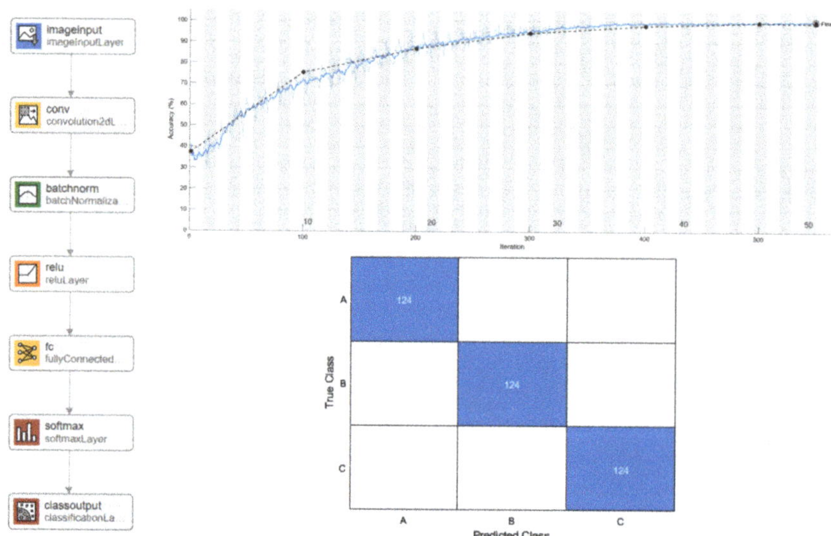

Figure 12. CNN structure (**left**), training process accuracy (**top right**) dark blue – training smoothed; light blue – training; dotted – validation; and test confusion matrix of the learnt CNN (**right bottom**).

5. Discussion

Fault diagnosis firstly requires the detection of a fault; secondly, it requires the location of the fault. The three-phase inverter studied in the 6S3P topology can have one switch (upper or lower) be unhealthy in one phase. An unhealthy condition means that the circuit is opened by a non-working transistor, but the diode is conducting normally. Such a situation is typical for transistor gate driver failures. In the methodology proposed in the first stage, the fault of the inverter is detected by a vector of features that are calculated for each phase current. The features are in the time and frequency domain and are calculated from samples that are collected in a short time window. After fault detection, more samples in a wider time window are examined. The phase currents i_a, i_b, and i_c collected in the time window are transformed into a matrix with a size of $16 \times 16 \times 3$, which corresponds to a 16×16 RGB image. The fault simulation allows for the preparation of the data acquisition. In total, six defects were investigated. For each transistor in phase, a fault was identified and labelled in the three classes corresponding to phases 'A', 'B', and 'C'. This approach allowed us to transform the collected currents into RGB images that were labelled by phase. The RGB image recognition was designed by selecting the CNN structure and performing the training process. The result of the localisation test (multi-class RGB image classification) with 100% accuracy confirms the appropriateness of this approach.

The approach of the RGB image localisation was compared with the build of the reference localisation classifiers based on the selected 27 features used for fault detection.

For all reference localisation classifiers, the time window size was equal to the 256 samples. Therefore, all features were recalculated in each window with 25 overlap samples. Each of the reference classifiers had a worse accuracy (lower than the 99.4%) than proposed approach. The confusion matrix and the accuracy percentage of the reference classifiers are presented in Table 1.

Table 1. Reference localisation classifiers.

Classifier Type	Validation Confusion Matrix	Validation Accuracy
Fine tree	1332 / 4 / 7 — 7 / 1330 / 6 — 9 / 17 / 1317	98.8%
Medium tree	1318 / 4 / 21 — 10 / 1213 / 120 — 18 / 12 / 1313	95.4%
Naive Bayes	1153 / 89 / 101 — 52 / 1176 / 115 — 137 / 27 / 1179	87.1%
SVM (support vector machine)	1264 / 59 / 20 — 263 / 1000 / 80 — (—) / 270 / 1073	82.8%
KNN (k-nearest neighbours)	1293 / 41 / 9 — 83 / 1158 / 102 — 71 / 18 / 1254	92%
Narrow neural network	1335 / 8 / (—) — 4 / 1334 / 5 — 4 / 4 / 1335	99.4%

The detection of the inter-turn short circuit for the PMSM phase was investigated in [82], where three different CNN structures were compared. However, the diagnosis system operated at three full periods of the phase current signal (500 samples), and the fault detection was performed within a time of 0.06 s. At a previous stage of the investigation, the fault detection of the inter-turn short circuit was obtained based on 200 samples, which was equivalent to 0.02 s of measurement [83]. Instead of using a CNN model, fuzzy logic can also be applied in PMSM fault diagnosis [84]; this approach requires an appropriate formulation of fuzzy rules by a specialist in the field of electrical drives. The single-power switch, open-circuit fault was detected 0.08 s after the fault occurrence. In [12], a CNN diagnosed three motor conditions (health motor, demagnetised motor, and motor with bearing fault) based on features extracted in the frequency domain. The drawback of conducting a conversion from time domain data to the frequency domain is that a long time window (large number of samples) is needed to achieve good resolution in the frequency domain, e.g., a time window of 1 s will lead to 1 Hz resolution. Therefore, a long time window is not appropriate for quickly detecting and localising a fault occurring in a 6S3P inverter. The diagnosis system with demagnetisation and semi-demagnetisation faults can use other domain characteristics by applying a DWT (discrete wavelet transform) [85], which requires the selection of the level of DWT decomposition and the choice of mother wavelet shape (one low-pass filter for approximation and one high-pass filter for detail). The suggested approach is to calculate the fault detection of open switch use features in the time and frequency domain in a short time window for fast fault disclosure. The advised data-driven method for fault localisation uses raw data and deep learning without the need for extraction features from the frequency or scale domain. The proposed approach operates around one period of the phase currents with faster fault detection based on a 128-sample window (corresponding to a time of 0.64 ms) and faster localisation of faulty inverter phase based on a 256-sample window (corresponding to a time of 1.28 ms).

6. Conclusions

Research was carried out at TRL (technology readiness level) 1 to validate the proof of concept. Further research will be performed to increase the TRL to higher levels to validate the rotary electric machine (electric drive) with a fault diagnosis system in the laboratory environment. At TRL 1, a single fault of an energo-electronic switch in the six-switch and three-phase (6S3P) topology was considered. This leads to considering one of six possible faults; however, faults of two switches in the same phase were tested. The 4S3P topology with modified SVPWM can operate properly with only one faulty phase, but not more. Multiple faults in different 6S3P phases are less probable. However, the authors are currently developing multiple fault diagnoses for all possible combinations (63 fault classes) of switch faults in a three-phase inverter. The preliminary research of those 63 classes provides less satisfactory results and needs more effort invested into them in future research. The authors at the current research stage considered an abrupt fault without an incipient or intermittent part. Fault detection should return an abnormal state; however, fault localisation will need a further extension for new types of faults; that extension could require a different approach or a retraining of a convolutional neural network with an extended phase current RGB images dataset. Another aspect of possible future research is an investigation of the power distribution system, with the detection and localisation of one of many SVPWM inverter connected in power grid. The issue of a short circuit or break circuit can be found in a hybrid circuit system [86] or vehicles with an internal power grid system, e.g., a car, aircraft [87], or ship.

In the event of a transistor fault in a three-phase inverter with 6S3P topology, it is possible to operate in a fault-tolerant manner after hardware reconfiguration to 4S3P. This approach can be used when one phase fails. The presented research with the proposed methodology in two steps allows for the proper fault detection and precise localisation of the faulty phase. Fault detection is the first step and is a trigger for the execution of the fault localisation part of the system. After fault detection, the control system can be shut down or the reference speed can be slightly reduced to slow down the system. In parallel, the localisation module can detect which phase of the inverter is faulty in order to switch one of the PMSM phases between capacitors. After hardware reconfiguration and modification of the SVPWM algorithm, the system can still operate normally under the constrained current conditions compared with a healthy state. The proposed multi-class classification of phase currents as RGB images provides satisfactory results at the current research stage.

Author Contributions: Feature and RGB image extraction, D.Ł.; fault detection and localisation classifier design, D.Ł.; two-stage fault diagnosis approach, D.Ł. and S.B.; faults selection of three-phase inverter, D.Ł. and S.B.; data generation of three-phase inverter faults, K.S. All authors have read and agreed to the published version of the manuscript.

Funding: This work has been funded by the National Science Centre, Poland (Grant No. 2015/17/N/ST7/03796).

Data Availability Statement: Not applicable.

Conflicts of Interest: The authors declare no conflict of interest.

References

1. Pan, X.; Wong, C.W.Y.; Li, C. Circular Economy Practices in the Waste Electrical and Electronic Equipment (WEEE) Industry: A Systematic Review and Future Research Agendas. *J. Clean. Prod.* **2022**, *365*, 132671. [CrossRef]
2. Rene, E.R.; Sethurajan, M.; Kumar Ponnusamy, V.; Kumar, G.; Bao Dung, T.N.; Brindhadevi, K.; Pugazhendhi, A. Electronic Waste Generation, Recycling and Resource Recovery: Technological Perspectives and Trends. *J. Hazard. Mater.* **2021**, *416*, 125664. [CrossRef]
3. Kan, Y.; Liu, H.; Yang, Y.; Wei, Y.; Yu, Y.; Qiu, R.; Ouyang, Y. Two Birds with One Stone: The Route from Waste Printed Circuit Board Electronic Trash to Multifunctional Biomimetic Slippery Liquid-Infused Coating. *J. Ind. Eng. Chem.* **2022**, *114*, 233–241. [CrossRef]
4. Ji, X.; Yang, M.; Wan, A.; Yu, S.; Yao, Z. Bioleaching of Typical Electronic Waste—Printed Circuit Boards (WPCBs): A Short Review. *Int. J. Environ. Res. Public Health* **2022**, *19*, 7508. [CrossRef]
5. Marinello, S.; Gamberini, R. Multi-Criteria Decision Making Approaches Applied to Waste Electrical and Electronic Equipment (WEEE): A Comprehensive Literature Review. *Toxics* **2021**, *9*, 13. [CrossRef]
6. Breque, M.; De Nul, L.; Petridis, A.; Directorate-General for Research and Innovation (European Commission). *Industry 5.0: Towards a Sustainable, Human Centric and Resilient European Industry*; Publications Office of the European Union: Luxembourg, 2021; ISBN 978-92-76-25308-2. [CrossRef]
7. Wang, X.; Mao, D.; Li, X. Bearing Fault Diagnosis Based on Vibro-Acoustic Data Fusion and 1D-CNN Network. *Measurement* **2021**, *173*, 108518. [CrossRef]
8. Liu, D.; Cheng, W.; Wen, W. Rolling Bearing Fault Diagnosis via STFT and Improved Instantaneous Frequency Estimation Method. *Procedia Manuf.* **2020**, *49*, 166–172. [CrossRef]
9. Miao, Y.; Zhang, B.; Li, C.; Lin, J.; Zhang, D. Feature Mode Decomposition: New Decomposition Theory for Rotating Machinery Fault Diagnosis. *IEEE Trans. Ind. Electron.* **2023**, *70*, 1949–1960. [CrossRef]
10. Han, T.; Ding, L.; Qi, D.; Li, C.; Fu, Z.; Chen, W. Compound Faults Diagnosis Method for Wind Turbine Mainshaft Bearing with Teager and Second-Order Stochastic Resonance. *Measurement* **2022**, *202*, 111931. [CrossRef]
11. Dhiman, H.S.; Deb, D.; Muyeen, S.M.; Kamwa, I. Wind Turbine Gearbox Anomaly Detection Based on Adaptive Threshold and Twin Support Vector Machines. *IEEE Trans. Energy Convers.* **2021**, *36*, 3462–3469. [CrossRef]
12. Wang, C.-S.; Kao, I.-H.; Perng, J.-W. Fault Diagnosis and Fault Frequency Determination of Permanent Magnet Synchronous Motor Based on Deep Learning. *Sensors* **2021**, *21*, 3608. [CrossRef]
13. Łuczak, D. Nonlinear Identification with Constraints in Frequency Domain of Electric Direct Drive with Multi-Resonant Mechanical Part. *Energies* **2021**, *14*, 7190. [CrossRef]
14. Brock, S.; Luczak, D.; Nowopolski, K.; Pajchrowski, T.; Zawirski, K. Two Approaches to Speed Control for Multi-Mass System with Variable Mechanical Parameters. *IEEE Trans. Ind. Electron.* **2016**, *64*, 3338–3347. [CrossRef]
15. Luczak, D. Mathematical Model of Multi-Mass Electric Drive System with Flexible Connection. In Proceedings of the 19th International Conference On Methods and Models in Automation and Robotics (MMAR), Miedzyzdroje, Poland, 2–5 September 2014; pp. 590–595.
16. Luczak, D.; Nowopolski, K. Identification of Multi-Mass Mechanical Systems in Electrical Drives. In Proceedings of the 2014 16th International Conference on Mechatronics—Mechatronika (ME), Brno, Czech Republic, 3–5 December 2014; pp. 275–282.
17. Wróbel, K.; Serkies, P.; Szabat, K. Model Predictive Base Direct Speed Control of Induction Motor Drive—Continuous and Finite Set Approaches. *Energies* **2020**, *13*, 1193. [CrossRef]

18. Szabat, K.; Wróbel, K.; Dróżdż, K.; Janiszewski, D.; Pajchrowski, T.; Wójcik, A. A Fuzzy Unscented Kalman Filter in the Adaptive Control System of a Drive System with a Flexible Joint. *Energies* **2020**, *13*, 2056. [CrossRef]
19. Nalepa, R.; Najdek, K.; Wróbel, K.; Szabat, K. Application of D-Decomposition Technique to Selection of Controller Parameters for a Two-Mass Drive System. *Energies* **2020**, *13*, 6614. [CrossRef]
20. Wicher, B.; Brock, S. Comparison of Robustness of Selected Speed Control Systems Applied for Two Mass System with Backlash. In *Advanced, Contemporary Control*; Bartoszewicz, A., Kabziński, J., Kacprzyk, J., Eds.; Springer International Publishing: Cham, Switzerland, 2020; pp. 1371–1382.
21. Guerra, R.H.; Quiza, R.; Villalonga, A.; Arenas, J.; Castaño, F. Digital Twin-Based Optimization for Ultraprecision Motion Systems With Backlash and Friction. *IEEE Access* **2019**, *7*, 93462–93472. [CrossRef]
22. Luczak, D.; Siwek, P.; Nowopolski, K. Speed Controller for Four-Mass Mechanical System with Two Drive Units. In Proceedings of the 2015 IEEE 2nd International Conference on Cybernetics (CYBCONF), Gdynia, Poland, 24–26 June 2015; pp. 449–454.
23. Hung, Y.-H.; Lee, C.-Y.; Tsai, C.-H.; Lu, Y.-M. Constrained Particle Swarm Optimization for Health Maintenance in Three-Mass Resonant Servo Control System with LuGre Friction Model. *Ann. Oper. Res.* **2022**, *311*, 131–150. [CrossRef]
24. Korendiy, V.; Kachur, O.; Gursky, V.; Gurey, V.; Maherus, N.; Kotsiumbas, O.; Havrylchenko, O. Kinematic and Dynamic Analysis of Three-Mass Oscillatory System of Vibro-Impact Plate Compactor with Crank Excitation Mechanism. *Vibroengineering PROCEDIA* **2022**, *40*, 14–19. [CrossRef]
25. Binder, D.; Bendrat, F.; Sourkounis, C. Model Predictive Control of a High Power Rolling-Mill Drive Considering Shaft Torque Constraints. In Proceedings of the IECON 2021—47th Annual Conference of the IEEE Industrial Electronics Society, Toronto, ON, Canada, 13–16 October 2021; pp. 1–7.
26. Łuczak, D. Mechanical Vibrations Analysis in Direct Drive Using CWT with Complex Morlet Wavelet. *Power Electron. Drives* **2023**, *8*, 65–73. [CrossRef]
27. Wang, B.; Cai, J.; Du, X.; Zhou, L. Review of Power Semiconductor Device Reliability for Power Converters. *CPSS Trans. Power Electron. Appl.* **2017**, *2*, 101–117. [CrossRef]
28. Manohar, S.S.; Sahoo, A.; Subramaniam, A.; Panda, S.K. Condition Monitoring of Power Electronic Converters in Power Plants—A Review. In Proceedings of the 2017 20th International Conference on Electrical Machines and Systems (ICEMS), Sydney, NSW, Australia, 11–14 August 2017; pp. 1–5.
29. Yang, S.; Xiang, D.; Bryant, A.; Mawby, P.; Ran, L.; Tavner, P. Condition Monitoring for Device Reliability in Power Electronic Converters: A Review. *IEEE Trans. Power Electron.* **2010**, *25*, 2734–2752. [CrossRef]
30. Spinato, F.; Tavner, P.J.; van Bussel, G.J.W.; Koutoulakos, E. Reliability of Wind Turbine Subassemblies. *IET Renew. Power Gener.* **2009**, *3*, 387–401. [CrossRef]
31. Lu, B.; Sharma, S.K. A Literature Review of IGBT Fault Diagnostic and Protection Methods for Power Inverters. *IEEE Trans. Ind. Appl.* **2009**, *45*, 1770–1777. [CrossRef]
32. Łuczak, D.; Siembab, K. Comparison of Fault Tolerant Control Algorithm Using Space Vector Modulation of PMSM Drive. In Proceedings of the 16th International Conference on Mechatronics—Mechatronika 2014, Brno, Czech Republic, 3–5 December 2014; pp. 24–31.
33. Siembab, K.; Zawirski, K. Modified Space Vector Modulation for Fault Tolerant Control of PMSM Drive. In Proceedings of the 2016 IEEE International Power Electronics and Motion Control Conference (PEMC), Varna, Bulgaria, 25–28 September 2016; pp. 1064–1071.
34. Łuczak, D.; Nowopolski, K.; Siembab, K.; Wicher, B. PMSM Laboratory Stand for Investigations on Advanced Structures of Electrical Drive Control. In Proceedings of the 2015 20th International Conference on Methods and Models in Automation and Robotics (MMAR), Miedzyzdroje, Poland, 24–27 August 2015; pp. 596–601.
35. Huang, W.; Du, J.; Hua, W.; Lu, W.; Bi, K.; Zhu, Y.; Fan, Q. Current-Based Open-Circuit Fault Diagnosis for PMSM Drives With Model Predictive Control. *IEEE Trans. Power Electron.* **2021**, *36*, 10695–10704. [CrossRef]
36. Wu, C.; Guo, C.; Xie, Z.; Ni, F.; Liu, H. A Signal-Based Fault Detection and Tolerance Control Method of Current Sensor for PMSM Drive. *IEEE Trans. Ind. Electron.* **2018**, *65*, 9646–9657. [CrossRef]
37. Jiang, L.; Deng, Z.; Tang, X.; Hu, L.; Lin, X.; Hu, X. Data-Driven Fault Diagnosis and Thermal Runaway Warning for Battery Packs Using Real-World Vehicle Data. *Energy* **2021**, *234*, 121266. [CrossRef]
38. Chang, C.; Zhou, X.; Jiang, J.; Gao, Y.; Jiang, Y.; Wu, T. Electric Vehicle Battery Pack Micro-Short Circuit Fault Diagnosis Based on Charging Voltage Ranking Evolution. *J. Power Sources* **2022**, *542*, 231733. [CrossRef]
39. Gao, S.; Xu, L.; Zhang, Y.; Pei, Z. Rolling Bearing Fault Diagnosis Based on SSA Optimized Self-Adaptive DBN. *ISA Trans.* **2022**, *128*, 485–502. [CrossRef]
40. Feng, Z.; Gao, A.; Li, K.; Ma, H. Planetary Gearbox Fault Diagnosis via Rotary Encoder Signal Analysis. *Mech. Syst. Signal Process.* **2021**, *149*, 107325. [CrossRef]
41. Ma, J.; Li, C.; Zhang, G. Rolling Bearing Fault Diagnosis Based on Deep Learning and Autoencoder Information Fusion. *Symmetry* **2022**, *14*, 13. [CrossRef]
42. Kim, M.S.; Yun, J.P.; Park, P. Deep Learning-Based Explainable Fault Diagnosis Model With an Individually Grouped 1-D Convolution for Three-Axis Vibration Signals. *IEEE Trans. Ind. Inform.* **2022**, *18*, 8807–8817. [CrossRef]
43. Zhang, X.; Zhao, Z.; Wang, Z.; Wang, X. Fault Detection and Identification Method for Quadcopter Based on Airframe Vibration Signals. *Sensors* **2021**, *21*, 581. [CrossRef] [PubMed]

44. Abbas, S.H.; Jang, J.-K.; Kim, D.-H.; Lee, J.-R. Underwater Vibration Analysis Method for Rotating Propeller Blades Using Laser Doppler Vibrometer. *Opt. Lasers Eng.* **2020**, *132*, 106133. [CrossRef]
45. Dutta, S.; Basu, B.; Talukdar, F.A. Classification of Motor Faults Based on Transmission Coefficient and Reflection Coefficient of Omni-Directional Antenna Using DCNN. *Expert Syst. Appl.* **2022**, *198*, 116832. [CrossRef]
46. Zhang, X.; Niu, H.; Hou, C.; Di, F. An Edge-Filter FBG Interrogation Approach Based on Tunable Fabry-Perot Filter for Strain Measurement of Planetary Gearbox. *Opt. Fiber Technol.* **2020**, *60*, 102379. [CrossRef]
47. Zhang, P.; Lu, D. A Survey of Condition Monitoring and Fault Diagnosis toward Integrated O&M for Wind Turbines. *Energies* **2019**, *12*, 2801. [CrossRef]
48. Wu, J.; Yang, Y.; Wang, P.; Wang, J.; Cheng, J. A Novel Method for Gear Crack Fault Diagnosis Using Improved Analytical-FE and Strain Measurement. *Measurement* **2020**, *163*, 107936. [CrossRef]
49. Fedorko, G.; Molnár, V.; Vasiľ, M.; Salai, R. Proposal of Digital Twin for Testing and Measuring of Transport Belts for Pipe Conveyors within the Concept Industry 4.0. *Measurement* **2021**, *174*, 108978. [CrossRef]
50. Pu, H.; He, L.; Zhao, C.; Yau, D.K.Y.; Cheng, P.; Chen, J. Fingerprinting Movements of Industrial Robots for Replay Attack Detection. *IEEE Trans. Mob. Comput.* **2022**, *21*, 3629–3643. [CrossRef]
51. Rafati, A.; Shaker, H.R.; Ghahghahzadeh, S. Fault Detection and Efficiency Assessment for HVAC Systems Using Non-Intrusive Load Monitoring: A Review. *Energies* **2022**, *15*, 341. [CrossRef]
52. Sabry, A.H.; Nordin, F.H.; Sabry, A.H.; Abidin Ab Kadir, M.Z. Fault Detection and Diagnosis of Industrial Robot Based on Power Consumption Modeling. *IEEE Trans. Ind. Electron.* **2020**, *67*, 7929–7940. [CrossRef]
53. Sánchez-Sutil, F.; Cano-Ortega, A.; Hernández, J.C. Design and Implementation of a Smart Energy Meter Using a LoRa Network in Real Time. *Electronics* **2021**, *10*, 3152. [CrossRef]
54. Wang, Z.; Tian, B.; Qiao, W.; Qu, L. Real-Time Aging Monitoring for IGBT Modules Using Case Temperature. *IEEE Trans. Ind. Electron.* **2016**, *63*, 1168–1178. [CrossRef]
55. Glowacz, A. Fault Diagnosis of Electric Impact Drills Using Thermal Imaging. *Measurement* **2021**, *171*, 108815. [CrossRef]
56. Al-Musawi, A.K.; Anayi, F.; Packianather, M. Three-Phase Induction Motor Fault Detection Based on Thermal Image Segmentation. *Infrared Phys. Technol.* **2020**, *104*, 103140. [CrossRef]
57. Li, Z.; Zhang, Y.; Abu-Siada, A.; Chen, X.; Li, Z.; Xu, Y.; Zhang, L.; Tong, Y. Fault Diagnosis of Transformer Windings Based on Decision Tree and Fully Connected Neural Network. *Energies* **2021**, *14*, 1531. [CrossRef]
58. Wang, Y.; Yang, M.; Li, Y.; Xu, Z.; Wang, J.; Fang, X. A Multi-Input and Multi-Task Convolutional Neural Network for Fault Diagnosis Based on Bearing Vibration Signal. *IEEE Sens. J.* **2021**, *21*, 10946–10956. [CrossRef]
59. Rauber, T.W.; da Silva Loca, A.L.; Boldt, F.d.A.; Rodrigues, A.L.; Varejão, F.M. An Experimental Methodology to Evaluate Machine Learning Methods for Fault Diagnosis Based on Vibration Signals. *Expert Syst. Appl.* **2021**, *167*, 114022. [CrossRef]
60. Meyer, A. Vibration Fault Diagnosis in Wind Turbines Based on Automated Feature Learning. *Energies* **2022**, *15*, 1514. [CrossRef]
61. Lee, J.-H.; Pack, J.-H.; Lee, I.-S. Fault Diagnosis of Induction Motor Using Convolutional Neural Network. *Appl. Sci.* **2019**, *9*, 2950. [CrossRef]
62. Cao, Y.; Sun, Y.; Xie, G.; Li, P. A Sound-Based Fault Diagnosis Method for Railway Point Machines Based on Two-Stage Feature Selection Strategy and Ensemble Classifier. *IEEE Trans. Intell. Transp. Syst.* **2022**, *23*, 12074–12083. [CrossRef]
63. Shiri, H.; Wodecki, J.; Ziętek, B.; Zimroz, R. Inspection Robotic UGV Platform and the Procedure for an Acoustic Signal-Based Fault Detection in Belt Conveyor Idler. *Energies* **2021**, *14*, 7646. [CrossRef]
64. Karabacak, Y.E.; Gürsel Özmen, N.; Gümüşel, L. Intelligent Worm Gearbox Fault Diagnosis under Various Working Conditions Using Vibration, Sound and Thermal Features. *Appl. Acoust.* **2022**, *186*, 108463. [CrossRef]
65. Yao, Y.; Wang, H.; Li, S.; Liu, Z.; Gui, G.; Dan, Y.; Hu, J. End-To-End Convolutional Neural Network Model for Gear Fault Diagnosis Based on Sound Signals. *Appl. Sci.* **2018**, *8*, 1584. [CrossRef]
66. Zhang, Z.; Li, J.; Song, Y.; Sun, Y.; Zhang, X.; Hu, Y.; Guo, R.; Han, X. A Novel Ultrasound-Vibration Composite Sensor for Defects Detection of Electrical Equipment. *IEEE Trans. Power Deliv.* **2022**, *37*, 4477–4480. [CrossRef]
67. Wang, W.; Xue, Y.; He, C.; Zhao, Y. Review of the Typical Damage and Damage-Detection Methods of Large Wind Turbine Blades. *Energies* **2022**, *15*, 5672. [CrossRef]
68. Maruyama, T.; Maeda, M.; Nakano, K. Lubrication Condition Monitoring of Practical Ball Bearings by Electrical Impedance Method. *Tribol. Online* **2019**, *14*, 327–338. [CrossRef]
69. Wakiru, J.M.; Pintelon, L.; Muchiri, P.N.; Chemweno, P.K. A Review on Lubricant Condition Monitoring Information Analysis for Maintenance Decision Support. *Mech. Syst. Signal Process.* **2019**, *118*, 108–132. [CrossRef]
70. Rizk, P.; Younes, R.; Ilinca, A.; Khoder, J. Wind Turbine Ice Detection Using Hyperspectral Imaging. *Remote Sens. Appl. Soc. Environ.* **2022**, *26*, 100711. [CrossRef]
71. Rizk, P.; Younes, R.; Ilinca, A.; Khoder, J. Wind Turbine Blade Defect Detection Using Hyperspectral Imaging. *Remote Sens. Appl. Soc. Environ.* **2021**, *22*, 100522. [CrossRef]
72. Meribout, M. Gas Leak-Detection and Measurement Systems: Prospects and Future Trends. *IEEE Trans. Instrum. Meas.* **2021**, *70*, 1–13. [CrossRef]
73. Li, Y.; Yu, Q.; Xie, M.; Zhang, Z.; Ma, Z.; Cao, K. Identifying Oil Spill Types Based on Remotely Sensed Reflectance Spectra and Multiple Machine Learning Algorithms. *IEEE J. Sel. Top. Appl. Earth Obs. Remote Sens.* **2021**, *14*, 9071–9078. [CrossRef]

74. Zhou, Q.; Chen, R.; Huang, B.; Liu, C.; Yu, J.; Yu, X. An Automatic Surface Defect Inspection System for Automobiles Using Machine Vision Methods. *Sensors* **2019**, *19*, 644. [CrossRef] [PubMed]
75. Yang, L.; Fan, J.; Liu, Y.; Li, E.; Peng, J.; Liang, Z. A Review on State-of-the-Art Power Line Inspection Techniques. *IEEE Trans. Instrum. Meas.* **2020**, *69*, 9350–9365. [CrossRef]
76. Davari, N.; Akbarizadeh, G.; Mashhour, E. Intelligent Diagnosis of Incipient Fault in Power Distribution Lines Based on Corona Detection in UV-Visible Videos. *IEEE Trans. Power Deliv.* **2021**, *36*, 3640–3648. [CrossRef]
77. Kim, S.; Kim, D.; Jeong, S.; Ham, J.-W.; Lee, J.-K.; Oh, K.-Y. Fault Diagnosis of Power Transmission Lines Using a UAV-Mounted Smart Inspection System. *IEEE Access* **2020**, *8*, 149999–150009. [CrossRef]
78. Ullah, Z.; Lodhi, B.A.; Hur, J. Detection and Identification of Demagnetization and Bearing Faults in PMSM Using Transfer Learning-Based VGG. *Energies* **2020**, *13*, 3834. [CrossRef]
79. Long, H.; Xu, S.; Gu, W. An Abnormal Wind Turbine Data Cleaning Algorithm Based on Color Space Conversion and Image Feature Detection. *Appl. Energy* **2022**, *311*, 118594. [CrossRef]
80. Kreutz, M.; Alla, A.A.; Eisenstadt, A.; Freitag, M.; Thoben, K.-D. Ice Detection on Rotor Blades of Wind Turbines Using RGB Images and Convolutional Neural Networks. *Procedia CIRP* **2020**, *93*, 1292–1297. [CrossRef]
81. Xie, T.; Huang, X.; Choi, S.-K. Intelligent Mechanical Fault Diagnosis Using Multisensor Fusion and Convolution Neural Network. *IEEE Trans. Ind. Inform.* **2022**, *18*, 3213–3223. [CrossRef]
82. Skowron, M.; Orlowska-Kowalska, T.; Kowalski, C.T. Application of Simplified Convolutional Neural Networks for Initial Stator Winding Fault Detection of the PMSM Drive Using Different Raw Signal Data. *IET Electr. Power Appl.* **2021**, *15*, 932–946. [CrossRef]
83. Skowron, M.; Orlowska-Kowalska, T.; Wolkiewicz, M.; Kowalski, C.T. Convolutional Neural Network-Based Stator Current Data-Driven Incipient Stator Fault Diagnosis of Inverter-Fed Induction Motor. *Energies* **2020**, *13*, 1475. [CrossRef]
84. Yan, H.; Xu, Y.; Cai, F.; Zhang, H.; Zhao, W.; Gerada, C. PWM-VSI Fault Diagnosis for a PMSM Drive Based on the Fuzzy Logic Approach. *IEEE Trans. Power Electron.* **2019**, *34*, 759–768. [CrossRef]
85. Kao, I.-H.; Wang, W.-J.; Lai, Y.-H.; Perng, J.-W. Analysis of Permanent Magnet Synchronous Motor Fault Diagnosis Based on Learning. *IEEE Trans. Instrum. Meas.* **2019**, *68*, 310–324. [CrossRef]
86. Xiao, C.; Yu, M.; Zhang, B.; Wang, H.; Jiang, C. Discrete Component Prognosis for Hybrid Systems Under Intermittent Faults. *IEEE Trans. Autom. Sci. Eng.* **2021**, *18*, 1766–1777. [CrossRef]
87. Yaramasu, A.; Cao, Y.; Liu, G.; Wu, B. Aircraft Electric System Intermittent Arc Fault Detection and Location. *IEEE Trans. Aerosp. Electron. Syst.* **2015**, *51*, 40–51. [CrossRef]

Article

Comparing Different Resonant Control Approaches for Torque Ripple Minimisation in Electric Machines

Thomas Steffen [1,*], Muhammad Saad Rafaq [2] and Will Midgley [3]

1 Department for Aeronautical and Automotive Engineering, Loughborough University, Loughborough LE11 3TU, UK
2 Wolfson School of Mechanical, Electrical and Manufacturing Engineering, Loughborough University, Loughborough LE11 3TU, UK
3 School of Mechanical and Manufacturing Engineering, University of New South Wales, Sydney, NSW 2052, Australia
* Correspondence: t.steffen@lboro.ac.uk or t.steffen@ieee.org; Tel.: +44-1509-227224

Abstract: Electric machines are highly efficient and highly controllable actuators, but they do still suffer from a number of imperfections. One of them is torque ripple, which introduces high frequency harmonics into the motion. One (cost- and performance-neutral) countermeasure is to apply control that counters the torque ripple. This paper compares several single-input single-output (SISO) control approaches for feedback control of torque ripple of a Permanent Magnet Synchronous Machine (PMSM). The baseline is PI (proportional-integral) control, which does not suppress torque ripple, and the most popular control approach is proportional-integral resonant (PIR) control. Both are compared to an advanced PIR controller (PIRA), frequency modulation, a mixed sensitivity design, and an iterative learning controller (ILC). The analysis demonstrates that PIR control, mixed sensitivity state feedback, and the modulating controller achieve identical behaviour. The choice between these three options is therefore dependent on preferences for the design methodology, or on implementation factors. The PIRA and the ILC on the other hand show more sophisticated behaviour that may be advantageous for certain applications, at the expense of higher complexity.

Keywords: electric machines; torque ripple; resonant control; modulation and demodulation control; mixed sensitivity loop shaping; iterative learning control

Citation: Steffen, T.; Rafaq, M.S.; Midgley, W. Comparing Different Resonant Control Approaches for Torque Ripple Minimisation in Electric Machines. *Actuators* **2022**, *11*, 349. https://doi.org/10.3390/act11120349

Academic Editor: Zhuming Bi

Received: 31 October 2022
Accepted: 25 November 2022
Published: 27 November 2022

Publisher's Note: MDPI stays neutral with regard to jurisdictional claims in published maps and institutional affiliations.

1. Introduction

This paper deals with problem of suppressing torque ripple in electric machines using feedback control. Torque ripple is a serious problem with electric machines, and it can come from a variety of sources, from cogging torque with stator teeth over harmonics in the back-EMF curve to defficiencies in the control scheme [1,2].

Different methods have been proposed to deal with this problem. Constructive measures are able to reduce torque ripple, but they often have a financial cost and a performance impact [3]. Feedforward control is commonly used in industry, where either a back-EMF curve different from a sinusoid is assumed [4], or a current is added based on the rotor position [5]. Any open loop approach will require extensive callibration work to be effective, and it is sensitive to product-to-product variaton [6,7].

For this reason, this paper follows a closed-loop control approach. It starts with a simplified control challenge of suppressing the torque ripple using a closed loop approach by controlling the voltage of the motor based on the speed measurement. In addition to a baseline proportional integral (PI) controller that is not able to address this challenge, four control laws are considered. The first is a conventional proportional-integral resonant controller (PIR), the second is a similar structure using a Mixed Sensitivity Loop Shaping approach. Then, a resonant controller is tested based on a demodulation and remodulation stage. Finally, an Interative Learning Controller (ILC) with a delay block is considered.

This covers the most common feedback controllers for torque ripple, but is limited by choice to linear approaches. Self-optimising controllers are excluded from consideration, because they always contain a non-linear element for the adaptation, and thus show more complex behaviour.

This work is novel in that for the first time it compares these different approaches to torque ripple minimisation starting from first principles. It is the first paper to prove that the resonant controller is exactly identical to a modulating controller. Furthermore, the detailed analytical and simulated comparison of controller simulations has not been seen before in the literature.

2. Materials and Methods

The aim of this paper is to compare different control approaches for reducing torque ripple in an electric machine algebraically, and then to compare their performance using simluations.

To keep the problem simple, a very basic model of the permanent magnet synchronous machine (PMSM) is used, which abstracts from the phase voltages and the inner control, instead focusing on the speed control loop. [8] A list of the symbols used can be found in Appendix A.

The plant is defined as a first order low pass filter capturing the key mechanical dynamics of the rotor:

$$G_P(s) = \frac{k_0}{\omega_P + s} \tag{1}$$

The input is the driving voltage for the motor (V_q), and the output is the shaft speed. In the interest of simplicity, time delays and the electrical time constant are ignored, but they could be added easily (and they may be relevant for the loop response at resonant frequency). The time constant of the first order low pass (PT1) is the result of the damping effect of the winding resistance acting on the inertia of the rotor. This model ignores the electrical time constant caused by the motor inductance in the interest of simplicity, but the same approach can be applied to more complex plant models.

The torque ripple is modelled as an external disturbance at a known frequency, acting onto the plant. This enables to stay within a linear framework, while a speed/position dependent model of the ripple would require a non-linear analysis. For constant speed operation or for moderately slow speed changes, this approach is perfectly sufficient. It leads to a structure as shown in Figure 1.

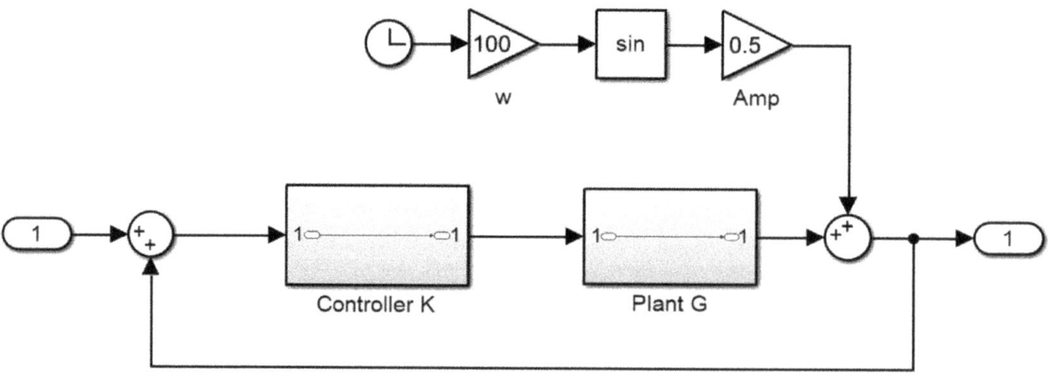

Figure 1. Basic Control Structure with Disturbance.

The disturbance could also be generated by a resonator without damping, with equal results. It is worth noting that this resonance is not connected to the control input, which means it does not become part of the feedback loop. This makes resonant control fundamentally different from the control of resonant plants, as discussed in [9].

This simple structure allows to compare different controllers for their ability to deal with the ripple torque, without making too many assumptions about the specific motor behaviour.

3. Methodology

3.1. Baseline Control

Single-input single-output (SISO) control design typically starts with a PI controller for a low pass plant, and we will follow this approach here.

Additionally, the corresponding PI controller is

$$G_{PI}(s; k_P, k_I) = k_P + \frac{k_I}{s} = \frac{k_P s + k_I}{s} = k_P \frac{s + \omega_{PI}}{s} \tag{2}$$

Conventional tuning would be based on cancellation of the dominant pole, which means that $\omega_{PI} = \omega_P$ and therefore the open loop chain is

$$G_0(s) = G_P G_{PI} = \frac{k_P k_0}{s} \tag{3}$$

The bandwidth is set via k_P, and a typical choice would be $k_P k_0 = 10$, leading to a closed-loop time constant of 0.1 s (and a control bandwidth of $\omega_C = 10$ rad/s). This would not be enough to make any difference to the ripple frequency.

It is possible to push the control bandwidth higher to achieve decent attenuation of the ripple frequency, but this would require a very high control gain. This may reduce the robustness of the controller, and it may introduce significant sensor noise into the control loop and therefore the system output. More targeted approaches to suppress the ripple will be presented below.

3.2. Resonant Control

The conventional approach to eliminating torque ripple is resonant control [10,11], which is based on a resonant second-order filter of the form

$$G_r(s; \omega_0, \zeta, a, b) = \frac{as + b}{s^2 + 2\zeta \omega_0 s + \omega_0^2} \tag{4}$$

This filter can be used in parallel with the existing controller

$$G_{PIR} = G_{PI} + G_r = k_p \frac{s + \omega_{PI}}{s} + \frac{as + b}{s^2 + 2\zeta \omega_0 s + \omega_0^2} \tag{5}$$

as shown in Figure 2, or in series

$$G_{PI} G_r = k_p \frac{s + \omega_{PI}}{s} \frac{as + b}{s^2 + 2\zeta \omega_0 s + \omega_0^2} \tag{6}$$

As long as there is clear separation between the operating frequencies of the PI branch and the resonant branch, both options produce very similar results for a single resonant frequency, with only some higher order differences that are typically negligible. However, if several ripple frequencies have to be eleminated, the parallel form has a distinct advantage, because they are treated independently. Although the problem considered here concerns only one frequency, the parallel form is used here for its easy of extension to several frequencies.

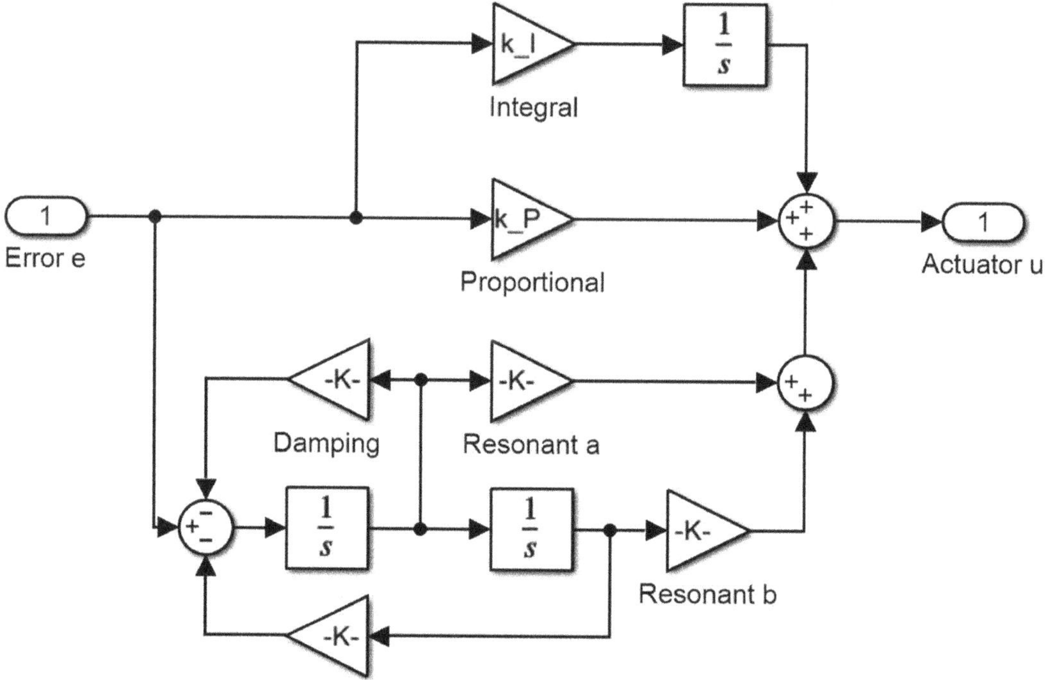

Figure 2. Structure of a PIR Controller.

Note that the phase at resonance is adjusted via the proportion of a and b, and unfortunately this has an impact on the amplitude response. This connection between phase and amplitude is the main complication when designing a PIR controller.

The operating principle of this filter is to bring up the loop gain above one at the resonance frequency ω_0 to reduce the loop sensitivity function at this frequency. The structure satisfies the internal model principle, which states that an effective controller has to include a model of the disturbance it is targeting, which the resonator in the PIR controller achieves. Unlike a normal controller, which has a limited bandwidth denoted by the unity open loop gain $|G_0(\omega)| = 1$, a resonant controller intersects this limit several times, leading to several discontinuous frequency ranges of operation. In other words, the resonant controller operates at low frequencies and at the resonant frequency, but not in the frequency range in between, where it has a very low gain.

The design of the resonant filters differs slightly between the parallel and series configuration. The parallel design usually follows two independent design processes for the two branches, where the other branch is treated as a small disturbance. This requires clear separation between the control bandwidth of both controllers, specifically $\omega_C \ll \omega_0$.

If the separation is not as clear, a serial design process is indicated, where one controller is designed first, and the second controller is designed for an extended plant including the first controller. This approach works for the series connection of the resonant controller, which can be interpreted as a second order phase lag filter.

One of the challenges of designing a resonant controller is to get the phase of the control action right. In theory, this can be tuned via the nominator coefficients a and b, but these also have an impact on the magnitude response.

Alternatively, an all-pass filter or a phase lead compensator can be added to delay the phase without affecting amplitude, while keeping $b = 0$. This leads to the following form, which has more consistent amplitude behaviour at the expense of another state:

$$G_{PIRA}\left(k_p, \omega_{PI}, \omega_0, \zeta, a, p, z\right) = G_{PI} + G_{rA} = k_p \frac{s + \omega_{PI}}{s} + \frac{s - z}{s - p}\frac{as}{s^2 + 2\zeta\omega_0 s + \omega_0^2} \tag{7}$$

This controller will be called PIRA (PIR with phase Advance). It presents significantly different behaviour from the basic PIR controller.

3.3. Mixed Sensitivity Design

Mixed Sensitivity Loop Shaping is an optimal controller design approach coming out of robust multiple input multiple output (MIMO) control. It is based on an extended plant model that contains both the plant model and weights to tune the optimisation problem for specific areas. Three weights are used: W_1 shapes the sensitivity function, W_2 reduces the control energy, and W_3 shapes the co-sensitivity function. The full structure is shown in Figure 3.

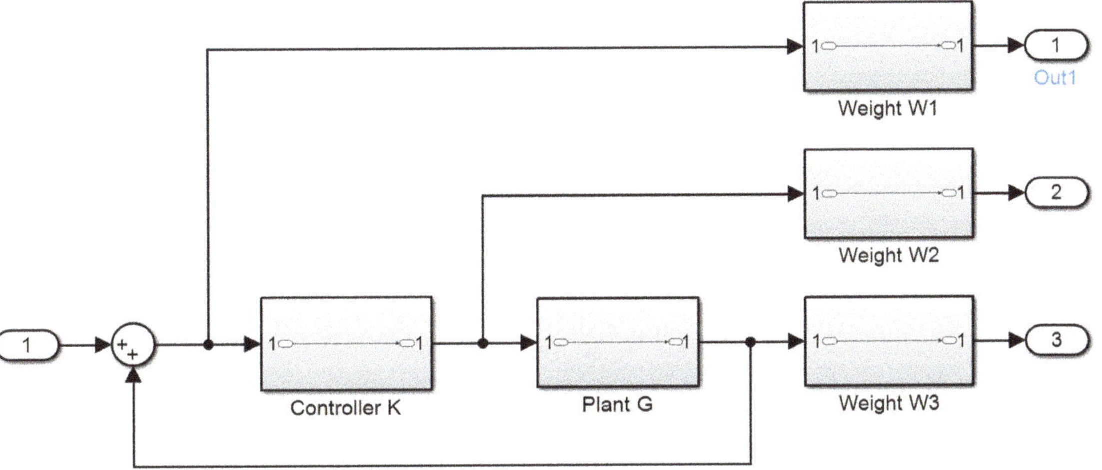

Figure 3. The Control Problem for Mixed Sensitivity Loop Shaping.

It is standard to use an integral weight in W_1 to force reference tracking, possibly in connection with a constant weight.

$$W_{1a}(s) = \begin{bmatrix} \frac{w_I}{s} \\ w_P \end{bmatrix} \tag{8}$$

and to start with a constant weight on the input

$$W_2(s) = 1 \tag{9}$$

For the suppression of torque ripple, an additional weight based on G_R is added to W_1 to respond to this ripple frequency:

$$W_{1b}(s) = G_r(s; \omega_0, \zeta, w_r, 0) = \frac{w_r s}{s^2 + 2\zeta\omega_0 s + \omega_0^2} \tag{10}$$

Together with the original plant, this leads to the following state space model for the extended plant (the bold symbols denote matrices and vectors):

$$\begin{aligned}
\dot{x} &= Ax + Bu \\
y &= C_y x \\
z &= C_z x + D_z u
\end{aligned}$$

(11)

with

$$A = \begin{bmatrix} -\omega_P & 0 & 0 & 0 \\ w_I & 0 & 0 & 0 \\ w_R & 0 & -2\zeta & -\frac{1}{\omega_0} \\ 0 & 0 & \frac{1}{\omega_0} & 0 \end{bmatrix}, B = \begin{bmatrix} k_P \\ 0 \\ 0 \\ 0 \end{bmatrix}$$

$$C_y = \begin{bmatrix} 1 & 0 & 0 & 0 \end{bmatrix}$$

$$C_z = \begin{bmatrix} 0 & 1 & 0 & 0 \\ 0 & 0 & 1 & 0 \\ 0 & 0 & 0 & 0 \end{bmatrix}, D_z = \begin{bmatrix} 0 \\ 0 \\ 1 \end{bmatrix}$$

(12)

The first state is the state of the original plant. The second state is the integral weight W_{1a}, and the last two states belong to W_{1b}.

Typical loop shaping designs are robust and therefore based on the H_∞ norm. However, for simplicity reasons, we will use the H_2 norm here, because it leads to a convex optimisation problem and a low order controller. A state feedback controller is designed following a standard LQR approach, with a control law of the structure

$$u = Kx$$

(13)

Since the weights are not physical, they have to be copied into the controller to complete the transfer function. In this example, all states are known (measured or simulated), and therefore no state observer is required. The resulting transfer function of the controller is

$$G_{MS}(s) = k_1 + \frac{k_2}{s} + \frac{k_3 s + k_4}{s^2 + 2\zeta\omega_0 s + \omega_0^2}$$

(14)

which shows exactly the same structure as the PIR controller. This means it will produce the same potential performance, just via a different design approach. The inclusion of the resonator from the weight in the extended plant into the controller again satisfies the internal model principle.

The advantages of the mixed sensitivity design are shared with most optimal control approaches: the user only has to specify the control objectives, and the optimal design process will synthesis a controller to achieve these. Stability is guaranteed, as is robustness (within limits).

3.4. Modulating Controller

A different approach found in the literature is to use modulation with a carrier frequency to inject a precisely designed signal to cancel out the torque. This approach originates from open-loop approaches of torque ripple compensation [12], and it adds an integral controller to make the cancellation signal adaptive [13]. A typical block diagram is shown in Figure 4.

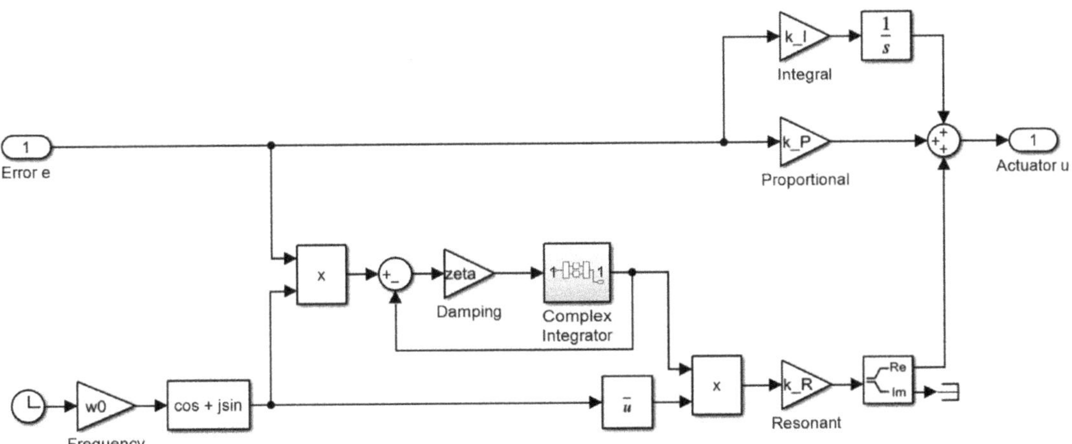

Figure 4. A Modulating Integral Controller.

Note that the integrator operates on a complex signal, and the standard integrator in Simulink does not support this, so it has to be replaced with a custom-made integrator block. It is possible to separate the sine and cosine and run them through separate integrators, but the complex signals make for a much easier analysis.

The transfer function of a modulating controller can be calculated using careful algebraic manipulation. We start with an input signal of

$$x_1(t) = \frac{1}{2}\left(\underline{X}_0 e^{j\omega t} + \underline{X}'_0 e^{-j\omega t}\right) \tag{15}$$

where $\underline{X}_0$ is the input phasor, and $\underline{X}'_0$ its conjugate. The modulation with the carrier

$$x_C(t) = e^{j\omega_0 t}$$

leads to

$$x_2(t) = x_1(t)x_C(t) = \frac{1}{2}\left(\underline{X}_0 e^{j(\omega_0+\omega)t} + \underline{X}'_0 e^{j(\omega_0-\omega)t}\right) \tag{16}$$

Note that this is a complex signal. For practical purposes, it may be split into a real and imaginary component, but this is not necessary for the analysis. A first order low pass filter is applied with the transfer function

$$G_L(s) = \frac{1}{s + \omega_1}$$

This has to be done separately for the two frequencies, leading to

$$x_3(t) = \frac{1}{2}\left(\underline{X}_0 \frac{1}{\omega_1 + j\omega_0 + j\omega} e^{j(\omega_0+\omega)t} + \underline{X}'_0 \frac{1}{\omega_1 + j\omega_0 - j\omega} e^{j(\omega_0-\omega)t}\right) \tag{17}$$

This signal is modulated back into the desired frequency by dividing by the carrier (or multiplying with its conjugate):

$$x_4(t) = \frac{x_3(t)}{x_C(t)} = k_r \frac{1}{2}\left(\underline{X}_0 \frac{1}{\omega_1 + j\omega_0 + j\omega} e^{j\omega t} + \underline{X}'_0 \frac{1}{\omega_1 + j\omega_0 - j\omega} e^{-j\omega t}\right) \tag{18}$$

This signal has imaginary components, and we are not interested in that. Instead, we only use the real part:

$$
\begin{aligned}
x_5(t) = \Re(x_4(t)) &= \tfrac{1}{4}\left(\underline{k}_r\underline{X}_0\tfrac{1}{\omega_1+j\omega_0+j\omega}e^{j\omega t} + \underline{k}_r\underline{X}_0'\tfrac{1}{\omega_1-j\omega_0-j\omega}e^{-j\omega t} + \underline{k}_r\!'\underline{X}_0\tfrac{1}{\omega_1-j\omega_0+j\omega}e^{j\omega t} + \underline{k}_r\underline{X}_0'\tfrac{1}{\omega_1+j\omega_0-j\omega}e^{-j\omega t}\right) \\
&= \tfrac{1}{2}\left(\underline{X}_0\frac{\Re(\underline{k}_R)(j\omega+\omega_1)-\Im(\underline{k}_R)\omega_0}{(j\omega)^2+2\omega_1 j\omega+\omega_1^2+\omega_0^2}e^{j\omega t} + \underline{k}_r\!'\underline{X}_0'\frac{\Re(\underline{k}_R)(j\omega+\omega_1)-\Im(\underline{k}_R)\omega_0}{(j\omega)^2-2\omega_1 j\omega+\omega_1^2+\omega_0^2}e^{-j\omega t}\right) = \Re\left(\underline{X}_0\frac{\Re(\underline{k}_R)(j\omega+\omega_1)-\Im(\underline{k}_R)\omega_0}{(j\omega)^2+2\omega_1 j\omega+\omega_1^2+\omega_0^2}e^{j\omega t}\right)
\end{aligned}
\tag{19}
$$

This establishes the transfer function of the modulating controller to be $G_M(s)$

$$
G_M(s) = \frac{\Re(\underline{k}_R)(j\omega+\omega_1)-\Im(\underline{k}_R)\omega_0}{s^2+2\omega_1 s+\omega_1^2+\omega_0^2}
\tag{20}
$$

Again, this transfer function is identical to the structure of the PIR controller, which means that the same control performance can be achieved. By matching the specific parameters, it can be matched exactly to the transfer of the resonant controller by adjusting ω_0 and $\underline{k}_r$ accordingly. Note that the phase can be adjusted using $\underline{k}_r$, and this has exactly the same effect as a and b in the PIR.

What may be remarkable is that there are no modulation artefacts, no harmonics, no cross-modulation: the output is again a pure sinusoid. This is because the second modulation is a division, and not a multiplication, and therefore the harmonics of the carrier do not appear in the output.

Despite being identical to the PIR in behaviour, there are three distinct advantages to this implementation of the modulating controller. The first one is that the actual controller works on a low bandwidth signal, so it could have lower sampling rate. The second is that the frequency is easy to adjust. For a variable speed motor, the modulating controller only needs the rotor position, which is always known for a synchronous machine. Finally, the compensation of phase lag in the plant is easy by just tuning the complex gain $\underline{k}_r$. This phasor can be frequency dependent, based on a simplified model of the transfer function of the plant.

3.5. Iterative Learning Control

Iterative Learning Control (ILC) is a control approach developed for cyclical reference signals, exploiting the fact that the cycles are always the same. By knowing the response to the previous cycle, the response to the next cycle can be planned without the causality constraints usually in effect. This approach can also be used to deal with cyclical disturbances, as long as the frequency is known [14,15].

The key element of an Iterative Learning Controller is a delay block e^{-Ts} that delays the signal by the duration of one period T. This allows the controller to respond to deviations during the previous cycle. It is often used in an iterative way such as

$$
\frac{1}{e^{-Ts}+1}
$$

which adds the new input signal to the output from the previous cycle, which creates a basic linear adaptation or learning capability.

A key advantage is that T can be adjusted either up or down to account for phase differences when going through the plant. An important feature of ILC is that it shows the typical "comb" pattern of the delay block in the amplitude plot of a closed loop, which means that it works for a fundamental frequency and all its harmonics. This can be an advantage, especially for electric machines that can show a range of harmonics (and DC or low frequency signals). At the same time, it can also be a concern, because the response at these frequencies cannot be tuned separately, and this can lead to unwanted interaction with high frequency plant dynamics.

Just as with the resonant filter, adjusting the phase of the ILC is not easy. It may be tempting to add a time delay to control the phase, but this tends to lead to large phase lags that affect the stability of the loop, especially given that the ILC has a significant steady state gain. A typical control structure would be as shown in Figure 5, and the resulting control law is

$$G_{ILC}(s; T, \zeta, k, p, z) = \frac{s - z}{s + p} \frac{k\zeta}{(1 - \zeta)e^{-Ts} + 1} \tag{21}$$

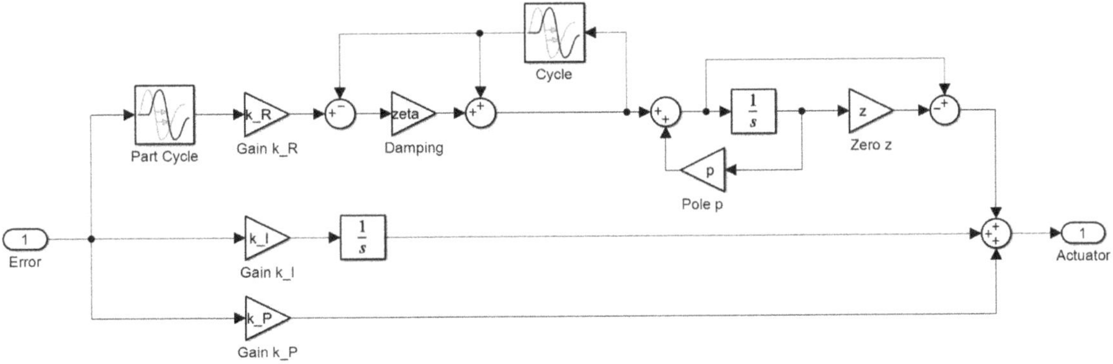

Figure 5. The Iterative Learning Control structure.

Note that this control law has essentially the same parameters as the advanced resonant controller, but the parametrisation may have to be different. This is because the resonant controller operates at all multiples of the resonant frequency, including the frequency 0. Thus it replaces the proportional action of the PI controller, but not the integral branch. This is mostly a consequence of this controller being significantly more powerful than the previous examples.

Especially with complex plant models, further filtering and compensation will be required to guarantee stability at all resonant frequencies. In this sense, the ILC controller is very similar to a high bandwidth PI controller, but it has the advantage of acting only at the relevant frequencies, keeping the magnification of sensor noise to a minimum.

4. Numerical Results

Unfortunately, an experimental rig was not available to the authors for real-world validation of the above analysis. However, work is progressing in this direction. Therefore, Simulink simulations are presented to give an initial and fair comparison between the different controllers, highlighting key features and demonstrating the correctness of the derivations above. The application of these controllers to an experimental rig is planned for a future paper. All files used to produce these results are available at https://github.com/LU-Centre-for-Autonomous-Systems/resonant_control/, version 1.0, published 27 November 2022.

4.1. Plant

All controllers are applied to the same plant, which is a simplified model of an electric motor, with the structure given in (1). The transfer function is a PT1 low pass filter with a time constant of 1 s, a gain of 1, and a ripple frequency of 100 rad/s.

$$G_P(s) = \frac{1}{1 + s} \tag{22}$$

A sinusoidal disturbance with a frequency of 100 rad/s and an amplitude of 1 is added to the output, this represents the effect of the ripple torque:

$$d(t) = \sin 100t \tag{23}$$

More complex plant models can be used, and they will impact on the design of the PI controller, but the impact on the resonant controller is mostly limited to the plant behaviour at the resonant frequency.

4.2. Controller Design

To keep the different control structures comparable, the same parameters are used, based on a mixed sensitivity design. The weights are defined as

$$W_{1a}(s) = \left[\begin{array}{c} \frac{10}{s} \\ \frac{50 \cdot 100s}{s^2 + 2 \cdot 0.05 \cdot 100s + 100^2} \end{array} \right] \tag{24}$$

$$W_2(s) = 1 \tag{25}$$

These weights already contain some important information about the controller. They reflect the ripple frequency of $\omega_0 = 100\,\text{rad/s}$, and they have a defined damping of $\zeta = 0.05$, which means that response to the ripple frequency is expected to be very focused, and it will take many periods for the controller to settle. The numerical gains 10 and 50 determine the importance of reference following and resonance suppression. Note that these gains are a lot more sensitive than R and Q in the LQ design, so it would not be appropriate to move in steps of factors of ten. Instead, changes by a factor of 2 were used to determine appropriate gains with an appropriate balance between integral and resonant action.

The mixed sensitivity H2 design leads to a state feedback controller law with

$$u = -\begin{bmatrix} 1 & 0.2 & -0.78 & 43 \end{bmatrix} x \tag{26}$$

and a transfer function for the resulting controller of

$$G_{MS}(s) = 43 + \frac{10}{s} + \frac{950s - 3.9 \cdot 10^5}{s^2 + 10s + 10^4} \tag{27}$$

The Bode Plot of this transfer function is shown in Figure 6. From this transfer function, all the controller parameters can be determined:

$$k_I = 10 \tag{28}$$

$$k_P = 43 \tag{29}$$

$$a = 950 \tag{30}$$

$$b = -3.9 \cdot 10^5 \tag{31}$$

and all five controller laws can then be calculated to be as similar in effect as possible:

$$G_{PI}(s) = 43 + \frac{10}{s} \tag{32}$$

$$G_{PIR}(s) = 43 + \frac{10}{s} + \frac{950s - 3.9 \cdot 10^5}{s^2 + 10s + 10^4} \tag{33}$$

$$G_{PIRA}(s) = 43 + \frac{10}{s} + \frac{s - 21}{s + 210} \frac{9300s}{s^2 + 10s + 10^4} \tag{34}$$

$$G_{MS}(s) = 43 + \frac{10}{s} + \frac{950s - 3.9 \cdot 10^5}{s^2 + 10s + 10^4} \tag{35}$$

$$G_{Mod}(s) = 43 + \frac{10}{s} + \frac{950s - 3.9 \cdot 10^5 s}{s^2 + 10s + 10^4} \tag{36}$$

$$G_{ILC}(s) = \frac{10}{s} + \frac{s + 21}{s + 210} \frac{930}{(1 - 0.05)e^{-0.01s} + 1} \tag{37}$$

The compensator parameters and the gains for G_{PIRA}, G_{Mod} and G_{ILC} are chosen such that the behaviour at the resonant frequency is identical, matching amplitude and phase. The ILC required a slightly different compensator to achieve stability, because the ILC branch has significant steady state gain. Note that the ILC does not have a proportional branch, because it already has proprotional action at low frequencies.

4.3. Open Loop Behaviour

Note that Simulink is unable to derive correct transfer functions for the modulating controller or the ILC, so all diagrams have been numerically derived using a FFT of an impulse response. Through careful choice of the excitation signal, the result should be accurate and free from noise. The phase is not unrolled, so discontinuities are to be expected.

The first observation is that despite different design approaches and different control structures, the PIR controller, the Mixed Sensitivity controller, and the modulating controller have exactly the same transfer function, and therefore the same effect. This is not surprising, given that they have the same structure of the control law, as demonstrated in the previous section, and the parameters were chosen specifically to match the behaviour. Obviously, it would be unsusual to find exactly the same parameters using different design methods, but this paper is about comparison of the potential of different control structures, not design methodologies, so in this case, the matched parameters are appropriate. The resulting behaviour is identical, and this is confirmed on the open loop Bode plot of the different resonant branches of the controllers in Figure 7. The PI branch is drawn out and plotted separately to focus on the behaviour of the resonant branch. Only one line is shown for both the resonant controller and the mixed sensitivity design, because the implementation is in fact idential, while the modulation controller has a completely different implementation, but with the same behaviour.

This plot also shows how different the PIRA controller is with a phase advance compensator. It shows a much stronger roll-off below the resonant frequency, which means that it interfers much less with the low frequency behaviour of the PI branch—but it does have slighty higher gains at higher frequencies, which could potentially affect the robustness.

Because the PIR, MS and modulating controller are indentical, only the PIR controller is considered further, and compared with the PIRA and ILC design. In terms of the design approach, the Mixed Sensitivity design has the advantage that it deals with phase and stability automatically, while these have to be considered as additional complications during the turning of the PIR controller.

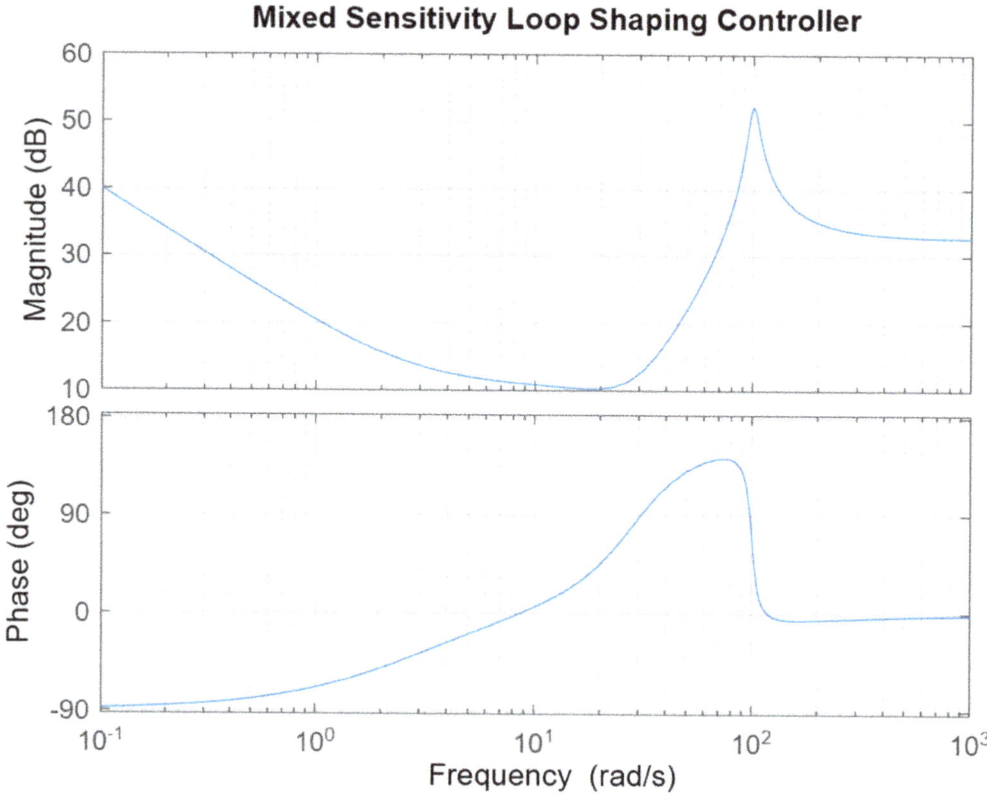

Figure 6. Bode Plot of the Mixed Sensitivity Design controller.

Finally, Figures 8 and 9 shows a comparison of the four different controllers (PI, PIR, PIRA and ILC) for the controller only and for the open loop chain. By including the plant model, this Bode Plot of the open loop chain allows an analysis of the stability and robustness of the different controllers.

The plots highlights both key similarities and differences between the approaches. The PI controller obviously does not address the ripple frequency, this is expected. All other controllers show how the loop gain goes above one (0 dB) again around the ripple frequency, and how the phase is well controlled in this second critical region. The PIR controller achieves the objective, but there is a very significant impact on the operation of the PI controller for low speed operation, which means that the PI controller may have to be retuned after adding the resonant branch. The PIRA avoids this, and preserves the correct operation of the PI branch while also achieving good suppression of the ripple frequency. The ILC finally is very similar to the PIR controller around the resonant frequency, but it deviates from all other controllers both at low frequencies and at harmonics of the ripple frequency, as expected from the theoretical analysis.

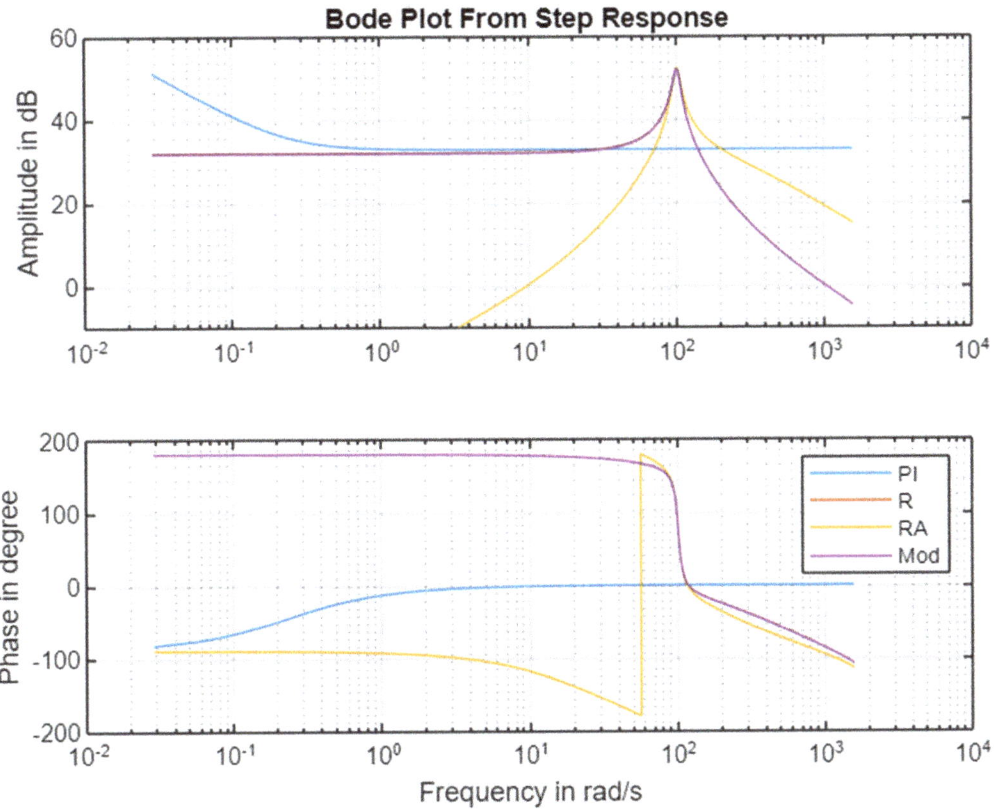

Figure 7. Bode Plot of PIR, PIRA and modulating controller.

The main conclusion is that apart from the PIRA approach, all controllers have interferences between the PI controller and the resonant part, despite the significant frequency separation. The PIRA is the only resonant controller that avoids this interference.

4.4. Closed Loop Behaviour

The closed loop behaviour is studied both in the frequency domain and in the time domain. The frequency domain is shown in the Bode plots of the complementary sensitivity function in Figure 10. This shows the transfer of disturbances of the reference on the error signal. The goal is to have a gain as low as possible for low frequencies (reference tracking), and at the ripple frequency, as this would be an indication of successful control. This view again demonstrates how similar the different controllers act around the ripple frequency, and how the resulting loop behaviour is very similar.

All controllers achieve this objective, with the expected exception of the PI controller, which does not address the ripple frequency. The plot also highlights how much influence the resonant controller (PIR) has at low frequencies—so the notion that it can be designed separately from the PI controller is not applicable here. The ILC performs reasonably well, but it is clearly quite different at low frequencies from the other controllers.

The time domain response is shown in Figure 11. This simulation is subject both to a reference jump at 0.5 s and to a sudden sinusoidal disturbance at 3 s, so it shows a combination of how the system deals with both. This diagrams shows very clearly that the resonant controllers affect the step response. The high frequency content in the step excites the resonance—leading both to some eratic movements and possibly some overshoot not present in the original pure PI controller. Both the PIRA and ILC at least maintain most of the low frequency step response, but it is still not as clean as the PI controller, and they all have some overshoot.

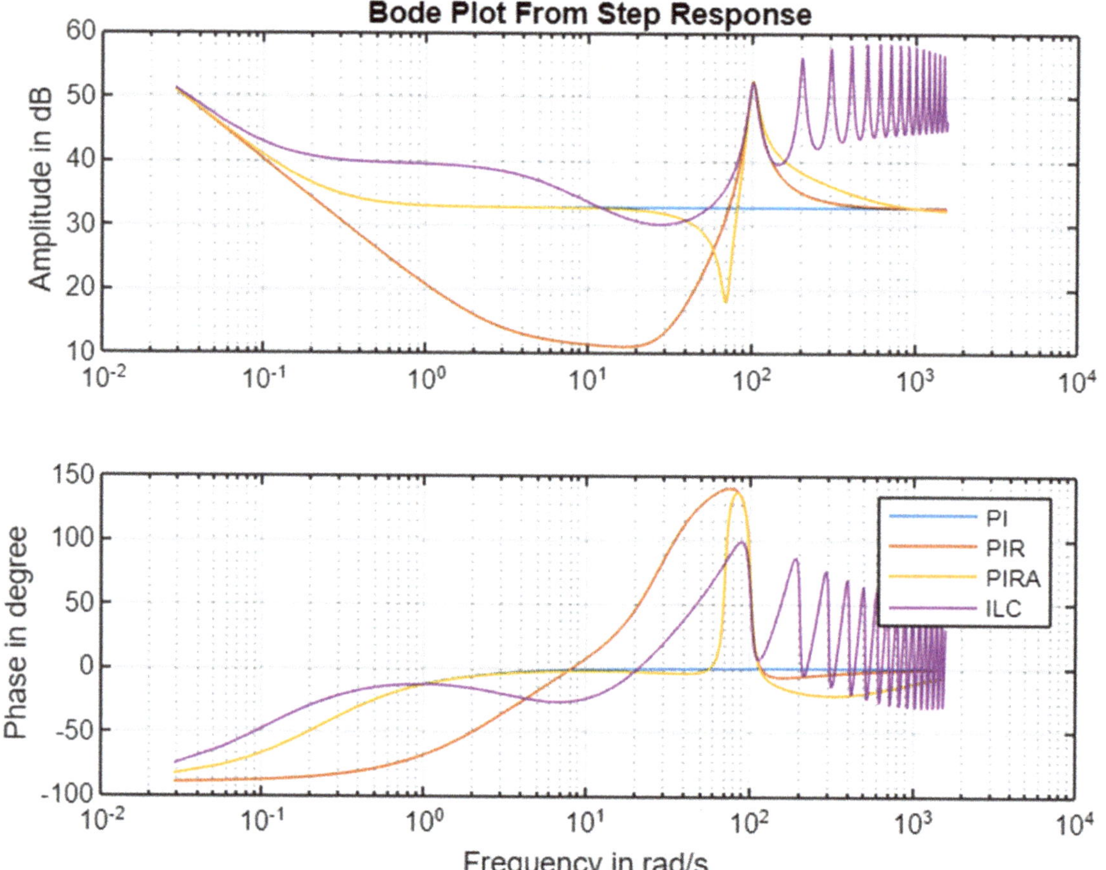

Figure 8. Comparing the Bode Plot of the four controllers.

To avoid this, it may be appropriate to apply a low pass filter to the reference input that limits high frequency content. Since that is a feed-forward filter, it does not affect stability, and it can be tuned seperately from the loop.

All controllers except the PI controller achieve good ripple suppression. The time domain view shows very nicely that the ripple suppression needs some time to settle and take effects. This is because the ripple suppression only applies to a very narrow frequency range, and the sudden onset of the ripple disturbance contains a much wider frequency range initially. Stationary behaviour is achieved after about 0.3 s, which means it is a bit slower than the settling time of the PI controller. Usually, fast settling is not required, and this would be considered more than acceptable.

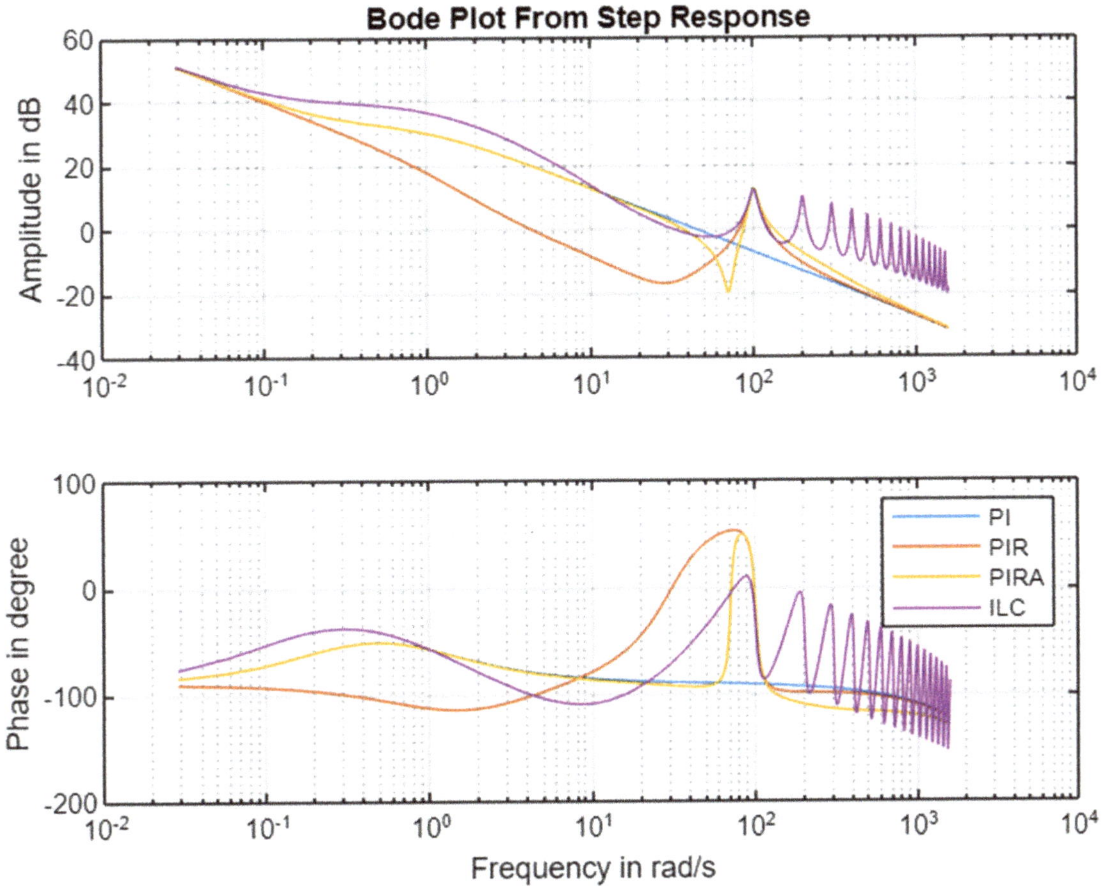

Figure 9. Comparing the Bode Plot of the open loop chain for the three controllers.

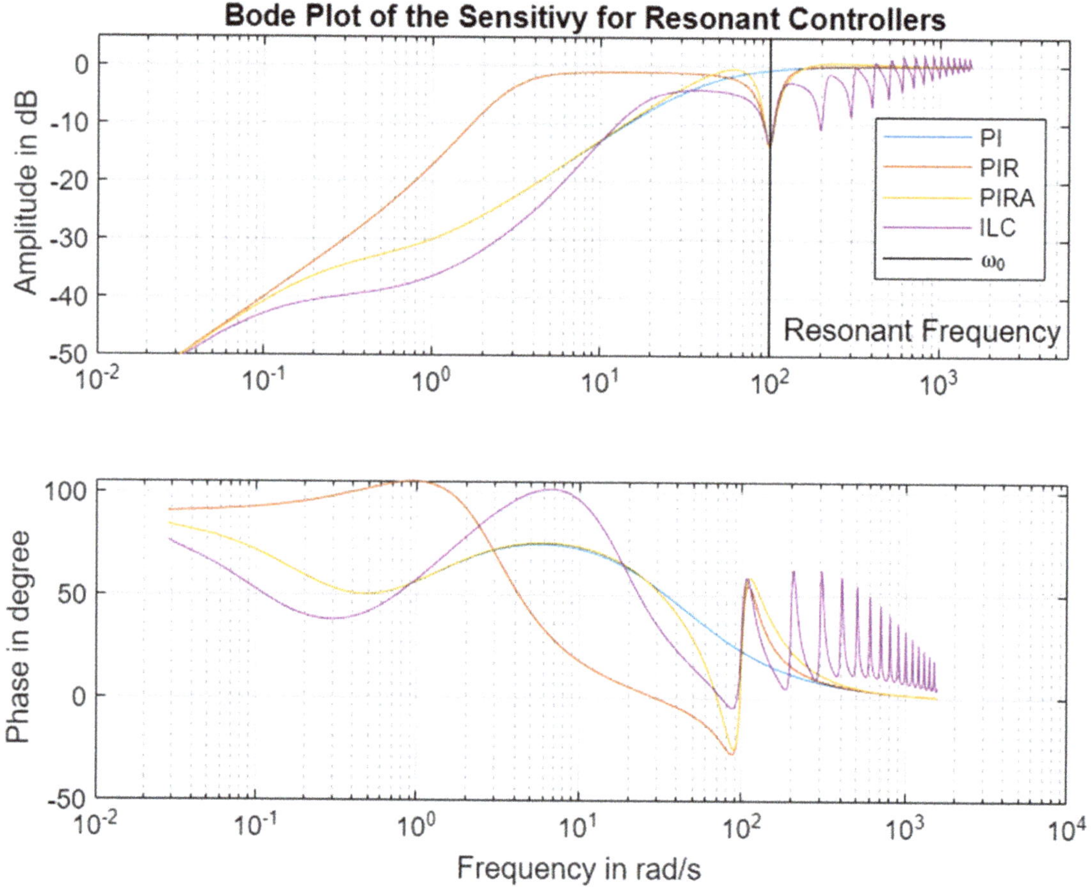

Figure 10. Bode Plot of the Closed Loop Sensitivity comparing all three controllers.

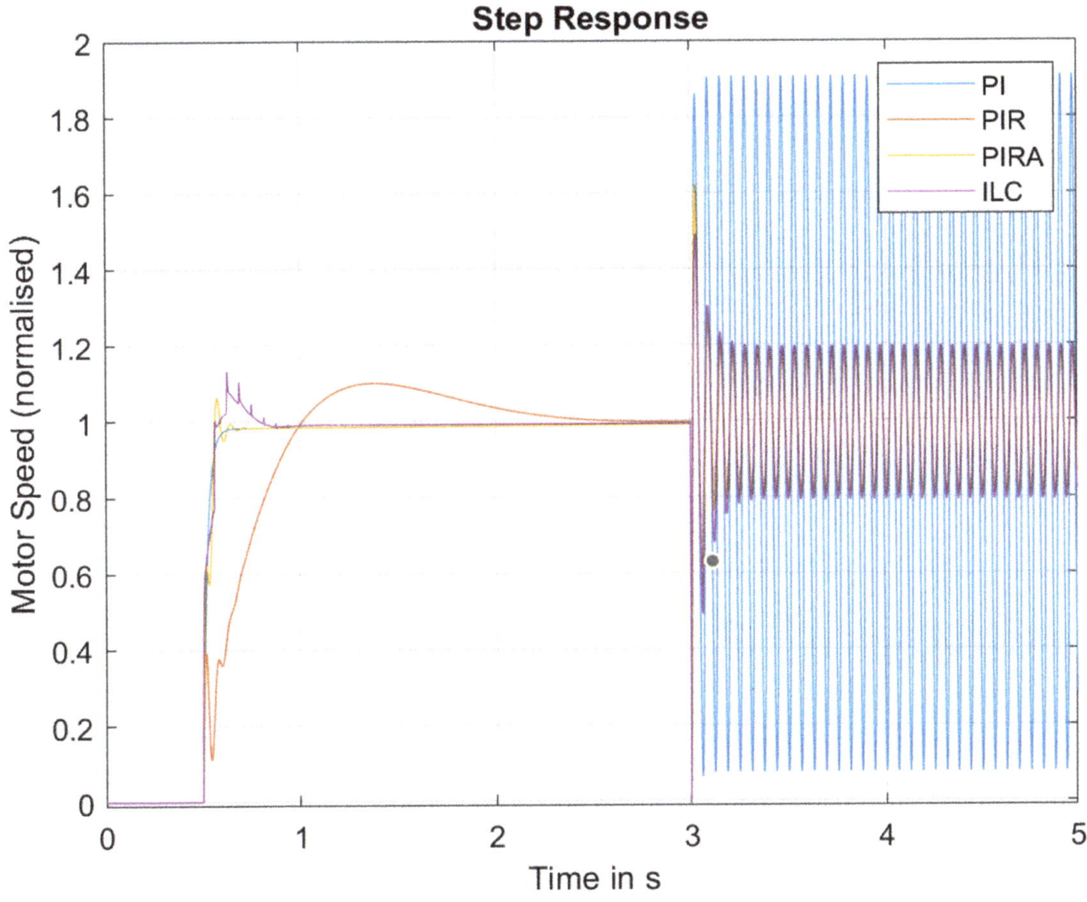

Figure 11. Closed Loop Step Response comparing all three controllers.

5. Discussion

The paper has compared three very different controllers for the control of torque ripple in an electric machine using a simplified plant model, as well as a baseline controller without this capability.

The most common controller is the resonant controller (PIR). It can be implemented in two very different ways, either using a resonanting branch, or using amplitude modulation with the resonant frequency—both show the same behaviour. While it is successful in addressing the problem, the design of this filter with its four parameters is not straightforward, because there is a very close interaction between the low frequency behaviour of the resonant branch and the PI branch. An open question in the literature is how to find and adjust the phase of the resonant branch.

The use of a Mixed Sensitivity Loop Shaping can help to find the right parameters of the resonant controller, based on a plant model and well chosen weighting function. This approach still has three parameters: the frequency, damping, and the desired suppression, but the phase is chosen implicitly to achieve the best results. The structure of the resulting controller is identical to the resonant controller, although more complex structures with additional filters can be used easily.

This interference can be avoided by adding an additional pass filter to shift the phase around the resonant frequency, as demonstrated in the PIRA. The phase needs to compensate the phase of the plant around the resonant frequency to achieve a stable sufficient of the torque ripple. The PIRA controller seems clearly superior for this application.

Finally, ILC is clearly a more powerful control structure, which is able to address several ripple frequencies at once, without designing a part of the control for each frequency to be targeted. However, stability needs to be achieved at each of those frequencies, and despite the inclusion of a generic filter, the form used here does not quite offer enough tuning parameters to be generally applicable to any plant model. This could be addressed by adding additional compensators or an approximated plant model inverse.

The modulating controller and the ILC later two structures also have the advantage that they could be run in a synchoronous fashion based on an angle measurement of the electric machine. That would make it easier to apply the same general solution at different motor speeds. However, the challenge is to achieve stability at all speeds, and thus further consideration of the plant dynamics is again necessary.

An overview of all the observations is shown in Table 1.

Table 1. Comparison between the different control approaches.

Control Method	Tuning	Implementation	Stability	Reference Tracking	Ripple Suppression
PI	Easy	Easy	Excellent	Excellent	None
PIR Mixed Sensitivity Modulating	Hard Automated Hard	Easy Easy Medium	Good Guaranteed Good	Marginal	Good (adjustable)
PIRA	Medium	Easy	Excellent	Very Good	Good
ILC	TBD	Medium	Tricky	Good	Good

6. Conclusions

For controlling torque ripple in electric machines, the standard approach is a proportional-integral resonant (PIR) controller. Four extensions have been considered here, that all offer notable advantages. The model-based Mixed Sensitivity Loop Shaping approach turns out to be a very convenient way of tuning the PIR controller, and it addresses the difficult question of how to best find the right phase. The modulating controller shows identical behaviour, but it allows direct tuning with a complex gain. Adding a filter leads to the PIRA controller, which provides much better separation between the PI branch and the resonant branch. Additionally, the Iterative Learning Controller (ILC) shows great promise, but requires further work to compensate plant dynamics at all relevant frequencies. The ILC and the modulating controller also lend themselves to synchronous operation based on a rotor angle position sensor.

This paper logically leads to three areas of further research. The first one is how to achieve ripple suppression at variable speed [16,17]. There are two obvious options: gain scheduling over frequency, or compensation of the plant dynamics (both based on an approximation of the plant transfer function). Both would lead to very different design approaches for the same problem. The second big question is how to measure the ripple, because the speed sensor is often not sufficiently accurate. A MIMO control approach may be indicated, using a microphone or an accelerometer to pick up the ripple. The final question is how to translate the design into the discrete-time domain, and how to apply it practically, because the implementation would typically be on a computer system, and the execution is time critical. A time-discrete design thus offers potentially superiour performance at frequencies close to the Nyquist limit.

Author Contributions: M.S.R. performed the initial research that gave the idea to this article. The conception of the comparison as well as the writing of the document and the simulations were predominantly performed by T.S., the first author. W.M. has contributed in an advisory role, providing feedback and hints for presentation. All authors have read and agreed to the published version of the manuscript.

Funding: This research was funded by the Advanced Propulsion Centre (APC) via UKRI Innovate UK under project ViVID, grant number 113210.

Institutional Review Board Statement: Not applicable.

Informed Consent Statement: Not applicable.

Data Availability Statement: All data is the result of simulation, and available on githib. No personal data has been used.

Conflicts of Interest: The authors declare no conflict of interest. The funders had no role in the design of the study; in the collection, analyses, or interpretation of data; in the writing of the manuscript; or in the decision to publish the results.

Appendix A. Abbreviations and Symbols

ILC—Iterative Learning Controller; MS—Mixed Sensitivity (loop shaping); PI—Proportional Integral (controller); PIR—Proportional Integral Resonant (controller); PIRA—Proportional Integral Resonant controller with phase Advance; PMSM—Permanent Magnet Synchronous Machine; j—Imaginary Unit $\sqrt{-1}$; $G(s)$—Transfer Function; t—Time in seconds; $u(t)$—Input; $x(t)$—State Vector; $y(t)$—Output; ω—Angular Frequency in rad/s.

References

1. Liu, C. Emerging Electric Machines and Drives—An Overview. *IEEE Trans. Energy Convers.* **2018**, *33*, 2270–2280. [CrossRef]
2. Lopez, C.A.; Jensen, W.R.; Hayslett, S.; Foster, S.N.; Strangas, E.G. A Review of Control Methods for PMSM Torque Ripple Reduction. In Proceedings of the 2018 23rd International Conference on Electrical Machines, ICEM 2018, Alexandroupoli, Greece, 3–6 September 2018; pp. 521–526. [CrossRef]
3. Petrov, I.; Ponomarev, P.; Alexandrova, Y.; Pyrhonen, J. Unequal Teeth Widths for Torque Ripple Reduction in Permanent Magnet Synchronous Machines With Fractional-Slot Non-Overlapping Windings. *IEEE Trans. Magn.* **2014**, *51*, 1–9. [CrossRef]
4. Holtz, J.; Springob, L. Identification and compensation of torque ripple in high-precision permanent magnet motor drives. *IEEE Trans. Ind. Electron.* **1996**, *43*, 309–320. [CrossRef]
5. Cho, H.-J.; Kwon, Y.-C.; Sul, S.-K. Torque Ripple-Minimizing Control of IPMSM With Optimized Current Trajectory. *IEEE Trans. Ind. Appl.* **2021**, *57*, 3852–3862. [CrossRef]
6. Khan, M.A.; Husain, I.; Islam, M.R.; Klass, J.T. Design of Experiments to Address Manufacturing Tolerances and Process Variations Influencing Cogging Torque and Back EMF in the Mass Production of the Permanent-Magnet Synchronous Motors. *IEEE Trans. Ind. Appl.* **2013**, *50*, 346–355. [CrossRef]
7. Dai, M.; Keyhani, A.; Sebastian, T. Torque ripple analysis of a permanent magnet brushless DC motor using finite element method. In Proceedings of the IEMDC 2001—IEEE International Electric Machines and Drives Conference, Cambridge, MA, USA, 17–20 June 2001; pp. 241–245. [CrossRef]
8. Gebregergis, A.; Chowdhury, M.H.; Islam, M.S.; Sebastian, T. Modeling of Permanent-Magnet Synchronous Machine Including Torque Ripple Effects. *IEEE Trans. Ind. Appl.* **2014**, *51*, 232–239. [CrossRef]
9. Pugi, L.; Reatti, A.; Corti, F. Application of modal analysis methods to the design of wireless power transfer systems. *Meccanica* **2019**, *54*, 321–331. [CrossRef]
10. Chuan, H.; Fazeli, S.M.; Wu, Z.; Burke, R. Mitigating the Torque Ripple in Electric Traction Using Proportional Integral Resonant Controller. *IEEE Trans. Veh. Technol.* **2020**, *69*, 10820–10831. [CrossRef]
11. Huang, M.; Deng, Y.; Li, H.; Wang, J. Torque Ripple Suppression of PMSM Using Fractional-Order Vector Resonant and Robust Internal Model Control. *IEEE Trans. Transp. Electrif.* **2021**, *7*, 1437–1453. [CrossRef]
12. Huang, C.L.; Yang, S.C. Torque Ripple Reduction for BLDC Permanent Magnet Motor Drive using DC-link Voltage and Current Modulation. In Proceedings of the 2021 IEEE International Future Energy Electronics Conference, IFEEC 2021, Taipei, Taiwan, 16–19 November 2021. [CrossRef]
13. Feng, G.; Lai, C.; Kar, N.C. A Closed-Loop Fuzzy-Logic-Based Current Controller for PMSM Torque Ripple Minimization Using the Magnitude of Speed Harmonic as the Feedback Control Signal. *IEEE Trans. Ind. Electron.* **2016**, *64*, 2642–2653. [CrossRef]
14. Liu, J.; Li, H.; Deng, Y. Torque Ripple Minimization of PMSM Based on Robust ILC Via Adaptive Sliding Mode Control. *IEEE Trans. Power Electron.* **2017**, *33*, 3655–3671. [CrossRef]

15. Qian, W.; Panda, S.; Xu, J. Speed Ripple Minimization in PM Synchronous Motor Using Iterative Learning Control. *IEEE Trans. Energy Convers.* **2005**, *20*, 53–61. [CrossRef]
16. Yansong, H.; Dejun, Y.; Yunhao, P. Research on Torque Ripple Suppression Strategy of PMSM under Variable Speed Condition. In Proceedings of the 2021 6th Asia Conference on Power and Electrical Engineering, ACPEE 2021, Chongqing, China, 8–11 April 2021; pp. 898–904. [CrossRef]
17. Tang, M.; Gaeta, A.; Formentini, A.; Zanchetta, P. A Fractional Delay Variable Frequency Repetitive Control for Torque Ripple Reduction in PMSMs. *IEEE Trans. Ind. Appl.* **2017**, *53*, 5553–5562. [CrossRef]

Article

Position Control of a Cost-Effective Bellow Pneumatic Actuator Using an LQR Approach

Goran Gregov [1], Samuel Pincin [1], Antonio Šoljić [1] and Ervin Kamenar [1,2,*]

1   Faculty of Engineering, University of Rijeka, Vukovarska 58, 51000 Rijeka, Croatia
2   Centre for Micro- and Nanosciences and Technologies, University of Rijeka, Radmile Matejčić 2, 51000 Rijeka, Croatia
*   Correspondence: ekamenar@riteh.hr; Tel.: +385-(0)51-651-585

Abstract: Today, we are witnessing an increasing trend in the number of soft pneumatic actuator solutions in industrial environments, especially due to their human-safe interaction capabilities. An interesting solution in this frame is a vacuum pneumatic muscle actuator (PMA) with a bellow structure, which is characterized by a high contraction ratio and the ability to generate high forces considering its relatively small dimensions. Moreover, such a solution is generally very cost-effective since can be developed by using easily accessible, off-the-shelf components combined with additive manufacturing procedures. The presented research analyzes the precision positioning performances of a newly developed cost-effective bellow PMA in a closed-loop setting, by utilizing a Proportional-Integral-Derivative (PID) controller and a Linear Quadratic Regulator (LQR). In a first instance, the system identification was performed and a numerical model of the PMA was developed. It was experimentally shown that the actuator is characterized by nonlinear dynamical behavior. Based on the numerical model, a PID controller was developed as a benchmark. In the next phase, an LQR that involves a nonlinear pregain term was built. The point-to-point positioning experimental results showed that both controllers allow fast responses without overshoot within the whole working range. On the other hand, it was discovered that the LQR with the corresponding nonlinear pregain term allows an error of a few tens of micrometers to be achieved across the entire working range of the muscle. Additionally, two different experimental pneumatic solutions for indirect and direct vacuum control were analyzed with the aim of investigating the PMA response time and comparing their energy consumption. This research contributes to the future development of the pneumatically driven mechatronics systems used for precise position control.

Keywords: bellow pneumatic muscle; soft actuator; position control; LQR control; PID control

Citation: Gregov, G.; Pincin, S.; Šoljić, A.; Kamenar, E. Position Control of a Cost-Effective Bellow Pneumatic Actuator Using an LQR Approach. *Actuators* **2023**, *12*, 73. https://doi.org/10.3390/act12020073

Academic Editor: Giorgio Olmi

Received: 30 December 2022
Revised: 7 February 2023
Accepted: 7 February 2023
Published: 9 February 2023

1. Introduction

Soft actuators are characterized by their flexible nature, light weight and high specific power [1–3]. They are frequently used in close proximity to humans [4,5], as components of rehabilitation devices [6,7], and as industrial devices, particularly for handling delicate objects [8,9]. They are, however, generally characterized by nonlinear dynamical behavior which makes controlling their displacement very complicated, especially if precise position control is aimed for. In our previous study [10], we presented the design, development and experimental assessment of a simple vacuum-driven bellow type pneumatic muscle actuator (PMA). The main idea behind the earlier research was to devise simple and cost-effective PMA using off-the-shelf components and additive manufacturing techniques. The developed and fabricated PMA had a high contraction ratio and a significant maximum force value with respect to its dimensions. As a soft actuator "skin", we used universal rubber cylindrical bellows, which are typically used as protection of the front suspension on motorcycles. Aside from representing the soft part of the system, the rubber bellow also ensures the spring effect for returning the PMA to its initial position. Furthermore, we place

3D-printed rigid rings along the bellow length to eliminate radial deformations of PMA and increase the value of force [10]. We analyzed two bellow PMAs with different dimensional and material properties with respect to different vacuum values and loading conditions. Experimental measurements and analyses of the maximum blocking force, displacement–velocity curve and sinusoidally forced motion responses were performed. Additional sinusoidally forced cyclic experiments were conducted to investigate the influence of fatigue, creep or relaxation of material. Upon the development of the PMA, we performed initial precision positioning tests and concluded that positioning in the micrometric range can be obtained. However, additional refinements are required since a significant overshoot and dead-band of the response can be observed in some cases. Furthermore, the positioning experiments in the previous research are performed without loading the system. These are some of the key points that will be examined in this study.

Due to its simple design and implementation in real mechatronic systems, the proportional-integral-derivative (PID) controller is one of the most used closed-loop controllers [11,12]. Using the PID controller for positioning of the artificial pneumatic muscle [13] resulted in a good control ability but the PID parameters have to be tuned additionally if working conditions change during the system's operation. The PID parameters can be tuned using well known methods or advanced numerical algorithms, such as neural networks [14,15], simulated annealing optimization algorithm [16] or fuzzy logic [17]. In the mentioned articles, the authors analyzed the positioning of the pneumatic actuators that operate with the positive pressure and the results showed the ability of achieving steady-state errors in the micrometric range. Furthermore, positive pressure is used to operate PMAs in the majority of research articles dealing with position control using the LQR approach [18]. In this article we analyzed the pneumatic actuator which works with negative pressure (vacuum). Using the PD and LQR controllers, the positioning analysis of the PMA with a similar working principle was conducted in [19]. The results showed that the accuracy (steady-state error) for both algorithms equals approximately 1.15 mm, and the LQR consumes less energy. A similar comparison of LQR and PID controllers for controlling pneumatic diaphragm valves was conducted in [20]. In all considered scenarios, the authors demonstrated the LQR's superiority over the PID controller. Besides the above-mentioned literature, the LQR approach has not been applied often in previous studies of PMA control [12]. Therefore, one of the main goals of the given research is to apply an LQR technique in addition to the widely used PID control as a benchmark algorithm.

Initially, the PID controller was developed, and its parameters were tuned by using Ziegler–Nichols method followed by trial-and-error refinement during the initial testing phase. Given the drawbacks of the previously mentioned tuning methods, we developed a numerical model of the system in the following step. The preliminary results of point-to-point positioning of the cost-effective bellow PMA with a developed PID controller revealed that there is no overshoot in the response; however, a relatively large steady-state error is present. Considering that the PID controller was unable to ensure optimal response, it was decided that the LQR approach will be applied in the following step to minimize steady-state error. Additionally, a mathematical model was developed and used as a base for building an LQR controller. Following preliminary experimental tests, the nonlinear pregain term was applied with the goal of improving positioning performances (i.e., eliminating steady-state error). The LQR approach resulted in better dynamical behavior, with a lower steady-state error and overshoot. However, a significant amount of dead-band was noticed at the beginning of each positioning cycle.

Therefore, in addition to indirect vacuum control, which achieves vacuum by controlling input pressure to the vacuum ejector, we investigate an additional pneumatic solution that allows direct vacuum control by controlling the vacuum value at the output of the ejector. These two solutions are used and compared with the goal of investigating the control properties of the used PMAs and the system's total energy consumption. Figure 1 graphically summarizes the approaches for obtaining precise positioning of the cost-effective bellow PMA using the two different pneumatic solutions.

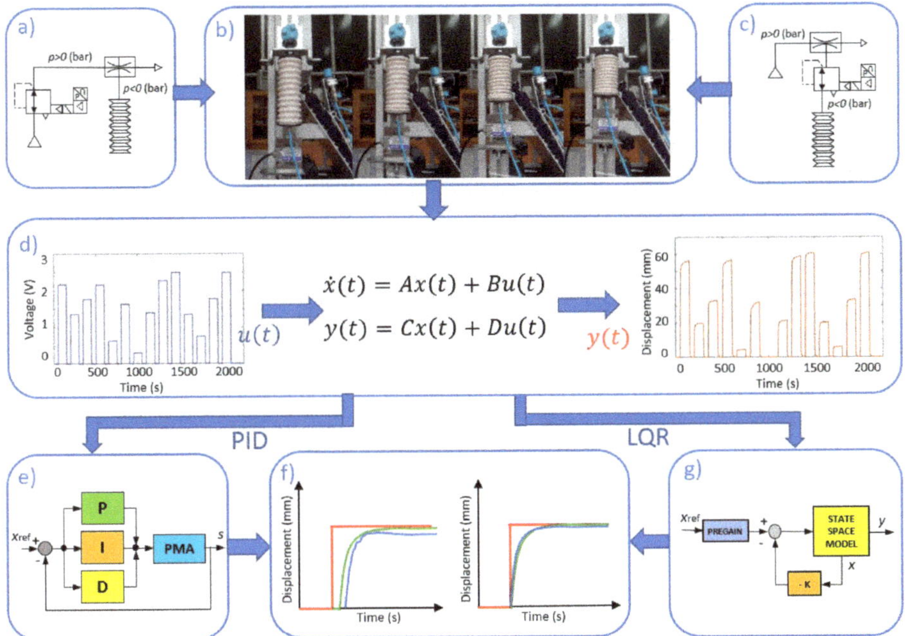

Figure 1. The development and experimental evaluation of PID and LQR controllers: (**a**) schematic view of the pneumatic setup for indirect vacuum control; (**b**) working principle of the bellow PMA; (**c**) schematic view of the pneumatic setup for direct vacuum control; (**d**) numerical model identification; (**e**) schematic model of the PID controller; (**f**) experimental assessment of position control for indirect (**left**) and direct (**right**) vacuum control; (**g**) simplified model of the LQR controller.

The remainder of the paper is structured as follows: Section 2 describes the materials and methods used, including a brief overview of the experimental apparatus and numerical modeling of PID and LQR controllers. In Section 3 presents experimental results of the positioning control under various working conditions and by using two distinct vacuum control systems. Finally, in Section 4 the main conclusions are drawn and future work is outlined.

2. Materials and Methods

This section introduces an experimental setup and briefly describes its main components. Furthermore, the steps in developing the numerical model of the bellow PMA, as well as PID and LQR tuning procedures using the MATLAB environment, are presented. Additionally, the development of PID and LQR controllers using LabVIEW and the preliminary experimental results of PMA positioning are presented.

2.1. Experimental Apparatus

The experiments were carried out at the Laboratory for Hydraulic and Pneumatic Systems at the Faculty of Engineering, University of Rijeka. The testbed (Figure 2) was built from widely used strut profiles and contains a Planet Air L-S50-25 compressor, a FESTO LR-MICRO-MA40-Q4 manual pressure regulator, FESTO VPPE-3-1/8-6-010 and FESTO VPPI-5L-G18-1V1H-V1-S1D proportional pressure regulators and a FESTO VN-05-H-T2-PQ1-VQ1-RQ vacuum generator. The measuring equipment used for control and data acquisition consists of National Instruments NI myRIO 1900 device, Schmalz VS VP8 SA M8-4 vacuum and pressure sensors, and a Burster 8713-100 potentiometric displacement sensor.

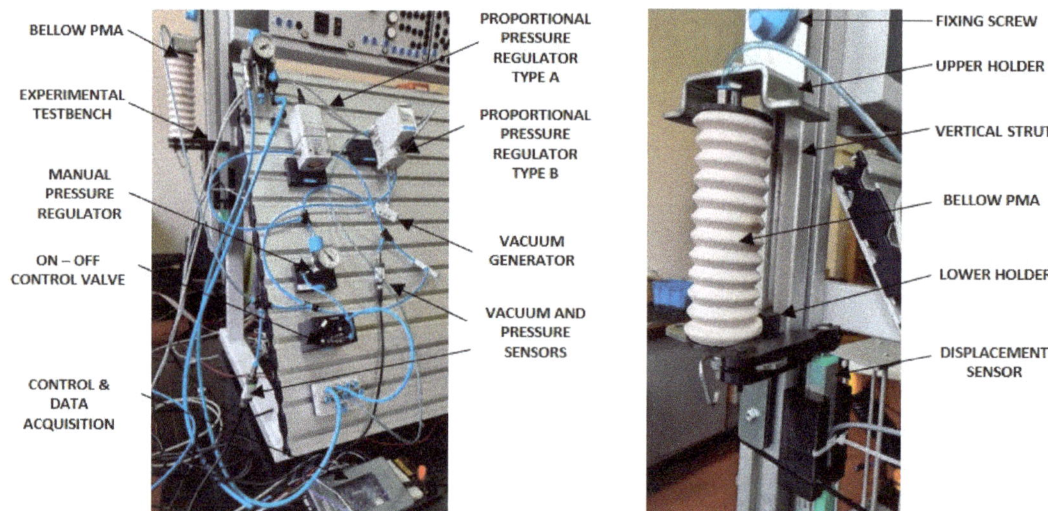

Figure 2. The main components of the experimental setup.

The bellow PMA, whose development is described in detail in [10], is anchored to an upper holder that is fixed to the profile, and a lower holder that allows sliding linear motion on the strut profile. Additionally, the movable part of the potentiometric displacement sensor is attached to the lower profile. The vacuum generator function is based on the Venturi principle and for the used FESTO vacuum generator, an input value of 0 to 4 bar results in a vacuum range of 0 to −0.9 bar. Therefore, the input value of the air pressure was adjusted by a manual pressure regulator to the maximum value of $p = 4$ bar. The user application for data collection and control was developed in the LabVIEW programming environment as a Virtual Instrument (VI).

In contrast to previous work [10], here we conducted additional analyses of two different pneumatic system configurations with different vacuum control approaches (Figure 1a,c). In the first configuration, we used a proportional pressure regulator which we call type A in the following text (FESTO VPPE-3-1/8-6-010), with an output range from 0 to 8 bar to control the input pressure in the vacuum ejector (indirect vacuum control). In the second system, we employed a proportional pressure regulator named type B (FESTO VPPE-3-1/8-6-010), with a pressure output range from −1 to 1 bar for directional vacuum control at the vacuum generator output. These two systems were used and compared with the aim of analyzing dynamical properties of the PMAs used. Finally, the total energy consumption of the system was compared in these two distinct cases. The direct vacuum control was assumed to have better dynamical properties (i.e., faster response times) than the indirect approach, but this comes at the expense of more supplied air. That is, direct vacuum control method requires continuous vacuum generation, whereas the indirect principle uses air only when the proportional pressure regulator is triggered. Furthermore, the influence of switching times of both used pressure regulators was investigated.

2.2. Numerical Modeling, PID Controller Development and Initial Experiments

With the aim of assessing the precision positioning parameters of the developed system, a widely used PID controller was employed first. Although a PID controller can ensure fast responses and a relatively low steady-state error, its response is generally characterized by some amount of overshoot.

The PID controller was custom-developed using the LabVIEW programming environment according to [21]. The algorithm is available on the author's Github repository [22]. Its parameters were tuned by using Ziegler–Nichols closed-loop method and additionally refined by the trial-and-error method in the initial testing phase. The final values of the PID gains that produced satisfactory responses were $K_P = 0.115$, $K_I = 0.015$ and $K_D = 0.001$. In the experimental phase, it was shown that increasing the reference value causes greater overshoot when such parameters are employed.

Taking into account the disadvantages of the mentioned tuning methods, in the next step we aimed to obtain a numerical model of the system. We selected the voltage of the proportional pressure regulator as the input and linear displacement of the PMA as the output. For the input signal, we employed step input signals with normally distributed amplitudes within a 0–2.5 V range. The upper limit was chosen since for this value, the system reaches its working range limit (0–61 mm) in the conditions with no external load. The sample time was set to 0.01 s. The duty cycle and period of the signal were set to 75% and 150 s, respectively. These parameters were chosen with respect to the dynamics of the open-loop system to account for the system's over-damped behavior, as well as to capture its exact dynamical behavior. That is, the duty-cycle and period were chosen with respect to the time constants of the system in both directions of motion. The input signal generator script written in MATLAB can be found in the author's Github repository [23].

The experiments with the system running in an open loop were performed using three different sets of randomly generated input signals. Figure 3a depicts an example of a typically obtained open-loop response. The MATLAB Identification Toolbox was then used to obtain the system's transfer function based on the experimental data. Given that the system was overdamped, it could be approximated as a first order system:

$$G(s) = \frac{10.6}{s + 0.5834} \tag{1}$$

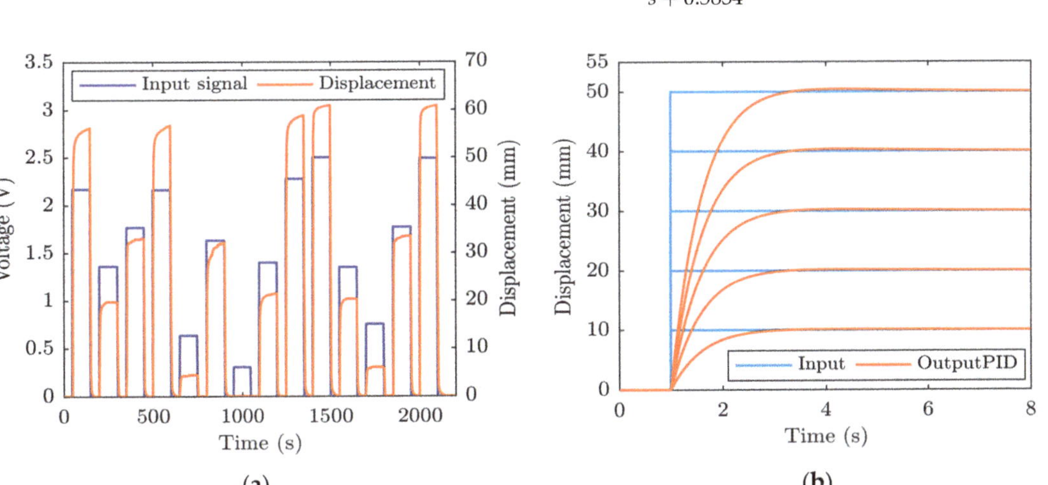

Figure 3. (a) The system's open-loop response for an input data set with normally distributed amplitudes in the 0–2.5 V range, corresponding to a displacement range of 0–61 mm. (b) Simulated responses of the PID controlled system for different reference values in SIMULINK.

It was found that the R^2 is equal to 0.655. After the model was built, a corresponding PID controller was found by using the SIMULINK PID tuning toolbox and the corresponding simulated responses to different set-point values were recorded, as shown in Figure 3b. It can be noticed that the steady-state error is nearly zero, while the overshoot of the system is negligible for smaller reference values, and very little for the larger references.

Due to the limitations of the PID implemented in SIMULINK, where the derivative component of error is filtered by using a low pass filter to avoid the effect of derivative kicks, we modified the derivative part in the implementation of the PID on a real-time hardware setup such that the derivative gain multiplies process value only [24]. Therefore, we adopted proportional (K_P = 0.16) and integral (K_I = 0.11) gains obtained in SIMULINK while we experimentally found a new derivative gain (K_D = 100) that provides a response with no or very little overshoot for different reference values within the working range. The PMA was then subjected to experimental tests without using an external load, as shown in Figure 4.

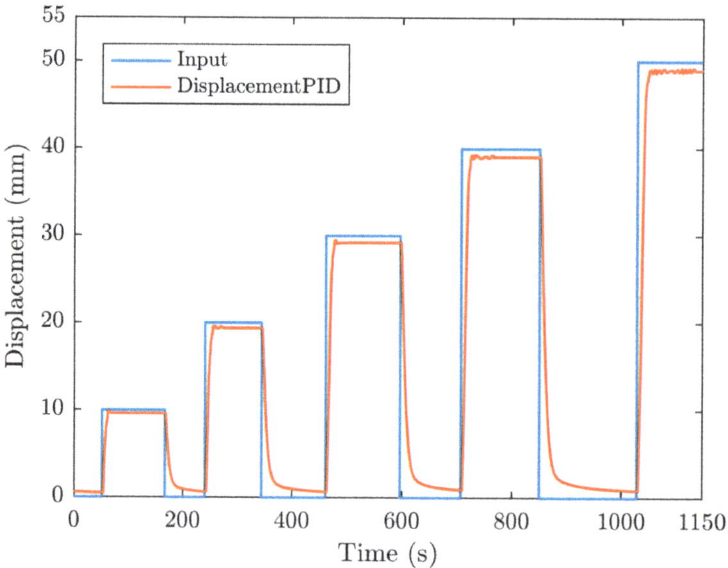

Figure 4. Positioning performance of the PMA controlled by PID.

It can be seen that the overshoot is negligible in all cases, which corresponds to the simulated responses (Figure 3b). A steady-state error from a control perspective (later in the text denoted simply as error), on the other hand, increases with muscle contraction, and can reach up to 1.5 mm for higher reference values. It is also noticeable that the PMA takes a longer time to return to its initial position (its maximum length). Furthermore, almost always, the muscle does not return to its initial position, because there is always a small amount of residual vacuum left in the body of the muscle. The aforementioned behavior is also influenced by the slower valve response time during its switch-off phase which is a result of the dynamics of the spring used to return the piston of the proportional pressure regulator to its initial position.

Given that the PID controller was unable to ensure optimal response, it was decided that the LQR approach will be investigated in the following steps with the attempt of minimizing steady-state error.

2.3. Numerical Modeling, LQR Controller Development and Initial Experiments

In order to apply an LQR approach to the considered system, additional sets of experiments were performed (using the parameters defined in Section 2.2) and the MATLAB Identification Toolbox was again employed to obtain the mathematical model of the system. In all cases, a sampling time of 5 ms was used. The second order state-space model in observable canonical form was, thus, developed. Linear displacement and velocity were used as states. The model was then used as a base for building an LQR controller. Based on previous research [10] and numerical simulations performed on the developed model, for the calculation of LQR gains, matrix Q was chosen to be an identity matrix multiplied by 10, while R was set to 0.001. By solving Riccati's equation [25], a vector of gains K = [0.11 0.01] was obtained. The LQR controller was then implemented in the

LabVIEW programming environment. Linear displacement was measured by employing a linear displacement sensor, and velocity was calculated from displacement by using shift registers in LabVIEW. The velocity was calculated in each consecutive time step as the difference between current and previous displacement values divided by sampling time. Finally, to reduce the effect of noise, the average velocity was calculated in real-time as the moving average of the previous five velocity values. Performed point-to-point positioning experiments allowed us to establish that a considerable steady-state error is present due to the inherent nonlinearity of the system, the compressibility of air and PMA's different behavior depending on the direction of motion, as discussed in Section 2.2. Therefore, a data-driven nonlinear pregain term that modifies the reference signal was determined experimentally and introduced into the system. Based on the performed experimental measurements, two 2nd order polynomials were determined each for one direction of motion. Table 1 shows the coefficients of each polynomial with respect to the direction of motion, whereas coefficient *a* multiplies the highest order member. These functions (Figure 5) were used to modify the reference signal so as to minimize the steady-state error. They were implemented in the LabVIEW environment by using a state-machine approach.

Table 1. Nonlinear pregain term polynomial coefficients for the type A pneumatic system.

Motion Direction	a	b	c
Forward	3.03×10^{-5}	-3.07×10^{-3}	1.161×10^{-1}
Backward	1.1×10^{-5}	-1.4×10^{-3}	8.057×10^{-2}

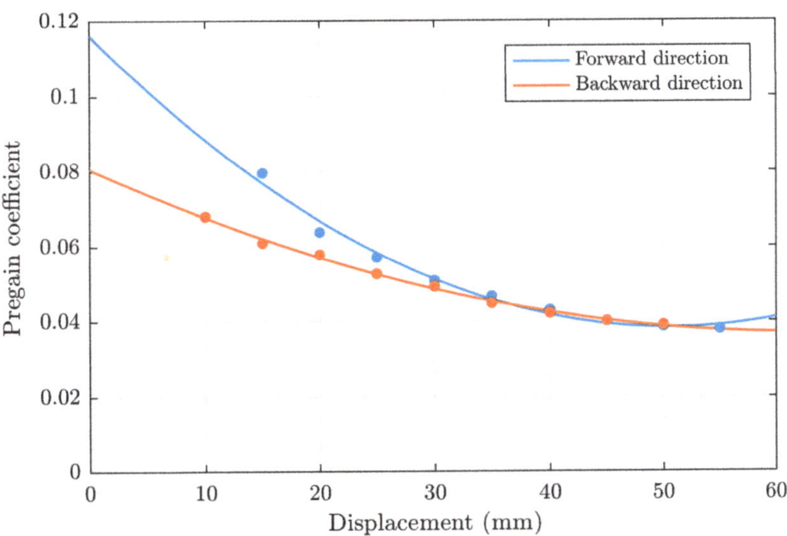

Figure 5. Controller pregain for the type A system represented as two 2nd order polynomials.

Figure 6 depicts the positioning performances of the developed LQR with pregain in no-load conditions. When the results are compared to those obtained by using the PID (Figure 4), it can be observed better dynamical behavior is obtained since the responses for each reference input are less jittery in close proximity to the steady state. Another clear advantage of using the LQR with pregain is a considerably lower steady-state error (0.03–0.14 mm) when compared to the results achieved with PID (0.1–1.5 mm).

Furthermore, in the case of LQR, overshoot is virtually eliminated. As in the case when PID is used, it can be seen that the muscle does not return exactly to the initial position, since there is always a small amount of residual vacuum left in the body while all experiments are performed without applying an external load.

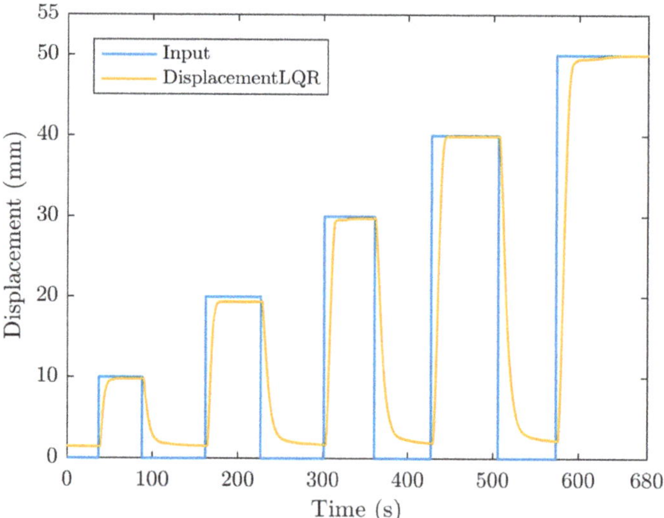

Figure 6. Positioning performances of the developed LQR controller.

3. Results

In this section, we compare the positioning performances of two different vacuum sources using PID and LQR control methods. First, we perform experiments with no external load, and then we add a constant load of 500 g to the system.

3.1. Position Performances of the Bellow PMA by Utilizing Indirect Vacuum Control

The PMA responses in no-load conditions are recorded for the PID and LQR approaches for different reference values (5, 10, 20, 30, 40 and 50 mm). Please keep in mind that, while it has been demonstrated that the working range of the employed PMA can be extended to approximately 60 mm [10], we limit the working range to 50 mm in this paper to avoid the contact of the inner supporting rings which can introduce additional nonlinearity to the setup.

As previously stated, in the case of indirect vacuum control (type A—please refer to Section 2.1 and Figure 1a), the proportional regulator varies the output pressure at the input rail of the used ejector, creating vacuum at its output. In this case, we experimentally confirmed that there is some dead-band present in the PMA response, as it can be observed from Figure 7. It can also be seen that the PMA has very similar dynamical behavior for both PID and LQR, though PID shows slightly faster dynamics for 5, 20 and 40 mm reference values. Steady-state error is, however, always considerably lower when LQR is used and this result is mainly limited by the used feedback sensor. In some cases, LQR outperforms PID by an order of magnitude, while the error is kept to a few tens of micrometers for all reference values. However, for both control typologies, there is a tendency for steady-state error to increase with higher reference values (except for the highest reference in case of LQR). When compared to similar research in this field [19], where error is measured in millimeters, this can be considered as a very good result, especially given that it was achieved on a highly nonlinear pneumatic system. Please note that the achieved positioning results are also limited by the properties of the used sensor (see also Section 2). Both controllers provide responses without overshoot, but the jitter effect is pronounced when PID is used. Similar to previous experiments, the muscle does not return exactly to the reference position, since all experiments are performed without the application of an external load. Table 2 summarizes the values of rising times and steady-state errors.

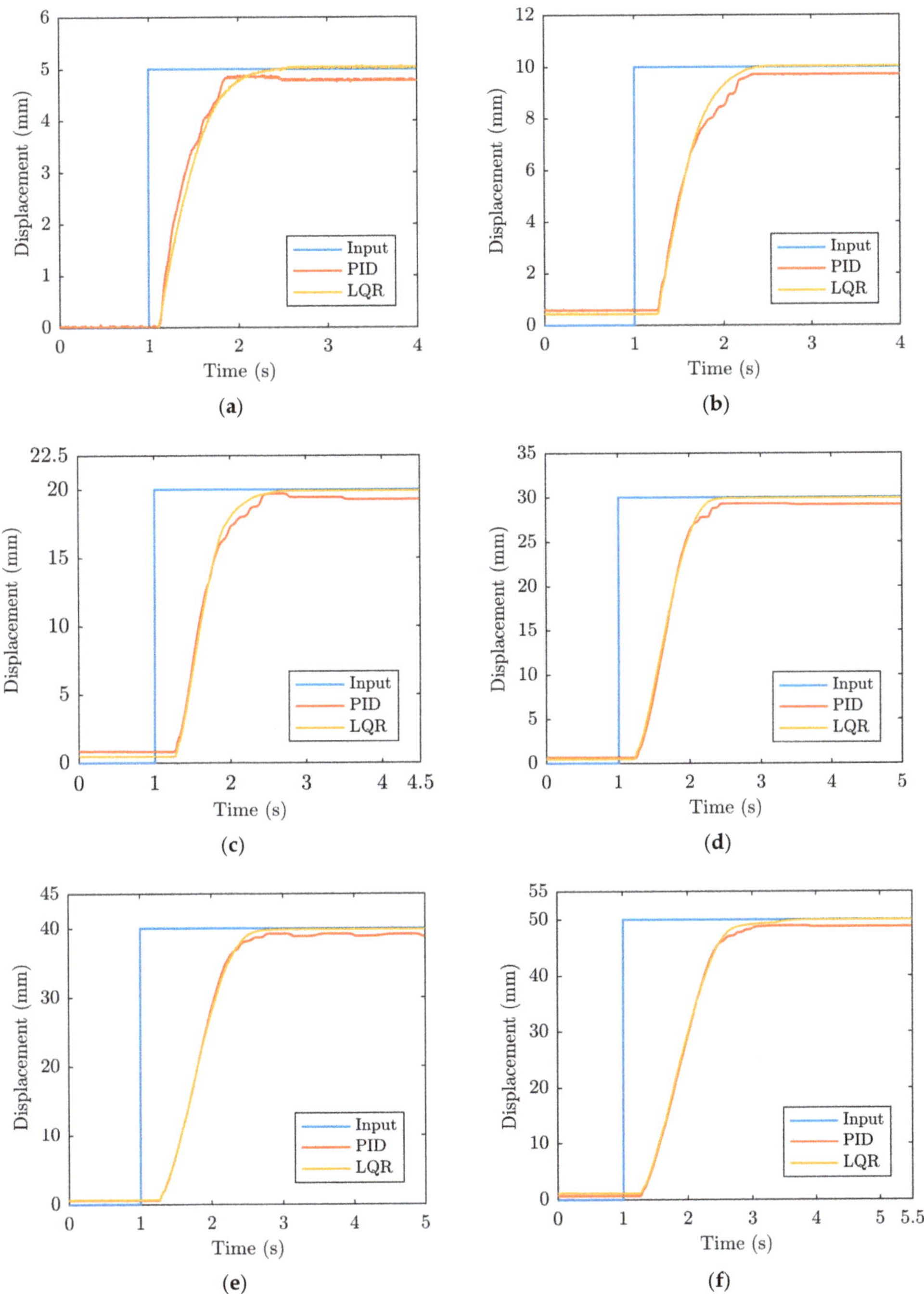

Figure 7. Results of PMA position control with PID (red lines) and LQR (yellow lines) for the type A pneumatic experimental setup: (**a**) 5 mm; (**b**) 10 mm; (**c**) 20 mm; (**d**) 30 mm; (**e**) 40 mm; (**f**) 50 mm reference.

Table 2. Rising time constants and steady-state errors for the type A pneumatic system for different set points.

Displacement (mm)	τ_s (s)		Steady-State Error (µm)	
	PID	**LQR**	**PID**	**LQR**
5	0.64	0.675	227	17
10	0.85	0.644	291	30
20	0.865	0.665	620	90
30	0.705	0.725	830	100
40	0.86	0.89	1000	140
50	1.01	1.03	1230	30

Given the fact that there is always a certain amount of dead-band present in the responses when indirect vacuum control is used, we analyze the behavior of the system by employing direct vacuum control in the following section.

3.2. Position Performances of the Bellow PMA by Utilizing Direct Vacuum Control

In this section, we conduct the experiments by using the type B pneumatic system (see Figure 1c), which allows for direct vacuum control at the output of the ejector. This configuration also allows for much faster valve switching (approximately 3 Hz frequency). Please keep in mind that one of the disadvantages of this system is that it consumes more energy due to the need for constant vacuum supply to the valve.

Since the system's hardware has been considerably modified, the optimal gains of both utilized controllers had to be adjusted. The PID parameters that were adopted are as follows: $K_P = 0.295$, $K_I = 0.035$ and $K_D = 3$. The vector with LQR gains, on the other hand, is calculated to be K = [0.2 0.002]. As with the type A system, we also calculate an additional pregain term in this case to allow for the elimination of the steady-state error. As shown in Table 3, the pregain term (Figure 8) is defined as a third order polynomial with coefficients that again depend on the motion direction.

Table 3. Nonlinear pregain term polynomial coefficients for the type B pneumatic system.

Motion Direction	a	b	c	d
Forward	-4.6×10^{-7}	5.724×10^{-5}	-2.442×10^{-1}	5.929×10^{-2}
Backward	-6×10^{-7}	7.65×10^{-5}	-3.11×10^{-3}	6.188×10^{-2}

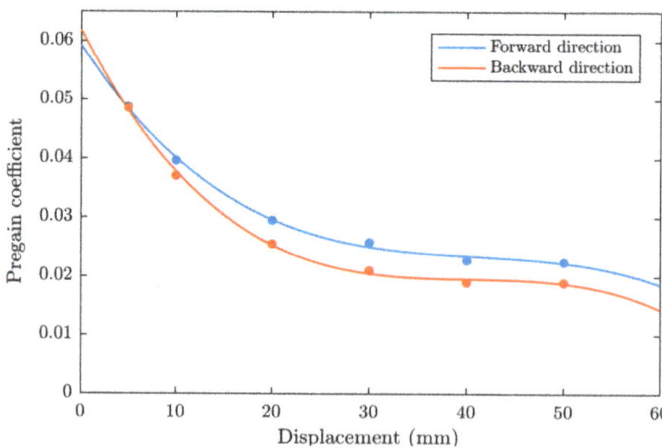

Figure 8. Controller pregain for type B system represented as two third order polynomials.

Figure 9 depicts and compares the experimental results for both control typologies when no external load is applied.

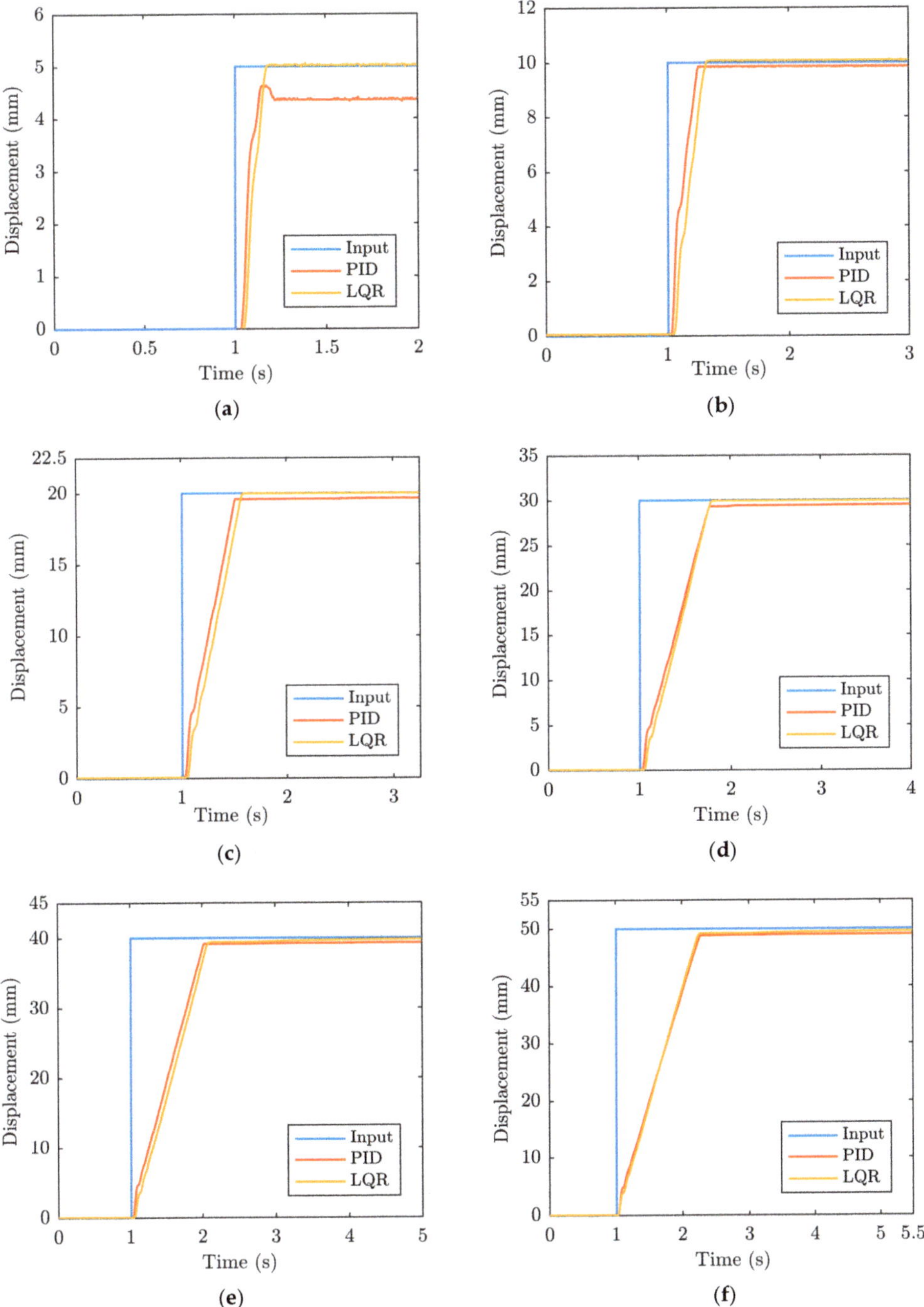

Figure 9. Results of PMA position control with PID (red lines) and LQR (yellow lines) for the second pneumatic experimental setup: (**a**) 5 mm; (**b**) 10 mm; (**c**) 20 mm; (**d**) 30 mm; (**e**) 40 mm; (**f**) 50 mm reference.

From the experimental results, it can be seen that the overshoot is not present when LQR is employed. For the lowest reference value, the PID controller induces an overshoot of approximately 6% (Figure 9a). Moreover, the previously observed dead-band is almost completely diminished in this case. Besides that, the faster switching time allows for a much faster overall system response for both PID and LQR controllers, which once more justifies the need of using the direct vacuum control principle if faster dynamics is desired. It can be noticed that the PMA has very similar dynamics for both control typologies, though PID again has slightly lower rising time constants for some reference values. Steady-state error is, however, always much lower in the case of LQR. LQR outperforms PID in terms of steady-state error by an order of magnitude in most cases (except for the 10 mm reference), while the error is a few tens of micrometers for all reference values. Table 4 summarizes the values of rising times and steady-state errors.

Table 4. Rising time constants and steady-state errors for the type B pneumatic system for different set points.

	τ_s (s)		Steady-State Error (μm)	
Displacement (mm)	PID	LQR	PID	LQR
5	0.095	0.1	642	17
10	0.185	0.205	144	80
20	0.42	0.44	380	10
30	0.65	0.625	520	50
40	0.86	0.84	730	20
50	1.05	1.005	960	50

On the other hand, when the results are compared to those of type A system (Table 5), it can be observed that when the PID controller is used, the rising time constant is considerably lower for smaller reference values and slightly higher (4%) for the highest reference value when type B system is considered. The steady-state error is significantly lower for almost all references, with the exception of the lowest reference value (5 mm) where it is much higher in the case of the system with type B vacuum control. This behavior can be attributed to the highly nonlinear behavior of the analyzed pneumatic muscle.

Table 5. Comparison of rising time constants and steady-state errors for indirect (**A**) and direct (**B**) vacuum control.

	τ_s		Steady-State Error	
Displacement (mm)	PIDB vs. PIDA	PIDB vs. PIDA	PIDB vs. PIDA	PIDB vs. PIDA
5	85% less	85% less	65% more	same
10	69% less	69% less	50% less	62% more
20	48% less	34% less	39% less	90% less
30	8% less	14% less	38% less	50% less
40	same	6% less	28% less	85% less
50	4% more	2% less	22% less	40% more

Rising time constants are lower in all cases when using the LQR controller, and this is especially noticeable for lower reference values. Except for the highest reference value, steady-state error is again much lower in almost all cases.

Finally, the experimental results obtained by employing an experimental system with direct vacuum control (type B) allowed establishing significantly better results from a dynamical point of view. This is especially evident taking into account the fact that the dead-band effect during PMA activations is almost eliminated. Moreover, if compared to type A, the dynamical response is much faster especially for lower reference values. The steady-state error is in the case of LQR again several tens of micrometers (20–80 μm).

In order to test the muscle in more realistic conditions, we assess the positioning performances of the loaded system in the final set of experiments. The system is given a constant weight of 500 g. The results of the positioning performances for the loaded system are shown in Figure 10, while the achieved dynamical performances are again evaluated in terms of rising time constants for each reference value, as shown in Table 6. The graphs show that positioning performances without overshoot in the case of LQR and with slight overshoot for some references in the case of PID, are achieved. In both cases, a very small dead-band value is obtained at the beginning of the actuation cycle. When the rising time constants are compared to those achieved in the previous experiments, it can be concluded that the values are very similar and only differ by about 10%.

However, when using a PID controller, the steady-state error is much higher when the system is loaded, and this is especially evident for the lower references. This means that if the loading conditions change, the PID parameters have to be tuned again [13]. This once more justifies the need for using more refined control typologies. When LQR with an additional pregain term is used, on the other hand, the steady-state error is a few tens of micrometers and it is not substantially influenced by external loading.

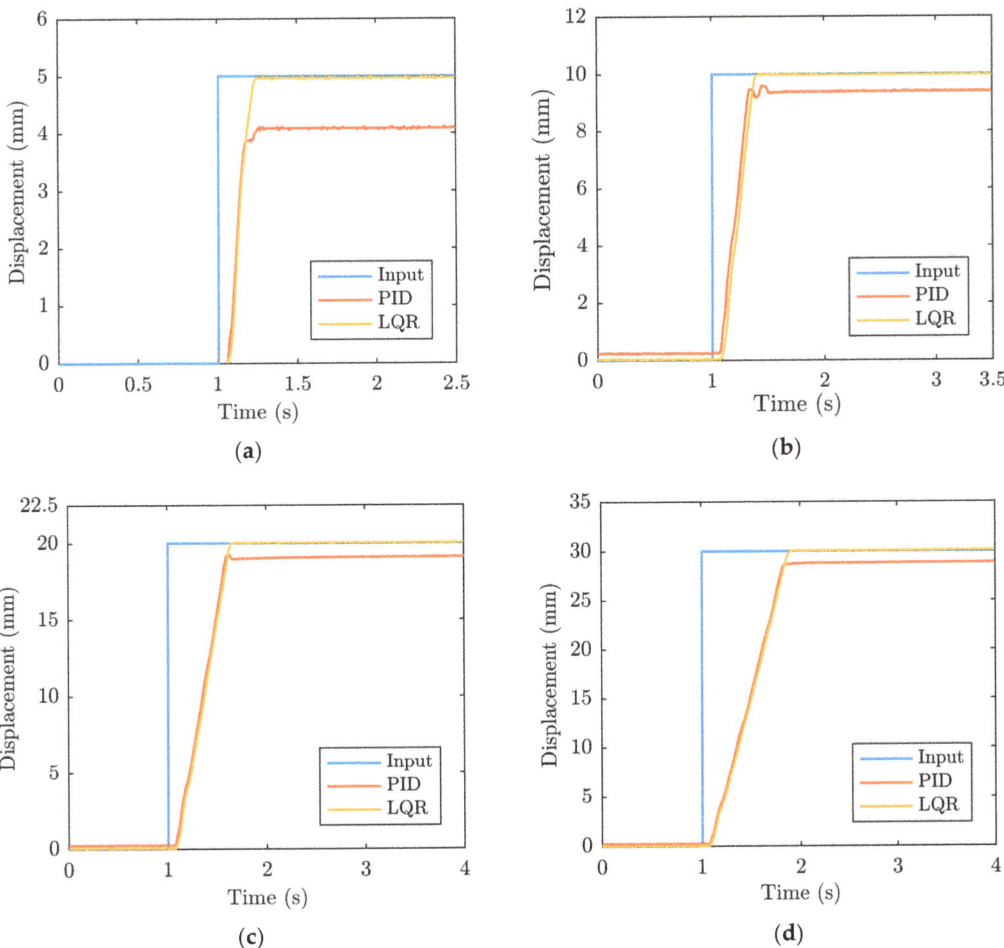

Figure 10. *Cont.*

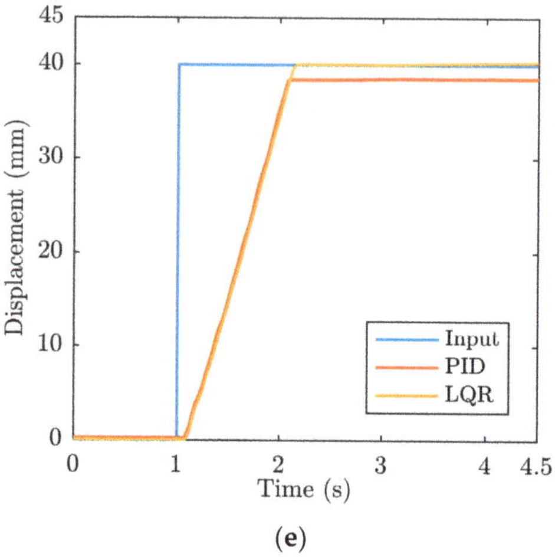

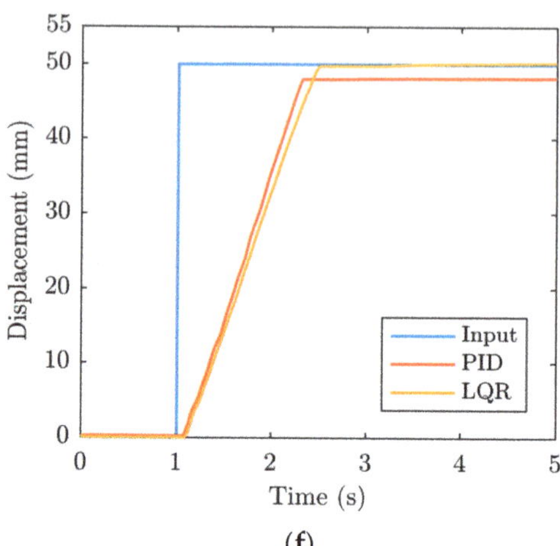

Figure 10. Results of PMA position control with PID (red lines) and LQR (yellow lines) for the second pneumatic experimental setup with 500 g load: (**a**) 5 mm; (**b**) 10 mm; (**c**) 20 mm; (**d**) 30 mm; (**e**) 40 mm; (**f**) 50 mm reference.

Table 6. Rising time constants and steady-state errors for the type B pneumatic system and a load of 500 g for different set points (in seconds).

Displacement (mm)	τ_s (s)		Steady-State Error (μm)	
	PID	**LQR**	**PID**	**LQR**
5	0.205	0.12	911	32
10	0.23	0.225	608	27
20	0.435	0.44	890	30
30	0.63	0.635	1150	40
40	0.83	0.83	1410	30
50	1.015	1.085	1790	70

Finally, we conducted energy consumption analyses for the pneumatic systems under consideration by measuring the time required for the pressure in the compressor's air reservoir to drop by 2 bar during the PMA operation. In both systems, the input pressure is held constant at 4 bar, and the control signal to the valves is sinusoidal with 0.01 Hz frequency. In these conditions, the total time measured was 870 and 264 s for type A (indirect vacuum control) and type B (direct vacuum control) systems, respectively. This allowed us to establish that the direct vacuum control system consumes 70% more compressed air than the indirect vacuum control approach. The direct vacuum control approach, however, allows better dynamical behavior, i.e., a faster response.

4. Conclusions

In the presented research, we demonstrated the possibility of precision positioning by using PID and LQR control approaches of a custom-designed bellow PMA actuator. Two distinct vacuum control configurations were investigated; a type A system that performs indirect vacuum control and a type B system that allows direct vacuum control. The data-driven numerical model of the system was developed using MATLAB. Both control algorithms were developed and tested in the simulation environment using the developed numerical model. The LabVIEW programming environment was then used to develop the algorithms for real-time control.

After initial evaluation of the system, both vacuum control systems were thoroughly evaluated by using PID and LQR control approaches. In almost all cases, the steady-state error of the LQR was an order of magnitude lower than that of the PID controller. The LQR approach also outperformed PID in terms of smaller overshoot. The response velocity for both control algorithms was comparable. When external load was added to the system, steady-state error increased significantly when the PID controller was used. LQR, on the other hand, successfully minimized error when the load was added and kept it within micrometric boundaries. A comparison of direct and indirect vacuum control approaches revealed that direct vacuum control allows significantly faster dynamical behavior. Furthermore, we performed energy consumption analyses for the pneumatic systems used and showed that the direct vacuum control consumes up to 70% more air than the indirect vacuum control approach.

Finally, the presented research demonstrated that the analyzed artificial muscle can achieve a positioning error of a few tens of micrometers (which is mainly limited by the used sensor) despite its highly nonlinear behavior. An additional design optimization could be performed to the muscle to refine the results across its entire working range.

Future research will concentrate on developing a 3-DOF motion (Stewart) platform with pneumatic muscle actuators and the LQR control approach described in this paper. Special attention will be given to the type of the 3D motion detection sensors during the design phase to ensure proper feedback.

Author Contributions: Conceptualization, G.G. and E.K.; methodology, G.G., E.K., A.Š. and S.P.; software, A.Š. and S.P.; validation, G.G., E.K., A.Š. and S.P.; formal analysis, A.Š. and S.P.; investigation, A.Š. and S.P.; resources, G.G., E.K., A.Š. and S.P.; data curation, A.Š.; writing—original draft preparation, G.G. and E.K.; writing—review and editing, G.G. and E.K.; visualization, G.G. and A.Š.; supervision, G.G. and E.K.; funding acquisition, G.G. and E.K. All authors have read and agreed to the published version of the manuscript.

Funding: This research received no external funding.

Institutional Review Board Statement: Not applicable.

Informed Consent Statement: Not applicable.

Data Availability Statement: Not applicable.

Conflicts of Interest: The authors declare no conflict of interest.

References

1. Su, H.; Hou, X.; Zhang, X.; Qi, W.; Cai, S.; Xiong, X.; Guo, J. Pneumatic soft robots: Challenge and benefits. *Actuators* **2022**, *11*, 92. [CrossRef]
2. Walker, J.; Zidek, T.; Harbel, C.; Yoon, S.; Strickland, F.S.; Kumar, S.; Shin, M. Soft robotics: A review of recent developments of pneumatic soft actuators. *Actuators* **2020**, *9*, 3. [CrossRef]
3. Majidi, C. Soft Robotics: A Perspective—Current Trends and Prospects for the Future. *Soft Robot.* **2014**, *1*, 5–11. [CrossRef]
4. Dong, W.; Wang, Y.; Zhou, Y.; Bai, Y.; Ju, Z.; Guo, J.; Gu, G.; Bai, K.; Ouyang, G.; Chen, S.; et al. Soft human-machine interfaces: Design, sensing and simulation. *Int. J. Intell. Robot. Appl.* **2018**, *2*, 313–338. [CrossRef]
5. Yu, M.; Cheng, X.; Peng, S.; Cao, Y.; Lu, Y.; Li, B.; Feng, X.; Znang, Y.; Wang, H.; Jiao, Z.; et al. A self-sensing soft pneumatic actuator with closed-loop control for haptic feedback wearable devices. *Mater. Des.* **2022**, *223*, 111149. [CrossRef]
6. Pan, M.; Yuan, C.; Liang, X.; Dong, T.; Liu, T.; Zhang, J.; Zou, J.; Yang, H.; Bowen, C. Soft Actuators and Robotic Devices for Rehabilitation and Assistance. *Adv. Intell. Syst.* **2022**, *4*, 2100140. [CrossRef]
7. Belforte, G.; Eula, G.; Ivanov, A.; Sirolli, S. Soft Pneumatic Actuators for Rehabilitation. *Actuators* **2014**, *3*, 84–106. [CrossRef]
8. Kamenar, E.; Črnjarić-Žic, N.; Haggerty, D.; Zelenika, S.; Hawkes, E.; Mezić, I. Prediction of the behavior of a pneumatic soft robot based on Koopman operator theory. In Proceedings of the 2020 43rd International Convention on Information, Communication and Electronic Technology, Opatija, Croatia, 28 September–2 October 2020.
9. Tanaka, J.; Ogawa, A.; Nakamoto, H.; Sonoura, T.; Eto, H. Suction pad unit using a bellows pneumatic actuator as a support mechanism for an end effector of depalletizing robots. *Robomech J.* **2020**, *7*, 2. [CrossRef]
10. Gregov, G.; Ploh, T.; Kamenar, E. Design, Development and Experimental Assessment of a Cost-Effective Bellow Pneumatic Actuator. *Actuators* **2022**, *11*, 170. [CrossRef]
11. Petre, I.M. Studies regarding the Use of Pneumatic Muscles in Precise Positioning Systems. *Appl. Sci.* **2021**, *11*, 9855. [CrossRef]
12. Xavier, M.S.; Fleming, A.J.; Yong, Y.K. Design and Control of Pneumatic Systems for Soft Robotics: A Simulation Aproach. *IEEE Robot. Autom. Lett.* **2021**, *6*, 5800–5807. [CrossRef]

13. Tang, T.F.; Chong, S.H.; Chan, C.Y.; Sakthivelu, V. Point-to-Point Positioning Control of a Pneumatic Muscle Actuated System Using Improved-PID Control. In Proceedings of the 2016 IEEE International Conference on Automatic Control and Intelligent Systems (I2CACIS), Selangor, Malaysia, 22 October 2016.
14. Al-Ibadi, A.; Nefti-Meziani, S.; Davis, S. Controlling of Pneumatic Muscle Actuators Systems by Parallel Structure of Neural Network and Proportional Controllers (PNNP). *Front. Robot. AI* **2020**, *7*, 115. [CrossRef]
15. Zhao, J.; Zhong, J.; Fan, J. Position Control of a Pneumatic Muscle Actuators Using RBF Neural Network Tuned PID Controller. *Math. Probl. Eng.* **2015**, *2015*, 810231. [CrossRef]
16. Scaff, W.; Horikawa, O.; Tsuzuki, M.S. Pneumatic Artificial Muscle Optimal Control with Simulated Annealing. *IFAC* **2018**, *51*, 333–338. [CrossRef]
17. Nuchkrua, T.; Leephakpreeda, T. Fuzzy Self-Tuning PID Control of Hydrogen-Driven Pneumatic Artificial Muscle Actuator. *J. Bionic Eng.* **2013**, *10*, 329–340. [CrossRef]
18. Šitum, Ž.; Herceg, S.; Bolf, N.; Željka, U.A. Design, Construction and Control of a Manipulator Driven by Pneumatic Artificial Muscles. *Sensors* **2023**, *23*, 776. [CrossRef] [PubMed]
19. Kazemi, S.; Hashem, R.; Stommel, M.; Cheng, L.K.; Xu, W. Experimental Study on the Closed-Loop Control of a Soft Ring-Shaped Actuator for In-Vitro Gastric Simulator. *IEEE/ASME Trans. Mechatron.* **2022**, *27*, 3548–3558. [CrossRef]
20. Conte, G.Y.C.; Marques, F.G.; Garcia, C. LQR and PID Control Design for a Pneumatic Diaphragm Valve. In Proceedings of the IEEE International Conference on Automation/XXIV Congers of the Chilean Association of Automatic Control (ICA-ACCA), Valparaiso, Chile, 22–26 March 2021.
21. Baćac, N.; Slukić, V.; Puškarić, M.; Štih, B.; Kamenar, E.; Zelenika, S. Comparison of different DC motor positioning control algorithms. In Proceedings of the 2014 37th International Convention on Information and Communication Technology, Electronics and Microelectronics (MIPRO), Opatija, Croatia, 26–30 May 2014; pp. 1654–1659.
22. Control_Algorithms. Available online: https://github.com/ekamenar/control_algorithms (accessed on 6 February 2023).
23. Signal_Generator. Available online: https://github.com/ekamenar/signal_generator (accessed on 6 February 2023).
24. Kamenar, E.; Zelenika, S. Micropositioning mechatronics system based on FPGA architecture. In Proceedings of the 2013 36th International Convention on Information and Communication Technology, Electronics and Microelectronics (MIPRO), Opatija, Croatia, 20–24 May 2013; pp. 125–130.
25. Levine, W.S. *The Control Handbook*; Three Volume Set; CRC Press: Boca Raton, FL, USA, 2018.

Article

Iterative Feedback Tuning of Model-Free Intelligent PID Controllers

Andrei Baciu and Corneliu Lazar *

Department of Automatic Control and Applied Informatics, "Gheorghe Asachi" Technical University of Iasi, 700050 Iasi, Romania
* Correspondence: clazar@ac.tuiasi.ro

Abstract: In the last decades, model-free control (MFC) has become an alternative for complex processes whose models are not available or are difficult to obtain. Among the model-free control techniques, intelligent PID (iPID) algorithms, which are based on the ultralocal model parameterized with the constant α and including a classical PID, are used in many applications. This paper presents a new method for tuning iPID controllers based on the iterative feedback tuning (IFT) technique. This model-free tuning technique iteratively optimizes the parameters of a fixed structure controller using data coming from the closed-loop system operation. First, the discrete transfer functions of the iPID are deduced, considering the first and second order derivatives of the output variable from the ultralocal model. Using the discrete transfer functions, the iPID controllers become the fixed structure type, and the IFT parameter tuning method can be applied. Thus, in addition to the classical gains of the PID algorithm, the value of the parameter α is also obtained, which is usually determined by trial-and-error. The performances of the IFT-tuned iPID controllers were experimentally tested and validated in real-time using Quanser AERO 2 laboratory equipment with a one degree of freedom (1-DOF) configuration.

Keywords: model-free control; intelligent PID controllers; iterative feedback tuning; data-driven tuning; pitch angle control; aerodynamic system control

1. Introduction

Due to current advanced hardware and software technologies, data acquisition, storage, computing, and communication allow for the processing of increased amounts of data online. In this way, a new method referred to as data-driven control (DDC) has been developed, by which the controllers are directly designed using input–output data collected from the control system without process modeling. Through the development of information technology, the industrial processes have become more complex, making their modeling difficult. An alternative to the model-based control method is DDC, which has rapidly developed in recent decades. Overviews on the model-based control and data-driven control techniques are presented in [1,2], while the DDC theory is covered in books such as [3–5]. At the same time, the data-driven tuning techniques were developed to optimize the controller parameters, distinguishing two classes based on the controller structure. The first class is intended for controllers with a fixed structure. Among these, the iterative feedback tuning (IFT) algorithm [6] iteratively optimizes the controller parameters by estimating the gradient of a cost function in relation to the input and output signals; the virtual reference feedback tuning (VRFT) method [7] is a direct data-driven method to optimize the controller parameters, which does not require iterations. The second class includes tuning methods that do not require knowledge of the regulator structure, such as iterative learning control [8] or model-free adaptive control [4].

One DDC method is the model-free control (MFC) algorithm, which was developed by Fliess and Join [9] and has been successfully applied in many fields. It is based on the

Citation: Baciu, A.; Lazar, C. Iterative Feedback Tuning of Model-Free Intelligent PID Controllers. *Actuators* 2023, 12, 56. https://doi.org/10.3390/act12020056

Academic Editor: Gary M. Bone

Received: 12 December 2022
Revised: 12 January 2023
Accepted: 25 January 2023
Published: 28 January 2023

use of the ultralocal model, which is valid for a short time interval and is on the embedded of a PID controller in the control law, which is called an intelligent PID (iPID). The tuning parameters of the iPID controller are the usual tuning gains of the PID controller, to which the non-physical constant parameter α from the ultralocal model is added. Considering the ultralocal model and closing the tuning loop with the iPID controller, a description of the control system is obtained through a linear differential equation of the control error, which only depends on the tuning gains of the PID, if a good estimate of the term including the unknown parts of the ultralocal model is available [8]. Hence, the conclusion from [8] is made, where tuning the iPID controller is easy to perform by enforcing the condition that the differential equation must have stable roots, thus canceling the steady-state error. Furthermore, in accordance with a condition from [9,10], the user will choose the parameter α. Although it seems easy to tune an iPID controller, the use of only the linear differential equation does not directly lead to achieving the tuning gains.

For tuning the iPID controllers, several methods have been reported in the literature. Starting from the tuning proposed in [8] based on the cancellation of the control error in the steady-state, in [11,12], domains were determined in the tuning gains space, for which the stability of the roots of the error differential equation is ensured, thus obtaining a zero steady-state error. The VRFT one-shot data-driven tuning algorithm can be used to tune the iPID controllers. Thus, in [13], the method is shown regarding how the VRFT algorithm can be used to tune the iP, iPI, and iPID model-free controllers, where the three tuned controllers are validated on modular servo system laboratory equipment. Another hybrid method is generated by mixing the model-free control with the sliding mode control (SMC) [14], resulting in a simple design method that ensures the stability of the tracking error dynamics specific to MFC. If the MFC in [14] is of the iPI type, the method in [15] presents the MFC-SMC with an iPD controller intended for the attitude and position control of a quadrotor.

The design methods above lead to the acquisition of the tuning gains of the iPID controllers without the possibility of finding a suitable value for α, which remains a constant design parameter to be chosen by the user according to [9,10]. However, other methods have been proposed to determine the parameter, such as in [16], where α is found using an algorithm based on the ultralocal model. Another approach considers α as a time-varying parameter for which the values are instantly estimated [17,18].

The data-driven IFT method for tuning controllers with a fixed structure is a gradient descent-based iterative algorithm presented in [6], are both theoretical aspects and applications regarding the tuning of controllers for mechanical and chemical systems. The tuning objective is to obtain the step response of the closed-loop system with a minimal settling time and small overshoot. The advantages of applying the IFT method for tuning PID controllers in relation to the classic methods are presented in [19]. To ensure the con-vergence for the initial parameters of the controller, [20] presents a solution based on a domain of attraction (DoA) approach. With respect to the regulatory control system, several solutions have proposed the use of the IFT algorithm. By using a special experiment, the authors of [21] describe the gradient estimate method based on a spectral analysis; in [22], the authors suggest the correlation-based IFT algorithm to reduce the influence of noise from the collected data. Solutions that do not require a special experiment to estimate the gradient related to the feedforward controller have also been proposed, such as those presented in [23,24]. A constrained IFT method was introduced in [25] for the robust cascade path-tracking control of a networked robot. For a model-free adaptive iterative learning controller, the IFT algorithm was used in [26] for controller parameters tuning. A first attempt of using the IFT algorithm for tuning a model-free iP controller intended for the speed control of an engine was presented in [27].

This paper proposes a new method for tuning the model-free iPID controllers based on the IFT data-driven algorithm. Being a data-driven approach, a process model is not required to tune the controller parameters; it only requires the sets of input–output data collected from the control system. Because the IFT algorithm is a method for tuning the parameters of a controller with a fixed structure, this study first determined a structure for the iPID controller using the connections between the iPID and PID discrete-time controllers. Only the first and second order derivatives related to the ultralocal model, which is the basis of the model-free control concept, were considered. Using the derivative approximation method based on the backward difference, the fixed structures of the iPID controllers were obtained, which were described by discrete transfer functions that were parameterized with a parameter vector. Based on these discrete transfer functions, behaviors that assimilated to those of conventional PID controllers were found, which allowed for the selection of intelligent controllers with behaviors like the PID ones and the removal of those with unusual behaviors in the control engineering process. Knowing the fixed structures of the iPID controllers, the IFT data-driven algorithm was applied, and the optimal tuning parameters of the controllers were obtained. The proposed tuning method based on IFT data-driven algorithms has the advantage of generating iPID tuning parameters related to the PID control law, but also the parameter α, which was usually chosen by the user. The new iPID controller tuning method based on the IFT algorithm is illustrated by real-time experiments that are aimed to validate the tuned controllers. The experiments were conducted with Quanser AERO 2 laboratory equipment in the one degree of freedom (1-DOF) configuration, with two DC motors for driving the propellers, for which the pitch angle was controlled.

This laboratory equipment, due to offering a wide range of system structures obtained through half-quad (pitch-locked, yaw-free), 1-DOF (yaw-locked, pitch-free), or 2-DOF con-figurations [28], was often used for testing and validating new control methods. For all configurations, the control with PID regulators is presented in [28]. New control algorithms for the 2-DOF configuration can also be found in the literature, such as adaptive neural control [29], approximation-based quantized state feedback tracking [30], observer-based sliding mode control [31], fuzzy neural networks [32], or adaptive attitude control with input and output quantization [33].

This paper proposes the following new contributions with respect to the state-of-the-art:

(i)   A new method of tuning model-free iPID controllers based on the IFT algorithm;
(ii) The calculation of both the tuning gains of the iPID controller and the parameter α;
(iii) The determination of a fixed structure for iPID controllers based on the connections between the iPID and PID discrete-time controllers to make it possible to apply the IFT tuning algorithm;
(iv) An analysis of the connections between the iPID and PID controllers; model-free control laws that correspond to some variants of the classical PID were determined, and those which are uncommon in the control engineering process were eliminated;
(v)   The tuning of the iPID controllers using the IFT algorithm was tested and validated experimentally on Quanser AERO 2 laboratory equipment.

The paper is organized as follows. Section 2 discusses the connections between the iPID and PID controllers, and based on them, the fixed structure of the iPID controller is presented. Section 3 first briefly introduces the IFT approach and then describes how to tune the iPID controller using the IFT algorithm. The experimental validation of the tuned iPID controllers is shown in Section 4, and Section 5 concludes the paper.

2. iPID Fixed Structure Based on Connections between the iPID and PID Discrete-Time Controllers

The IFT technique is a model-free technique used to iteratively optimize the parameters of a fixed-order controller using data coming from the closed-loop system operation. Applying the IFT technique for tuning an iPID controller first requires the determination of a fixed structure for this type of controller, which is parametrized by a vector $\rho \in \mathbb{R}^n$. For this reason, the discrete form of the iPID controllers has been related to the classical PID, which has a fixed structure.

The MFC method is designed around the concept of the ultralocal model [9], which is obtained from signals of the system and is available for a very short time frame. The equation that describes the ultralocal model, with notations from the operational calculus, is:

$$s^v Y(s) = F(s) + \alpha U(s),$$ (1)

where Y is the output signal of the system, U is the control signal, α is a parameter chosen by the practitioner, and F is a function which contains all of the unmodelled components of the ultralocal model. The derivative order v of y is usually chosen as 1 or 2 [9]. The control law iPID is defined as:

$$U(s) = \frac{1}{\alpha}\left(s^v \widetilde{Y}(s) - F(s) + \left(K_p + \frac{K_i}{s} + K_d s\right)E(s)\right),$$ (2)

where $\widetilde{Y}$ is the reference signal, $E = \widetilde{Y} - Y$ is the control error, and K_p, K_i, and K_d are the tuning gains of a classical PID controller.

The discrete-time ultralocal model is obtained using the backward difference approximation, as suggested in [34], by replacing s with $(1 - z^{-1})/T_s$, where T_s is the sampling period, resulting in:

$$\left(\frac{1 - z^{-1}}{T_s}\right)^v y_k = F_k + \alpha u_k.$$ (3)

Using Equation (3), the estimation of the unmodelled components of the ultralocal model is defined as:

$$\hat{F}_k = \left(\frac{1 - z^{-1}}{T_s}\right)^v y_k - \alpha u_{k-1}.$$ (4)

In the following, the estimation error between the actual F_k and its estimation $\hat{F}_k$ is considered negligible.

Applying the backward difference approximation to Equation (2) and using the estimation from Equation (4), the discrete-time iPID control law is obtained:

$$u_k = \frac{1}{\alpha(1 - z^{-1})}\left(\left(\frac{1 - z^{-1}}{T_s}\right)^v e_k + \left(K_p + K_i\frac{T_s}{1 - z^{-1}} + K_d\frac{1 - z^{-1}}{T_s}\right)e_k\right).$$ (5)

From Equation (5), the discrete-time classical PID control law is easily deduced:

$$u_k = \left(K_p + K_i\frac{T_s}{1 - z^{-1}} + K_d\frac{1 - z^{-1}}{T_s}\right)e_k.$$ (6)

Next, the connections between the discrete-time iPID and PID controllers will be investigated, considering the two recommended values for the derivative order v.

First, for $v = 1$, from Equation (5) after a simple calculation, the following form of the iPID$_1$ regulator results:

$$u_k = \frac{1}{\alpha}\left(\frac{1 + K_d}{T_s}e_k + \frac{K_p}{T_s}\frac{T_s}{1 - z^{-1}}e_k + \frac{K_i}{T_s}\frac{T_s^2}{(1 - z^{-1})^2}e_k\right),$$ (7)

which highlights a classical PI-I^2 type controller. The subscript of iPID$_1$ indicates the derivative order $v = 1$.

Using Equation (7), the following equations are obtained for the intelligent controllers iP$_1$, iPI$_1$, and iPD$_1$, by setting the appropriate tuning parameters. Thus, for $K_i = K_d = 0$, the iP$_1$ control law is found:

$$u_k = \frac{1}{\alpha}\left(\frac{1}{T_s}e_k + \frac{K_p}{T_s}\frac{T_s}{1 - z^{-1}}e_k\right),$$ (8)

which corresponds to a classical PI controller. If only $K_d = 0$, the result is the iPI_1 control law:

$$u_k = \frac{1}{\alpha}\left(\frac{1}{T_s}e_k + \frac{K_p}{T_s}\frac{T_s}{1-z^{-1}}e_k + \frac{K_i}{T_s}\frac{T_s^2}{(1-z^{-1})^2}e_k\right),$$ (9)

which has the form of a PI-I^2 controller, as in the case of the $iPID_1$ controller, but with different tuning parameters. Setting $K_i = 0$, the iPD_1 control law is obtained:

$$u_k = \frac{1}{\alpha}\left(\frac{1+K_d}{T_s}e_k + \frac{K_p}{T_s}\frac{T_s}{1-z^{-1}}e_k\right),$$ (10)

which is a classical PI controller, as in the case of the iP_1 controller, but with different tuning parameters. When analyzing the comparisons between the $iPID_1$ and classical PID controllers for $\nu = 1$, it is discovered that the intelligent controllers behave like PI or PI-I^2. For the iP_1 and iPD_1 controllers that behave like PI, by applying the Z transform to Equations (8) and (10), the discrete transfer function results:

$$G_{iP\&D_1}(z^{-1}) = \frac{q_0 + q_1 z^{-1}}{1 - z^{-1}},$$ (11)

whose tuning parameters grouped in the vector $\rho = [q_0\ q_1]^T$ are determined according to the type of the intelligent controller, as follows:

$$iP_1:\ q_0 = \frac{1+T_s K_p}{\alpha T_s}, q_1 = -\frac{1}{\alpha T_s},$$ (12)

$$iPD_1:\ q_0 = \frac{1+K_d+\alpha T_s K_p}{\alpha T_s}, q_1 = -\frac{1+K_d}{\alpha T_s}.$$ (13)

Applying the Z transform to Equations (7) and (9), the discrete transfer function of the iPI_1 and $iPID_1$ controllers that behave like PI-I^2 is obtained:

$$G_{iPI\&D_1}(z^{-1}) = \frac{q_0 + q_1 z^{-1} + q_2 z^{-2}}{1 - 2z^{-1} + z^{-2}},$$ (14)

with the following tuning parameters grouped in the vector $\rho = [q_0\ q_1\ q_2]^T$:

$$iPI_1:\ q_0 = \frac{K_p T_s + 1}{\alpha T_s}, q_1 = -\frac{K_p T_s + K_i T_s^2 + 2}{\alpha T_s}, q_2 = \frac{1}{\alpha T_s},$$ (15)

$$iPID_1:\ q_0 = \frac{K_p T_s + K_i T_s^2 + K_d + 1}{\alpha T_s}, q_1 = -\frac{K_p T_s + 2K_d + 2}{\alpha T_s}, q_2 = \frac{1+K_d}{\alpha T_s}$$ (16)

In conclusion, for $\nu = 1$, the intelligent controllers iP_1 and iPD_1 behave as a PI controller, having a fixed structure given by the discrete transfer function from Equation (11), parametrized by a vector $\rho \in \mathbb{R}^2$; the intelligent controllers iPI_1 and $iPID_1$ behave like a PI-I^2 controller with a fixed-order structure, given by the discrete transfer function from the Equation (14), parametrized by a vector $\rho \in \mathbb{R}^3$.

For $\nu = 2$, using Equation (5), the $iPID_2$ control law is obtained:

$$u_k = \frac{1}{\alpha}\left(\frac{K_d}{T_s}e_k + \frac{K_p}{T_s}\frac{T_s}{1-z^{-1}}e_k + \frac{1}{T_s}\frac{1-z^{-1}}{T_s}e_k + \frac{K_i}{T_s}\frac{T_s^2}{(1-z^{-1})^2}e_k\right),$$ (17)

which highlights a classical PID-I^2 controller. The subscript of $iPID_2$ indicates the derivative order $\nu = 2$.

Similar to the case of $v = 1$, by setting the tuning parameters in Equation (17), the control laws iP$_2$, iPI$_2$, and iPD$_2$ for $v = 2$ are obtained. Therefore, for $K_i = K_d = 0$, the iP$_2$ control law is achieved:

$$u_k = \frac{1}{\alpha}\left(\frac{K_p}{T_s}\frac{T_s}{1-z^{-1}}e_k + \frac{1}{T_s}\frac{1-z^{-1}}{T_s}e_k \right),$$ (18)

with a classical ID controller behavior, which is uncommon in the control engineering. Considering only $K_d = 0$, the control law iPI$_2$ is obtained:

$$u_k = \frac{1}{\alpha}\left(\frac{K_p}{T_s}\frac{T_s}{1-z^{-1}}e_k + \frac{1}{T_s}\frac{1-z^{-1}}{T_s}e_k + \frac{K_i}{T_s}\frac{T_s^2}{\left(1-z^{-1}\right)^2}e_k \right),$$ (19)

which is equivalent to a classical ID-I^2 controller, also unused in the control engineering. Setting $K_i = 0$, the iPD$_2$ control law results in:

$$u_k = \frac{1}{\alpha}\left(\frac{K_d}{T_s}e_k + \frac{K_p}{T_s}\frac{T_s}{1-z^{-1}}e_k + \frac{1}{T_s}\frac{1-z^{-1}}{T_s}e_k \right),$$ (20)

having a classical PID behavior.

Considering only the iPD$_2$ and iPID$_2$ controllers and applying the Z transform to Equations (17) and (20), which are related to the two controllers, the iPD$_2$ discrete transfer function is obtained:

$$G_{iPD_2}\left(z^{-1}\right) = \frac{q_0 + q_1 z^{-1} + q_2 z^{-2}}{1 - z^{-1}},$$ (21)

and, respectively, for iPID$_2$:

$$G_{iPID_2}\left(z^{-1}\right) = \frac{q_0 + q_1 z^{-1} + q_2 z^{-2} + q_3 z^{-3}}{1 - 2z^{-1} + z^{-2}}.$$ (22)

The parameter vectors for the two controllers and the expressions of the tuning parameters are:

$$\text{iPD}_2 : \rho = [q_0\ q_1\ q_2]^T; \text{ with}: q_0 = \frac{K_p T_s^2 + K_d T_s + 1}{\alpha T_s^2}, q_1 = -\frac{K_d T_s + 2}{\alpha T_s^2}, q_2 = \frac{1}{\alpha T_s^2},$$ (23)

$$\text{iPID}_2 : \rho = [q_0\ q_1\ q_2\ q_3]^T; \text{ with}: q_0 = \frac{1 + K_p T_s^2 + K_d T_s}{\alpha T_s^2},$$
$$q_1 = -\frac{3 + K_p T_s^2 - K_i T_s^3 + 2K_d T_s}{\alpha T_s^2}, \ q_2 = \frac{2 + K_d T_s}{\alpha T_s^2}, q_3 = -\frac{1}{\alpha T_s^2}.$$ (24)

Analyzing for the case of $v = 2$ in the comparisons between the iPID$_2$ and the classic PID controllers, it turned out that the iP$_2$ and iPI$_2$ controllers behave similarly to equivalent classical controllers, which is unusual in the control engineering. The iPD$_2$ and iPID$_2$ controllers, PID and PID-I^2 type behaviors, and fixed structures expressed by the transfer functions from Equations (21) and (22) were found, which were parameterized by the vectors ρ given by Equations (23) and (24).

The connections analyzed between the iPID and PID showed that both for $v = 1$ and for $v = 2$, the intelligent controllers have either an integrator or an integrator supplemented by a double integrator. It also resulted that a derivative behavior of the iPID controllers was only obtained for $v = 2$.

The parameters vector ρ will be determined using the model-free IFT approach. Because the tuning parameters K_p, K_i, K_d, and α are obtained based on the ρ vector, only the iP$_1$ and iPI$_1$ controllers were chosen for $v = 1$; this is because from Equations (12) and (15), unique solutions for the tuning gains and α are obtained. At the same time, for the case of $v = 1$, the neglected controllers iPD$_1$ and iPID$_1$ have the same type of behavior as the iP$_1$

and iPI$_1$ controllers; however, from Equations (13) and (16), unique solutions for the tuning parameters and α are not obtained.

In the following, we will consider $\nu = 1$ for the controllers iP$_1$ and iPI$_1$, having the fixed structure described by the transfer functions G_{iP_1} and G_{iPI_1}, given by Equations (11) and (14); we will also consider $\nu = 2$ for the controllers iPD$_2$ and iPID$_2$ with the fixed structure described by the transfer functions G_{iPD_2} and G_{iPID_2}, provided by Equations (21) and (22).

3. Model-Free IFT for iPID Controllers

3.1. Iterative Feedback Tuning Approach

It is considered a linear time-invariant discrete-time process with unknown dynamics controlled by a linear discrete-time controller with a fixed-order structure described by a discrete transfer function $G_c(\rho)$ and parameterized with the ρ vector. For this controller, the minimization of some norm of the error between the achieved and the desired response over the controller parameter vector ρ:

$$e_k(\rho) = y_k(\rho) - \tilde{y}_k, \tag{25}$$

is imposed as a design objective. For IFT, the following quadratic criterion is used:

$$J(\rho) = \frac{1}{2N}\left\{\sum_{k=1}^{N}(L_y e_k(\rho))^2 + \lambda\sum_{k=1}^{N}(L_u u_k(\rho))^2\right\}, \tag{26}$$

where $y_k(\rho)$ represents the output of the system after using the set of parameters ρ for the controller, $\tilde{y}_k$ represents the reference signal of the control system, and $u_k(\rho)$ is the control signal for the used set of parameters ρ. L_y and L_u are filters that weighted the output and control signals, λ is a penalty factor, and N is the number of samples. The cost function from Equation (26) will be minimized by an optimal parameter vector:

$$\rho^* = \arg\min_{\rho} J(\rho). \tag{27}$$

A necessary condition to obtain the minimum is the cancellation of the derivative of $J(\rho)$ with respect to the vector parameters ρ. Considering for simplicity $L_y = L_u = 1$, the derivative of $J(\rho)$ becomes:

$$\frac{\partial J}{\partial \rho}(\rho) = \frac{1}{N}\left[\sum_{k=1}^{N}e_k(\rho)\frac{\partial e_k}{\partial \rho}(\rho) + \frac{\lambda}{N}\sum_{k=1}^{N}u_k(\rho)\frac{\partial u_k}{\partial \rho}(\rho)\right]. \tag{28}$$

If the gradient $\frac{\partial J}{\partial \rho}(\rho)$ could be computed using Equation (28), the solution of the equation $\frac{\partial J}{\partial \rho}(\rho) = 0$ can be obtained in an iterative way, as follows:

$$\rho_{i+1} = \rho_i - \gamma_i R_i^{-1}\frac{\partial J}{\partial \rho}(\rho_i), \tag{29}$$

where γ_i is a positive real scalar that determines the step size, whereas R_i is a symmetric positive defined matrix used to update the direction. To obtain the negative gradient descent direction, the identity matrix is usually used. Furthermore, using the data collected from the control system, the matrix defined by:

$$R_i = \frac{1}{N}\sum_{k=1}^{N}\left(\left[\frac{\partial y_k}{\partial \rho}(\rho_i)\right]\left[\frac{\partial y_k}{\partial \rho}(\rho_i)\right]^T + \lambda\left[\frac{\partial u_k}{\partial \rho}(\rho_i)\right]\left[\frac{\partial u_k}{\partial \rho}(\rho_i)\right]^T\right). \tag{30}$$

can be employed [6].

Because the exact gradient is not known, it is replaced in Equation (29) by an unbiased estimate $\frac{\partial \hat{J}}{\partial \rho}(\rho_i)$. To determine the unbiased estimate $\frac{\partial \hat{J}}{\partial \rho}(\rho_i)$, the following quantities are needed:

(i)   The signals $e_k(\rho_i)$ and $u_k(\rho_i)$;

(ii)   The estimation of the gradients $\frac{\partial e_k}{\partial \rho}(\rho_i)$ and $\frac{\partial u_k}{\partial \rho}(\rho_i)$;

(iii) Unbiased estimates of the products $e_k \frac{\partial e_k}{\partial \rho}(\rho_i)$ and $u_k \frac{\partial u_k}{\partial \rho}(\rho_i)$.

These quantities are obtained through closed-loop experiments with the controller, having the transfer function G_c parameterized with the ρ vector.

From Equation (25), it follows:

$$\frac{\partial e_k}{\partial \rho}(\rho_i) = \frac{\partial y_k}{\partial \rho}(\rho_i), \tag{31}$$

and thus, the estimations from (ii) refer to the output and input signals.

Denoting the closed loop transfer function with G_0 and the sensitivity function with S, the following expressions of the derivatives of the output and input signals were obtained in [6]:

$$\begin{aligned}
\frac{\partial y_k}{\partial \rho}(\rho_i) &= \frac{1}{G_c(\rho_i)} \frac{\partial G_c}{\partial \rho}(\rho_i) G_0(\rho_i)(r - y_k(\rho_i)) \\
\frac{\partial u_k}{\partial \rho}(\rho_i) &= \frac{1}{G_c(\rho_i)} \frac{\partial G_c}{\partial \rho}(\rho_i) S(\rho_i)(r - y_k(\rho_i))
\end{aligned} \tag{32}$$

At iteration i, with the fixed controller $G_c(\rho_i)$ being known, its derivative can be calculated. With the controller $G_c(\rho_i)$ in the closed loop, based on two experiments of length N denoted by *(a)* and *(b)*, the derivatives from Equation (32) are estimated. For the first experiment *(a)*, the reference of the control loop is $\widetilde{y}$, resulting in:

$$\begin{aligned}
\widetilde{y}_i^{(a)} &= \widetilde{y} \\
y_k^{(a)}(\rho_i) &= G_0(\rho_i)\widetilde{y} \\
u_k^{(a)}(\rho_i) &= S(\rho_i)G_c(\rho_i)\widetilde{y}
\end{aligned} \tag{33}$$

The second experiment *(b)* is a special one. The reference is $\widetilde{y} - y_k^{(a)}(\rho_i)$, and the output/input signals are:

$$\begin{aligned}
\widetilde{y}_i^{(b)} &= \widetilde{y} - y_k^{(a)}(\rho_i) \\
y_k^{(b)}(\rho_i) &= G_0(\rho_i)\left(\widetilde{y} - y_k^{(a)}(\rho_i)\right) \\
u_k^{(b)}(\rho_i) &= S(\rho_i)G_c(\rho_i)\left(\widetilde{y} - y_k^{(a)}(\rho_i)\right)
\end{aligned} \tag{34}$$

The superscript *(a)* and *(b)* used in Equations (33) and (34) indicate that the signals were obtained during experiment *(a)* and *(b)*.

Based on the signals obtained in the two experiments, the estimation of the output/input gradients is determined as it is suggested in [6]:

$$\begin{aligned}
est\left[\frac{\partial y_k}{\partial \rho}(\rho_i)\right] &= \frac{1}{G_c(\rho_i)} \frac{\partial G_c}{\partial \rho}(\rho_i) y_k^{(b)}(\rho_i), \\
est\left[\frac{\partial u_k}{\partial \rho}(\rho_i)\right] &= \frac{1}{G_c(\rho_i)} \frac{\partial G_c}{\partial \rho}(\rho_i) u_k^{(b)}(\rho_i).
\end{aligned} \tag{35}$$

With the estimations from the Equation (35) obtained based on the two experiments, the unbiased estimate $\frac{\partial \hat{J}}{\partial \rho}(\rho_i)$ of the cost function can be computed with:

$$\frac{\partial \hat{J}}{\partial \rho}(\rho_i) = \frac{1}{N}\left\{\left[\sum_{k=1}^{N} e_k(\rho_i) est\left[\frac{\partial y_k}{\partial \rho}(\rho_i)\right] + \frac{\lambda}{N}\sum_{k=1}^{N} u_k(\rho_i) est\left[\frac{\partial u_k}{\partial \rho}(\rho_i)\right]\right]\right\}. \tag{36}$$

The same estimates available after the two experiments are used to estimate the matrix R_i, defined by (30), which becomes:

$$\hat{R}_i = \frac{1}{N}\sum_{k=1}^{N}\left(est\left[\frac{\partial y_k}{\partial \rho}(\rho_i)\right]est\left[\frac{\partial y_k}{\partial \rho}(\rho_i)\right]^T + \lambda est\left[\frac{\partial u_k}{\partial \rho}(\rho_i)\right]est\left[\frac{\partial u_k}{\partial \rho}(\rho_i)\right]^T\right). \tag{37}$$

Using the estimates given by Equations (36) and (37), the iterative calculation of the tuning parameters with Equation (29) turns to:

$$\rho_{i+1} = \rho_i - \gamma_i \hat{R}_i^{-1}\frac{\partial \hat{J}}{\partial \rho}(\rho_i). \tag{38}$$

The following algorithm gives the steps of the IFT approach:

1. In the first iteration, the tuning parameters of the controller are chosen in such a way as to obtain a stable response in the closed loop.
2. At each iteration i, two experiments of length N are performed:
   (a) The reference $\tilde{y}$ is applied to the control system, and the input/output signals are collected $\left\{u_k^{(a)}(\rho_i), y_k^{(a)}(\rho_i)\right\}_{k=\overline{1,N}}$;
   (b) The reference $\tilde{y} - y_k^{(a)}(\rho_i)$ is applied to the control system, and the input/output signals are collected $\left\{u_k^{(b)}(\rho_i), y_k^{(b)}(\rho_i)\right\}_{k=\overline{1,N}}$;
3. The estimates for the gradients of $u_k(\rho_i)$ and $y_k(\rho_i)$ are calculated by Equation (35);
4. Using the estimates from Step 3, the gradient estimates for the cost function $\frac{\partial \hat{J}}{\partial \rho}(\rho_i)$ and for the matrix $\hat{R}_i$ are determined with Equations (36) and (37), respectively;
5. Based on the estimates from Equations (36) and (37), the new vector of parameters ρ_{i+1} is calculated with Equation (38);
6. Repeat Step 2–5 until the optimal vector of parameters ρ^* is obtained.

The IFT algorithm presented in this section can be applied for the calculation of gradient approximations in the case of nonlinear systems from the collected experimental data [35]. The solution proposed by Hjalmarsson in [6,35] is to generate the true gradient using the linear time-varying system obtained by linearizing the nonlinear system around the system trajectory under normal operating conditions. In this context, the same IFT procedure as the one used for linear systems can be applied.

3.2. IFT of iPID Controllers

For tuning the iPID controllers with the IFT approach, the discrete transfer functions obtained in the previous Section for $v = 1$ (G_{iP_1} and G_{iPI_1}) and $v = 2$ (G_{iPD_2} and G_{iPID_2}) will be considered:

$$\begin{aligned}
G_{iP_1}(z^{-1}) &= \frac{q_0+q_1 z^{-1}}{1-z^{-1}}, \\
G_{iPI_1}(z^{-1}) &= \frac{q_0+q_1 z^{-1}+q_2 z^{-2}}{1-2z^{-1}+z^{-2}}, \\
G_{iPD_2}(z^{-1}) &= \frac{q_0+q_1 z^{-1}+q_2 z^{-2}}{1-z^{-1}}, \\
G_{iPID_2}(z^{-1}) &= \frac{q_0+q_1 z^{-1}+q_2 z^{-2}+q_3 z^{-3}}{1-2z^{-1}+z^{-2}}.
\end{aligned} \tag{39}$$

Based on these transfer functions, for the estimations of the output and input quantities with Equation (35), the derivatives of the transfer functions, weighted with their inverses, are firstly calculated for each type of iPID controller, as follows:

- iP$_1$ controller

$$\frac{1}{G_{iP_1}}\frac{\partial G_{iP_1}}{\partial q_0} = \frac{1}{q_0+q_1z^{-1}},$$
$$\frac{1}{G_{iP_1}}\frac{\partial G_{iP_1}}{\partial q_1} = \frac{z^{-1}}{q_0+q_1z^{-1}}. \tag{40}$$

- iPI_1 controller

$$\frac{1}{G_{iPI_1}}\frac{\partial G_{iPI_1}}{\partial q_0} = \frac{1}{q_0+q_1z^{-1}+q_2z^{-2}},$$
$$\frac{1}{G_{iPI_1}}\frac{\partial G_{iPI_1}}{\partial q_1} = \frac{z^{-1}}{q_0+q_1z^{-1}+q_2z^{-2}},$$
$$\frac{1}{G_{iPI_1}}\frac{\partial G_{iPI_1}}{\partial q_2} = \frac{z^{-2}}{q_0+q_1z^{-1}+q_2z^{-2}}. \tag{41}$$

- iPD2 controller

$$\frac{1}{G_{iPD_2}}\frac{\partial G_{iPD_2}}{\partial q_0} = \frac{1}{q_0+q_1z^{-1}+q_2z^{-2}},$$
$$\frac{1}{G_{iPD_2}}\frac{\partial G_{iPD_2}}{\partial q_1} = \frac{z^{-1}}{q_0+q_1z^{-1}+q_2z^{-2}},$$
$$\frac{1}{G_{iPD_2}}\frac{\partial G_{iPD_2}}{\partial q_2} = \frac{z^{-2}}{q_0+q_1z^{-1}+q_2z^{-2}}. \tag{42}$$

- iPID2 controller

$$\frac{1}{G_{iPID_2}}\frac{\partial G_{iPID_2}}{\partial q_0} = \frac{1}{q_0+q_1z^{-1}+q_2z^{-2}+q_3z^{-3}},$$
$$\frac{1}{G_{iPID_2}}\frac{\partial G_{iPID_2}}{\partial q_1} = \frac{z^{-1}}{q_0+q_1z^{-1}+q_2z^{-2}+q_3z^{-3}},$$
$$\frac{1}{G_{iPID_2}}\frac{\partial G_{iPID_2}}{\partial q_2} = \frac{z^{-2}}{q_0+q_1z^{-1}+q_2z^{-2}+q_3z^{-3}},$$
$$\frac{1}{G_{iPID_2}}\frac{\partial G_{iPID_2}}{\partial q_3} = \frac{z^{-3}}{q_0+q_1z^{-1}+q_2z^{-2}+q_3z^{-3}}. \tag{43}$$

Applying the above algorithm, the optimal parameter vectors ρ^* are obtained for each individual iPID controller, and based on them, the tuning parameters K_p, K_i, K_d, and α are calculated using Equations (15), (16), (23), and (24), with the results being summarized in Table 1.

Table 1. Tuning parameters.

	K_p	K_i	K_d	α
iP_1	$-\dfrac{q_0+q_1}{q_1 T_s}$			$-\dfrac{1}{q_1 T_s}$
iPI_1	$\dfrac{q_0-q_2}{q_2 T_s}$	$-\dfrac{q_0+q_1+q_2}{q_2 T_s^2}$		$\dfrac{1}{T_s q_2}$
iPD_2	$\dfrac{q_0+q_1+q_2}{q_2 T_s^2}$		$-\dfrac{q_1+2q_2}{q_2 T_s}$	$\dfrac{1}{q_2 T_s^2}$
$iPID_2$	$\dfrac{q_2-q_0+q_3}{q_3 T_s^2}$	$-\dfrac{q_0+q_1+q_2}{q_3 T_s^3}$	$-\dfrac{q_2+2q_3}{q_3 T_s}$	$-\dfrac{1}{q_3 T_s^2}$

4. Case Studies—Real-Time Control of Pitch Angle for Aero 2 Laboratory Equipment

To test and validate the iPID controllers tuning using the IFT algorithm, we considered a real-time application for controlling the pitch angle for the dual-motor aerospace system Aero 2, provided by Quanser. The system has two rotors that can be used for various aerospace applications, with the possibility to control the pitch and yaw angles as suggested in [28–33]. To control the pitch angle, the 1-DOF configuration was chosen for the Aero 2 platform, which locks the yaw and has the pitch as free. This configuration, represented in Figure 1, was used for real-time experiments, having both propellers horizontal and ensuring a variation of the pitch angle of 90° degrees, 45° degrees for each side.

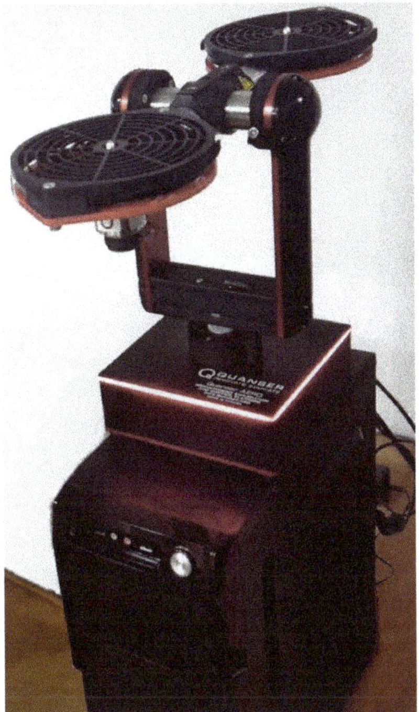

Figure 1. Quanser Aero 2 laboratory equipment with a 1-DOF configuration.

Single-ended optical shaft encoders were used to measure the pitch of the Aero body and the angular speed of the DC motors [28]. The pitch encoder is 2880 counts per revolution in quadrature mode, and the angular speed encoder is a counter that returns the number of the encoder counts every second.

The scheme of the 1-DOF Aero 2 plant for the pitch angle control is shown in Figure 2, where M_b is the aero body mass, D_t is the thrust displacement, D_m is the center of mass displacement, and θ is the pitch angle to the equilibrium position; F_0 and F_1 are the thrust forces generated by propellers that are driven by two DC motors, and g is the gravitational acceleration. To introduce nonlinearities, an asymmetry was created by using a single propeller operating in both positive and negative thrust and by shifting the center of mass towards thruster 1.

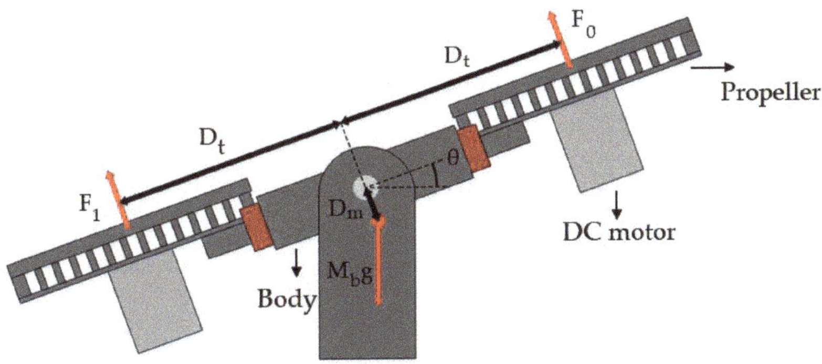

Figure 2. The one degree of freedom Aero 2 plant.

Each propeller was driven by a DC motor, its supply voltage v_m being considered the input variable, and the output variable is the pitch angle θ. The plant model was obtained based on the DC motor and propeller models.

The DC motor dynamics are described by [28]:

$$\begin{aligned}
L_m \frac{di_m}{dt} &= v_m - R_m i_m - e_b \\
J_m \frac{d\omega_m}{dt} &= \tau_m - \tau_d \\
e_b &= k_e \omega_m \\
\tau_m &= k_m i_m
\end{aligned} \tag{44}$$

where i_m is the motor current, R_m and L_m are the resistance and inductances of the rotor, e_b is the back electromotive force (emf), J_m is the rotor inertia, ω_m is the motor angular speed, τ_m is the motor torque, τ_d is the drag torque and k_e and k_m are the motor back EMF and torque constant. Neglecting the rotor inductance, an assumption valid for low-power motors, the following model of the DC motor results from Equation (44):

$$J_m \frac{d\omega_m}{dt} = -\frac{k_e k_m}{R_m}\omega_m + \frac{k_m}{R_m}v_m - \tau_d \tag{45}$$

The dynamic behavior of the propeller system can be modeled by taking into account [28] and [29,31]:

$$\begin{aligned}
(J_p + M_b D_m^2)\ddot{\theta} + D_p\dot{\theta} + K_{sp}\theta + M_b g D_m \cos\theta &= \tau_p \\
\tau_p &= K_{pp}\omega_m D_t
\end{aligned} \tag{46}$$

where J_p is the moment of inertia related to the pitch motion, D_p is the viscous damping coefficient, K_{sp} is the stiffness, K_{pp} is the force thrust gain relative to the rotor speed, D_t is the distance from the pivot point to the center of the rotor, and τ_p is the torque acting on the pitch axis, created by the thrust force $F = K_{pp}\omega_m$.

Considering the models of the two components of the plant, a cascade control structure was chosen, with a faster inner loop for controlling the motor angular speed ω_m and an outer loop for controlling the pitch angle θ. In this way, the disturbance introduced by the drag torque τ_d will be rejected. The cascade control structure is a special one, whose scheme is represented in Figure 3.

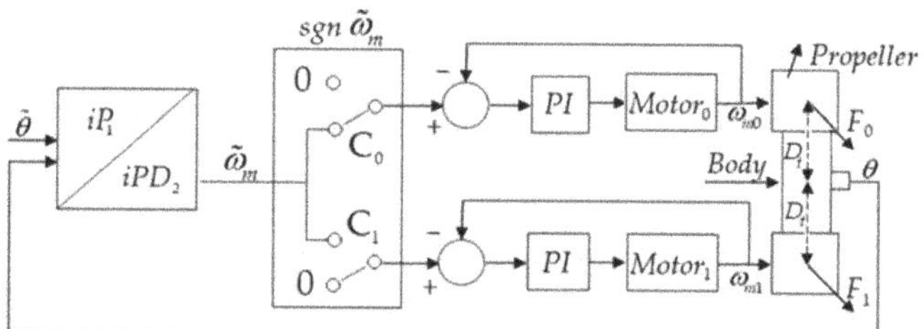

Figure 3. Control scheme for pitch angle with MFC controllers.

The C_0 and C_1 switches, which depend on the sign of the angular velocity reference $\tilde{\omega}_m$, have been introduced to activate a single motor. Thus, for $\tilde{\omega}_m > 0$, the related internal propeller loop that generates the thrust force F_0 will come into operation, and for $\tilde{\omega}_m < 0$, it will be the propeller that generates the thrust force F_1. The presence of the two switches, through which a single inner loop is activated depending on the sign of $\tilde{\omega}_m$, introduces a nonlinearity in the control system. Another nonlinearity from Equation (46) is created by the asymmetric placement of the center of gravity, as seen in Figure 2.

For each inner loop, a PI controller described by [28]:

$$u = -k_p \omega_m + k_i(\widetilde{\omega}_m - \omega_m) \tag{47}$$

was designed, starting from the simplified model (45) of the DC motor and considering the overshoot σ and peak time t_p as desired performances, as in [28]. Applying the Laplace transform to the DC motor model (45), a first-order transfer function is obtained with gain $k = 1/k_e$ and time constant $\tau = J_m k_m k_e / R_m$. Using the plant transfer function, in [28], the following tuning parameters of the PI controller were obtained based on a pole placement method:

$$\begin{aligned} k_p &= (2\zeta\omega_n\tau - 1)/k \\ k_i &= \omega_n^2\tau/k \end{aligned} \tag{48}$$

where the damping ratio ζ and natural frequency ω_n are calculated based on the desired performances. Employing the DC motor parameters and the desired performances from [28], the following tuning parameters of the inner loop PI controller were obtained: $k_p = 0.1553$, $k_i = 2.3176$.

For the outer loop, two model-free controllers were tested to validate the proposed design method based on the data-driven IFT algorithm. The first one tested was an iP$_1$ controller with a PI-type behavior, as shown in Section 2, which ensures a zero steady-state error; the second one was an iPD$_2$ with a PID-type behavior, for which the influence of the derivative component was analyzed. The structures of the two model-free controllers are depicted in Figure 4.

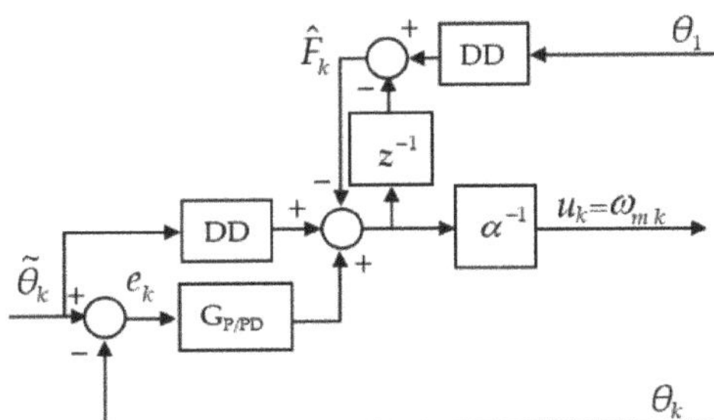

Figure 4. Model-free controller structure.

In the model-free controller structure, the Discrete Derivative block DD computes the discrete-time derivative using the backward difference approximation; the $G_{P/PD}$ block implements the discrete transfer function of the P or PD controller, resulting in the iP$_1$ or iPD$_2$ controllers. The sample time $T_s = 0.002$ s for both controllers was considered due to the small value of the time constants of the inner loop.

The outer loop plant is formed by the closed-loop transfer function of the inner loop, to which the propeller model from Equation (46) is added. Although the closed-loop model of the inner loop is known by designing with the pole placement method, the propeller model is considered as totally unknown. The design of the model-free controller was conducted only based on the data collected from the control system without using any outer loop plant model.

The Aero 2 system from Quanser was connected with a PC, where the control scheme from Figure 2 was implemented using MATLAB/Simulink 2021b software tools.

For tuning the two model-free controllers, the IFT algorithm from Section 3 was used, which is based on two closed-loop experiments and the collection of input and output data related to the real-time experiments. The proposed aero pitch controller tuning methodology is the following:

1. In the first step of the IFT algorithm, the tuning parameters of the two controllers that lead to a stable response must be found. These parameters were determined as in [11] and [12] based on the characteristic polynomial of the error, for which it was required that the roots be stable, thus cancelling the error in the steady-state. For iP_1, it follows from [11] that the roots of the characteristic polynomial of the error are stable if $K_p \in (0, 2/T_s)$. Applying the Jury test, the domain in the plane of gains parameters, where the roots of the characteristic polynomial are stable, was obtained in [12] for the iPD_2 controller, defined by:

$$
\begin{aligned}
&T_s^2 K_p > 0 \\
&4 - 2T_s K_d + T_s^2 K_p > 0 \\
&\left|1 + T_s^2 K_p - T_s K_d\right| < 1
\end{aligned}
\tag{49}
$$

For the two model-free controllers, the initial tuning parameters were chosen using values from the stability domains of the roots. With the tuning parameters thus determined, an experiment is carried out, and if undesirable performances are obtained, proceed to Step 2 of the IFT algorithm.

2. At each iteration i, two experiments of length $N = 30/T_s$ are performed:

    (a) The reference $\widetilde{\theta} = 0.2$ rad is applied to the pitch control system, and the input/output signals are collected $\left\{ u_k^{(a)}(\rho_i), \theta_k^{(a)}(\rho_i) \right\}_{k=\overline{1,N}}$;

    (b) The reference $\widetilde{\theta} - \theta_k^{(a)}(\rho_i)$ is applied to the control system, and the input/output signals are collected $\left\{ u_k^{(b)}(\rho_i), \theta_k^{(b)}(\rho_i) \right\}_{k=\overline{1,N}}$.

3. Based on the data $\left\{ u_k^{(b)}(\rho_i), \theta_k^{(b)}(\rho_i) \right\}_{k=\overline{1,N}}$ collected in the special experiment (b), the estimates of the gradients for $u_k(\rho_i)$ and $\theta_k(\rho_i)$ and the estimates of the gradient of the cost function $\frac{\widehat{\partial J}}{\partial \rho}(\rho_i)$ and for the matrix $\hat{R}_i$, described by (37), are computed.

4. Using the estimates from Step 3, the new vector of parameters ρ_{i+1} is calculated; based on the vector, the tuning parameters are determined using Table 1.

5. With the tuning parameters determined in Step 4, the performances of the pitch control system are tested in a real-time experiment.

6. If the performances obtained with the tuning parameters from Step 5 are not satisfactory, restart the tuning procedure by repeating Step 2–5 until the desired performances are reached.

The penalty factor λ from Equation (26) and the step size γ_i from Equation (29) were determined through a trial-and-error procedure, while the matrix R_i was considered the identity matrix I_n, one of possibilities mentioned by [6] and [36]. To test the performances of the control system for each iteration i of the tuning procedure, the ITAE (integral time-weighted absolute error) criterion defined by:

$$
ITAE_i = \sum_{k=1}^{N} kT_s |e_k|
\tag{50}
$$

was used, where kT_s is the discrete-time expressed in seconds. This criterion is helpful because the objective is to have a smooth response without oscillations around the reference value.

4.1. Control of Pitch Angle with iP_1 Controller

Based on the stability condition of the roots of the characteristic polynomial, the $iP_{1\text{-}0}$ controller with the tuning parameters from Table 2 were chosen for the initial step. Using step size $\gamma_i = 0.0003$, penalty factor $\lambda_i = 0.001$, and matrix $R_i = I_2$, four iterations of the IFT algorithm have been created, resulting in the controllers $iP_{1\text{-}1,2,3}$ and finally the fourth controller, iP$_1$-IFT. The stopping point was determined based on the optimum performance criterion defined by Equation (50). The responses of the iP$_1$ controllers are shown in Figure 4, where the unsatisfactory performances of the iP$_{1\text{-}0}$ controller are visible. The IFT tuning procedure was applied in order to improve the performances. The initial response of the control system and other several intermediate responses results obtained with the iP$_{1\text{-}1,3}$ and iP$_1$-IFT controllers during the tuning procedure are shown in Figure 5.

Table 2. Tuning parameters for the iP$_1$ controllers and the ITAE criteria values.

Iteration	Initial Values		ITAE	Final Values	
1	$K_{p-0} = 17.5$ $\alpha_0 = 28$	$q_{0-0} = 18.4821$ $q_{1-0} = -17.8571$	$ITAE_0 =$ $43{,}517$	$K_{p-1} = 17.352$ $\alpha_1 = 27.9959$	$q_{0-1} = 18.4795$ $q_{1-1} = -17.8597$
2	$K_{p-1} = 17.352$ $\alpha_1 = 27.9959$	$q_{0-1} = 18.4795$ $q_{1-1} = -17.8597$	$ITAE_1 =$ $43{,}428$	$K_{p-2} = 17.1917$ $\alpha_2 = 27.9971$	$q_{0-2} = 18.4768$ $q_{1-2} = -17.8625$
3	$K_{p-2} = 17.1917$ $\alpha_2 = 27.9971$	$q_{0-2} = 18.4768$ $q_{1-2} = -17.8625$	$ITAE_2 =$ $43{,}419$	$K_{p-3} = 17.0322$ $\alpha_3 = 27.9871$	$q_{0-3} = 18.4739$ $q_{1-3} = -17.8654$
4	$K_{p-3} = 17.0322$ $\alpha_3 = 27.9871$	$q_{0-3} = 18.4739$ $q_{1-3} = -17.8654$	$ITAE_3 =$ $43{,}205$	$K_{p-4} = 16.8649$ $\alpha_3 = 27.9825$	$q_{0-4} = 18.4710$ $q_{1-4} = -17.8683$
Final	$K_{p-IFT} = 16.8649$ $\alpha_{IFT} = 27.9825$	$q_{0-IFT} = 18.4710$ $q_{1-IFT} = -17.8683$	$ITAE_4 =$ $43{,}175$		

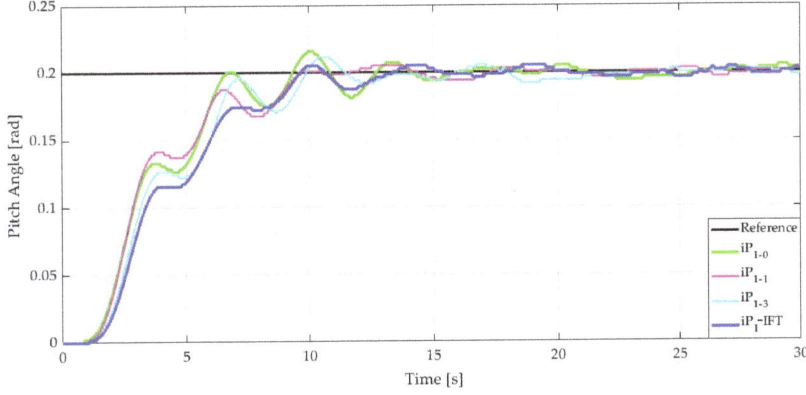

Figure 5. Control system responses with the iP$_1$ controller during the tuning with the IFT algorithm.

In Figure 5, the system responses obtained with the iP$_1$ controllers are represented by considering the tuning parameters found using the tuning methodology. The optimal values have been obtained in the fourth step. The responses from the figure show that during the application of the IFT algorithm, the oscillations of the output signal become more damped and thus, the settling time decreases.

The parameters ρ_i and the values of the tuning parameters of the iP$_1$ controllers, which were determined with the relationships from Table 1, are presented in Table 2 together with the values of the ITAE criteria for each iteration.

In Table 2, the values of the ITAE criterion decrease with the increase in the number of iterations, indicating the performances improvement.

The iP$_{1\text{-}0}$ and iP$_1$-IFT controllers are used in a comparative experiment, which involves modifications of the reference signal, with different steps, in both directions of the pitch axis. The result of the comparison is depicted in Figure 6, from which the good behavior of the iP$_1$-IFT controller in relation to iP$_{1\text{-}0}$ is shown.

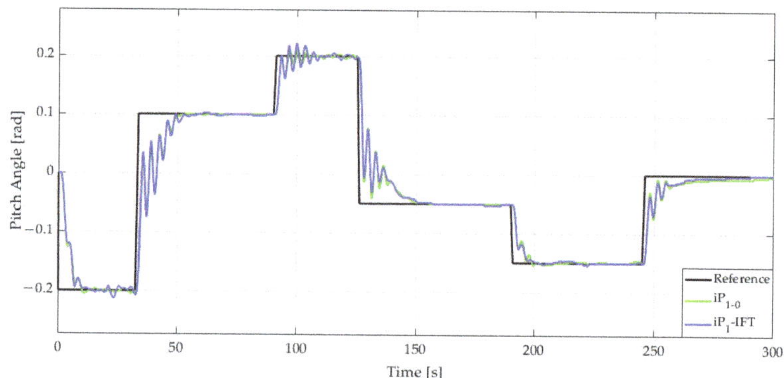

Figure 6. Comparison between $iP_{1\text{-}0}$ and iP_1-IFT controllers for the pitch angle variation experiments.

The control signals for the $iP_{1\text{-}0}$ and iP_1-IFT controllers during the comparison experiments are presented in Figure 7. Therein represented are the reference angular speed of the motors $\widetilde{\omega}_m$ and, at the same time, the responses of the two inner loops generated by the C_0 and C_1 switches from Figure 3, depending on $\text{sgn}\,\widetilde{\omega}_m$. When $\widetilde{\omega}_m > 0$, the reference generated by the iP_1 controller of the outer loop will be applied through C_0 to the inner loop related to the motor_0, and the inner loop of the motor_1 will have the reference equal to zero through switch C_1. For $\widetilde{\omega}_m < 0$, the switches change their position and thus, the inner loop related to the motor_1 will receive the reference; for motor_0, the reference will be zero. The operation of the switches according to $\text{sgn}\,\widetilde{\omega}_m$ introduces a nonlinearity in the plant model related to the outer loop.

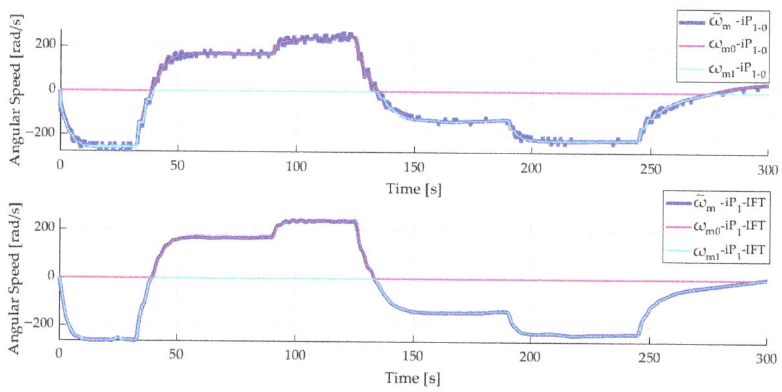

Figure 7. Control signals of $iP_{1\text{-}0}$ and iP_1-IFT controllers and the controlled angular speeds for the pitch angle variation experiments.

From Figure 7, an oscillating operation is observed for the DC motors, which actuate the propellers for the $iP_{1\text{-}0}$ controller, which can cause the wear of the Aero 2 system. These oscillations disappear in the case of the iP_1-IFT controller.

4.2. Control of Pitch Angle with the iPD_2 Controller

As shown in the conclusions from Section 2, the iPD_2 controller will introduce a derivative component that accelerates the control loop and leads to a faster response of the system, but at the same time amplifying the high-frequency signals. A similar procedure with the iP_1 case presented in Section 4.1 was created, starting with a stable response of the system with an $iPD_{2\text{-}0}$ controller that was now determined based on the Jury test. For the iPD_2 controller, five iterations of the IFT algorithm have been achieved with step size $\gamma_i = 0.000025$, penalty factor $\lambda_i = 0.1$, and matrix $R_i = I_3$. The responses obtained during the tuning phase with the initial controller $iPD_{2\text{-}0}$, with some intermediate controllers $iPD_{2\text{-}2,3}$ and the final controller iPD_2-IFT, are presented in Figure 8.

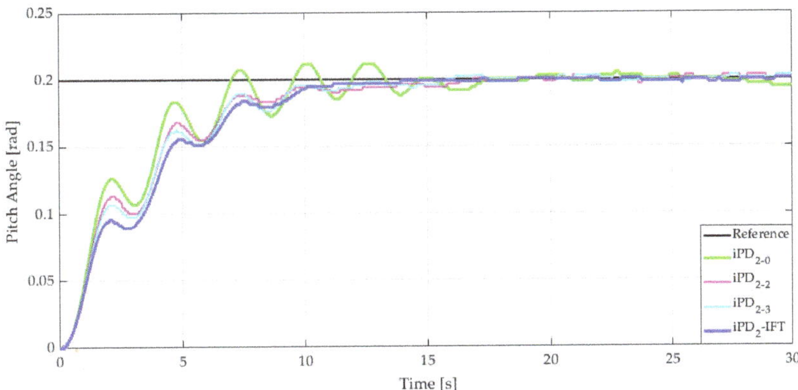

Figure 8. Control system responses with the iPD_2 controller during tuning with the IFT algorithm.

The system controlled with $iPD_{2\text{-}0}$ shows oscillations with high amplitude, which are reduced after five iterations of the IFT algorithm. This gives the system a high range, where it can reach the reference for the pitch angle. The $iPD_{2\text{-}5}$ from the final iteration, noted as iPD_2-IFT, was chosen due to the obtained performances, which have been evaluated according to the ITAE criterion from Equation (50). The damping of the oscillations during the tuning phase leads to the decrease in the settling time, which becomes smaller than the one obtained with the iP_1-IFT controller in Figure 5 due to the derivative component. In Table 3, the parameters ρ_i and the values of the tuning parameters of the iPD_2 controllers are summarized together with the ITAE criteria values. As Table 3 shows, the values of the ITAE criterion are decreasing from the initial response $iPD_{2\text{-}0}$ to the fifth iteration of the IFT algorithm due to the improvement of the performances.

To analyze the behavior of the tuned controllers, a comparison between the initial $iPD_{2\text{-}0}$ and the final version of iPD_2-IFT for the step variations of the reference signal is made, and the results are presented in Figure 9.

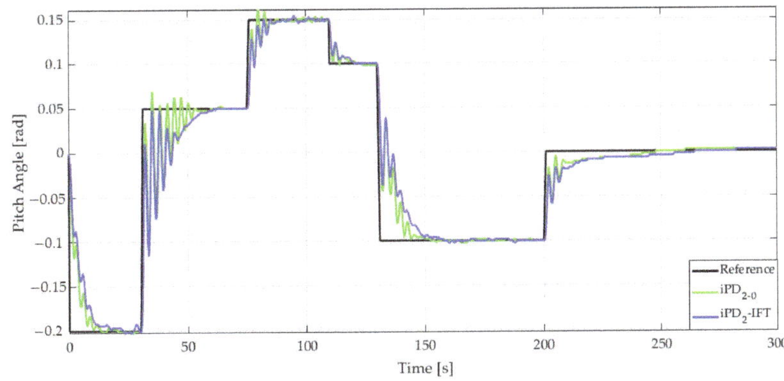

Figure 9. Comparison between the $iPD_{2\text{-}0}$ and iPD_2-IFT controllers for the pitch angle variations.

Table 3. Tuning parameters for the iPD$_2$ controllers and the ITAE criteria values.

Iteration	Initial Values		ITAE	Final Values	
1	$K_{p-0} = 20$ $K_{d-0} = 20$ $\alpha_0 = 24$	$q_{0-1} = 10{,}834.17$ $q_{1-1} = -21{,}250.00$ $q_{2-1} = 10{,}416.67$	$ITAE_0 =$ 1401	$K_{p-1} = 19.0842$ $K_{d-1} = 20.0019$ $\alpha_1 = 24$	$q_{0-1} = 10{,}834.15$ $q_{1-1} = -21{,}250.01$ $q_{2-1} = 10{,}416.65$
2	$K_{p-1} = 19.0842$ $K_{d-1} = 20.0019$ $\alpha_1 = 24$	$q_{0-1} = 10{,}834.15$ $q_{1-1} = -21{,}250.01$ $q_{2-1} = 10{,}416.65$	$ITAE_1 =$ 1369	$K_{p-2} = 18.0803$ $K_{d-2} = 20.0039$ $\alpha_2 = 24.0001$	$q_{0-2} = 10{,}834.14$ $q_{1-2} = -21{,}250.03$ $q_{2-2} = 10{,}416.64$
3	$K_{p-2} = 18.0803$ $K_{d-2} = 20.0039$ $\alpha_2 = 24.0001$	$q_{0-2} = 10{,}834.14$ $q_{1-2} = -21{,}250.03$ $q_{2-2} = 10{,}416.64$	$ITAE_2 =$ 1317	$K_{p-3} = 17.0496$ $K_{d-3} = 20.0060$ $\alpha_3 = 24.0001$	$q_{0-3} = 10{,}834.13$ $q_{1-3} = -21{,}250.04$ $q_{2-3} = 10{,}416.63$
4	$K_{p-3} = 17.0496$ $K_{d-3} = 20.0060$ $\alpha_3 = 24.0001$	$q_{0-3} = 10{,}834.13$ $q_{1-3} = -21{,}250.04$ $q_{2-3} = 10{,}416.63$	$ITAE_3 =$ 1295	$K_{p-4} = 16.0102$ $K_{d-4} = 20.0081$ $\alpha_4 = 24.0001$	$q_{0-4} = 10{,}834.11$ $q_{1-4} = -21{,}250.06$ $q_{2-4} = 10{,}416.61$
5	$K_{p-4} = 16.0102$ $K_{d-4} = 20.0081$ $\alpha_4 = 24.0001$	$q_{0-4} = 10{,}834.11$ $q_{1-4} = -21{,}250.06$ $q_{2-4} = 10{,}416.61$	$ITAE_4 =$ 1276	$K_{p-5} = 15.1076$ $K_{d-5} = 20.0099$ $\alpha_5 = 24.0002$	$q_{0-4} = 10{,}834.10$ $q_{1-4} = -21{,}250.07$ $q_{2-4} = 10{,}416.60$
Final	$K_{p-IFT} = 15.1076$ $K_{d-IFT} = 20.0099$ $\alpha_{IFT} = 24.0002$	$q_{0-IFT} = 10{,}834.10$ $q_{1-IFT} = -21{,}250.07$ $q_{2-4IFT} = 10{,}416.60$	$ITAE_5 =$ 1269		

The figure reveals a similar behavior with the previous one presented in Figure 8 for the tuning phase, with the iPD$_{2-0}$ controller having larger oscillations than the iPD$_2$-IFT when the reference is modified. This aspect is important with respect to achieving a high interval of movement on the pitch axis. The initial set of parameters is limited to variations of angle under 0.25 radians (around 15°) because the oscillations introduced make the system unstable. Considering this limitation, a different reference for the pitch angle is used compared with the iP$_1$ one depicted in Figure 6. The control signals for the iPD$_2$ controllers are shown in Figure 10 with the reference control signal $\widetilde{\omega}_m$ and controlled signals ω_{m0}, ω_{m1}.

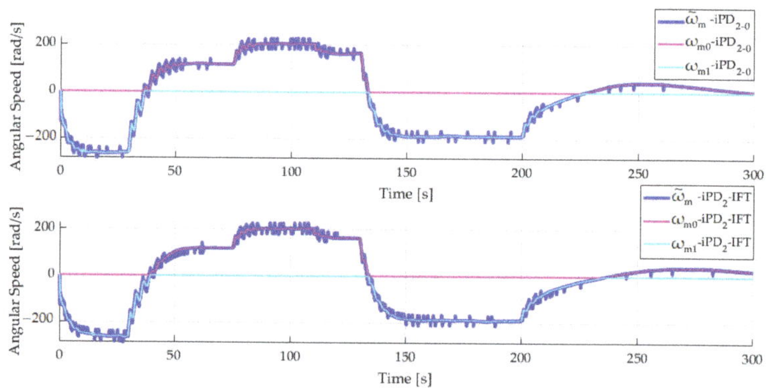

Figure 10. Control signals of the iPD$_2$ and iPD$_2$-IFT controllers and the controlled angular speeds for the pitch angle variations experiment.

The reference angular speed of the motors $\widetilde{\omega}_m$ for the experiment presented in Figure 9 can take positive or negative values, as Figure 10 reveals. Taking into account the sign of $\widetilde{\omega}_m$, the C_0 and C_1 switches from Figure 3 make it possible to use motor$_0$ for the positive values and motor$_1$ for the negative values of the reference angular speed signal.

4.3. Control of Pitch Angle with iP$_1$-IFT and iPD$_2$-IFT Controllers

The controllers tuned with the IFT algorithm that were presented in Sections 4.1 and 4.2 are used for a performance comparison with the step variations of the reference signal. The reference profile from Figure 6 used for the iP$_1$ comparison was considered for the experiments against the iPD$_2$-IFT controller. Moreover, it is necessary to analyze which type of MFC controller can offer better results for the pitch angle control of an Aero System. The results are depicted in Figure 11.

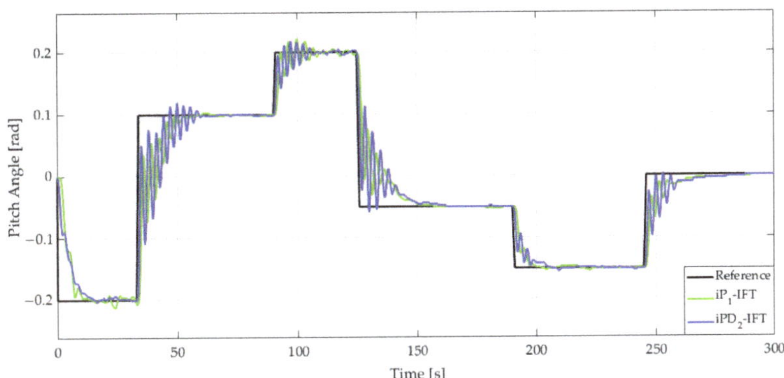

Figure 11. Comparison between the iP$_1$-IFT and iPD$_2$-IFT controllers for the pitch angle variation experiments.

The comparison reveals a response with larger oscillations for iPD$_2$-IFT in relation to iP$_1$-IFT to the step variations of the reference signal. However, due to the derivative component, iPD$_2$-IFT will generate a shorter settling time. The control signals obtained from the comparative experiment are depicted in Figure 12 together with the angular speeds of the two DC motors.

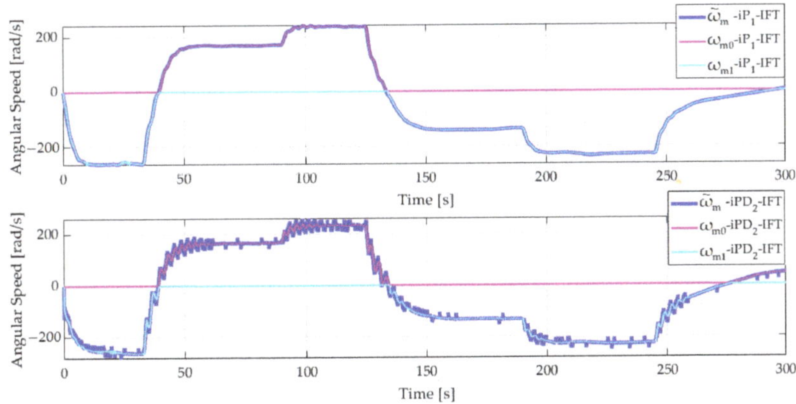

Figure 12. Control signals of iP$_1$-IFT and iPD$_2$-IFT and the controlled angular speeds for the pitch angle variation experiments.

The control signal of the iP$_1$-IFT controller is smooth compared with that of the iPD$_2$-IFT controller, which exhibits oscillations due to the amplification of the high-frequency signals from the control system by the derivative component. These oscillations propagate at the output of the inner loops and finally at the output of the control system, determining less satisfactory performances of the control system with iPD$_2$-IFT compared with those obtained using iP$_1$-IFT.

5. Conclusions

In this paper, a new method for tuning model-free iPID controllers based on the IFT data-driven algorithm is described. Considering that the IFT algorithm can only be used for the tuning controllers with a fixed structure, such a structure was first determined for the iPID controllers. Thus, starting from the connections between the iPID and PID discrete-time controllers, a fixed structure of the iPID controller was obtained, described by an equivalent transfer function that were parametrized with the parameters vector. Moreover, analyzing the behaviors of the iPID controllers in relation to the PID control laws, two intelligent controllers were established for $v = 1$ and $v = 2$, respectively, with the others being eliminated because they had unusual behaviors in the control engineering. The IFT algorithm could be applied by having the iPID controllers with a fixed structure, whose transfer functions were used to estimate the gradients of the input/output variables. At the same time, based on two experiments in the closed loop, the input and output signals were collected, which together with the above derivative estimates allowed for the estimation of the gradients of the cost function and the matrix used to update the direction. Finally, the parameters grouped in vector ρ of the fixed structure of the controller were determined; based on these parameters, the tuning gains and parameter α were found. The methodology for tuning the iPID controllers based on the IFT algorithm was tested and experimentally validated using a real-time application for controlling the pitch angle of the dual-motor Aero 2 aerospace system, which was provided by Quanser.

The main advantages of the proposed tuning method based on the IFT algorithm consist of the following: the determination of both the gain parameters related to the PID algorithm and the α parameter; the IFT tuning of the iPID parameters does not require knowledge of the plant model; and the gradients of the quadratic control criterion can be easily estimated based on the data collected after the two experiments. Among the disadvantages, it is mentioned that the selection of the starting values of the iPID tuning parameters must be conducted in such a way so as to obtain a stable response and the two experiments must be performed to determine the tuning parameters at each step. The initial choice of tuning parameters, along with finding the related values of the step size and the penalty factor of the control effort through a trial-and-error procedure, are the main design challenges.

For future research, some techniques will be considered for tuning the parameters related to the step size and penalty factor of the control effort as well as the extending of the proposed method to multivariable systems.

Author Contributions: Conceptualization and methodology, C.L. and A.B.; software, A.B.; validation, formal analysis, investigation, resources, writing—original draft preparation, data curation, and investigation, A.B. and C.L.; writing—review and editing, visualization, supervision, and project administration, C.L. and A.B. All authors have read and agreed to the published version of the manuscript.

Funding: This research received no external funding.

Data Availability Statement: Data available from the authors upon request.

Conflicts of Interest: The authors declare no conflict of interest.

References

1. Hou, Z.-S.; Wang, Z. From model-based control to data-driven control: Survey, classification and perspective. *Inf. Sci.* **2013**, *235*, 3–35. [CrossRef]
2. Hou, Z.-S.; Chi, R.; Gao, H. An Overview of Dynamic-Linearization-Based Data-Driven Control and Applications. *IEEE Trans. Ind. Electron.* **2017**, *64*, 4076–4090. [CrossRef]
3. Bazanella, A.S.; Campestrini, L.; Eckhard, D. *Data-Driven Controller Design: The H2 Approach*, 1st ed.; Springer: Dordrecht, The Netherlands, 2012.

4. Hou, Z.; Jin, S. *Model Free Adaptive Control: Theory and Applications*, 1st ed.; CRC Press: Boca Raton, FL, USA, 2016.
5. Precup, R.-E.; Roman, R.-C.; Safaei, A. *Data-Driven Model-Free Controllers*, 1st ed.; CRC Press: Boca Raton, FL, USA, 2021.
6. Hjalmarsson, H.; Gevers, M.; Gunnarsson, S.; Lequin, O. Iterative feedback tuning: Theory and applications. *IEEE Control Syst. Mag.* **1998**, *18*, 26–41.
7. Guardabassi, G.O.; Savaresi, S.-M. Virtual reference direct design method: An off-line approach to data-based control system design. *IEEE Trans. Autom. Control* **2000**, *45*, 954–959. [CrossRef]
8. Chi, R.; Liu, X.; Zhang, R.; Hou, Z.; Huang, B. Constrained data-driven optimal iterative learning control. *J. Process Control* **2017**, *55*, 10–29. [CrossRef]
9. Fliess, M.; Join, C. Model free control. *Int. J. Control.* **2013**, *86*, 2228–2252. [CrossRef]
10. Join, C.; Fliess, M.; Chaxel, F. Model-Free Control as a Service in the Industrial Internet of Things: Packet loss and latency issues via preliminary experiments. In Proceedings of the 28th Mediterranean Conference on Control and Automation (MED2020), Saint-Raphaël, France, 16–18 September 2020; pp. 299–306.
11. Baciu, A.; Lazar, C. Model Free Speed Control of Spark Ignition Engines. In Proceedings of the 23rd International Conference on System Theory, Control and Computing (ICSTCC), Sinaia, Romania, 9–11 October 2019; pp. 480–485.
12. Baciu, A.; Lazar, C. Model-Free iPD Control Design for a Complex Nonlinear Automotive System. In Proceedings of the 24th International Conference on System Theory, Control and Computing (ICSTCC), Sinaia, Romania, 8–10 October 2020; pp. 868–873.
13. Roman, R.-C.; Radac, M.-B.; Precup, R.-E.; Petriu, E.M. Virtual Reference Feedback Tuning of Model-Free Control Algorithms for Servo Systems. *Machines* **2017**, *5*, 25. [CrossRef]
14. Precup, R.-E.; Radac, M.-B.; Roman, R.-C. Model-free sliding mode control of nonlinear systems: Algorithms and experiments. *Inf. Sci.* **2017**, *381*, 176–192. [CrossRef]
15. Wang, H.; Ye, X.; Tian, Y.; Christov, N. Attitude Control of a Quadrotor Using Model Free Based Sliding Model Controller. In Proceedings of the 20th International Conference of Control Systems and Science, Bucharest, Romania, 27–29 May 2015; pp. 149–154.
16. Polack, P.; d'Andréa-Novel, B.; Fliess, M.; de La Fortelle, A.; Menhour, L. Finite-Time Stabilization of Longitudinal Control for Autonomous Vehicles via a Model-Free Approach. *IFAC-PapersOnLine* **2017**, *50*, 12533–12538. [CrossRef]
17. Carrillo, F.-J.; Rotella, F. Some contributions to estimation for model-free control. *IFAC-PapersOnLine* **2015**, *48*, 150–155. [CrossRef]
18. Yaseen, A.A.; Bayart, M. A Model-Free Approach to Networked Control System with Time-Varying Communication Delay. *IFAC PapersOnLine* **2018**, *51*, 558–563. [CrossRef]
19. Lequin, O.; Gevers, M.; Mossberg, M.; Bosmans, E.; Triest, L. Iterative Feedback Tuning of PID parameters: Comparison with classical tuning rules. *Control Eng. Pract.* **2003**, *11*, 1023–1033. [CrossRef]
20. Bazanella, A.S.; Gevers, M.; Miskovic, L.; Anderson, B.D.O. Iterative minimization of h2 control performance criteria. *Automatica* **2008**, *44*, 2549–2559. [CrossRef]
21. Kammer, L.C.; Bitmead, R.R.; Bartlett, P.L. Direct iterative tuning via spectral analysis. *Automatica* **2002**, *36*, 2549–2559.
22. Karimi, A.; Miskovic, L.; Bonvin, D. Iterative correlation-based controller tuning. *Int. J. Adapt. Control Signal Process.* **2004**, *18*, 1410–1415. [CrossRef]
23. Masuda, S. Iterative controller parameters tuning using gradient estimate of variance evaluation. In Proceedings of the International Conference on Advanced Mechatronic Systems ICAMechS, Shiga, Japan, 26–28 August 2019; pp. 789–794.
24. Takano, Y.; Masuda, S. Iterative Feedback Tuning for regulatory control systems with measurable disturbances. *IFAC Paper-sOnLine* **2021**, *54*, 245–250. [CrossRef]
25. Xie, Y.; Jin, J.; Tang, X.; Ye, B.; Tao, J. Robust Cascade Path-Tracking Control of Networked Industrial Robot Using Constrained Iterative Feedback Tuning. *IEEE Access* **2019**, *7*, 8470–8482. [CrossRef]
26. Wu, W.; Li, D.; Meng, W.; Zhuo, J.; Liu, Q.; Ai, Q. Iterative Feedback Tuning-based Model-Free Adaptive Iterative Learning Control of Pneumatic Artificial Muscle, In Proceedings of the 2019 IEEE/ASME International Conference on Advanced Intelligent Mechatronics (AIM). Hong Kong, China, 8–12 July 2019; pp. 954–959.
27. Baciu, A.; Lazar, C.; Caruntu, C.-F. Iterative Feedback Tuning of Model-Free Controllers. In Proceedings of the 25th International Conference on System Theory, Control and Computing (ICSTCC), Iasi, Romania, 20–23 October 2021; pp. 467–472.
28. *Quanser AERO 2 User Manual*; Quanser Inc.: Markham, ON, Canada, 2022.
29. Wu, B.; Wu, J.; Zhang, J.; Tang, G.; Zhao, Z. Adaptive Neural Control of a 2DOF Helicopter with Input Saturation and Time-Varying Output Constraint. *Actuators* **2022**, *11*, 336. [CrossRef]
30. Kim, B.M.; Yoo, S.J. Approximation-Based Quantized State Feedback Tracking of Uncertain Input-Saturated MIMO Nonlinear Systems with Application to 2-DOFHelicopter. *Mathematics* **2021**, *9*, 1062. [CrossRef]
31. Lambert, P.; Reyhanoglu, M. Observer-based sliding mode control of a 2-DoF Helicopter system. In Proceedings of the IECON 2018 44th Annual Conference of the IEEE Industrial Electronics Society, Washington, DC, USA, 21–23 October 2018.
32. Khanesar, M.A.; Kayacan, E. Controlling the pitch and yaw angles of a 2-DOF helicopter using interval type-2 fuzzy neural networks, Studies in Systems. *Decis. Control* **2015**, *24*, 349–370.
33. Schlanbusch, S.M.; Zhou, J.; Schlanbusch, R. Adaptive Attitude Control of a Rigid Body with Input and Output Quantization. *IEEE Trans. Ind. Electron.* **2022**, *69*, 8296–8305. [CrossRef]
34. Atkinson, K.E. *An Introduction to Numerical Analysis*, 2nd ed.; John Wiley and Sons: New York, NY, USA, 1989.

35. Hjalmarsson, H. Control of nonlinear systems using iterative feedback tuning. In Proceedings of the American Control Conference, Philadelphia, PA, USA, 26 June 1998; pp. 2083–2087.
36. Solari, G. Iterative Model-Free Controller Tuning. Ph.D. Thesis, Université Catholique de Louvain, Ottignies-Louvain-la-Neuve, Belgium, 2005.

 *actuators*

MDPI

Article

A Robust Model Predictive Control for Virtual Coupling in Train Sets

Jesus Felez *, **Miguel Angel Vaquero-Serrano** and **Juan de Dios Sanz**

Mechanical Engineering Department, Universidad Politécnica de Madrid, Jose Gutierrez Abascal 2, 28006 Madrid, Spain
* Correspondence: jesus.felez@upm.es

Abstract: In recent decades, the demand for rail transport has been growing steadily and faces a double problem. Not only must the transport capacity be increased, but also a more flexible service is needed to meet the real demand. Both objectives can be achieved through virtual coupling (VC), which is an evolution of the current moving block systems. Trains under VC can run much closer together, forming what is called a virtually coupled train set (VCTS). In this paper, we propose an approach in which virtual coupling is implemented via model predictive control (MPC). For this purpose, we define a robust controller that can predict, based on a dynamic model of the train, the state of the system at later moments of time and make the appropriate control decisions. A robust MPC (RMPC) is obtained by introducing two uncertain variables. The first uncertain variable is added to the acceleration equation of the dynamic model, while the second uncertain variable is used to define the uncertainty in the train positioning. To test the RMPC for virtual coupling, two simulation cases are performed for a metro line, analysing the influence of both the uncertainties. In all cases, the results obtained show a safer operation of the virtual coupling without significantly affecting the service.

Keywords: railway; virtual coupling; optimal control; model predictive control; robust MPC

Citation: Felez, J.; Vaquero-Serrano, M.A.; de Dios Sanz, J. A Robust Model Predictive Control for Virtual Coupling in Train Sets. *Actuators* **2022**, *11*, 372. https://doi.org/10.3390/act11120372

Academic Editor: Patrick Lanusse

Received: 26 October 2022
Accepted: 8 December 2022
Published: 10 December 2022

Publisher's Note: MDPI stays neutral with regard to jurisdictional claims in published maps and institutional affiliations.

1. Introduction

In recent decades, the demand for transport has been progressively increasing in both the public and private sectors. The railroad, as an essential, efficient, and sustainable service to reduce congestion in other modes of transport, has been affected by this tendency [1] and must, therefore, face a double problem. On the one hand, it is necessary to increase the transport capacity to meet this demand. In this case, the solutions proposed involve reducing the minimum intervals between trains instead of building expensive new lines. On the other hand, it is necessary to make the service more flexible by reducing the operating costs and maintaining the quality of service required to always meet the real demand.

An interesting solution to address both problems is virtual coupling between trains [2] forming a virtually coupled train set (VCTS), which is an evolution of the current moving block systems. These systems, such as Communication-Based Train Control (CBTC) and European Train Control System (ETCS) Level 3, require that a train be able to brake and stop before the last known position of the preceding train. However, this assumption is conservative as it does not consider the speed and braking capability of the preceding train. Thus, in this context, based on the principle of spacing trains in a relative braking distance, as well as the basic principles of autonomous vehicles and platoon cars, a new concept has emerged: a train convoy, virtual coupling, or virtually coupled train sets (VCTSs) [2].

The authors in [3] presented initial estimations showing that virtual coupling can reduce the distance between trains by 64% compared to the European Train Control System (ETCS) Level 2 and by 43% compared to the ETCS Level 3, thereby increasing the line capacity significantly.

The VCTS has two main advantages. First, the VCTS significantly improves the current capacity limit imposed by automatic train protection (ATP) by allowing trains to run safely over a shorter distance. This significantly reduces the costs of an additional capacity, as it uses mainly on-board equipment rather than major infrastructure changes, such as the installation of new tracks or signalling systems.

Second, besides improving the capacity, the VCTS concept aims, above all, to benefit from operational flexibility and robustness by enabling an interoperability between different rail vehicles and replacing sensitive mechanical couplings. The VCTS enables a more flexible operation of the trainsets since it allows for the circulation of a train set as if it were mechanically coupled, but with no real physical connection. This allows for a more flexible service under an optimized infrastructure by connecting or disconnecting units from the train set in a fast and efficient way. This fact is especially relevant, e.g., in situations with two lines featuring shared tracks in some sections, but different terminal stations and in stations with less traffic, as seen in [4,5]. The main idea of this second advantage is twofold: to be able to operate with smaller configurations adapted to the traffic needs and to increase flexibility to serve heterogeneous demands, also enabling an interoperability between different rail vehicles.

In this paper, we propose an approach in which virtual coupling is implemented via model predictive control (MPC) for solving the control problems of the VCTS. MPC is an advanced control method that can control a system while satisfying a set of constraints. One of the main advantages of this system is that it can be formulated in a simple way since its dynamics and constraints have a real physical meaning. On the other hand, the response of the current control system is also optimized based on the future predictions, so this feature allows the MPC to anticipate future events and, consequently, take the necessary control actions in advance.

Due to this prediction capability, MPC is one of the most widely used control techniques in autonomous vehicle driving and vehicle platoons, as noted in [6–8], and its results form the basis and foundation for most of the research on virtual coupling control in railroads.

The accuracy of the prediction and, therefore, the control actions taken depends on the accuracy of the dynamic model used by the controller, requiring a balance between the complexity of the controller formulation and the computational load required. In this sense, there are two traditional approaches to deal with errors in the control model [9].

The first approach formulates a controller that does not consider the possible errors committed using a nominal MPC (NMPC). Subsequently, the behaviour of this controller under the existence of these errors is studied using a robustness analysis. If the behaviour is still acceptable, the errors can be ignored, and the nominal MPC will be considered adequate.

The second approach formulates an MPC controller that considers the possible errors made within its own formulation, i.e., synthesizing what is called a robust MPC (RMPC). This approach is typically used when an accuracy is required or when a dynamic disturbance or error has a large impact on the behaviour of the controller.

There are several works that use an NMPC to address the work at hand. In this way, the authors in [10] provided a proof of concept for the VCTS by introducing a specific operating mode within the ERTMS/ETCS standard specification and defining a coupling control algorithm accounting for the time-varying delays affecting the communication links. Likewise, the authors in [11] developed a distributed nominal MPC for high-speed trains, evaluated its stability, and derived sufficient feasibility and stability conditions for a platoon of up to two trains by designing the terminal controller and invariant set. Additionally, for high-speed trains, the authors in [12] proposed a control strategy focused on both the local and string stability under variant manoeuvres in high-speed scenarios using an analytical algorithm based on Pontryagin's maximum principle. Similarly, the authors in [13,14] applied an NMPC in metro lines.

However, there are a few examples in the literature of RMPC being applied to the VCTS. For example, the authors in [15] implemented a robust event-triggered model predictive control based on the equations of longitudinal train dynamics and a communication switching topology characterized as a Markov chain. However, this study only focused on the communication problems with a constant-speed leader and, therefore, did not test the changing operating conditions as a function of the speed and track parameters. In addition, the authors in [16] proposed a robust gap controller based on sliding mode control with a nonlinear train model featuring uncertainties. However, as in the previous study, the uncertainties were limited to the consideration of a small Davis formula and track resistances. On the other hand, the authors in [17] presented consensus-based robust cooperative control schemes for both homogeneous and heterogeneous train platoons by using robust strictly negative–imaginary controllers considering the network topology to track a predefined motion reference. Recently, the authors in [18] developed an alternative way to manage the uncertainties using a tube-based MPC, but also expressed the need to incorporate uncertainties in the train dynamics and consider and study factors such as the weather conditions, positioning errors, and communication delays.

Finally, the authors in [19] formulated a nominal MPC, showing its application in metro lines. According to this study, VCTSs can be implemented through two different control architectures: centralized control and decentralized control. In the former, a single controller makes the trains cooperate to optimize the overall platoon strategy. In the latter, each train has its own individual controller, and, as a result, each train optimizes its own strategy according to the trajectory estimation of the preceding vehicle. The authors in [19] also expressed the need to improve the nominal MPC through uncertainties and the development of a robust MPC.

The main objective and contribution of this work is the development of a decentralized RMPC for virtual coupling in railways, considering the effects of adhesion loss and positioning errors. The research in [19] serves as a starting point for this work, from which the design of a nominal MPC is taken. Unlike this previous work, the implemented robust MPC considers the external disturbances with uncertainties that consider the previously mentioned positioning errors and a loss of the adhesion during braking. In this way, the proposed controller is compared with a nominal MPC in different simulations under the different applied disturbances. The results will be studied from an operational point of view, i.e., by analysing the distance between the trains and the overall behaviour of a two-train VCTS.

Ultimately, the objective of this work is to improve the line capacity under the condition of virtual coupling. In this sense, we consider that the VCTS fulfils its function and that all the components of the VCTS work as connected when all of them arrive and stop at the same time at the same station with a minimum delay time between them. In this way, the convoy operationally behaves as a single train since, from the users' point of view, all the VCTSs' components arrive at the station at the same time and behave as if they were a single complete train. Although when the convoy is running the distance between the components increases due to the safety conditions which have been set, we consider that the convoy preserves its integrity, operating as a single train, when the communication between trains is maintained and the above-mentioned condition of all the components arriving at the station at the same time is respected. The figures included in the simulations section illustrate this phenomenon.

The remainder of this paper is organized as follows. Section 2 presents the dynamic model. Section 3 formulates the design of the RMPC. Section 4 describes the simulation cases developed and presents their corresponding results. Finally, Section 5 includes the conclusions of this work.

2. System Dynamics

The model defining the train motion of this work is based on the principles of longitudinal train dynamics (LTD). Then, the train is considered as a point mass with one degree of freedom, where the traction/brake system, rolling resistances, air intake, aerodynamic drag, and slope and curve resistances are applied. The dynamic equations considered are as follows:

$$\dot{s} = v \tag{1}$$

$$\dot{v} = \left(-a - bv - cv^2 - F_e + F\right)/M + w^a \tag{2}$$

$$\dot{F} = (u - F)/\tau \tag{3}$$

where s (m) and v (m/s) denote the position and train speed, u (N) is the controlled driving/braking force, F (N) is the integrated driving/braking force, F_e (N) is the resistance force due to the track, τ is the inertial lag of the longitudinal dynamics, M (kg) is the train's mass, a (N) is a term that includes the rolling resistance plus the bearing resistance, b (Ns/m) is a coefficient related to the air intake, c (Ns^2/m^2) is the aerodynamic coefficient, and w^a (m/s^2) represents the uncertainty in the acceleration contemplated in the robust control.

The resistance force F_e includes two components, F_g and F_R, which are defined as follows:

$$F_e = F_g + F_R \tag{4}$$

$$F_g = -Mg \times slope \tag{5}$$

$$F_R = -M \times 6/R \tag{6}$$

where F_g (N) is the component of the gravity force due to the slope of the track, the *slope* (m/m) is the slope of the track, g (m/s^2) is the acceleration of gravity, F_R (N) is the resistance in the curve, and R (m) is the radius of the curve.

The values of the *slope* and R depend on the line profile and the position s of the train on the line and are, therefore, known at each time. Figure 1 shows as an example of the value of the *slope* and R used in this work.

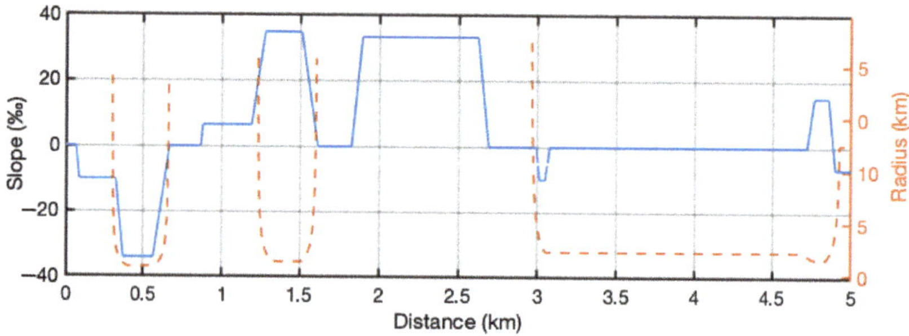

Figure 1. Slope and radius considered in the simulation scenario.

Starting from the dynamic equations, a linearized set of equations at the operating point will be used to handle uncertainties in the RMPC optimization algorithm [20].

Thus, the vector of states X and outputs Y, that will be used for a linearization, are defined as follows:

$$X = \begin{bmatrix} s \\ v \\ F \end{bmatrix} \tag{7}$$

$$Y = \begin{bmatrix} u \cdot v \\ v^2/2a^e \end{bmatrix} \tag{8}$$

where a^e (m/s^2) is the maximum possible deceleration of the controlled train.

By linearizing Equations (1)–(3), these equations are transformed into (9)–(11), thereby formulating the linearized dynamics as the first-order Taylor polynomial of the nonlinear model as follows:

$$U = [u], \qquad W = \begin{bmatrix} 0 \\ w^a \\ 0 \end{bmatrix} \tag{9}$$

$$\dot{X} = \dot{X}_0 + A(X - X_0) + B(U - U_0) + W \tag{10}$$

$$Y = Y_0 + C(X - X_0) + D(U - U_0) \tag{11}$$

and

$$A = \begin{bmatrix} 0 & 1 & 0 \\ 0 & -(B + 2Cv_0)/M & 1/M \\ 0 & 0 & -1/\tau \end{bmatrix} \qquad B = \begin{bmatrix} 0 \\ 0 \\ 1/\tau \end{bmatrix}$$

$$C = \begin{bmatrix} 0 & u_0 & 0 \\ 0 & v_0/a_f & 0 \end{bmatrix} \qquad D = \begin{bmatrix} v_0 \\ 0 \end{bmatrix}$$

where v_0 (*m/s*) and u_0 (*N*) corresponds to the state v and the control force u at the operating point, respectively; X_0, U_0, and Y_0 are the state vector X, the decision vector U, and the output vector Y at the operating point, respectively; and $\dot{X}_0$ is Equations (1)–(3) evaluated at the operating point.

In the above Equations (1)–(11), the states s, v, F; the controlled driving/braking force u; and the uncertainty w are time dependent. Therefore, X and Y are also time dependent. The matrices A, B, C, D are constant, but v_0 and u_0 depend on the operating point at which the equations are linearised. In addition, the values of *slope* and R are known at each time and dependent on the line profile and train location on the line, i.e., the train position s, as shown in Figure 1. The remaining variables in (1)–(11) are constant parameters that depend on the train's characteristics.

3. MPC Controller Design

3.1. Control Architecture

We consider a convoy (Figure 2) composed of a leader and n followers, all of them with length L. The superscript indicates the train, with 0 for the leader and i, where $i = 1$, ... , n, for the followers.

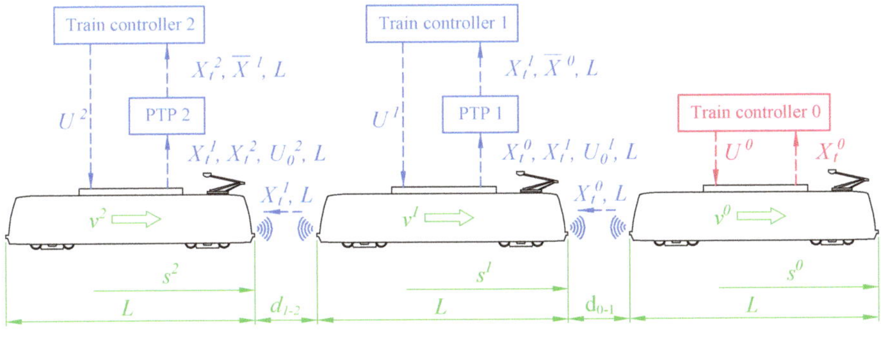

Figure 2. Control architecture.

We consider a decentralized VCTS control problem, with independent controllers for the leader and each follower. Figure 2 represents this control architecture. In this figure, index 0 represents the leader; therefore, X_t^0 represents the current measured states for the leader; v^0 represents the leader's speed; s^0 is the leader's current position; and U^0 represents the vector containing the driving/braking force for the leader. X_t^i represents the current measured states for the follower i in the platoon, which is virtually coupled to its previous train; v^i and s^i represent, respectively, the speed and current position of the follower i; and U^i represents the driving/braking force for the follower i. As seen in Section 2 for the linearized control model, U_o^i is the value of U^i at the operating point.

The end-front distance $d_t{}^i$ between two consecutive trains at any time is calculated using (12), where L is the train length:

$$d_t{}^i = s_t{}^{i-1} - s_t{}^i - L. \tag{12}$$

Furthermore, $\overline{X}^{i-1}$ represents the predicted states for the preceding train $(i-1)$ calculated by the follower's Preceding Train Predictor (PTP) module, which will be described in Section 3.5. For the followers, we design an MPC robust controller that ensures a safety and control efficiency while considering the parameter uncertainties.

With the proposed control strategy, the leader will track a given speed curve, and the followers will guarantee a safe minimum distance between trains and the string stability of the VCTS with parameter uncertainties. Consequently, as seen in Figure 2, the follower needs information about the preceding train to ensure its safety. This information ($X_t^{i-1}(t)$ and L) is used by the follower's controller for its calculations because the predicted states for the preceding train $i-1$ are calculated using the PTP embedded in each follower i based on the information received from the preceding train.

3.2. Leader Controller

When using a decentralized control architecture, the leader can operate under any control and signalling method. Therefore, the leader can use any conventional control method based on Automatic Train Control, Communication-Based Train Control, or the ETCS.

For simplicity, for the leading train, we use a driving mechanism based on an Automatic Train Control system that tracks a given speed curve. This speed curve is obtained using a Dynamic Programming (DP) approach (Figure 3).

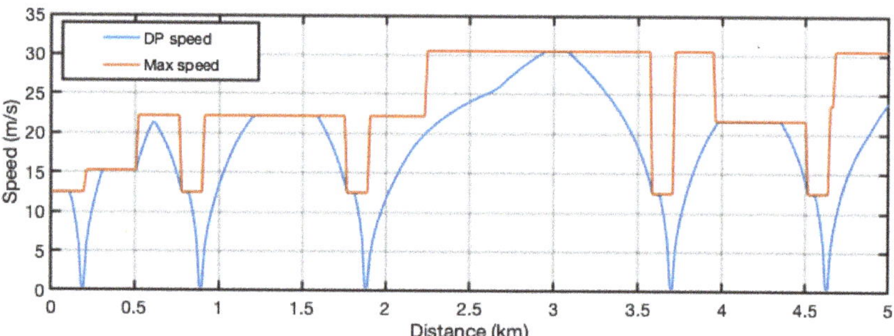

Figure 3. Maximum driving speed in compliance with speed limits.

Because the role of the leader is to set the VCTSs' movement policy, we use the DP approach to precompute the reference behaviour of the leader. Then, the result of the DP establishes the general policy followed by the train's convoy. This general control policy can have different objective functions such as minimizing the energy consumption or maximizing the convoy speed. In this paper, we use the second policy with an optimal speed profile that finds the maximum velocity permitted by the speed limitations imposed by the line operation, thereby satisfying the speed constraints at all times.

Figure 3 shows the speed profile obtained for the leader, corresponding to the convoy policy of the maximum possible velocity allowed by the speed limitations established for the line's operation. In this figure, the curve labelled "Max speed" represents the maximum limitations set in the line design. In this figure, it can be seen where the stations are located and that the speed at these stations is limited to 12.5 m/s (45 km/h). The other speed limitations depend on the design of the railway line.

On the other hand, the curve labelled "DP speed" represents the maximum speed at which it is possible to run without exceeding at any time the speed allowed on the line and represents the result of the DP calculation. On this curve, the stopping point of the leading train at the end of each station can also be seen. A detailed explanation of the implementation of this DP approach can be found in [19].

3.3. Follower Controller

For the controller design, a model predictive control (MPC) approach is used. The MPC optimizes over a finite time horizon but implements only the current time window of the finite horizon optimization problem solution.

For the MPC formulation, a prediction horizon $[t, t + N_p]$ with origin at time t is considered. The notation $x_{t+k|t}$ represents the state vector at time $t + k$ predicted at time t, obtained from the current state $x_{t|t} = x(t) \equiv x_t$, while $u_{\cdot|t} = \left[u_{t|t}, \ldots, u_{t+N_p-1|t} \right]$ denotes the unknown input variables (or inputs) to be optimized.

The RMPC for virtual coupling is obtained by introducing two uncertain variables.

The first uncertain variable is w^a (m/s^2) and is added to the acceleration equation of the dynamic model, as seen in (9).

The second uncertain variable w^p (m) is considered in the virtual coupling specific constraints and the follower cost functions, and it is used to define the uncertainty in the positioning, as will be seen later.

As a result of introducing uncertain variables, the optimization problem for the follower i is formulated as the *min–max* problem given in (13):

$$\min_{u^i_{\cdot|t} \ w^{a,i}_{\cdot|t}, \ w^{p,i}_{\cdot|t}} \max J^i \left(X^i_{k|t}, u^i_{k|t}, w^{a,i}_{k|t}, w^{p,i}_{k|t} \right). \tag{13}$$

The controller's objective for each follower is to try to keep the trains running as close together as possible by minimising the difference in the distance between trains $d_{k|t}$ and a desired separation d_{des} while maintaining safe conditions. Therefore, we can formulate the following optimization problem:

$$J^i = \sum_{k=1}^{k=N_p} K_D \left\| \frac{d^i_{k+1|t} + w^{p,i}_{k+1|t} - d_{des}}{d_{des}} \right\|_1 \tag{14}$$

$$d^i_{k|t} = \bar{s}^{i-1}_{k|t} - s^i_{k|t} - L$$

subject to:

$$X^i_{k+1|t} = X^i_{k|t} + t_s \dot{X}^i_{k|t} \tag{15}$$

$$X^i_{t|t} = X^i_t \tag{16}$$

$$0 \leq v^i_{k|t} \leq v_{lim}\left(s^i_{k|t} \right) \tag{17}$$

$$-j_{max} \leq j^i_{k|t} \leq j_{max} \tag{18}$$

$$j^i_{k|t} = \left(u^i_{k+1|t} - u^i_{k|t} \right) / M / \Delta t$$

$$-Ma^i_{br} \leq u^i_{k|t} \leq Ma^i_{dr} \tag{19}$$

$$-P^i_{br} \leq Y^i_{1k|t} \leq P^i_{dr} \tag{20}$$

$$\forall k = t, \ldots, t + N_p - 1$$

$$0 \leq v^i_{t+N_p|t} \leq v^{i*}_{DP}\left(s^i_{t+N_p|t}\right) \tag{21}$$

$$d_{min} \leq w^{p,i}_{k|t} + d^i_{k|t} \tag{22}$$

$$\forall k = t, \ldots, t + N_p - 1$$

$$d_{min} \leq w^{p,i}_{k|t} + d^i_{t+N_p|t} \tag{23}$$

$$d_{min} \leq w^{p,i}_{k|t} + d^i_{t+N_p|t} + \frac{\left(\overline{v}^{i-1}_{t+N_p|t}\right)^2}{2a^s} - Y^i_{2k|t} \tag{24}$$

$$w_{a,min} \leq w^{a,i}_{k|t} \leq w_{a,max} \tag{25}$$

$$w_{p,\,min} \leq w^{p,i}_{k|t} \leq w_{p,\,max}. \tag{26}$$

Superscript i is related to the corresponding follower.

In the cost function J^i (14), $K_D \geq 0$ is a dimensionless coefficient representing the weight that penalizes the deviation from the desired distance d_{des}, and $\overline{v}^{i-1}_{t+N_p|t}$ is the predicted speed of the preceding train, which is pre-calculated according to the PTP module that will be described later in Section 3.5.

The equations in (15) represent the train dynamics updates for the model obtained in (9)–(11) according to the t_s integration time step. The initial state is set in (16).

Equation (17) corresponds to the velocity constraint v_{lim} (Max speed in Figure 3).

Equation (18) represents the jerk constraint established for u^i.

Equations (19) and (20) represent the input constraints, including the maximum driving/braking force and the maximum power on traction/braking, with $Y^i_{jk|t}$ representing the estimated j component of vector Y at instant k calculated at time t according to (11).

Equation (21) represents a terminal constraint. The upper speed limit $v^{i*}_{DP}(m/s)$ is obtained by pre-calculating the maximum speed at which the train can run while respecting the speed limits set for the line (DP speed in Figure 3). This terminal constraint is used because the train must always respect the speed limit, and there are areas above the blue line in Figure 3 where if the train runs under these conditions, it may not respect the speed limits at future times.

Equation (22) represents the safety condition, stating that the distance between the trains must be greater than d_{min} at any time.

Equations (23) and (24) are the terminal constraints. These constraints are imposed to ensure that the controller is recursively feasible and safe (see Section 3.6). These equations guarantee that the follower can come to a complete stop without collision by applying a maximum service deceleration a^s when the preceding train performs emergency braking with the maximum deceleration a^e. Note that the terminal constraint (24) functions as an operational constraint, while the terminal constraint (23) works as a safety constraint. The safe train-to-train distance d_{min} is used to define a constraint to avoid a collision if the leader brakes. The distance d_{des} is used in the cost function as a target to maintain a safe desired driving distance greater than the minimum safe distance d_{min}. Finally, constraint (25) bounds the acceleration uncertainty w^a and (26) bounds the second uncertain variable w^p (m), which addresses the distance error from the preceding train. Both constraints are handled through the traditional method of robust optimization based on the duality theory (for more details, see [20]).

The first uncertain variable w^a (m/s^2) is added to the dynamic model's acceleration equation, as seen in (2). By means of this uncertainty, we can consider different phenomena. For instance, using this variable, an additional resistance due to the start resistance can be taken into account, thus solving the lack of accuracy of the Davis formula at speeds below 3 m/s. In addition, this variable can counteract the additional resistance introduced by the tunnel factor and the longitudinal component of lateral wind's impact. Moreover, this variable can avoid the risky situations produced by perturbation, such as a loss of adhesion during braking. This variable also enables the correction of the modelling errors produced by incorrect estimations of the Davis coefficients. The specific values that limit this uncertain variable will be discussed in Section 4.1.

Finally, the second uncertain variable w^p (m) addresses the positioning error that can influence the estimation of the follower's distance from the preceding train. The specific values that limit this uncertain variable will be discussed in Section 4.2.

3.4. Control Loop

For each train, the resulting optimal states and inputs of (13)–(26) are denoted as follows:

$$X_t^{i\,*} = \begin{pmatrix} X_{t|t}^{i\,*} & X_{t+1|t}^{i\,*} & \cdots & X_{t+N_p|t}^{i\,*} \end{pmatrix}^T$$
$$u_t^{i\,*} = \begin{pmatrix} u_{t|t}^{i\,*} & u_{t+1|t}^{i\,*} & \cdots & u_{t+N_p-1|t}^{i\,*} \end{pmatrix}^T. \tag{27}$$

Since constraints (16), (17), and (21) are space-dependent, to avoid numerical problems and reduce the computational burden, they can be estimated a priori by $\bar{s}_{t+k|t}^i$:

$$\overline{X}_{k+1|t}^i = \overline{X}_{k|t}^i + t_s \dot{\overline{X}}_{k|t}^i \tag{28}$$

$$\forall k = t, \ldots, t + N_p - 1$$

$$\overline{u}_{k|t}^i = u_{k|t-1}^{i\,*} \tag{29}$$

$$\forall k = t, \ldots, t + N_p - 2$$

$$\overline{u}_{t+N_p-1|t}^i = u_{t+N_p-2|t-1}^{i\,*} \tag{30}$$

$$\overline{X}_{t|t}^i = X_{t|t}^i \tag{31}$$

where $u_{k|t-1}^{i\,*}$ is the input predicted from (27) at the previous time step $t - 1$ with $u_{k|t-1}^{i\,*} = 0 \; \forall k = 0, \ldots, N_p - 2$.

To close the loop, the first input is applied to system (9)–(11) during the time interval $[t, t + 1)$:

$$u_t^i = u_{t|t}^{i\,*}. \tag{32}$$

At the next time step $t + 1$, a new optimal problem in the form of (13)–(26) based on a new state measurement will be solved over a shifted horizon.

3.5. Preceding Train Predictor

To establish the control loop, it is necessary to estimate a prediction of the preceding train's movement (train $i - 1$). This task is performed by the module, called the Preceding Train Predictor (PTP), which is described in this section.

Since both vehicles are equipped with a communication module, and intervehicular communication is available, the prediction task is executed by the follower based on the information received from the preceding train. For the prediction, the preceding train $i - 1$ communicates to the follower its position $s_{t|t}^{i-1}$, speed $v_{t|t}^{i-1}$, and the driving/braking force $F_{t|t-1}^{i-1\,*}$. In this way, the distance between trains $d_{k|t}^i$ is determined by the difference in the positions of the two trains. On the other hand, and as a redundancy, the follower has a sensor that can measure the distance to the front train. Therefore, for the estimation of the position of the preceding train, the worst case is considered, which corresponds to the lower of the two calculated values above.

Then, the MPC assumes that the preceding train's information can be predicted over a short horizon N_p. The corresponding expressions for the prediction of $\bar{s}_{k|t}^{i-1}$ and $\bar{v}_{k|t}^{i-1}$ are

$$\overline{X}_{t|t}^{i-1} = \begin{pmatrix} s_{t|t}^{i-1} & v_{t|t}^{i-1} & F_{t|t-1}^{i-1*} \end{pmatrix} \tag{33}$$

$$\overline{X}^{i-1}{}_{k+1|t} = \overline{X}^{i-1}{}_{k|t} + t_s \, \dot{\overline{X}}^{i-1}{}_{k|t} \tag{34}$$

$$\forall k = t+1, \ldots, t + N_p - 1,$$

where

$$\overline{u}_{k|t}^{i-1} = u_{k|t-1}^{i-1*} \tag{35}$$

$$\forall k = t, \ldots, t + N_p - 2$$

$$\overline{u}_{t+N_p-1|t}^{i-1} = u_{t+N_p-2|t-1}^{i-1*}. \tag{36}$$

Here, $\bar{s}_{t|t}^{i-1}$ and $\bar{v}_{t|t}^{i-1}$ are the measured position and velocity of the leader at t, respectively; $F_{t|t-1}^{i-1*}$ is the force at time t predicted from (27) at the previous time step $t-1$ initialized with $F_{0|t-1}^{i-1*} = 0$; and $u_{k|t-1}^{i-1*}$ is the input predicted from (27) at the previous time step $t-1$, with $u_{k|t-1}^{i-1*} = 0 \ \forall k = 0, \ldots, N_p - 2$.

In (34), $\dot{X}^i{}_{k|t}$ is calculated from the expressions of (9)–(11), particularly (10), and the disturbance w^a is considered to be the lower limit $w_{a,min}$, which is the most unfavourable situation corresponding to the highest risk of a collision between the two consecutive trains.

3.6. Min-Max Recursive Feasibility

To ensure the recursive feasibility of the RMPC, we define set $\mathbb{C}$ as the robust control invariant set. According to [21], set $\mathbb{C}$ is said to be a robust control invariant set for the system (9)–(11) subject to constraints (13)–(25) if

$$X(t) \in \mathbb{C} \implies \exists [u] \in \mathbb{U} \text{ such that system } (4) \in \mathbb{C}, \ \forall w(t) \in \mathbb{W} \ \forall t \geq 0 \tag{37}$$

where $\mathbb{U}$ is the set of the valid decision variable values given by (18)–(20), and $\mathbb{W}$ is the set that bounds the uncertain variables, i.e., the set given by (25) and (26).

As in [22], set $\mathbb{C}$ can be introduced by the constraints (21)–(23), which define a robust control invariant set because (21) ensures a proper decision of the train's controller according to the track profile, while (22)–(24) ensure a safe reaction to the emergency braking of the preceding train. In other words, these constraints ensure that there exists an output value for the control variables that satisfies (37).

4. Simulations

This section compares the results obtained with the nominal controller developed in [19] with those obtained with the robust controller developed in this paper. To test the robust MPC for the VCTS, two simulation cases were used for a metro line. The first simulation case analyses the influence of an uncertainty in acceleration w^a, while the second one studies the uncertainty in the train's positioning w^p.

The parameters considered for the rolling stock can be found in Table A1 of Appendix A. The profile of the selected line can be found in Figure 1, and its corresponding speed profile can be found in Figure 3.

The simulations were performed by comparing the results between an NMPC (without uncertainties) and an RMPC, where uncertainties are considered.

Simulations were performed in MATLAB and Yalmip [23] on a computer with an i7-1165G7 processor at 2.8 GHz and with 8 GB of RAM. For the different simulations, we determined that the average CPU computation time of the solver was 0.17 s for the RMPC. Moreover, the total CPU computation time was found to be 302 s, which is lower than the total simulated time (345 s). Therefore, the RMPC presents promising results for a hardware implementation that needs to be solved in real time.

4.1. Simulation 1: Uncertainty in Acceleration

At the end of Section 3.3, we introduced what the first uncertain variable w^a (m/s^2) can represent. Now, from a practical point of view in the train operation, we bound the values of the acceleration uncertainty (25) according to Section 3.3.

The start resistance of a train on a track with a null slope is 50 N/t, i.e., 0.05 m/s^2. Since this resistance is applied against the movement (negative), -0.05 m/s^2 is established as the lower limit $w_{a,min}$. Even though this value is lower because the Davis formula (2) is still applied in our controller, it allows for the simultaneous consideration of additional small resistances. For instance, the maximum errors in the estimation of the Davis coefficients for passenger trains are small (below 0.01 m/s^2 for this train) and would be equally compensated by the controller thanks to this setting. Therefore, this value corrects the modelling errors introduced by the Davis formula: both above 3 m/s (Davis coefficients) and lower speeds (start resistance and Davis coefficients).

Additional resistances under 0.05 m/s^2 will also be correctly compensated at speeds above 3 m/s, which is the case for an external perturbation, such as the resistance introduced by either the tunnel factor or the longitudinal component of a lateral wind's impact.

Conversely, $w_{a,max}$ is established to counteract an extreme loss of adhesion of 10% during braking. If a friction factor of 0.1 is considered, this loss will produce a reduction in the deceleration of 0.1 m/s^2. This reduction works as an additional traction force during braking. Therefore, 0.15 m/s^2 is established as the upper bound $w_{a,max}$ as a conservative approximation.

Among all the external perturbations, only the extreme perturbation of a 10% loss of adhesion will be studied in the simulation cases of this paper, but these results can be extrapolated to any of the acceleration uncertainty effects mentioned above.

In Simulation 1, the behaviour of the uncertain variable w^a has been assessed and corresponds to a disturbance involving a loss of adhesion at the entrance of the third station. This situation is extremely unfavourable since the disturbance was applied only to the follower, and the train in front was considered to continue braking without any loss of adhesion.

Figures 4–6 represent the behaviour of the convoy when the follower experiences a 10% loss of adhesion during braking.

The upper plot in Figure 4 represents the time–velocity curve and shows the speed of the leader, which is the same for both the nominal and robust controllers. Moreover, in acceleration (i.e., the increasing parts of the curves), the leader has a slightly higher velocity than the followers, thus increasing the distance between the trains. However, in braking (i.e., the decreasing part of the curve), when the velocity of the leader decreases, the follower becomes closer as it has a higher speed, always maintaining the safe distance imposed by the constraints. It is also possible to observe the stop at the stations.

In Figure 4, it can also be seen how the integrity of the convoy is maintained, as the two trains stop at the station at almost the same time. This effect can be seen in the time–velocity plot, where both trains in the convoy stop at the stations at almost the same time. The plot below shows the distance between the leader and the follower. Here, the most critical moment occurs at the entrance of the stations, where the leader stops, and the follower must also stop maintaining a safe distance.

The curves labelled "no adhesion loss" include the distance between the leader and the follower for the NMPC and RMPC when there is no disturbance, i.e., no adhesion loss. These curves show a very similar behaviour in both cases.

If the behaviour of the robust controller is compared with that of the nominal controller without disturbances, it can always be observed that the distance between two consecutive trains is always larger in the case of the robust controller. This result is logical since the controller contemplates the presence of uncertainty in its own design. In this case, the minimum distance between the trains is obtained at the station entrance, when the leader has stopped. It can also be seen that in the NMPC, this distance is 5 m, while the RMPC is more conservative and maintains a distance of approximately 8 m.

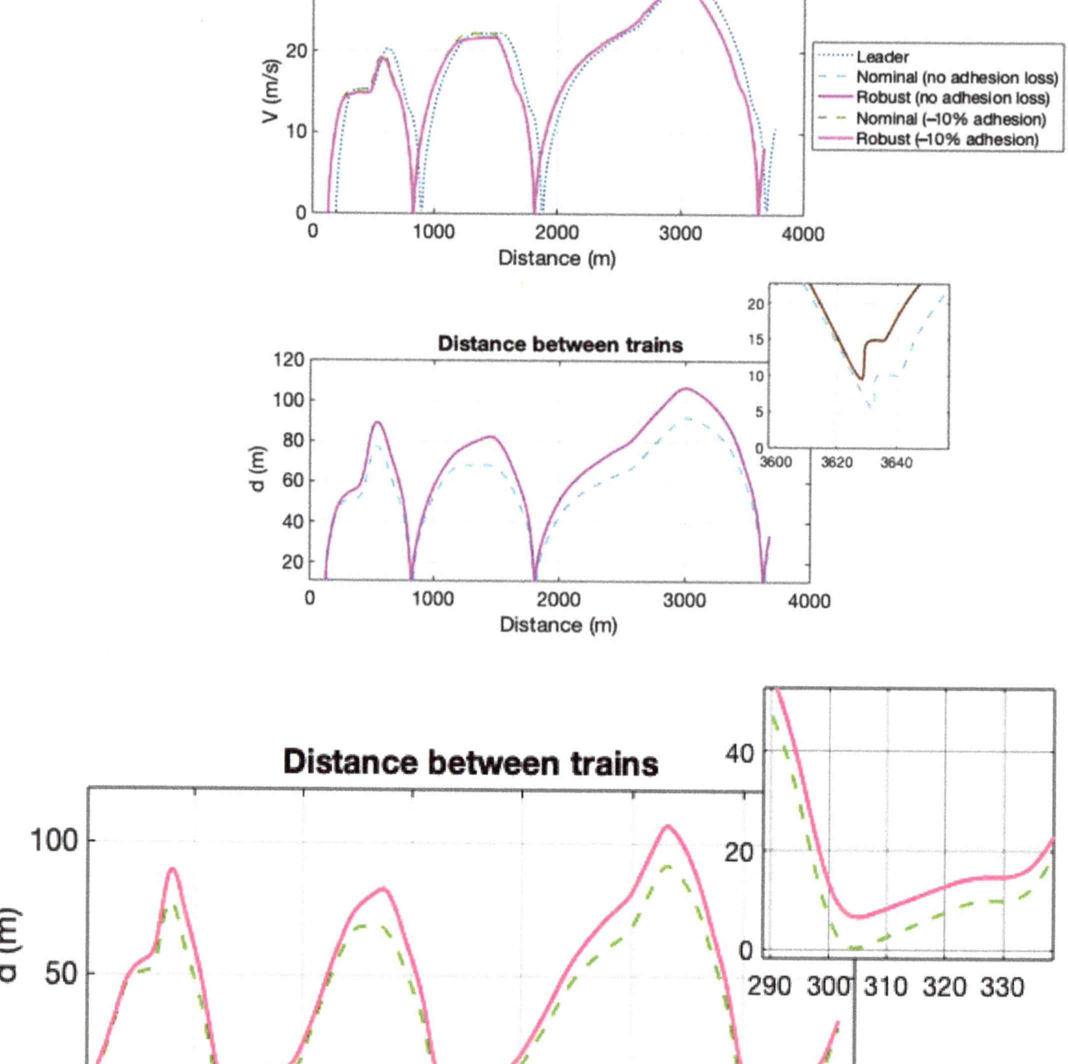

Figure 4. Results of simulation 1: 10% adhesion loss. Variables versus time.

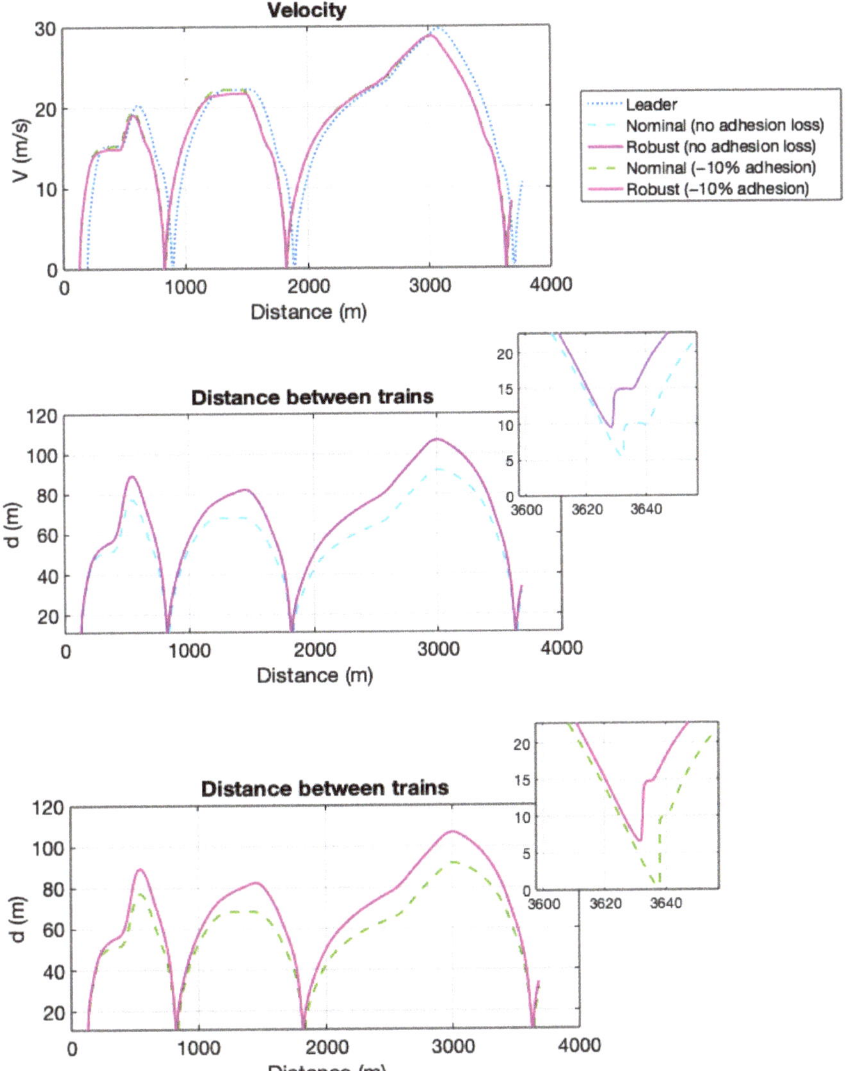

Figure 5. Results of simulation 1: 10% adhesion loss. Variables versus distance travelled.

However, in the case of a loss of adhesion, differences are observed. In the simulation, a 10% loss of adhesion is considered in the entrance area of the third station. In this case, while the RMPC can maintain a minimum distance of 8 m between the trains at the entrance to the station, a distance of 0.5 m is obtained with the NMPC. This situation seriously compromises safety because a collision is about to occur, and the safety conditions for which the nominal control was designed are not respected.

Figure 5 represents the speed and distance between the trains versus the distance travelled. How and when the trains stop at the station is accurately visualized in this figure. It can also be seen that the front train stops at the end of the station, and the rear train stops somewhat further back at the desired distance. Additionally, Figure 6 represents the driving/force obtained as a controlled variable.

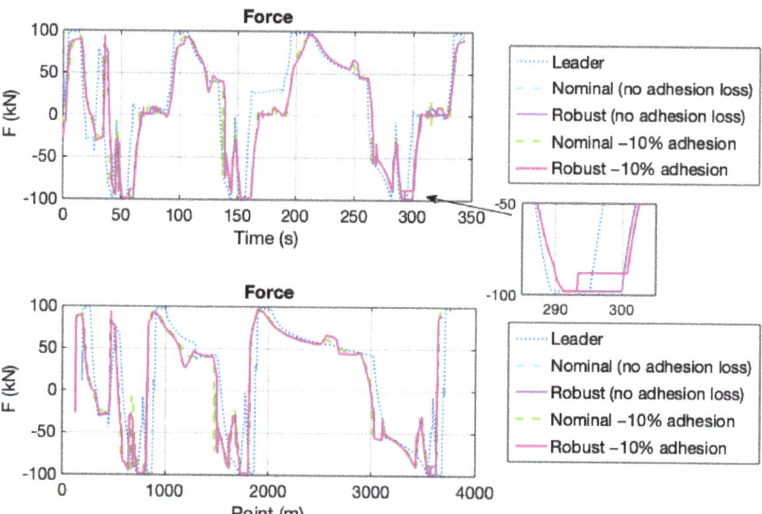

Figure 6. Results of simulation 1: 10% adhesion loss. Force vs. time.

4.2. Simulation 2: Uncertainty in Positioning Error

At the end of Section 3.3, we introduced the second uncertain variable w^p (m), which represents the positioning error that can influence the estimation of the follower's distance from the preceding train. Now, from a practical point of view in the train operation, we bound the values of the positional uncertainty (26) according to Section 3.3.

The positioning error is produced by two components. To begin with, ATO (Automatic Train Operation) controllers have a specified maximum positioning error e_p of ±0.5 m [24]. Additionally, a velocity sensor usually has a maximum error e_v of 1% of the real speed. If a metro line with a 110 km/h maximum speed is considered, the velocity error will be ±0.3 m/s.

Hence, the maximum distance error produced by the follower predictor appears at the end of the prediction horizon (38), where the error is more critical in the constraints. In (38), this result is multiplied by two to consider an extreme situation in which the preceding train is nearer, and the follower is farther. Since this distance error implies that the preceding train might be nearer, this value is considered in the lower bound. Therefore, -3.5 m is estimated as the lower bound of $w_{p,\,min}$ as a conservative approximation of (38). Nevertheless, 0 m is set as the upper bound $w_{p,\,max}$ in order to avoid a distance reduction above virtual coupling conditions, which could end in a collision between the trains:

$$d_{error} = 2\left(e_p + t_s N_p e_v\right) = 3.4 \text{ m}. \tag{38}$$

This simulation focuses on testing the uncertain variable w^p, which is applied to test the performance of the system when there is noise in the positioning information used by the follower. In this simulation, two sinusoidal signals are introduced in the positioning and speed information of the leading train (s^l and v^l). These signals have a period of 90 s and amplitudes of 0.8 m and 0.6 m/s, respectively. In addition, a random noise is superimposed on the previous wave. This noise is uniformly distributed and has an amplitude of 0.001 in each signal (m and m/s, respectively).

The simulation was performed in both the nominal MPC and robust MPC. The results presented in Figure 5 under the label "errors" correspond to the simulation with disturbances, while the results without the label "errors" correspond to the simulation without disturbances.

Figure 7 shows that an error in the estimation of the position of the preceding train, e.g., due to a failure in the odometry or noise in the position information of the preceding train, can cause a collision with the NMPC (green dashed line in Figure 7). This collision can occur at the end of braking from high speeds, i.e., near stations.

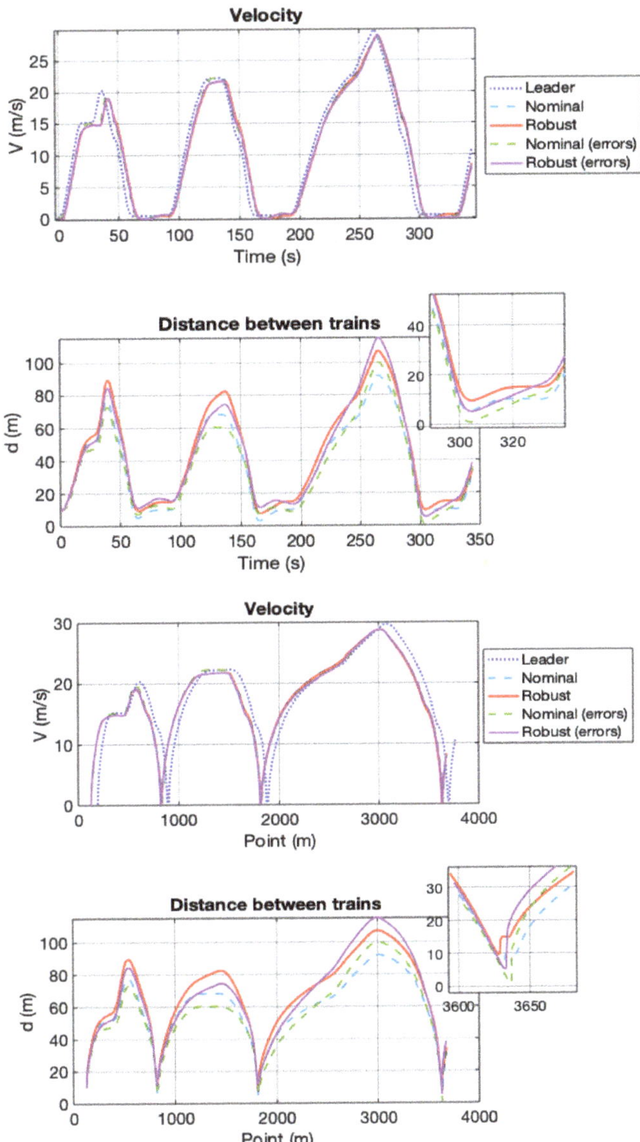

Figure 7. Simulation results 2: Errors in the preceding train information.

The results of the simulations show that in the case of the NMPC, the minimum distance reached with this disturbance was found to be 0.8 m near the 300 s instant of the simulation time.

In contrast, in the case of the RMPC, these risk situations are avoided (the solid magenta line in Figure 7). Additionally, Figure 8 represents the driving/force obtained as a controlled variable. In this simulation, the RMPC was found to achieve a minimum safe distance of 5.1 m, whereas the NMPC presented 0.8 m in the same situation.

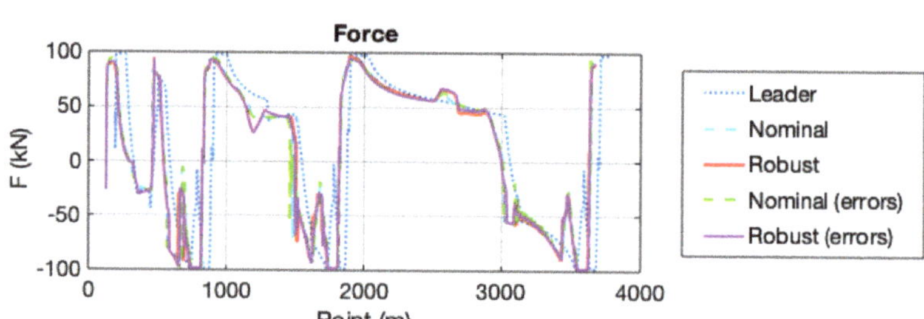

Figure 8. Results of simulation 2: errors in the preceding train information. Force vs. time.

The most important result of this simulation is that the nominal MPC control system tends to lead to risky situations when the environment in which it operates is not a nominal one. This statement is corroborated in Figure 7, which illustrates that, at the station entrance, trains approach up to an absolute distance of 0.8 m, which is an undesirably risky situation.

However, the robust controller results in a minimum distance of 5.1 m compared to 7.5 m for an operation without disturbances. Nevertheless, despite the unfavourable situation, a distance greater than the specified odometry error (3.5 m) is maintained at 48.6%, thereby ensuring a safe traffic flow. Furthermore, with respect to the NMPC, the minimum distance achieved by the RMPC in this unfavourable situation is almost six times greater. In other words, the NMPC reduces the distance in this situation by 83% due to its lack of robustness, which leads to risky situations.

5. Conclusions

In this paper, a decentralized RMPC for the VCTS in railroads was developed. Two types of uncertainties were included in the controller. We compared the proposed RMPC controller with an NMPC in different simulations that consider the positioning errors and a loss of adhesion during braking.

The result is a more conservative controller than a nominal controller, which is nevertheless capable of guaranteeing collision safety in situations which the nominal controller is not able to overcome, such as under a 10% loss of adhesion during braking and in the case of errors in the position information of the front train.

However, it is possible to improve the estimation of the virtual coupling predictor. In future research, fewer conservative solutions could be proposed based on the leader's braking estimates calculated from the last known information instead of using only the last known information directly in the controller.

On the other hand, the decentralized control architecture is acceptable as an advanced railway signalling system that improves the current train operations but not from the perspective of the VCTSs' concept, since in the proposed solution, the leader sets the VCTSs' movement policy but is not affected by the behaviour of the followers. Therefore, in the near future, we will use these results to design a robust centralized MPC for the VCTSs' strategy where all the elements of the VCTS can interact.

Author Contributions: Conceptualization, J.F., M.A.V.-S. and J.d.D.S.; methodology, J.F. and M.A.V.-S.; software, J.F. and M.A.V.-S.; investigation, J.F., M.A.V.-S. and J.d.D.S.; writing—original draft preparation, J.F. and M.A.V.-S.; writing—review and editing, J.F. and M.A.V.-S.; supervision, J.F. All authors have read and agreed to the published version of the manuscript.

Funding: This work was supported in part by the H2020 ECSEL EU Project Intelligent Secure Trustable Things—InSecTT (www.insectt.eu, (accessed on 7 December 2022)) receiving funding from the ECSEL Joint Undertaking (JU) under grant agreement No 876038, and in part by the Spanish Science and Innovation Ministry—State Research Agency under Grant PCI2020-112126/AEI/10.13039/501100011033. The JU receives support from the European Union's Horizon 2020 research and innovation programme and Austria, Sweden, Spain, Italy, France, Portugal, Ireland, Finland, Slovenia, Poland, Netherlands, Turkey. The document reflects only the author's view and Commission is not responsible for any use that may be made of the information it contains.

Institutional Review Board Statement: Not applicable.

Informed Consent Statement: Not applicable.

Data Availability Statement: Not applicable.

Conflicts of Interest: The authors declare no conflict of interest.

Appendix A

Table A1. Parameters used in the simulation.

Parameter	Value	Parameter	Value
M (kg)	99.972×10^3	t_s (s)	0.2
L (m)	54.9	H_p	20
a (N)	1216.13	v_{max} (m/s)	30.6
b (N/(m/s))	117.39	j_{max} (m/s^3)	0.98
c (N/(m/s)^2)	2.97	d_{des} (m)	10
τ (s)	0.7	d_{min} (m)	5
Ma_{br}, Ma_{dr} (N)	150.0×10^3	$w_{a,min}$ (m/s^2)	−0.05
P_{br}, P_{dr} (W)	1.584×10^6	$w_{a,max}$ (m/s^2)	0.15
a^e (m/s^2)	1.25	$w_{p,\,min}$ (m)	−3.5
a^s (m/s^2)	1	$w_{p,\,max}$ (m)	0

References

1. ERRAC. Rail 2050 Vision. 2017. Available online: https://errac.org/wp-content/uploads/2019/03/122017_ERRAC-RAIL-2050.pdf (accessed on 30 September 2022).
2. Mitchell, I.; Goddard, E.; Montes, F.; Stanley, P.; Muttram, R.; Coenraad, W.; Poré, J.; Andrews, S.; Lochman, L. ERTMS Level 4, Train Convoys or Virtual Coupling. IRSE News. 2016. Available online: https://webinfo.uk/webdocssl/irse-kbase/ref-viewer.aspx?Refno=1882928268&document=ITC%20Report%2039%20Train%20convoys%20and%20virtual%20coupling.pdf (accessed on 7 October 2022).
3. Quaglietta, E.; Wang, M.; Goverde, R. A multi-state train-following model for the analysis of virtual coupling railway operations. *J. Rail Transp. Plan. Manag.* **2020**, *15*, 100195. [CrossRef]
4. X2Rail-3 Grant agreement ID: 826141. Advanced Signalling, Automation and Communication System (IP2 and IP5)—Prototyping the future by means of capacity increase, autonomy and flexible communication. *Cordis database.* Available online: https://cordis.europa.eu/project/id/826141 (accessed on 7 October 2022).
5. Schumann, T. Increase of capacity on the shinkansen high-speed line using virtual coupling. *Int. J. Transp. Dev. Integr.* **2017**, *1*, 666–676. [CrossRef]
6. Zheng, Y.; Li, S.E.; Li, K.; Borrelli, F.; Hedrick, J.K. Distributed Model Predictive Control for Heterogeneous Vehicle Platoons Under Unidirectional Topologies. *IEEE Trans. Control. Syst. Technol.* **2016**, *25*, 899–910. [CrossRef]
7. Guanetti, J.; Kim, Y.; Borrelli, F. Control of connected and automated vehicles: State of the art and future challenges. *Annu. Rev. Control.* **2018**, *45*, 18–40. [CrossRef]
8. Lan, J.; Zhao, D. Min-Max Model Predictive Vehicle Platooning With Communication Delay. *IEEE Trans. Veh. Technol.* **2020**, *69*, 12570–12584. [CrossRef]

9. Bemporad, A.; Morari, M. Robust model predictive control: A survey. In *Robustness in Identification and Control*; Springer: London, UK, 2007; pp. 207–226. [CrossRef]
10. Di Meo, C.; Di Vaio, M.; Flammini, F.; Nardone, R.; Santini, S.; Vittorini, V. ERTMS/ETCS Virtual Coupling: Proof of Concept and Numerical Analysis. *IEEE Trans. Intell. Transp. Syst.* **2019**, *21*, 2545–2556. [CrossRef]
11. Liu, Y.; Liu, R.; Wei, C.; Xun, J.; Tang, T. Distributed Model Predictive Control Strategy for Constrained High-Speed Virtually Coupled Train Set. *IEEE Trans. Veh. Technol.* **2021**, *71*, 171–183. [CrossRef]
12. Liu, Y.; Zhou, Y.; Su, S.; Xun, J.; Tang, T. An analytical optimal control approach for virtually coupled high-speed trains with local and string stability. *Transp. Res. Part C Emerg. Technol.* **2021**, *125*, 102886. [CrossRef]
13. Luo, X.; Liu, H.; Zhang, L.; Xun, J. A Model Predictive Control Based Inter-Station Driving Strategy for Virtual Coupling Trains in Railway System. In Proceedings of the 2021 IEEE International Intelligent Transportation Systems Conference (ITSC), Indianapolis, IN, USA, 19–22 September 2021; pp. 3927–3932. [CrossRef]
14. Luo, X.; Tang, T.; Liu, H.; Zhang, L.; Li, K. An Adaptive Model Predictive Control System for Virtual Coupling in Metros. *Actuators* **2021**, *10*, 178. [CrossRef]
15. Zhao, H.; Dai, X.; Zhang, Q.; Ding, J. Robust Event-Triggered Model Predictive Control for Multiple High-Speed Trains With Switching Topologies. *IEEE Trans. Veh. Technol.* **2020**, *69*, 4700–4710. [CrossRef]
16. Park, J.; Lee, B.-H.; Eun, Y. Virtual Coupling of Railway Vehicles: Gap Reference for Merge and Separation, Robust Control, and Position Measurement. *IEEE Trans. Intell. Transp. Syst.* **2020**, *23*, 1085–1096. [CrossRef]
17. Li, C.; Wang, J.; Shan, J.; Lanzon, A.; Petersen, I.R. Robust Cooperative Control of Networked Train Platoons: A Negative-Imaginary Systems' Perspective. *IEEE Trans. Control. Netw. Syst.* **2021**, *8*, 1743–1753. [CrossRef]
18. Liu, Y.; Zhou, Y.; Su, S.; Xun, J.; Tang, T. Control strategy for stable formation of high-speed virtually coupled trains with disturbances and delays. *Comput. Civ. Infrastruct. Eng.* **2022**, 1–19. [CrossRef]
19. Felez, J.; Kim, Y.; Borrelli, F. A Model Predictive Control Approach for Virtual Coupling in Railways. *IEEE Trans. Intell. Transp. Syst.* **2019**, *20*, 2728–2739. [CrossRef]
20. Löfberg, J. Automatic robust convex programming. *Optim. Methods Softw.* **2012**, *27*, 115–129. [CrossRef]
21. Borrelli, F.; Bemporad, A.; Morari, M. *Predictive Control for Linear and Hybrid Systems*, 1st ed.; Cambridge University Press: Cambridge, UK, 2017. [CrossRef]
22. Turri, V.; Kim, Y.; Guanetti, J.; Johansson, K.H.; Borrelli, F. A model predictive controller for non-cooperative eco-platooning. In Proceedings of the 2017 American Control Conference (ACC), Seattle, WA, USA, 24–26 May 2017; pp. 2309–2314. [CrossRef]
23. Löfberg, J. YALMIP: A toolbox for modeling and optimization in MATLAB. In Proceedings of the CACSD Conference, New Orleans, LA, USA, 2–4 September 2004; pp. 284–289.
24. IEEE. IEEE 1474.1-2004 Standard for Communications-Based Train Control (CBTC). Performance and Functional Requirements. 2004. Available online: https://standards.ieee.org/ieee/1474.1/3552/ (accessed on 7 October 2022).

Article

Time-Optimal Current Control of Synchronous Motor Drives

Václav Šmídl *, Antonín Glac and Zdeněk Peroutka

Research and Innovation Centre for Electrical Engineering (RICE), University of West Bohemia,
30100 Plzeň, Czech Republic
* Correspondence: vsmidl@fel.zcu.cz

Abstract: We are concerned with the problem of fast and accurate tracking of currents in the general synchronous drive. The problem becomes complicated with decreasing available voltage, which is common in high-speed and field weakening regimes. The existing time-optimal controllers rely on a simplified model, ignoring stator resistance and differences in inductances. We derive a solution for the general model considering all parameters and show how the parameters affect the current trajectory. One simplifying assumption had to be made, but we show in simulation that it has a negligible impact on accuracy. The simplification allowed for the design of a feed-forward controller that has a low computational cost and can be easily implemented in realtime. We provide experimental validation of the controller on the developed IPMSM drive prototype of the rated power of 4.5 kW using conventional industrial DSP. The controller is compared to conventional PI and deadbeat solutions, demonstrating that the time-optimal controller can reach the required setpoint four times faster than the competitors at the field weakening regime of the drive. The proposed feed-forward control can be seen as a universal building block that can be combined with existing feedback controllers and observers and thus incorporated into existing control solutions.

Keywords: interior permanent magnet synchronous motor (IPMSM); maximum torque per ampere (MTPA); maximum torque per current (MTPC); dead beat control, predictive control

Citation: Šmídl, V.; Glac, A.; Peroutka, Z. Time-Optimal Current Control of Synchronous Motor Drives. *Actuators* **2023**, *12*, 15. https://doi.org/10.3390/act12010015

Academic Editor: Shuxiang Dong

Received: 30 November 2022
Revised: 16 December 2022
Accepted: 21 December 2022
Published: 29 December 2022

1. Introduction

Control of synchronous motor drives is traditionally decomposed into nested loops where the current loop is the fastest of them. A prominent example is the conventional field-oriented control of using a cascade of PID controllers [1]. Since the time constants of the current loops are relatively short, the optimal operating points of the drive are usually given as inputs to the current control loop [2]. The optimal current setpoints are computed for steady-state operations, using well-known results such as the maximum torque per ampere MTPA curve [3]. Even in the calculation of the MTPA, the stator resistance of the drive is often neglected to achieve simpler solutions. This restriction has been recently relaxed by designing optimal steady-state references for the full electric model of a synchronous motor drive [4].

Optimization of the steady state operation is typically considered for permanent operation of the drive, with application to pumps or ventilation, hence the most common concern of optimization is efficiency [4,5]. However, this optimization is incomplete for highly dynamic operations such as robotic actuators or manipulators, where high precision of set-point tracking is required. In such applications, the set-point trajectory is often known in advance, allowing optimization of the feed-forward part of the controller. In this paper, we focus on feed-forward optimization of the current loop. All other aspects of the control, such as set-point design [4], feed-back part of the control [6], and disturbance rejection [7,8] will be briefly commented at the end of the paper.

Computation of the optimal feed-forward trajectories for transients can be achieved numerically, using model predictive control [9]. However, even many predictive control solutions often use steady-state solutions [10,11] or decomposition into speed and current loop [12,13]. While MPC can also be utilized on a higher level with beneficial results [14,15], its application to demanding current control transients is not successful. The reason is the hard constraint on the input voltage. Due to this constraint, a very long prediction horizon is required to obtain the optimal current trajectory with a completely different trajectory for shorter horizons [16]. The results of long-horizon optimization are counter-intuitive, since the optimal current profile is increasing the tracking error at the beginning of the transient. The reason is that it will reach the set-point in a shorter time than if it tried to decrease the error. This solution can be obtained numerically with very high computational cost. While it is possible to transform the numerical solution to explicit model predictive control [12,17], any change in motor parameters requires recomputing of the expensive solution.

In this paper, we address this problem using time-optimal control (TOC) methods [18]. Minimizing the time to convergence can be treated using general-purpose methods [19]; however, some problems allow for an analytical solution. Specifically, we will be formalizing the current control problem in continuous time, where it can be addressed using Pontryagin's principle of maxima. This technique has been applied to the current control of PMSM drive problem, which has been studied in [20,21] using highly simplifying assumptions, such as neglected resistance. The solution of the full model problem is conceptually known [22,23]; however, its evaluation for the full model is non-trivial. It can be found, e.g., by dynamic programming [24], which is computationally demanding. An extension considering resistance has been presented in [25]; however, without considering different inductances in the direct and quadrature axis. In essence, the TOC serves to generate current setpoints for low level controllers such as deadbeat [2,26].

The contributions of our paper are as follows:

1. We review existing time-optimal current control methods.
2. We derive the explicit formula for the general case of a synchronous motor drive model, including stator resistance and different d- and q-axis inductances.
3. We provide simplification of the exact formula that allows real-time evaluation of the solution without a significant increase in computational complexity in comparison to previous approaches.
4. We demonstrate the advantage of the new proposed general formula over the previous solutions in torque-controlled IPMSM drive in simulations with various parameters.
5. The solution was implemented in real-time digital signal processor and experimentally validated on a drive prototype of the rated power of 4.5 kW.

The paper is organized as follows. The review of previous approaches is summarized in Section 2. The general formula of time-optimal control is derived in Section 3. A sensitivity study of the proposed control to parameters of the drive is performed in Section 4 using simulation. The experimental comparison with a deadbeat controller and the PI controller is reported in Section 5.

2. Review of Existing Approaches

2.1. Mathematical Model of Synchronous Motor Drive

Mathematical model of a synchronous motor drive in the dq reference frame linked to a rotor flux linkage vector is generally described by stator flux dynamics [4]

$$\dot{\psi}_s(t) = -R_s i_s(t) - \omega(t) J \psi_s(t) + u_s(t), \tag{1}$$

where ψ_s denotes the vector of stator flux in the dq reference frame, $\omega(t)$ is the electrical rotor speed, R_s is the stator winding resistance, J is the rotation matrix $J = [0, -1; 1, 0]$ and $u_s(t)$ is the stator voltage vector in the dq reference frame. We will assume that the flux can be (at least locally) approximated by linear formula

$$\psi_s(t) = L_s i_s(t) + \psi_{\text{pm}}, \tag{2}$$

where we assume a diagonal matrix of inductances $L_s = [L_{s,d}, 0; 0, L_{s,q}]$ and flux excited by the permanent magnets on the rotor $\boldsymbol{\psi}_{\mathrm{pm}} = [\psi_{\mathrm{pm},d}; \psi_{\mathrm{pm},q}]$. In this work, we consider $\boldsymbol{\psi}_{\mathrm{pm}}$ to be a general constant vector, which can be specialized for particular drives by setting its elements to zero [4]. For example, $\psi_{\mathrm{pm},q} = 0$ for PMSM and PM-enhanced RSM, $\psi_{\mathrm{pm},d} = 0$ for PM-assisted RSM, and $\boldsymbol{\psi}_{\mathrm{pm}} = [0; 0]$ for RSM drive. We will derive all equations for the general synchronous motor model.

Since we are concerned with drives that have a much faster current response than mechanical dynamics, we will consider the rotor speed ω to be constant during the transient. The state variable is thus the vector of fluxes $x = \boldsymbol{\psi} = [\psi_{s,d}; \psi_{s,q}]$, with the control variable $u = u_s$ and with dynamics

$$\dot{x}(t) = Ax(t) + u(t) + q, \tag{3}$$

where

$$A = -R_s L_s^{-1} - \omega J, \qquad\qquad q = R_s L_s^{-1} \boldsymbol{\psi}_{\mathrm{pm}}.$$

The trajectory of the system is thus

$$x(t) = e^{tA}\left(x(0) + \int_0^t e^{-sA}(u(s) + q)ds \right). \tag{4}$$

2.2. Time-Optimal Control (TOC)

Time-optimal control is a special case of optimal control where the aim is to reach the desired state x_{des} from the original state x_0 in the minimum possible time τ. The well-known Pontryagin principle of maxima [18] states necessary conditions for the optimal trajectory x^* and optimal control u^* via two dynamic equations

$$\dot{x}^*(t) = \frac{\partial H}{\partial p}(t) = Ax^*(t) + u^*(t) + q(t),$$

$$\dot{p}(t) = -\frac{\partial H}{\partial x}(t) = -A^T p(t),$$

and two boundary conditions $x(0) = x_0$ and $x(\tau) = x_{\mathrm{des}}$. Here, $H(p(t), x(t), u(t)) = 1 + p(t)^T(Ax(t) + u(t) + q)$ is the Hamiltonian of the system, and p is known as the costate (adjoint) variable satisfying

$$p(t) = e^{-tA^T} p_0. \tag{5}$$

The optimal control $u(t)$ has to satisfy for almost every $t \in [0, \tau]$ the maximum principle

$$H(p(t), x^*(t), u^*(t)) = \max(H(p(t), x^*(t), u(t)))$$
$$\text{s.t. } ||u(t)|| \leq \overline{U}. \tag{6}$$

Under the circular voltage constraint (6), the optimal control has the form

$$u(t) = \overline{U}\frac{p(t)}{||p(t)||}. \tag{7}$$

Substituting (7) into (4), we obtain

$$x(t) = e^{tA}\left(x_0 + \int_0^t \left[\overline{U}\frac{e^{-sA}e^{-sA^T}p_0}{\sqrt{p_0 e^{-sA}e^{-sA^T}p_0}} + e^{-sA}q \right] ds \right). \tag{8}$$

Solving (8) together with constraint $x(\tau) = x_{\mathrm{des}}$ for τ and p_0 yields the required solution. Note that we may normalize the costate by setting $||p_0|| = 1$. The complexity of the solution depends highly on the model matrix A.

2.3. TOC with Neglected Stator Resistance

To our best knowledge, all previous approaches such as [20,21,24], were designed with neglected stator resistance $R_s \approx 0$. It is understandable because this assumption is very convenient and allows us to greatly simplify the problem, since $A = -\omega J$ and $q = 0$. Then,

$$e^{tA} = \begin{bmatrix} \cos \omega t & \sin \omega t \\ -\sin \omega t & \cos \omega t \end{bmatrix}, \quad e^{-sA}e^{-sA^T} = I,$$

Using assumption $||p_0|| = 1$, Equation (8) becomes

$$x(t) = e^{tA}\left(x_0 + \overline{U}p_0 t\right). \tag{9}$$

Substituting $t = \tau$ and using the terminal condition $x(\tau) = x_{\text{des}}$ in (9) yields $p_0 = \frac{1}{\overline{U}\tau}\left(e^{-\tau A}x_{\text{des}} - x_0\right)$. Since p_0 has to satisfy the normalization condition $||p_0|| = 1$, we obtain an implicit equation:

$$||e^{-\tau A}x_{\text{des}} - x_0||^2 = (\overline{U}\tau)^2, \tag{10}$$

where the only free variable is τ. Finding the solution of (10) can be done using, e.g., the bisection method.

3. Time-Optimal Flux Control Considering Stator Resistance

Since the assumption of negligible resistance is not valid in many cases, we now use the study solution of the time-optimal control (7) for more general cases. First, we will analyze the case of equal inductances $L_{s,d} \approx L_{s,q}$, which is a reasonable assumption for surface mounted PMSM. The more general case is studied subsequently.

3.1. TOC with Equal Stator Inductances

Under the assumption of equal stator inductances in the d and q axes, $L_s = L_{s,d} = L_{s,q}$ the system (3) has the form

$$A = -\rho I - \omega J, \quad q = \rho \psi_{\text{pm}},$$

where $\rho = R_s/L_s$, yielding

$$e^{tA} = e^{-\rho t}\begin{bmatrix} \cos \omega t & -\sin \omega t \\ \sin \omega t & \cos \omega t \end{bmatrix}, \quad e^{-sA}e^{-sA^T} = e^{2\rho t}I, \tag{11}$$

then, Equation (8) becomes

$$x(t) = e^{tA}\left(x_0 + \int_0^t \left[e^{\rho s}\overline{U}p_0 + e^{-sA}q\right]ds\right) \tag{12}$$

$$= e^{tA}\left(x_0 + A^{-1}(I - e^{-tA})q + \frac{\overline{U}}{\rho}(e^{\rho t} - 1)p_0\right). \tag{13}$$

This yields an explicit form for the initial condition

$$p_0 = \frac{\rho}{\overline{U}(e^{\rho t} - 1)}\left(e^{-tA}x(t) - x_0 - A^{-1}(I - e^{-tA})q\right) \tag{14}$$

which once again has to be normalized and together with $x(\tau) = x_{\text{des}}$ yield

$$||e^{-tA}x_{\text{des}} - x_0 - A^{-1}(I - e^{-At})q||^2 = \left(\frac{\overline{U}}{\rho}(e^{\rho t} - 1)\right)^2. \tag{15}$$

When multiplying this formula by $e^{-\rho\tau}$, the vector on the left-hand side can be simplified and the whole formula results in

$$0 = \left(\frac{\overline{U}}{\rho}(1 - e^{-\rho\tau})\right)^2 - \left(-\alpha_1 \cos\omega\tau + \alpha_2 \sin\omega\tau + \beta_1 e^{-\rho\tau}\right)^2$$

$$- \left(-\alpha_2 \cos\omega\tau - \alpha_1 \sin\omega\tau + \beta_2 e^{-\rho\tau}\right)^2. \quad (16)$$

where

$$\alpha_1 = \psi_{\mathrm{des},d} - \frac{\rho\psi_{\mathrm{pm},d} + \omega\psi_{\mathrm{pm},q}}{\rho^2 + \omega^2}\rho, \qquad \beta_1 = \psi_{0,d} - \frac{\omega\psi_{\mathrm{pm},q} + \rho\psi_{\mathrm{pm},d}}{\rho^2 + \omega^2}\rho,$$

$$\alpha_2 = \psi_{\mathrm{des},q} + \frac{\omega\psi_{\mathrm{pm},d} - \rho\psi_{\mathrm{pm},q}}{\rho^2 + \omega^2}\rho, \qquad \beta_2 = \psi_{0,q} + \frac{\omega\psi_{\mathrm{pm},d} - \rho\psi_{\mathrm{pm},q}}{\rho^2 + \omega^2}\rho. \quad (17)$$

This is one equation in the time variable and thus of complexity comparable to that of (10). The only increase is due to the need to evaluate the exponential function. However, accurate approximations for its evaluation are available [27].

3.2. Different d, q Inductances

The assumption of the previous section is no longer valid in drives that have significantly different inductances. Since we use flux as the main variable, different inductances influence only the stator resistance dependent terms of Equation (3) via $R_s L_{s,d}^{-1}$ and $R_s L_{s,q}^{-1}$. For analytical convenience, we denote $\rho = \frac{1}{2}R_s(L_{s,d}^{-1} + L_{s,q}^{-1})$ and $\delta = \frac{1}{2}R_s(L_{s,d}^{-1} - L_{s,q}^{-1})$. Then, (3) has the form

$$A = -\begin{bmatrix} \rho + \delta & 0 \\ 0 & \rho - \delta \end{bmatrix} - \omega J. \quad (18)$$

The form of the system trajectory via the matrix exponential (4) then depends on the relation between δ and ω. Specifically, if $\omega > \delta$, the trajectory is periodic in the base of sin and cos functions just like (11), but with a different frequency $\sqrt{\omega^2 - \delta^2}$. When $\delta > \omega$, the trajectory becomes aperiodic with parameter $\sqrt{\delta^2 - \omega^2}$. When $\omega = \delta$, the trajectory becomes a straight line. To achieve compact notation for all cases, we introduce auxiliary variables:

$$c_1 = \sqrt{|\omega^2 - \delta^2|},$$
$$c_2 = \delta^2 - \omega^2 - \rho^2,$$

and auxiliary functions $\sigma(t)$ and $\mu(t)$, defined as:

$$\begin{cases} \sigma(t) = \dfrac{\sin(c_1 t)}{c_1} & \mu(t) = \cos(c_1 t) & \text{for } |\delta| < |\omega|, \\[2mm] \sigma(t) = \dfrac{\sinh(c_1 t)}{c_1} & \mu(t) = \cosh(c_1 t) & \text{for } |\delta| > |\omega|, \\[2mm] \sigma(t) = t & \mu(t) = 1 & \text{for } |\delta| = |\omega|. \end{cases} \quad (19)$$

Note that the last equation is a limit of both former cases for $c_1 \to 0$.

Using auxiliary functions (19), the system dynamics can be written as

$$e^{tA} = e^{-\rho t}\begin{bmatrix} \mu(t) - \delta\sigma(t) & \omega\sigma(t) \\ -\omega\sigma(t) & \mu(t) + \delta\sigma(t) \end{bmatrix}. \quad (20)$$

In this case, the norm of the costate, $\|p_0\|$, required in (7) is rather complex and analytical expression of (5) becomes intractable. Therefore, we will proceed with the same normalization as in the case of PMSM, which is now an approximation:

$$e^{-sA}e^{-sA^T} \approx e^{2\rho t}, \tag{21}$$

validity of this approximation will be tested in simulation.

Under this approximation, the solution of the costate equation becomes formally equal to (14), giving an implicit equation very similar to (16):

$$0 = \left(\frac{\overline{U}}{\rho}(1 - e^{-\rho\tau})c_2\right)^2 - \left(\alpha_1\mu(\tau) + \alpha_2\sigma(\tau) + \beta_1 e^{-\rho\tau}\right)^2$$
$$- \left(\alpha_3\mu(\tau) + \alpha_4\sigma(\tau) + \beta_2 e^{-\rho\tau}\right)^2. \tag{22}$$

where the only difference is in different constants

$$\begin{aligned}
\alpha_1 &= c_2\psi_{\text{des},d} - \delta q_1 + \omega q_2 + q_1\rho, \\
\alpha_2 &= q_1\omega^2 - q_1\delta^2 + \delta c_2\psi_{\text{des},d} + \delta q_1\rho - \omega c_2\psi_{\text{des},q} - \omega q_2\rho, \\
\alpha_3 &= c_2\psi_{\text{des},q} + \delta q_2 - \omega q_1 + q_2\rho, \\
\alpha_4 &= q_2\omega^2 - q_2\delta^2 - \delta c_2\psi_{\text{des},q} - \delta q_2\rho + \omega c_2\psi_{\text{des},d} + \omega q_1\rho, \\
\beta_1 &= -\omega q_2 - c_2\psi_{0,d} + \delta q_1 - q_1\rho, \\
\beta_2 &= -\delta q_2 - c_2\psi_{0,q} + \omega q_1 - q_2\rho.
\end{aligned} \tag{23}$$

Note that since these constants are independent of the time, they can be computed only once before the bisection method.

When the time of the transient τ is known, it is substituted to (14) to obtain p_0, which is then substituted to (7) to obtain $u^*(0)$:

$$u_d^* = \frac{\rho}{(1 - e^{-\rho\tau})c_2}\left(\alpha_1\mu(\tau) + \alpha_2\sigma(\tau) + \beta_1 e^{-\rho\tau}\right), \tag{24}$$

$$u_q^* = \frac{\rho}{(1 - e^{-\rho\tau})c_2}\left(\alpha_3\mu(\tau) + \alpha_4\sigma(\tau) + \beta_2 e^{-\rho\tau}\right). \tag{25}$$

3.3. Implementation Details

3.3.1. Time Discretization

The proposed controller is designed in continuous time; however, its implementation will be done in discrete time. Let us denote the time instant at which we will apply the controller by time index u_t. The time required to compute the solution prevents the use of the measurement at the same time. Hence, we assume that we have only one-step delayed measurement, i_{t-1}. Therefore, we have to perform delay compensation, i.e., the initial state x_0 is calculated using one-step ahead prediction (2). This value and the requested state x_{req} are substituted into the implicit Equation (16). The solution of this equation using the bisection method is a scalar value τ^*, which is substituted into (24) and (25), which is the time-optimal control law.

3.3.2. Multiple Extremes in Bisection

To use the bisection method, we need to define the minimum and maximum of the search interval such that the function in these points have opposite signs. A suitable choice of the lower bound is zero, where the implicit function (22) is always positive. The upper bound is more problematic. One possible choice is setting the upper limit to the maximum expected time of a transient (e.g., 512 sampling periods, Δt). This setting becomes problematic at a very high speed regime in the field weakening operation. At this region, the current trajectory starts circulating and may cross the requested value multiple times in a short window. The implicit function then has multiple roots, see Figure 1, and we need to select the one closest to zero. An optimal solution would be to derive an analytic upper bound on it. However, in the experimental evaluation, we use a simple heuristic solution. Specifically, we choose $t_{max} = 256\Delta t$, but the first evaluation is not made in the middle of the interval but at point $10\Delta t$. All subsequent evaluations are done using the standard bisection algorithm. We found that 20 steps of the bisection algorithm for solving Equation (22) are sufficient.

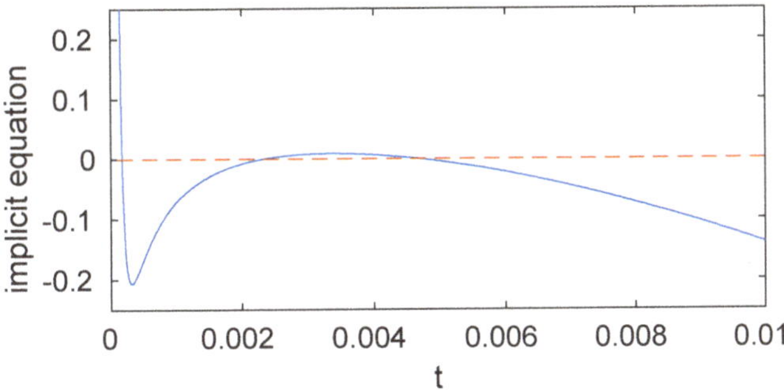

Figure 1. An example of multiple roots in the implicit Equation (22).

3.3.3. Small Step Solution

Another problem arises when the trajectory becomes too close to the requested value and $\tau < \Delta t$. In such a case, we compute the reference value for the PWM using the conventional deadbeat solution, which arises from the discrete-time version of model (3)

$$\frac{x_{t+1} - x_t}{\Delta t} = Ax_t + u_t + q,$$

where it is assumed that the requested value x^* can be reached in one step, yielding

$$u_t^{DB} = \frac{x^* - x_t}{\Delta t} - Ax_t - q. \tag{26}$$

When the state x_t is close to the requested value, the deadbeat solution is within the voltage limit. However, it becomes suboptimal when the amplitude of the solution $|u_t^{DB}|$ is higher than $\overline{U}$. This motivates the proposed algorithm for evaluation of the current control u_t^{opt}, Algorithm 1.

Algorithm 1 Time-optimal current control algorithm.

Input: measured current vector i_{t-1}, requested current i_{des}

1.   transform measured current to flux $x_{t-1} = \psi_{t-1}$, and $x_{des} = \psi_{des}$ using (2),
2.   evaluate delay compensation x_t using Euler approximation of (3),
3.   compute the deadbeat solution u_t^{DB} using (26),
4.   **if** $|u_t^{DB}| < \overline{U}$,
        assign $u_t^{opt} = u_t^{DB}$
    else
        find τ bisecting (16),
        evaluate u_t^{opt} using (24) and (25),
5.   compute modulation signal from u_t^{opt}.

The deadbeat solution alone is often used for current control [2]. However, in such a case, it has to be truncated to the voltage limit. Typical solution is

$$
u_t^{\mathrm{DBt}} = \begin{cases} u_t^{\mathrm{DB}} & \text{if } |u_t^{\mathrm{DB}}| < \overline{U}, \\ \frac{\overline{U}}{|u_t^{\mathrm{DB}}|} u_t^{\mathrm{DB}} & \text{otherwise.} \end{cases}
\tag{27}
$$

which we will use for comparison with the proposed TOC. The key difference between DB and TOC is the situation when the available voltage cannot satisfy the requirement in one step (i.e., the "else" part of step 4 in Algorithm 1).

Note that both algorithms are feed-forward controllers, relying on model correctness. Potentially this may lead to a small steady-state error, which can be compensated by an additional feedback controller. However, we have not experienced any steady state error in our experiments.

4. Simulations

In this section, we study the difference between the proposed solution and previous solutions that neglect stator resistances and different inductances on the time-optimal trajectory. First, we use the parameters of our experimental rig and then we study the sensitivity of the solution to parameter variations. All simulations were done in Matlab, without the use of any toolbox, since the simulation is fully determined by elementary arithmetic operations.

4.1. Nominal Parameters

Our experimental rig has the parameters

$$
\begin{aligned}
R_s &= 1.8 \,\Omega, & \psi_{pm,d} &= 0.438 \text{ Wb}, \\
L_{s,d} &= 14.0 \text{ mH}, & \Delta t &= 100 \text{ µs}, \\
L_{s,q} &= 19.3 \text{ mH}, & U_{dc} &= 450 \text{ V}.
\end{aligned}
\tag{28}
$$

We compare three different methods for time-optimal flux control: (i) control with neglected resistance [20,23], (ii) solution with equal inductances, where the inductance is set to mean value of the true machine parameters, $L_s = 0.5(L_{s,d} + L_{s,q})$, and (iii) the proposed method with full model parameters. The methods are compared on a step change of the requested stator current from zero to $i_d^* = -3\text{A}, i_q^* = 14\text{A}$. The current trajectories of different control strategies are displayed in Figure 2.

In two modes, open-loop (left) and closed-loop (right). The open loop trajectory applies the control strategy designed at the origin. In the closed loop, the strategy is recomputed at every sampling period. For comparison, the deadbeat control (27) is also computed and displayed in the right column of Figure 2. Note that all time-optimal control strategies are relatively close to each other. This indicates that the effect of resistance on the trajectory is low. In contrast, the deadbeat controller (27) takes a different path. The difference is in the time to reach the setpoints. While for $\omega = 10$ rad/s all controllers reach the setpoint in 16 sampling periods, for $\omega = 400$ rad/s, all time-optimal controllers reach the setpoint in 46 sampling periods ($46\Delta t$), but the deadbeat reaches the setpoint in $131\Delta t$.

4.2. Sensitivity Study

The difference between all versions of the time-optimal control becomes more obvious when the ratio ρ between the resistance and inductance is higher. This is simulated by setting smaller values of the inductances, with $L_{s,d} > L_{s,q}$, namely $L_{s,d} = 5\text{ mH}, L_{s,q} = 3\text{ mH}$. The resulting current trajectories for all tested algorithms and step change of the requested stator current from zero to $[5, 30]\text{A}$ are displayed in Figure 3.

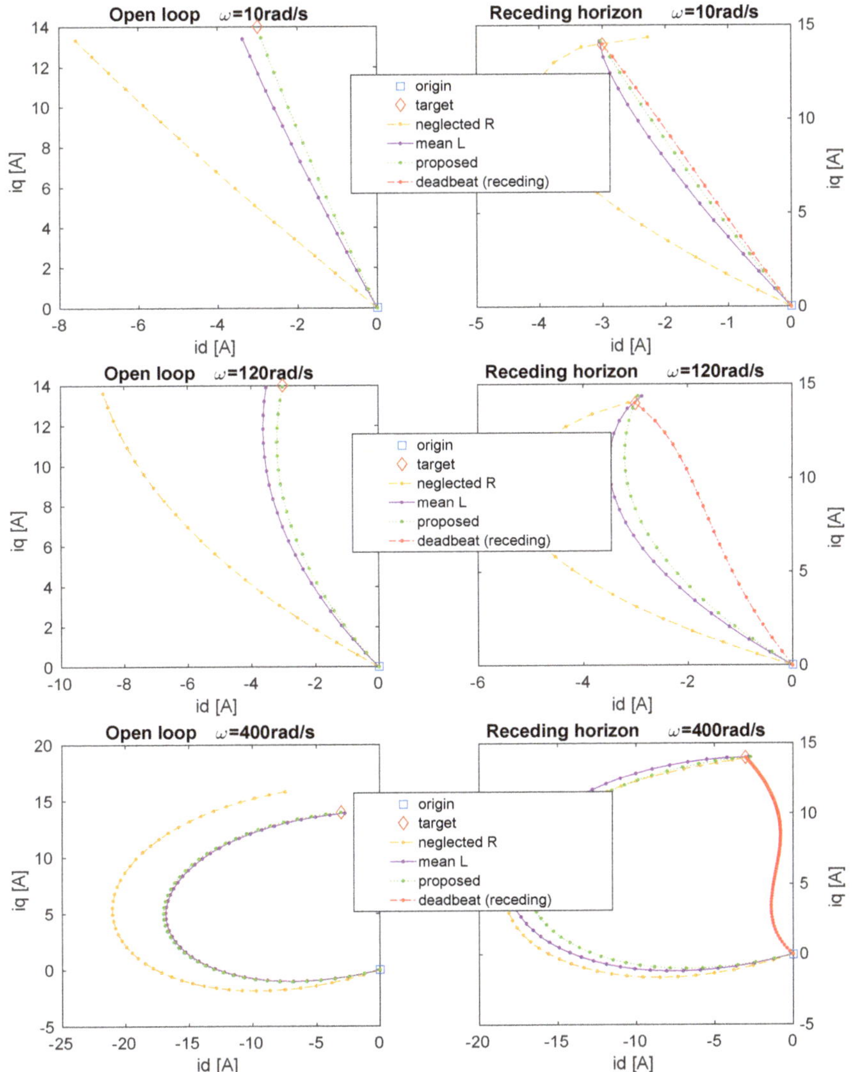

Figure 2. Comparison of studied versions of time-optimal control and deadbeat control on stator current reference step change from zero to $[-3, 14]$A for $\omega = 10$ rad/s (**top** row), $\omega = 120$ rad/s (**middle** row), and $\omega = 400$ rad/s (**bottom** row). The difference of the solutions in the origin is visible in the open-loop strategy (**left**), its impact on the receding horizon reevaluation in each sampling period (**right**), sampling times are denoted by dots on the full lines.

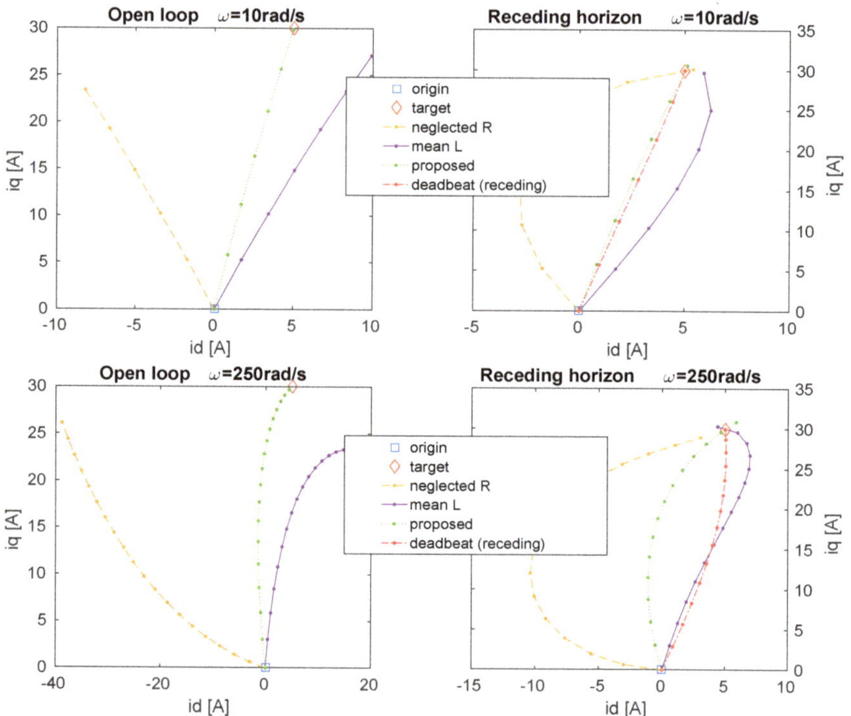

Figure 3. Comparison of studied versions of time-optimal control and deadbeat control on current reference step change from zero to $[5, 30]$A for modified parameters of the system with lower inductance at $\omega = 10$ rad/s (**top** row), and $\omega = 250$ rad/s (**bottom** row). The difference of the solutions in the origin is visible in open-loop strategy (**left**), its impact on the receding horizon reevaluation in each sampling period (**right**), sampling times are denoted by dots on the full lines.

In this case, simplifications of the time-optimal trajectory differ from the proposed solution. This is most obvious on the open-loop results, where only the proposed solution reaches the requested value, but the strategy with neglected resistance goes far to the left (negative i_d) and the one with averaged inductances far to the right (positive i_d). This tendency is corrected in the closed-loop due to recalculations of the trajectory in each time step. However, even with a correctly reached target, the strategy with neglected resistance generates a trajectory with very low i_d currents, and consequently reaches the setpoint in the longest time ($15\Delta t$). The strategy with averaged inductances takes a path with positive i_d currents and reaches the setpoint in $15\Delta t$. The proposed strategy is at low speed almost equivalent to the deadbeat controller, and reaches the setpoint in $14\Delta t$ equally with the deadbeat.

An important conclusion is that even for these parameters, the open-loop trajectory of the proposed control reaches the target setpoint. This implies that the approximation proposed in (21) is sufficiently accurate, in contrast to the simplified solutions.

4.3. Discussion of Results

The results of the controllers significantly differ for different parameters. When the stator resistance is low, it can be neglected in the time-optimal controllers, and the resulting controller yields in high speeds significantly faster transients than the deadbeat controller. However, when the effect of the resistance is higher, the TOC with neglected resistance provides a poor solution that is actually slower than the deadbeat solution and yields higher Joule losses. In such a case, the proposed TOC with resistance yields a better solution. The only computationally slow operation is the sinh function, which needs to be computed for $\omega < \delta$. If computational cost is a concern, it is safe to use the deadbeat controller in this operating regime and switch to TOC only for higher speeds.

5. Experimental Results for IPMSM

A laboratory prototype of the IPMSM drive with the same parameters as in the simulation (28) was used to verify the approach experimentally. The test rig is displayed in Figure 4.

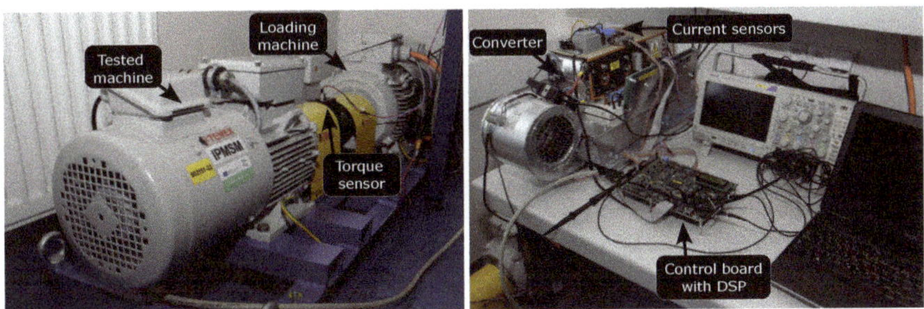

Figure 4. Photo of the test rig with the controlled IPMSM and loading induction machine (**left**), and the controlled converter, current sensors and the control board (**right**).

The rated power of the IPMSM drive is 4.5 kW, rated voltage 400 V, rated current 12.47 A rms, and rated speed 1500 rpm. The IPMSM drive is equipped with 12 bit absolute angular position encoder LARM ARC 405, torque sensor Burster 8661, voltage transducer LEM LV 25-P for converter dc-link voltage measurement, and current transducers LA 55-P for measurement of the stator phase currents. The switching frequency of the voltage-source converter supplying IPMSM is 10 kHz.

The current control strategies will be used to follow setpoint designed by optimal steady-state solution that respects stator resistance and different inductances as described in [4]. The full block diagram is displayed in Figure 5.

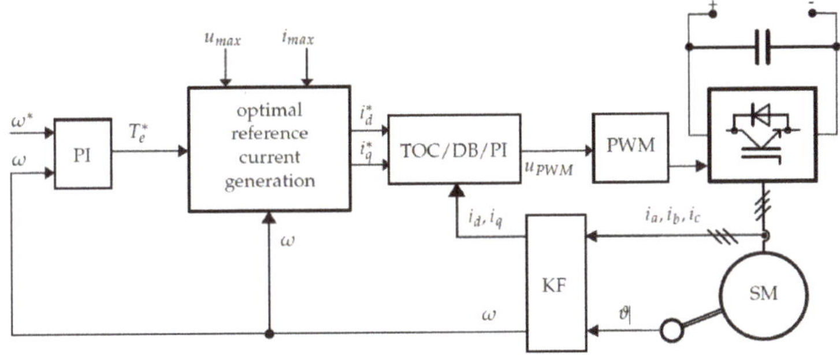

Figure 5. Block diagram of the tested closed loop controller.

This follows the conventional cascade structure, where the speed of the drive is controlled by a PI controller (or any other controller) yielding the requested torque, T_e^*. The requested stator current vector in the dq reference frame with elements i_d^* and i_q^* is calculated from the requested torque by an optimization scheme that minimizes Joule losses in the steady-state [4]. Specifically, the optimal setpoint is calculated as the intersection of the torque curve with the MTPA curve or with the field weakening (FW) curve. The method presented in [4] extends previous approaches (e.g., [2]) by explicit consideration of stator resistance. The effect of the stator resistance on the FW curve is visualized in Figure 6.

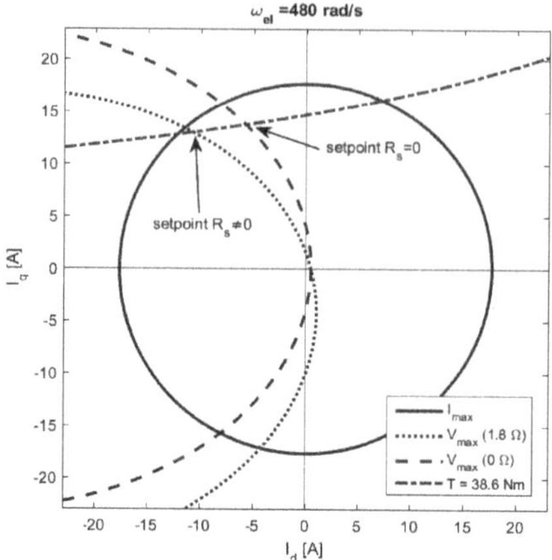

Figure 6. Visualization of optimal setpoints for rotor speed $\omega = 400$ rad/s. The optimal setpoint lies on the intersection of the FW curve with the iso-torque curve (T = 38.6 Nm). FW curve with correct resistance ($R_s = 1.8\ \Omega$) is compared to curve with neglected resistance ($R_s = 0\ \Omega$).

Note that the FW curve that does not consider that the resistance is symmetric around the x-axis, while the curve that considers that it is not. This means that the i_d current in the real drive has to be lower than predicted using the simplified FW curve. This mismatch has been achieved by using "safety" coefficient ζ, [2], $\overline{U} = \zeta U_{dc}/2$ where U_{dc} is the dc-link voltage. In real experiments, the safety coefficient for the simplified approach has to be set to $\zeta = 0.7$ to achieve good performance. The improved solution of [4] allows us to use higher safety coefficient, $\zeta = 0.9$ was found to be sufficient in our case. This improvement is achieved at a higher computational cost due to the need to solve roots of fourth-order polynomials.

In the experiments, the current control loop is either the proposed time-optimal controller (TOC), the standard truncated deadbeat controller (DB) from (27), or the FOC using the conventional PI controllers

$$u_{d,t}^* = k_p e_{d,t} + k_i \sum_{i=1} e_{t-i}, \quad e_{d,t} = (i_{d,t} - i_{d,t}^*),$$

and analogously for the q axis. Both the TOC and the deadbeat controllers are tuning free, but the performance of the PI controller depends on the choice of coefficients k_p and k_i (tuning). We have tuned the PI controller to yield the best overall performance. It is possible to re-tune the controller to obtain better results at one operating point, however, at the cost of deteriorating performance at another. However, the behavior of the control remains very similar for different tuning due to the fact that it is a pure feedback controller. The computational times of the control loop building-blocks are given in Table 1.

Table 1. Computational times of the control algorithm on DSP TMS320F28377S.

Computational Time	μs
data acquisition + KF	6.8
setpoint optimization	41.2
delay compensation + DB	1.5
TOC coefficient evaluation	1.5
TOC bisection for sin/sinh	18/28
modulation signals	5.8
total	74.8/84.8

The execution time of the TOC depends on the rotor speed. For lower speeds, the base functions are the hyperbolic functions (sinh,cosh), which are more expensive, while for higher speeds, the base functions are the sin and cos functions, which are cheaper.

Since the TOC controller is based on the assumption of the perfect state knowledge, we use the Kalman filter (block KF in Figure 5) to reconstruct the speed, position, and stator currents from the position and current measurements.

The performance of the controller is tested on a testing profile displayed in Figure 7.

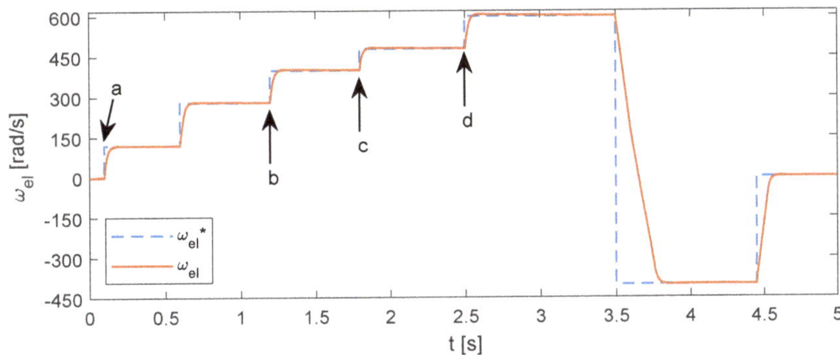

Figure 7. Speed profile of the experimental evaluation. Speed request is displayed in a dashed blue line, the actual speed in a red full line. Letters a, b, c and d indicate moments that are analyzed in detail in Figure 8.

Which is composed of multiple step changes of the requested rotor speed. Since the step change is high, the PI controller of the speed generates a request for the maximum possible torque. The optimization routine provides current setpoints corresponding to the best option at the given operating point. For low-speed operations, it selects points on the maximum torque per ampere curve. In the field weakening regime, it computes optimal currents for field weakening. References on the torque and currents remain almost constant for many sampling periods since the time-constant of the speed is much longer than that of the currents. To visualize the differences between the proposed controllers, we provide details of the current transient at the time of the step change at multiple operating points. These points are at different speeds $\omega = 0, 280, 400, 480$ for acceleration request and $\omega = 600$ rad/s for speed reversal request, see Figure 7 for illustration.

The resulting torque and current trajectories at selected operating points are displayed in Figure 8 in two modes: (i) time trajectories visualize the temporal convergence of the trajectories to setpoints displayed by dashed lines in the right subplots, and (ii) current d-q plane trajectories to the setpoints displayed by crosses in the left subplots. Note that at zero speed, the current trajectories of all controllers are very similar and equally fast. This is understandable since the TOC trajectory is almost a straight line at very low speeds. It starts deviating from a straight line with a higher speed, as can be noted for case (b) in Figure 8.

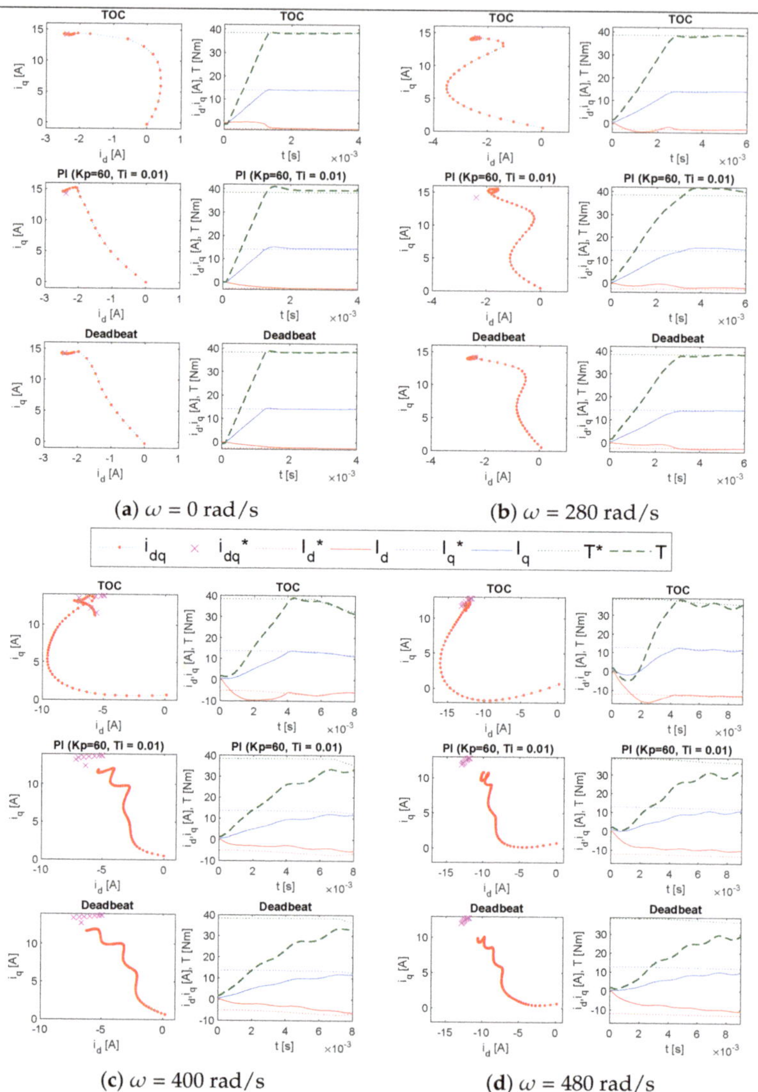

Figure 8. Details of torque and current trajectories at selected moments—(**a**–**d**)—of the experimental testing profile. For each transient, we plot the resulting trajectories for TOC controller (**top** row), PI controller (**middle** row), and deadbeat controller (**bottom** row). Trajectories of the currents in the *d-q* coordinates are displayed in the left subfigures with the requested current values denoted by crosses. Trajectories in the time domain are displayed in the right subfigures. The requested values of the torque and current are marked by dashed lines.

At $\omega = 280$ rad/s, the field weakening limit crosses the MTPA line, and the reference of the *d* current moves to negative values. At this moment, the TOC controller is able to reach the requested torque in 2.5 ms and remain stable at this value. The PI controller reaches the requested torque around the same time, but only due to the fact that it reaches the torque curve at different current values than requested. In an attempt to reach the requested currents, the torque request overshot, and it takes a while to settle it. The DB controller reaches the requested torque later, around 3 ms, but it is also able to remain stable at this value. Even more obvious is the difference at $\omega = 400$ rad/s and $\omega = 480$ rad/s, where the TOC controller is able to reach the requested torque in 4 ms, contrary to the other controllers, that are not able to reach it in twice that time. Note that the TOC controller starts the transient with decreasing i_d current at the cost of decreasing the i_q current. This strategy yields a decrease of the torque for a very short time. However, this short time decrease allows a much faster increase of the torque in the second part of the transient. This demonstrates the ability of the TOC controller to optimize on a very long horizon. Performance of all controllers at the speed reversal moment at $\omega = 600$ rad/s is again comparable.

A summary of the settling times for all transients is provided in Table 2. The settling time was measured as the time after which the measured current is within 5% of the requested value.

Table 2. Comparison of settling time of different controllers for the transients in Figure 8. All times are given in milliseconds.

	i_d			i_q		
Step at Speed	**TOC**	**PI**	**DB**	**TOC**	**PI**	**DB**
$\omega = 0$ rad/s	1.8	1.8	1.8	1.3	4.7	1.3
$\omega = 280$ rad/s	2.8	10.0	3.1	2.7	9.5	3.0
$\omega = 400$ rad/s	4.5	11.2	8.1	4.1	17.6	8.6
$\omega = 480$ rad/s	4.0	22.5	16.0	4.3	18.6	17.5

As expected, the TOC has the most visible benefits at high speeds. While minimization of the speed of the transient is its primary objective, we would like to point out that the current trajectories of the TOC controller also exhibit lower oscillations. This is propagated into the torque trajectory, which is followed by TOC with lower oscillations than those provided by the PI and deadbeat controllers.

6. Conclusions

We have proposed a time-optimal control strategy for current control of the general synchronous motor drive, that considers also stator resistance and differences in stator inductances in the direct and quadrature axis. We have shown that the base functions of the optimal trajectory differ from those of the previous simplified solutions. We have derived a bisection-based optimization algorithm for the evaluation of the more accurate control strategy, which has only a minor increase in computational demands compared to previous simplified solutions.

The proposed strategy was able to reach the requested current in the minimum possible time. The difference to conventional PI or DB controllers is negligible at lower speeds; however, it provides significantly faster transients at higher speeds. On the testing prototype, TOC achieved a four times faster settling time than the PI and DB controllers at rotor speed of 480 rad/s. This is caused by the fact that TOC is a feed-forward controller increasing the tracking error at the beginning of the transient, which is not natural for a feedback controller.

While the controller is designed to minimize the time of the transient, we have found that it also provides more stable tracking of the required torque. The use of this control may thus benefit applications that require fast and accurate torque tracking such as high dynamic servo drives for robotics or manipulators.

Note that the proposed feed-forward strategy can be combined with any feed-back strategy, even with the conventional PI cascade. We believe that the proposed controller can be used as a universal building block complementing existing solutions. Exploring all potential benefits of the approach in combinations with different feedback solutions and different observers is left for future study.

Author Contributions: Methodology, V.Š. and L.A.; software, L.A. and A.G.; validation, A.G.; data curation, A.G.; writing—original draft preparation, V.Š., L.A. and A.G.; writing—review and editing, Z.P. and L.A. All authors have read and agreed to the published version of the manuscript.

Funding: This research was funded by the Ministry of Education, Youth and Sports of the Czech Republic under the project OP VVV Electrical Engineering Technologies with High-Level of Embedded Intelligence, CZ.02.1.01/0.0/0.0/18 069/0009855, project OP VVV Research Center for Informatics, CZ.02.1.01/0.0/0.0/16 019/0000765, and by UWB Student Grant Project no. SGS-2021-021.

Institutional Review Board Statement: Not applicable.

Informed Consent Statement: Not applicable.

Data Availability Statement: Data sharing not applicable.

Conflicts of Interest: The authors declare no conflict of interest. The funders had no role in the design of the study; in the collection, analyses, or interpretation of data; in the writing of the manuscript, or in the decision to publish the results.

References

1. Vas, P. *Vector Control of AC Machines*; Oxford University Press: Oxford, UK, 1990; Volume 22.
2. Preindl, M.; Bolognani, S. Optimal State Reference Computation With Constrained MTPA Criterion for PM Motor Drives. *IEEE Trans. Power Electron.* **2015**, *30*, 4524–4535. [CrossRef]
3. Preindl, M.; Bolognani, S. Model Predictive Direct Torque Control With Finite Control Set for PMSM Drive Systems, Part 1: Maximum Torque Per Ampere Operation. *IEEE Trans. Ind. Inform.* **2013**, *9*, 1912–1921. [CrossRef]
4. Eldeeb, H.; Hackl, C.M.; Horlbeck, L.; Kullick, J. A unified theory for optimal feedforward torque control of anisotropic synchronous machines. *Int. J. Control* **2017**, 1–30. [CrossRef]
5. Glac, A.; Šmídl, V.; Peroutka, Z.; Hackl, C.M. Dependence of IPMSM Motor Efficiency on Parameter Estimates. *Sustainability* **2021**, *13*, 9299. [CrossRef]
6. Jung, J.W.; Leu, V.Q.; Do, T.D.; Kim, E.K.; Choi, H.H. Adaptive PID speed control design for permanent magnet synchronous motor drives. *IEEE Trans. Power Electron.* **2014**, *30*, 900–908. [CrossRef]
7. Sira-Ramírez, H.; Linares-Flores, J.; García-Rodríguez, C.; Contreras-Ordaz, M.A. On the control of the permanent magnet synchronous motor: An active disturbance rejection control approach. *IEEE Trans. Control. Syst. Technol.* **2014**, *22*, 2056–2063. [CrossRef]
8. Humaidi, A.J.; Hameed, A.H. PMLSM position control based on continuous projection adaptive sliding mode controller. *Syst. Sci. Control Eng.* **2018**, *6*, 242–252. [CrossRef]
9. Rodriguez, J.; Kazmierkowski, M.; Espinoza, J.; Zanchetta, P.; Abu-Rub, H.; Young, H.; Rojas, C. State of the Art of Finite Control Set Model Predictive Control in Power Electronics. *IEEE Trans. Ind. Inform.* **2013**, *9*, 1003–1016. [CrossRef]
10. Preindl, M.; Bolognani, S. Model predictive direct torque control with finite control set for pmsm drive systems, Part 2: Field weakening operation. *IEEE Trans. Ind. Inform.* **2013**, *9*, 648–657. [CrossRef]
11. Preindl, M.; Bolognani, S. Model predictive direct speed control with finite control set of PMSM drive systems. *IEEE Trans. Power Electron.* **2013**, *28*, 1007–1015. [CrossRef]
12. Mariethoz, S.; Domahidi, A.; Morari, M. High-bandwidth explicit model predictive control of electrical drives. *IEEE Trans. Ind. Appl.* **2012**, *48*, 1980–1992. [CrossRef]
13. Gao, S.; Wei, Y.; Zhang, D.; Qi, H.; Wei, Y. A Modified Model Predictive Torque Control with Parameters Robustness Improvement for PMSM of Electric Vehicles. *Actuators* **2021**, *10*, 132. [CrossRef]
14. Fehér, M.; Straka, O.; Šmídl, V. Model predictive control of electric drive system with L1-norm. *Eur. J. Control* **2020**, *56*, 242–253. [CrossRef]
15. Li, L.; Zhou, W.; Bi, X.; Sun, X.; Shi, X. Second-Order Model-Based Predictive Control of Dual Three-Phase PMSM Based on Current Loop Operation Optimization. *Actuators* **2022**, *11*, 251. [CrossRef]
16. Šmídl, V.; Mácha, V.; Janouš, Š.; Peroutka, Z. Analysis of cost functions and setpoints for predictive speed control of PMSM drives. In Proceedings of the 2016 18th European Conference on Power Electronics and Applications (EPE'16 ECCE Europe), Karlsruhe, Germany, 5–9 September 2016; pp. 1–8.
17. Besselmann, T.; Lofberg, J.; Morari, M. Explicit MPC for LPV Systems: Stability and Optimality. *IEEE Trans. Autom. Control* **2012**, *57*, 2322–2332. [CrossRef]
18. Clarke, F.H. *Functional Analysis, Calculus of Variations and Optimal Control*; Springer: Berlin/Heidelberg, Germany, 2013.
19. He, F.; Huang, Q. Time-Optimal Trajectory Planning of 6-DOF Manipulator Based on Fuzzy Control. *Actuators* **2022**, *11*, 332. [CrossRef]
20. Bianchi, N.; Bolognani, S.; Zigliotto, M. Time optimal current control for PMSM drives. In Proceedings of the IEEE 2002 28th Annual Conference of the Industrial Electronics Society, IECON 02, Seville, Spain, 5–8 November 2002; Volume 1, pp. 745–750.

21. Konghirun, M.; Xu, L. A fast transient-current control strategy in sensorless vector-controlled permanent magnet synchronous motor. *IEEE Trans. Power Electron.* **2006**, *21*, 1508–1512. [CrossRef]
22. Li, S.; Xu, L. Minimum-time flux-linkage transition for PMSM control with minimum number of inverter switching. In Proceedings of the ICEMS'2001, Fifth International Conference on Electrical Machines and Systems (IEEE Cat. No.01EX501), Shenyang, China, 18–20 August 2001; Volume 2, pp. 1207–1210.
23. Xu, L.; Li, S. A fast response torque control for interior permanent-magnet synchronous motors in extended flux-weakening operation regime. In Proceedings of the IEMDC 2001, IEEE International Electric Machines and Drives Conference (Cat. No.01EX485), Cambridge, MA, USA, 17–20 June 2001; pp. 33–36.
24. Lee, J.S.; Lorenz, R.D.; Valenzuela, M.A. Time-optimal and loss-minimizing deadbeat-direct torque and flux control for interior permanent-magnet synchronous machines. *IEEE Trans. Ind. Appl.* **2014**, *50*, 1880–1890. [CrossRef]
25. Šmídl, V.; Janouš, Š.; Peroutka, Z.; Adam, L. Time-optimal current trajectory for predictive speed control of PMSM drive. In Proceedings of the 2017 IEEE International Symposium on Predictive Control of Electrical Drives and Power Electronics (PRECEDE), Pilsen, Czech Republic, 4–6 September 2017; pp. 83–88.
26. Zhang, Z.; Jing, L.; Wu, X.; Xu, W.; Liu, J.; Lyu, G.; Fan, Z. A deadbeat PI controller with modified feedforward for PMSM under low carrier ratio. *IEEE Access* **2021**, *9*, 63463–63474. [CrossRef]
27. Schraudolph, N.N. A fast, compact approximation of the exponential function. *Neural Comput.* **1999**, *11*, 853–862. [CrossRef] [PubMed]

Article

Implementation and Control of a Wheeled Bipedal Robot Using a Fuzzy Logic Approach

Chun-Fei Hsu [1,*], Bo-Rui Chen [2] and Zi-Ling Lin [1]

[1] Department of Electrical Engineering, Tamkang University, No. 151, Yingzhuan Rd., Tamsui Dist., New Taipei City 25137, Taiwan
[2] Department of Electrical and Computer Engineering, Institute of Electrical and Control Engineering, National Yang Ming Chiao Tung University, Hsinchu 30010, Taiwan
* Correspondence: fei@ee.tku.edu.tw

Abstract: This study designs and implements a wheeled bipedal robot (WBR) that combines the mobility of wheeled robots and the dexterity of legged robots. The designed WBR has extra knee joints to maintain body balance when encountering uneven terrain. Because of the robot's highly nonlinear, dynamic, unstable, and under-actuated nature, an intelligent motion and balance controller (IMBC) based on a fuzzy logic approach is proposed to maintain the balance of the WBR while it is standing and moving on the ground. It should be emphasized that the proposed IMBC system does not require prior knowledge of system dynamics and the controller parameters are tuned using the qualitative aspects of human knowledge. Furthermore, a 32-bit microcontroller that has memory, programmable I/O peripherals, and a processor core is used to implement the IMBC method. Finally, moving and rotating, height-changing, posture-keeping, and "one leg on slope" movement scenarios are tested to demonstrate the feasibility of the proposed IMBC system. The experimental results show that, by using the proposed IMBC system, the WBR can not only balance and move well both on flat ground and in complex terrain but also extend each leg independently to maintain body balance.

Keywords: fuzzy control; balance control; movement control; wheeled bipedal robot

Citation: Hsu, C.-F.; Chen, B.-R.; Lin, Z.-L. Implementation and Control of a Wheeled Bipedal Robot Using a Fuzzy Logic Approach. *Actuators* **2022**, *11*, 357. https://doi.org/10.3390/act11120357

Academic Editor: Matteo Cianchetti

Received: 29 October 2022
Accepted: 30 November 2022
Published: 2 December 2022

Publisher's Note: MDPI stays neutral with regard to jurisdictional claims in published maps and institutional affiliations.

1. Introduction

Automation has become an integral part of all kinds of work, touting the benefits of not only reducing personnel costs but also maintaining the quality of the work done. In the past, industrial robots and human workers worked independently of each other in the workspace; however, robots and human workers will need to interact more directly in the future. Mechanisms for robot motion can be divided into two main domains: wheel-based approaches [1–3] and leg-based approaches [4–6]. Wheeled robots can change their position easily and quickly and their cost is lower; however, they cannot handle uneven terrain. Legged robots can travel almost anywhere but require heavy computational processing to actuate such complex movements. As they can achieve the advantages of both wheel- and leg-based designs, wheel-legged robots have attracted much attention [7–12]. The wheel-legged robot can move fast on flat ground and pass over uneven terrain, thereby improving the robot's adaptability in application scenarios. It also allows the robot's height to be changed to avoid collisions when it needs to pass under low obstacles.

Both academia and industry have conducted extensive research on self-balancing two-wheeled robots [13–15]. These have become an important member of the family of autonomous service robots, thanks to their good movement speed, load capacity, and terrain adaptability, and the concept has been successfully applied to transport robots [16] and hospitality robots [17,18]. Considering the diversity of locations, the wheeled bipedal robot (WBR) [19,20], which has two legs and two wheels (with the wheels installed on each foot), has more degrees of freedom than self-balancing two-wheeled robots [13–18] in the vertical direction. By achieving a higher movement speed and a larger movement range, it is designed to be flexible even in the variable terrain of indoor environments. The most famous WBRs are "Handle" from Boston Dynamics and "Ascento" from ETH Zurich, which can move quickly on flat terrain, go down stairs, and jump through obstacles. However, because there are only two contact points between the robot and the ground, the ability to balance plays a key role in the practical application of WBRs.

Recently, there is increasing interest in studying the design, modeling, and control of WBRs. Traditional control schemes have been employed to solve the balance and height control issues [21–32]. A linear quadratic regulator (LQR) controller was designed to implement stabilization and driving control, and a proportional-integral-derivative (PID) controller was designed to synchronize the two leg motors in order to change the height of the robot [21]. However, these control gains needed to be pre-constructed using a tuning procedure. A cascade PID controller was proposed as the movement control system of a WBR [22]. Although it could achieve satisfactory balancing motion performance, it did not take into account how the WBR could pass over rough terrain. An easy-to-implement LQR-based control system was proposed [23–26]. However, the control performance degrades in the case of WBR operating point changes. An inverse dynamics controller was designed to consider the full dynamics of the WBR without neglecting any nonlinear dynamics terms [27]. However, these system uncertainties are unavoidable in a real WBR. A whole-body control frame was proposed for a WBR so that it could not only accelerate and decelerate along the forward direction and balance dynamically but also withstand disturbances more efficiently [28–32]. However, its control performance was highly dependent on the accuracy of the WBR's dynamic model.

The parameters of the model-based controller in recent studies [21–32] have had to be manually adjusted to ensure effective control when the physical parameters of the WBR change, such as robot height and payload weight. To overcome this shortcoming as well as solve the WBR balance problem, intelligent control schemes have been proposed [33–35]. A cascade PID controller was proposed for the path tracking of a WBR in which a reinforcement learning method automatically obtained optimal gains for the control law in a simulation [33]. Although the obtained controller gains could be directly applied in real-world scenarios without any re-training, the effect of robot height variation on the control response was not considered. A learning-based solution was proposed based on adaptive optimal control, where the controller parameters could be learned directly from the input-state data in the Gazebo robotic simulator [34]. However, it required heavy computational processing. An intelligent adaptive sliding-mode controller was proposed for the trajectory tracking and stability of a WBR, where a fuzzy-basis-function-network was used to approximate the system dynamics online [35]. However, the controller did not consider the effect of WBR height variation on the system response.

Fuzzy control (FC) using linguistic information can model the qualitative aspects of human knowledge and possesses several advantages, such as robustness, being model-free, and being a rule-based algorithm [36–38]. Most of the operations in an FC use error and change-of-error as the fuzzy input variables, where the rules table is constructed in a two-dimensional input space, which cause difficulties in implementation. It can be found that the rule tables have the skew–symmetry property, where the absolute magnitude of the control input is proportional to the distance from its main diagonal line in the normalized input space. To reduce the complexity of the implementation, a single-input fuzzy controller (SFC) was proposed based on the sliding-mode control scheme, where a new fuzzy input variable (called the signed distance) was derived [39–42]. Compared to FC system, the total number of fuzzy rules used in SFC system is greatly reduced, so the control rules can be easy to generate and adjust. However, due to the linguistic expression of the FC and SFC systems, it has not been easy to guarantee the stability and robustness of the closed-loop control systems.

This study proposes a WBR that consists of a body and two legs with wheels where each leg can be independently extended and retracted by driving a motor mounted on the hip of the body. It has the advantages of both leg and wheel movement, with good maneuverability and a high load capacity. The height of the proposed WBR can be adjusted through the realization of a four-bar linkage. Since the four-bar linkage approximates the linear motion of the wheels perpendicular to the ground, the WBR can traverse simple terrain in the same manner as wheeled robots and adjust its posture to the shape of uneven ground.

To overcome the highly dynamic and under-actuated nature of the proposed WBR's behavior, this study proposes an intelligent motion and balance control (IMBC) system that comprises fuzzy movement balancing control (FMBC), fuzzy yaw steering control (FYSC), and fuzzy roll balancing control (FRBC). The FMBC, based on a dual-loop control structure, is designed to cope with the under-actuated nature problem of the WBR and allow the WBR to move at the desired velocity command while maintaining balance. The FYSC is a compensated speed control signal for the WBR that can rotate about its vertical axis, and the FRBC extends each leg independently to change the robot's height and altitude when driving over challenging terrain. Since the proposed IMBC method was designed based on the SFC method, it does not require complex analysis or heavy computational processing. A low-cost 32-bit STM32F446RE microcontroller is used to implement the control method for the WBR. Finally, four experimental scenarios are tested to explore the stability performance of the proposed IMBC system. The experimental results show that the proposed IMBC system is feasible and effective for robot motion on both flat and complex terrain.

The contribution of this study can be summarized as follows: (1) The mechanical structure of the WBR, which has two legs and two wheels (with the wheels installed on each foot), is designed and implemented. (2) An experimental setup is presented for the WBR using a low-cost 32-bit STM32F446RE microcontroller. (3) The IMBC system is shown to be able to maintain the balance of the WBR while standing and moving on the ground. (4) The IMBC system is shown to be able to independently control two leg motors to change the robot's height and attitude and allow the WBR to move over challenging terrain.

The remainder of this study is outlined as follows. In Section 2, the design and dynamic model of the WBR is presented. In Section 3, the structure of the IMBC system is presented. Experimental tests are presented in Section 4, and conclusions are given in Section 5.

2. System Structure and Modeling of the WBR

2.1. Mechanical Design

Figure 1 shows the mechanical structure of the WBR: Figure 1a shows the leg components and Figure 1b shows the kinematic models. The developed WBR, which consists of a body and two legs with wheels, is a promising way to improve the performance of bipedal robots in terms of flexibility, movement speed, and energy efficiency. By adding wheels to the ends of the legs, each leg can be independently extended and retracted by driving a leg motor mounted on the hip of the body. The torsion springs installed in the inner joints (see Figure 1a) are used to counteract the weight of the WBR itself, which can reduce the control effort of the leg motors when driving or standing to improve the overall efficiency of the robot. The total height of the WBR can be adjusted between 26 cm and 33 cm through the realization of a four-bar linkage.

In order to illustrate the relationship between the leg length of the WBR and the angle of the leg motor, part of the right leg is considered as shown in Figure 1b. The length between the pin joint and the inner joint, L_{kR}, can be found according to the Cosine law as follows:

$$L_{kR} = \sqrt{L_1^2 + L_2^2 - 2L_1 L_2 \cos \theta_{kR}} \tag{1}$$

where L_1 is the length between the pin joint and the knee joint, L_2 is the length between the pin joint and the leg joint, and θ_{kR} is the leg motor angle of the right leg. The torsion spring angle θ_{tR} and the length L_R of the right leg of the WBR can be found as follows:

$$\theta_{tR} = \pi - \cos^{-1}\left(\frac{L_1^2 + L_{kR}^2 - L_2^2}{2L_1 L_{kR}}\right) - \cos^{-1}\left(\frac{L_3^2 + L_{kR}^2 - L_1^2}{2L_3 L_{kR}}\right) \tag{2}$$

$$L_R = \sqrt{L_1^2 + L_4^2 - 2L_1 L_4 \cos(\theta_{tR})} \tag{3}$$

where L_3 is the length between the knee joint and the inner joint and L_4 is the length between the inner joint and the wheel motor. The height of the right side of the WBR is given as:

$$h_R = L_R + R \tag{4}$$

where R is the wheel radius. Similarly, the height of the left side of the WBR, h_L, can be obtained using the leg motor angle of the left leg θ_{kL} as follows:

$$h_L = \sqrt{L_1^2 + L_4^2 - 2L_1 L_4 \cos(\theta_{tL})} + R \tag{5}$$

in which

$$\theta_{tL} = \pi - \cos^{-1}\left(\frac{L_1^2 + L_{kL}^2 - L_2^2}{2L_1 L_{kL}}\right) - \cos^{-1}\left(\frac{L_3^2 + L_{kL}^2 - L_1^2}{2L_3 L_{kL}}\right) \tag{6}$$

$$L_{kL} = \sqrt{L_1^2 + L_2^2 - 2L_1 L_2 \cos\theta_{kL}} \tag{7}$$

where θ_{tL} is the torsion spring angle and L_{kL} is the length between the pin joint and the inner joint of the left leg of the WBR, respectively. Thus, the joint angles are adjusted to produce a change in the lengths of both sides of the robot (h_R and h_L) when the two leg motors are driven (θ_{kR} and θ_{kL}).

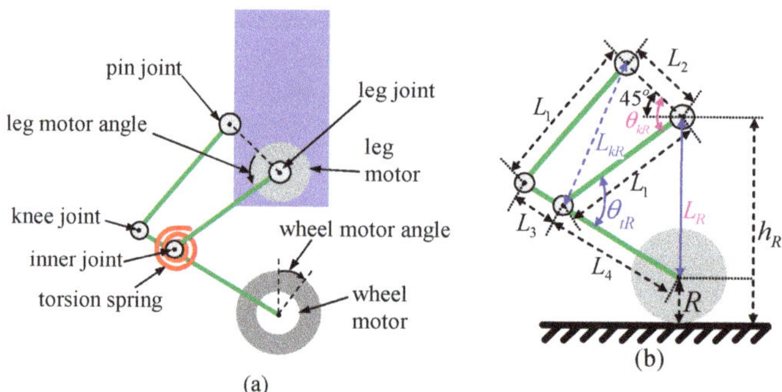

Figure 1. Mechanical structure. (**a**) Leg components. (**b**) Kinematic model of the right leg.

2.2. Hardware and Software

Figure 2 shows the photographs of the WBR: Figure 2a shows the WBR platform and Figure 2b shows the hardware connection diagram. The hardware connection diagram is composed of a STM32F446RE microcontroller, a MPU 6050, two wheel motors with optical encoders, and two leg motors. The MPU 6050, consisting of a 3-axis accelerometer and a 3-axis gyroscope, is located at the center of the robot body, the STM32F446RE microcontroller is placed on the upper body, and the motor drives are located on the lower body. The leg motor is a Dynamixel motor (MX-64) made by ROBOTIS, which can provide enough power for a leg to adjust the height of the robot. The wheel motor is a geared DC motor (GM25-370) made by ChiHai MOTOR, which can provide smooth stabilization and counteract disturbances to the tilt angle of the system. An encoder is installed behind the wheel motor to provide information on the robot's translational velocity and position. The MPU 6050 is used to measure the roll and pitch angles of the robot tilt. The STM32F446RE microcontroller, which is coded by Keil C, is used as a platform for the implementation of control algorithms; it has a Cortex-M4 core, a processor clock frequency of up to 180 MHz, and a rich set of IO ports for connecting to various sensors and motors. The microcontroller, leg motors, wheel motors, and remaining electronic devices are powered by a 3-cell LiPo Battery.

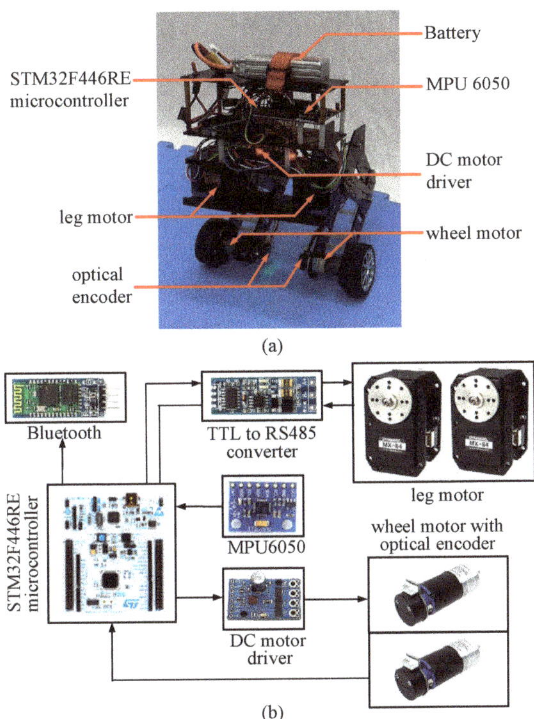

Figure 2. Photographs of the WBR. (**a**) WBR platform. (**b**) Hardware connection diagram.

2.3. Modeling

In order to manipulate the WBR, it is necessary to build a mathematical model that allows the robot and the designed controller to successfully achieve the desired control goals. Figure 3 shows the coordinate systems of the WBR and Appendix A lists all the symbols used and their definitions. The robot's position can be measured as $\theta_w = \frac{1}{2}(\theta_R + \theta_L)$ and the robot's body yaw angle ψ can be given as:

$$\psi = \frac{R}{2W}(\theta_R - \theta_L) \tag{8}$$

where θ_R is the rotary angle of the right wheel motor and θ_L is the rotary angle of the left wheel motor. The dynamic model of the WBR for the pitch axis can be described as [43]:

$$\mathbf{E}\begin{bmatrix} \ddot{\theta}_w \\ \ddot{\theta} \end{bmatrix} + \mathbf{F}\begin{bmatrix} \dot{\theta}_w \\ \dot{\theta} \end{bmatrix} + \mathbf{G}\begin{bmatrix} \theta_w \\ \theta \end{bmatrix} = \mathbf{H}\begin{bmatrix} v_L \\ v_R \end{bmatrix} + \mathbf{D} \tag{9}$$

where $\mathbf{E}$ is the inertia matrix, $\mathbf{F}$ is the matrix of centripetal and Coriolis forces, $\mathbf{G}$ is the gravity matrix, $\mathbf{H}$ is the gain matrix, $\mathbf{D}$ is the lump of uncertainties matrices (including the external disturbances and parameter uncertainties), and v_L and v_R are the input voltages of the left and right wheel motors, respectively. These matrices can be found by considering the limits $\theta \to 0$, $\sin\theta \to \theta$ and $\cos\theta \to 1$ as:

$$\mathbf{E} = \begin{bmatrix} (2m+M)R^2 + 2J_w + 2J_m & MLR - 2J_m \\ MLR - 2J_m & ML^2 + J_\psi + 2J_m \end{bmatrix} \tag{10}$$

$$\mathbf{F} = 2\begin{bmatrix} \dfrac{K_t K_b}{R_m} & -\dfrac{K_t K_b}{R_m} \\ -\dfrac{K_t K_b}{R_m} & \dfrac{K_t K_b}{R_m} \end{bmatrix} \tag{11}$$

$$\mathbf{G} = \begin{bmatrix} 0 & 0 \\ 0 & -MgL \end{bmatrix} \tag{12}$$

$$\mathbf{H} = \begin{bmatrix} \dfrac{K_t}{R_m} & \dfrac{K_t}{R_m} \\ -\dfrac{K_t}{R_m} & -\dfrac{K_t}{R_m} \end{bmatrix} \tag{13}$$

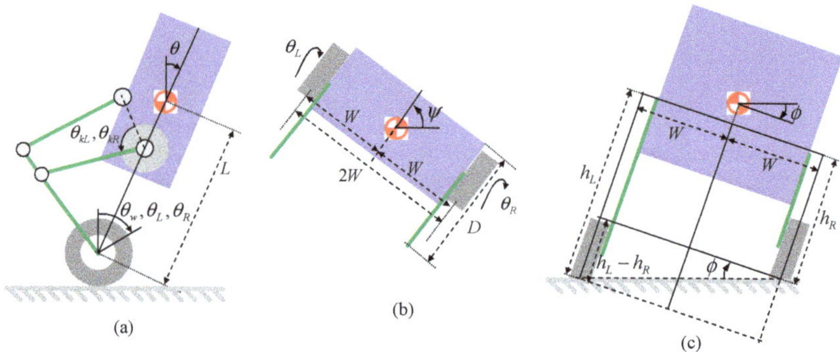

Figure 3. Generalized coordinates of the system. (**a**) Side view. (**b**) Top view. (**c**) Front view.

Equation (9) can be divided into two subsystems: one is the velocity control subsystem and the other is the balancing control subsystem. The velocity control subsystem is used to allow the WBR to move at the desired velocity command, and the balancing control subsystem is used to maintain the WBR's balance. Meanwhile, the dynamic model of the WBR for the yaw axis can be expressed as [43]:

$$\ddot{\psi} = -\frac{J}{I} + \begin{bmatrix} -\dfrac{K}{I} & \dfrac{K}{I} \end{bmatrix} \begin{bmatrix} v_L \\ v_R \end{bmatrix} \tag{14}$$

where $I = 2mW^2 + J_\phi + \frac{2W^2}{R^2}(J_w + J_m)$, $J = \frac{2W^2}{R^2}\frac{K_t K_b}{R_m}$, and $K = \frac{W}{R}\frac{K_t}{R_m}$. The robot body roll angle can be given as:

$$\phi = \tan^{-1}\left(\frac{h_L - h_R}{2W}\right) \tag{15}$$

When the robot's two legs are on the ground. From (9), (14) and (15), it shows the nonlinear and under-actuated nature of the WBR system. Since these robot parameters and the uncertainties matrix are unknown or perturbed in real WBR applications, a model-free control scheme should be designed that not only accelerates and decelerates along the forward direction and balances dynamically but is also robust to uncertainties.

3. Controller Design for the WBR

A dual-loop control structure was proposed to cope with the under-actuated nature problem of a self-balancing two-wheeled robot moving and turning while maintaining the robot's balance [43]. Extending the design ideas, this study proposes an IMBC system for the WBR, as shown in Figure 4. It comprises three controllers: FMBC, FYSC, and FRBC. It is similar to [43], where the FMBC designed using a dual-loop control structure allows the WBR to move at the desired velocity command while maintaining balance and the FYSC can control the WBR to rotate about its vertical axis. Unlike [43], the FRBC is a novel design that independently controls two leg motors to change the robot's height and altitude and allows the WBR to travel over challenging terrain.

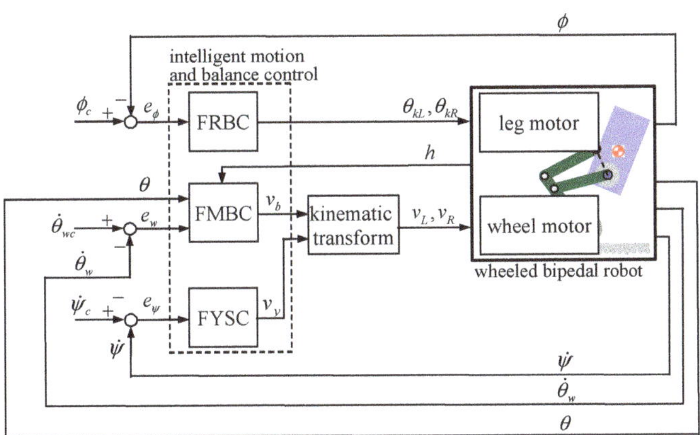

Figure 4. The block diagram of the IMBC system for the WBR.

3.1. FMBC Design

The FMBC should control the WBR to move at the desired velocity command and remain balanced even in the height variation of the WBR. A dual-loop control scheme is utilized, where the outer-loop controller (for the velocity control subsystem) is used to calculate the desired robot pitch angle for tracking velocity commands and the inner-loop controller (for the balancing control subsystem) is applied to keep the body balanced with the desired robot pitch angle. Thus, the advantage of the FMBC scheme lies in its simplicity and robust performance, compared to some other existing schemes [23–32]. According to the velocity command $\dot{\theta}_{wc}$ given by the users, an outer-loop sliding index is designed as [43]:

$$s_w = e_w + \lambda_1 \int_0^t e_w d\tau \tag{16}$$

where $e_w = \dot{\theta}_{wc} - \dot{\theta}_w$ is the translational velocity error and λ_1 is a positive constant. In order to converge the translational velocity error of the robot to zero quickly and have an excellent system response such as a faster rising time and smaller overshoot, each fuzzy rule of the outer-loop controller is given as follows:

$$\text{IF } s_w \text{ is } F_{sw}^i, \text{ THEN } \theta_c \text{ is } F_{\theta c}^i, \text{ for } i = 1, 2, \ldots, 7 \tag{17}$$

where F_{sw}^i are the labels of the fuzzy sets, $F_{\theta c}^i$ are the singleton control actions, and θ_c is the desired robot pitch angle. The fuzzy rules base of the outer-loop controller is given in Table 1, where the fuzzy labels are negative big (NB), negative medium (NM), negative small (NS), zero (ZO), positive small (PS), positive medium (PM), and positive big (PB), as shown in Figure 5a, where the membership functions of the fuzzy sets are given in a triangular form. The fuzzy rules in Table 1 are constructed using the idea that the outer-loop sliding index s_w can quickly reach zero via a time-consuming trial-and-error tuning procedure. The adopted outer-loop sliding index reduces the number of fuzzy rules, thus reducing the complexity of adjustment. It shows that the translational velocity error e_w will converge to zero when the outer-loop sliding index converges to zero. The outer-loop controller allows the WBR to move at the desired velocity command; thus, the velocity control subsystem can be asymptotically stable under the action of the outer-loop controller.

Table 1. The fuzzy rules for the FMBC.

s_w	NB	NM	NS	ZO	PS	PM	PB
$F_{\theta c}^i$	-15.0	-8.5	-5.4	0.0	5.4	8.5	15.0

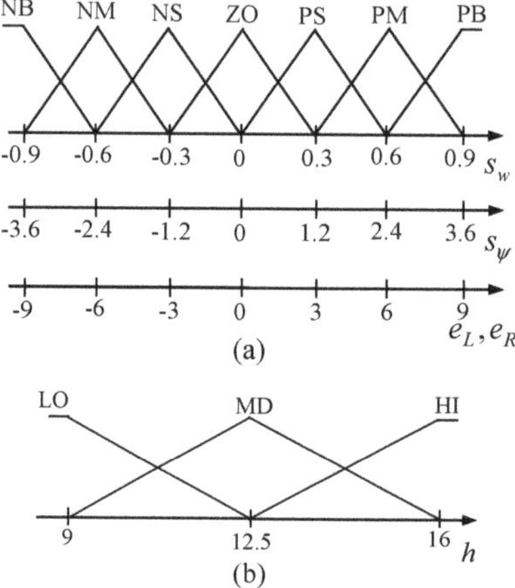

Figure 5. Membership functions. (**a**) FMBC and FRBC; (**b**) FYSC.

To cope with the system uncertainties caused by the changes in robot height, each fuzzy rule of the inner-loop controller is given as follows:

$$\text{IF } h \text{ is } F_h^j, \text{ THEN } v_b \text{ is } \mathbf{k}_j^T \mathbf{e_\theta}, \text{ for } j = 1, 2, \ldots, 3 \tag{18}$$

where the robot height can be measured as $h = \frac{1}{2}(h_R + h_L)$, F_h^j are the labels of the fuzzy sets, $e_\theta = \theta - \theta_c$ is the error of the body pitch angle, $\mathbf{e_\theta} = [e_\theta, \dot{e}_\theta]^T$ is the error vector, and $\mathbf{k}_j$ is the control gains vector of the inner-loop controller. The fuzzy labels used in this study are low (LO), medium (MD), and high (HI), as shown in Figure 5b, where the membership functions of the fuzzy sets are given in a triangular form. It is known that the outer-loop controller obtains the virtual angle command by defining the outer-loop sliding index and the inner-loop controller tracks the virtual angle command by defining the robot angle error. Thus, the balancing control subsystem can be asymptotically stable under the action of the inner-loop controller.

3.2. FYSC Design

FYSC is designed to control the WBR so that it can rotate along its vertical axis. The speed control signals applied to the right or left driving wheels are obtained as follows:

$$v_L = v_b + v_y \tag{19}$$

$$v_R = v_b - v_y \tag{20}$$

where v_b is the FMBC output and v_y is the FYSC output. According to the velocity command $\dot{\psi}_c$ given by the users, a yawing sliding index is designed as [43]:

$$s_\psi = e_\psi + \lambda_2 \int_0^t e_\psi d\tau \tag{21}$$

where the yaw speed error $e_\psi = \dot{\psi}_c - \dot{\psi}$ and λ_2 are positive constants. Each fuzzy rule of the FYSC is given as follows:

$$\text{IF } s_\psi \text{ is } F^i_{s\psi}, \text{ THEN } v_y \text{ is } F^i_{vy}, \text{ for } i = 1, 2, \ldots, 7 \tag{22}$$

where $F^i_{s\psi}$ are the labels of the fuzzy sets as shown in Figure 5a and F^i_{vy} are the singleton control actions. The fuzzy rules base of the FYSC is shown in Table 2, where it is constructed using the idea that the yawing sliding index s_ψ can quickly reach zero via a time-consuming trial-and-error tuning procedure. The adopted yaw sliding index reduces the number of fuzzy rules, thus reducing the complexity of adjustment. It shows that the yaw speed error e_ψ will converge to zero when the yaw sliding index converges to zero. The FYSC allows the WBR to rotate along its vertical axis at the desired velocity command. Thus, it can be asymptotically stable under the action of the FYSC.

Table 2. The fuzzy rules for the FYSC.

s_ψ	NB	NM	NS	ZO	PS	PM	PB
F^i_{vy}	-0.8	-0.3	-0.1	0.0	0.1	0.3	0.8

3.3. FRBC Design

When the ground is rough or the robot's legs are not of equal length, the FRBC is designed to control the WBR and maintain its horizontal posture and height by adjusting the length of its legs. Due to the hardware limitation, the WBR's leg length cannot be extended or shortened indefinitely. In the proposed WBR, the length of its legs can be adjusted between $h_{max} = 161.1$ mm and $h_{min} = 93.7$ mm through the realization of a four-bar linkage. When the leg length of the WBR is within the limitation, the FRBC is designed for posture control, and when the leg length of the WBR is on the limitation, the FRBC is designed for height control. First, considering the robot height is between h_{max} and h_{min}, the FRBC is designed to control the WBR to maintain its horizontal posture. In the posture controller design, a rolling sliding index is designed as:

$$s_\phi = \dot{\phi} + \lambda_3 \phi \tag{23}$$

where λ_3 is a positive constant. Each fuzzy rule of the posture controller is given as follows:

$$\text{IF } s_\phi \text{ is } F^i_{s\phi}, \text{ THEN } \Delta\theta_{kL} \text{ is } F^i_{\phi L} \text{ and } \Delta\theta_{kR} \text{ is } F^i_{\phi R}, \text{ for } i = 1, 2, \ldots, 7 \tag{24}$$

where $F^i_{s\phi}$ are the labels of the fuzzy sets as shown in Figure 5a and F^i_L and F^i_R are the singleton control actions. The fuzzy rules base of the posture controller is shown in Table 3a, where it is constructed using the idea that the rolling sliding index s_ϕ can quickly reach zero via a time-consuming trial-and-error tuning procedure. The adopted rolling sliding index reduces the number of fuzzy rules, thus reducing the complexity of adjustment. It shows that the roll angle of the robot body ϕ will converge to zero when the rolling sliding index converges to zero. Thus, the FRBC allows the WBR to control two leg motors independently so as to maintain its horizontal posture.

Secondly, given that the height of either leg of the robot is at the limitations h_{max} or h_{min}, the FRBC is designed to adjust the length of its legs to maintain its height. In order to ensure that the WBR can be kept at the setting height h_{set}, each fuzzy rule of the height controller can be designed as follows:

$$\text{IF } e_L \text{ is } F^i_{eL}, \text{ THEN } \Delta\theta_{kL} \text{ is } F^i_{hL} \tag{25}$$

$$\text{IF } e_R \text{ is } F^i_{eR}, \text{ THEN } \Delta\theta_{kR} \text{ is } F^i_{hR} \tag{26}$$

where $e_L = h_L - h_{set}$ and $e_R = h_R - h_{set}$ are the height errors, F_{eL}^i and F_{eR}^i are the labels of the fuzzy sets as shown in Figure 5a, and F_{hL}^i and F_{hR}^i are the singleton control actions. The fuzzy rules base is shown in Table 3b, where it is constructed using the idea that the height errors (e_L and e_R) can quickly reach zero by a time-consuming trial-and-error tuning procedure. This allows the WBR to control two leg motors independently in order to change the height of the robot while it is asymptotically stable under the action of FRBC. Thus, the motor angle of the left leg and the right leg are updated as follows:

$$\theta_{kL}(N+1) = \theta_{kL}(N) + \Delta\theta_{kL} \tag{27}$$

$$\theta_{kR}(N+1) = \theta_{kR}(N) + \Delta\theta_{kR} \tag{28}$$

where N denotes the number of iterations.

Table 3. The fuzzy rules for the FRBC.

			(a) posture control				
s_ϕ	**NB**	**NM**	**NS**	**ZO**	**PS**	**PM**	**PB**
$\Delta\theta_{kL}$	−2.0	−1.5	−0.7	0.0	0.7	1.5	2.0
$\Delta\theta_{kR}$	2.0	1.5	0.7	0.0	−0.7	−1.5	−2.0
			(b) height control				
e_L, e_R	**NB**	**NM**	**NS**	**ZO**	**PS**	**PM**	**PB**
$\Delta\theta_{kL}, \Delta\theta_{kR}$	−10	−7.5	−3.0	0.0	3.0	7.5	10

4. Experimental Results

The controller parameters of the IMBC system were selected as $\lambda_1 = 0.01$, $\lambda_2 = 0.125$, $\lambda_3 = 0.5$, $\mathbf{k}_1 = [130, 10]^T$, $\mathbf{k}_2 = [125, 10]^T$, and $\mathbf{k}_3 = [120, 10]^T$, considering the stability requirements and possible operating conditions. Generally, these control parameters require some trial-and-error tuning procedures to determine. To show the effectiveness of the IMBC system, a comparison between the cascade PID control [22] and the proposed IMBC system is presented. Four experimental scenarios (namely moving and rotating, height-changing, posture-keeping, and one leg on slope movement) were tested to explore the stability performance of the controllers. Figure 6 shows four photographs of the experiments, and the experimental results are discussed below.

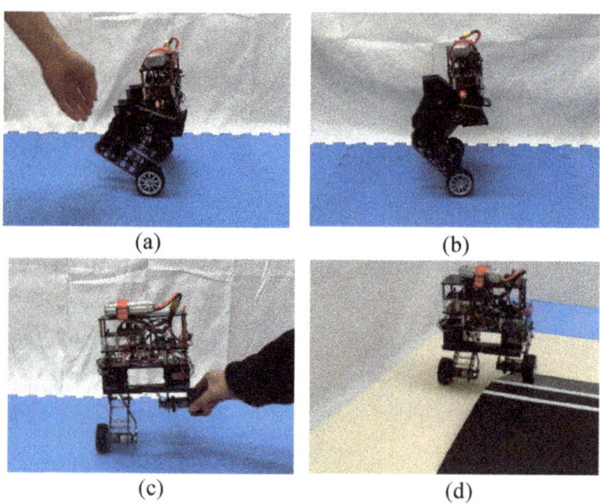

Figure 6. Photographs of the experiments: (**a**) moving and rotating; (**b**) height-changing; (**c**) posture-keeping; (**d**) one leg on slope movement.

4.1. Moving and Rotating Scenario

In this scenario, the WBR started by balancing in its initial posture at the setting height and staying at the setting height. The tests were then carried out, including self-balancing, station-keeping, anti-interference, moving, and rotating tests. The experiment was performed by executing three actions to demonstrate the flexibility of the WBR. Two external forces were pushed on both sides of its body by human hands at approximately the 3rd and 6th second. The WBR was then ordered to go forward for 3 s at the 12th second and turn right in place for 3 s at the 17th second. The experimental results of the moving and rotating scenario are shown in Figure 7a,b for the cascade PID control [22] and the IMBC system, respectively. These figures show that both the cascade PID control and the IMBC system are sufficiently robust against external disturbances and can perfectly perform tasks such as moving and rotating in place while maintaining good balance.

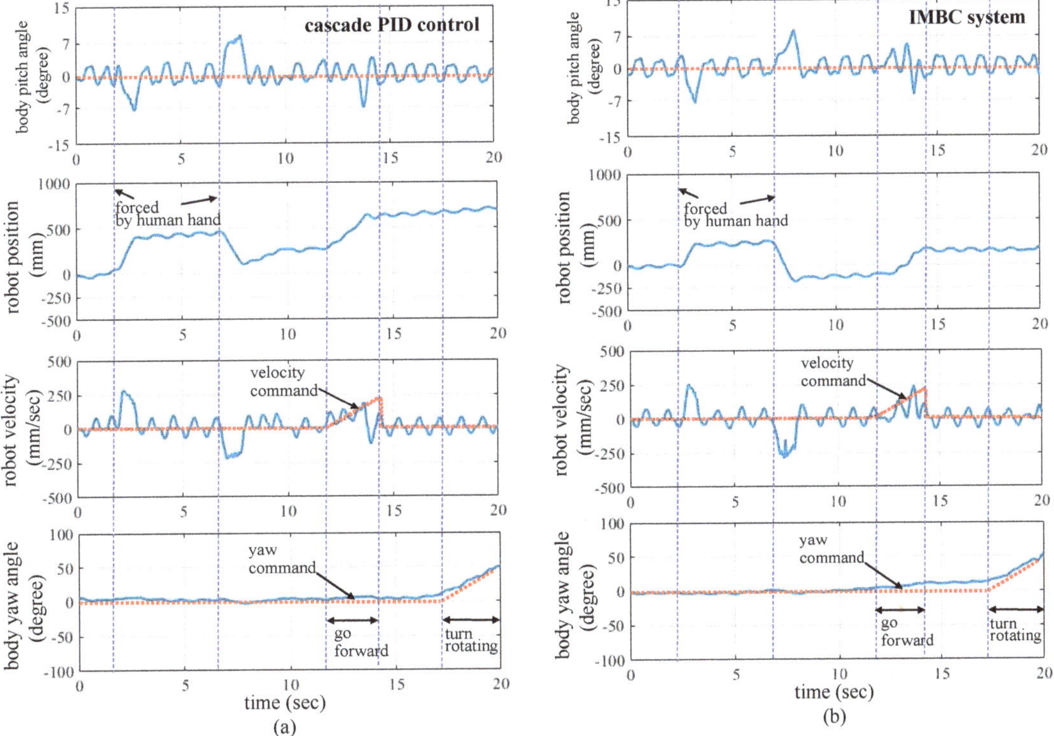

Figure 7. Results of moving and rotating. (**a**) Cascade PID control; (**b**) the IMBC system.

4.2. Height-Changing Scenario

In this scenario, the WBR started balancing in its initial posture at the minimum height. The experiment was performed by increasing the height of the robot to the maximum height starting at approximately the 2nd second and then changing the height of the robot to the minimum height at approximately the 4th second. The experimental results of the height-changing scenario are shown in Figure 8a,b for the cascade PID control [22] and the IMBC system, respectively. It can be seen that when the robot height command is changed, the length of both legs increases or decreases at the same time. Comparing Figure 8a,b shows that the WBR was able to stay in place without generating significant motion when using the proposed IMBC system. However, when using the cascade PID control, the robot could not stay in its original position, owing to the change in height of its center of gravity.

This result demonstrates that the proposed IMBC system can effectively overcome the effects of system uncertainties caused by robot height variations.

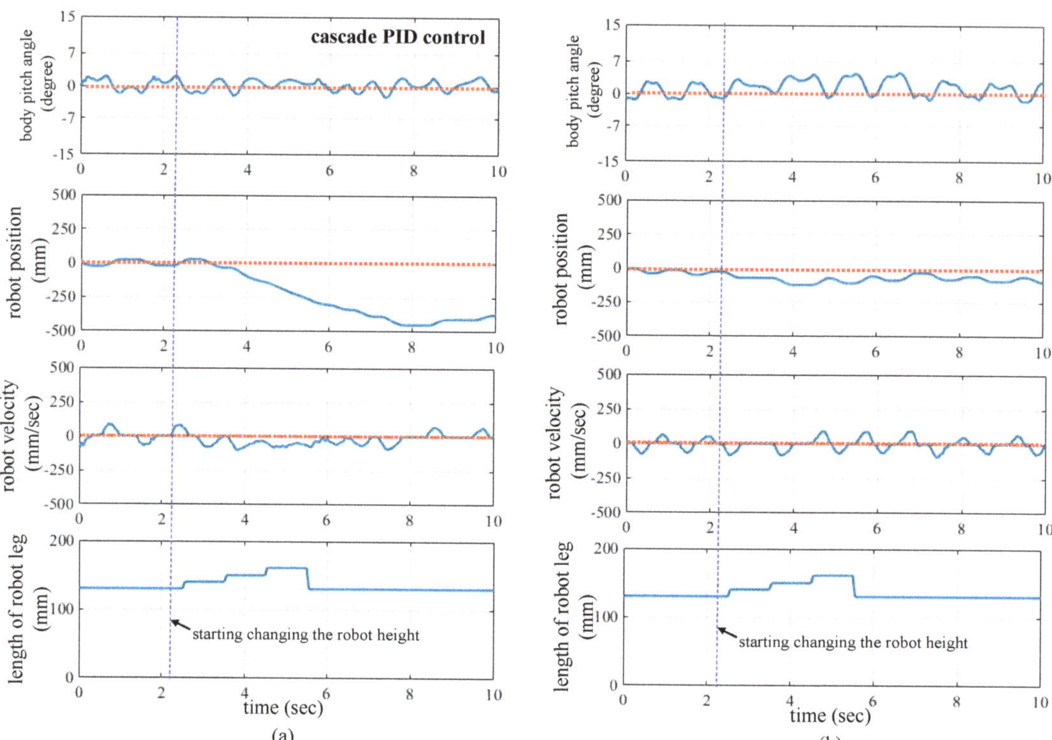

Figure 8. Results of height changing. (**a**) Cascade PID control; (**b**) the IMBC system.

4.3. Posture-Keeping Scenario

The robustness of the IMBC system was tested when one wheel was slightly off the ground. The purpose of the experiment was to test whether the leg strategy controller could maintain the robot's height when it was disturbed by human hands. The experiment was performed by lifting the right foot of the WBR by hand at approximately the 2nd second and putting it down at approximately the 6th second, and then lifting the left foot of the WBR by hand at approximately the 12th second and putting it down at approximately the 16th second. The experimental results of the posture-keeping scenario are shown in Figure 9a,b for the cascade PID control [22] and the IMBC system, respectively. They show that the proposed IMBC system can maintain the horizontal posture of the WBR during the leg lifting process, and when the single leg on one side of the leg lift reaches its limit, the other side can be extended or retracted to compensate for the error of the horizontal angle. Because the cascade PID controller cannot independently control two leg motors to change the robot's attitude, the horizontal posture of the WBR cannot be maintained during the leg lifting process.

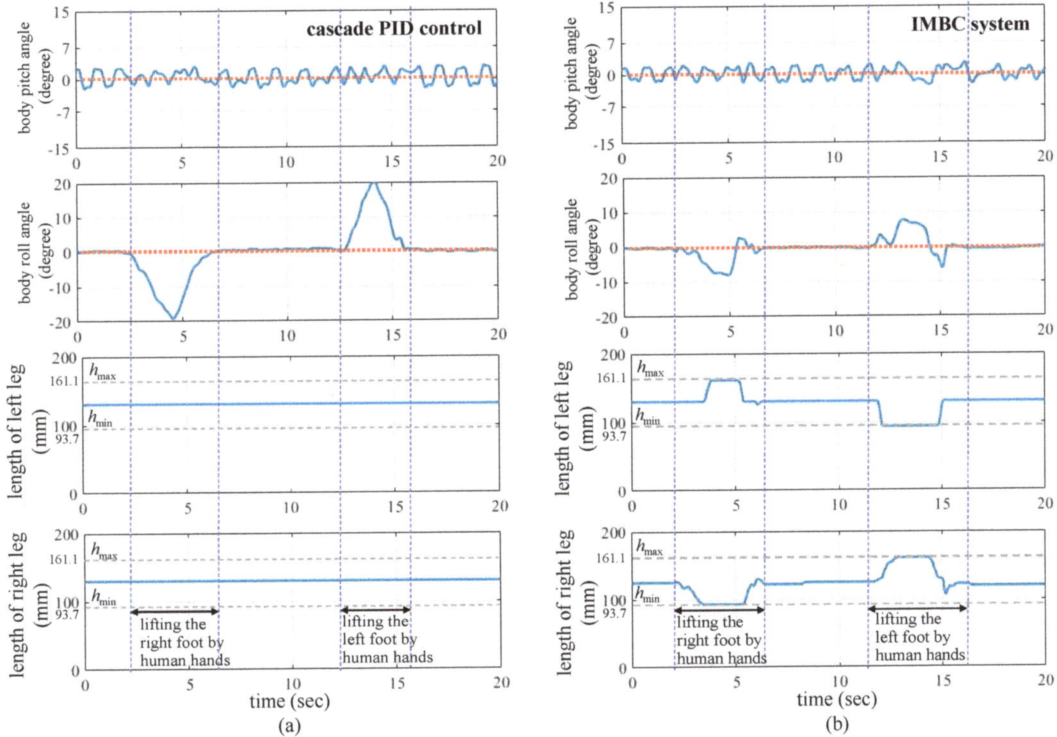

Figure 9. Results of body balancing. (**a**) Cascade PID control; (**b**) the IMBC system.

4.4. One Leg on Slope Movement Scenario

The WBR moved on a small slope of approximately 12° on one side and 8° on the other side, with the left leg on the small slope and the right leg on the flat ground. The purpose of the experiment was to test whether the leg strategy controller could keep the robot horizontal when it was disturbed by terrain. The experiment had the robot move forward at approximately the 5th second with the left foot on the ramp but the right foot on the ground and then move backward to its original position at approximately the 7th second. The experimental results of the posture-keeping scenario are shown in Figure 10a,b for the cascade PID control [22] and the IMBC system, respectively. These results show that the proposed IMBC system can adjust the height of the left and right legs in time to avoid the robot falling over as a result of a change in height of the left and right leg positions. However, the cascaded PID controller cannot adjust the height of the left and right legs independently, so when the WBR is disturbed by terrain changes, its balance can easily fail.

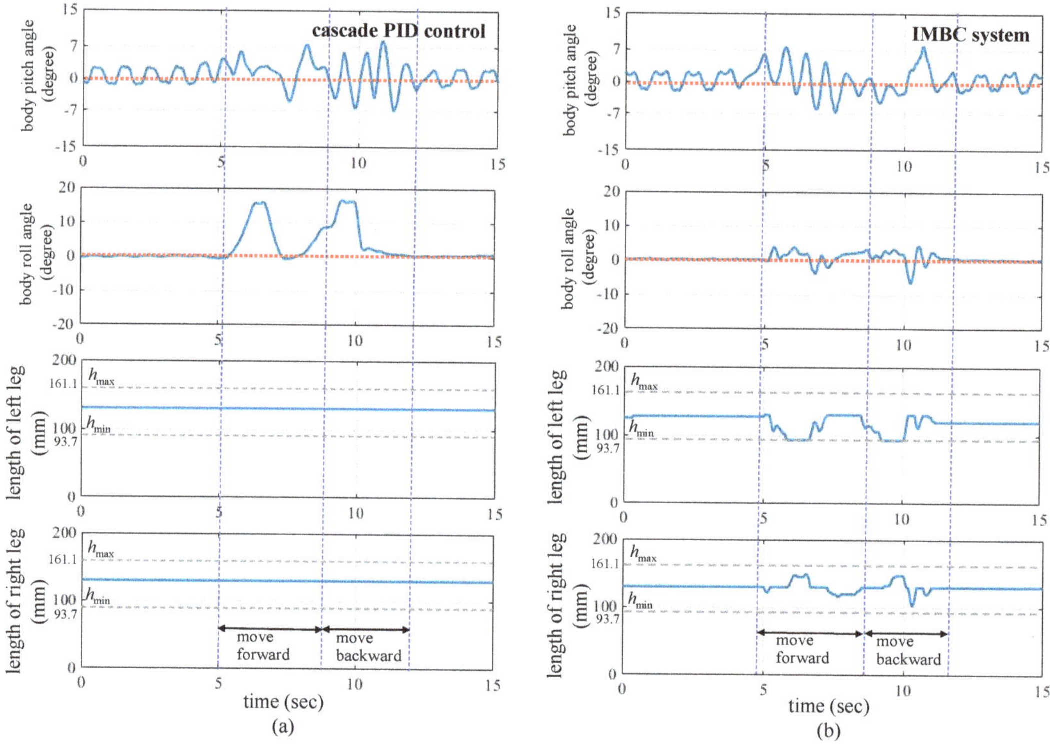

Figure 10. Results of one leg on slope moving. (**a**) Cascade PID control; (**b**) the IMBC system.

5. Conclusions

Due to the non-linear and under-actuated nature of the wheeled bipedal robot (WBR) system, this study presented an intelligent motion and balance control (IMBC) method based on the fuzzy logic approach. It should be emphasized that the IMBC system allows for stability and drive control and does not require the same system dynamics and system parameters of a WBR. Besides the simplicity of design, many other advantages are offered by the proposed control method (such as robustness against external disturbance and parameters variation). Four experimental scenarios, namely moving and rotating, height-changing, posture-keeping, and one leg on slope movement, were tested to explore the stability performance of the IMBC system for the WBR. The experimental results showed that the proposed IMBC system is capable of providing appropriate control outputs to maintain the body balance of the robot, as well as the motion and steering control of the robot, even under variable environmental terrain and with changing robot height. Meanwhile, the video showing the IMBC system for WBR can be found at https://www.youtube.com/watch?v=AR6yP99sSjs (accessed on 20 November 2022). However, the IMBC system depends to a large extent on the expert's knowledge or on trial and error. It was not easy to guarantee the stability and robustness of the closed-loop control systems.

Future works in this study may focus on: (1) Giving a more systematic way to apply external forces to provide a more specific description of performance comparisons. (2) Utilizing Type-2 fuzzy systems [44] to handle the rule uncertainties when it is hard to exactly determine the grade of membership functions. (3) Auto-tuning a fuzzy controller [45,46], whose fuzzy rules can be tuned online without requiring time-consuming adjustments, is developed to ensure the stability of the closed-loop system. (4) Detecting dynamic obstacle and road features using LiDAR and CCD images to develop an autonomous navigation framework [47,48] that interacts with humans.

Author Contributions: Conceptualization, C.-F.H. and B.-R.C.; Methodology, C.-F.H.; Software, B.-R.C. and Z.-L.L.; Validation, B.-R.C. and Z.-L.L.; Formal Analysis, B.-R.C. and Z.-L.L.; Investigation, C.-F.H. and Z.-L.L.; Resources, C.-F.H.; Data Curation, B.-R.C.; Writing—Original Draft Preparation, C.-F.H. and B.-R.C.; Writing—Review and Editing, C.-F.H. and B.-R.C.; Visualization, B.-R.C.; Supervision, C.-F.H.; Project Administration, C.-F.H.; Funding Acquisition, C.-F.H. All authors have read and agreed to the published version of the manuscript.

Funding: The study was funded by the Ministry of Science and Technology of Republic of China under Grant MOST 108-2221-E-032-039-MY2.

Data Availability Statement: Not applicable.

Acknowledgments: We thank the anonymous reviewers for providing critical comments and suggestions that improved the manuscript.

Conflicts of Interest: The authors declare no conflict of interest.

Appendix A

θ	body pitch angle
ψ	body yaw angle
ϕ	body roll angle
θ_w	robot position
θ_R	rotary angle of the right wheel motor
θ_L	rotary angle of the left wheel motor
R	wheel radius
W	half of body width
L	distance of CoG from the wheel axle
m	wheel weight
M	body weights
J_W	wheel inertia moment
J_θ	body pitch inertia moment
g	gravity acceleration
R_m	wheel motor resistance
K_t	wheel motor torque constant
K_b	wheel motor back electromotive force coefficient
J_m	wheel motor inertia moment
L_1	length between the pin joint and the knee joint
L_2	length between the pin joint and the leg joint
L_3	length between the knee joint and the inner joint
L_4	length between the inner joint and the wheel motor
h	robot height
h_L	height of the left side of the WBR
h_R	height of the right side of the WBR
θ_{kL}	rotary angle of the left leg motor
θ_{kR}	rotary angle of the right leg motor

References

1. McGinn, C.; Cullinan, M.F.; Otubela, M.; Kelly, K. Design of a terrain adaptive wheeled robot for human-orientated environments. *Auton. Robot.* **2019**, *43*, 63–78. [CrossRef]
2. Xu, S.S.D.; Huang, H.C.; Kung, Y.C.; Chu, Y.Y. A networked multirobot CPS with artificial immune fuzzy optimization for distributed formation control of embedded mobile robots. *IEEE Trans. Ind. Inform.* **2020**, *16*, 414–422. [CrossRef]
3. Pang, L.; Cao, Z.; Yu, J.; Guan, P.; Chen, X.; Zhang, W. A robust visual person-following approach for mobile robots in disturbing environments. *IEEE Syst. J.* **2020**, *14*, 2965–2968. [CrossRef]
4. Tazaki, Y.; Murooka, M. A survey of motion planning techniques for humanoid robots. *Adv. Robot.* **2020**, *34*, 1370–1379. [CrossRef]
5. Li, T.H.S.; Kuo, P.H.; Cheng, C.H.; Hung, C.C.; Luan, P.C.; Chang, C.H. Sequential sensor fusion-based real-time LSTM gait pattern controller for biped robot. *IEEE Sens. J.* **2021**, *21*, 2241–2255. [CrossRef]
6. Lee, J.; Bakolas, E.; Sentis, L. An efficient and direct method for trajectory optimization of robots constrained by contact kinematics and forces. *Auton. Robot.* **2021**, *45*, 135–153. [CrossRef]
7. Viragh, Y.; Bjelonic, M.; Bellicoso, C.D.; Jenelten, F.; Hutter, M. Trajectory optimization for wheeled-legged quadrupedal robots using linearized zmp constraints. *IEEE Robot. Autom. Lett.* **2019**, *4*, 1633–1640. [CrossRef]
8. Medeiros, V.S.; Jelavic, E.; Bjelonic, M.; Siegwart, R.; Meggiolaro, M.A.; Hutter, M. Trajectory optimization for wheeled-legged quadrupedal robots driving in challenging terrain. *IEEE Robot. Autom. Lett.* **2020**, *5*, 4172–4179. [CrossRef]
9. Sun, J.; You, Y.; Zhao, X.; Adiwahono, A.H.; Chew, C.M. Towards more possibilities: Motion planning and control for hybrid locomotion of wheeled-legged robots. *IEEE Robot. Autom. Lett.* **2020**, *5*, 3723–3730. [CrossRef]
10. Li, J.; Wang, J.; Peng, H.; Hu, Y.; Su, H. Fuzzy-torque approximation-enhanced sliding mode control for lateral stability of mobile robot. *IEEE Trans. Syst. Man Cybern. Syst.* **2022**, *52*, 2491–2500. [CrossRef]
11. He, J.; Sun, Y.; Yang, L.; Sun, J.; Xing, Y.; Gao, F. Design and control of TAWL-a wheel-legged rover with terrain-adaptive wheel speed allocation capability. *IEEE/ASME Trans. Mechatron.* **2022**, *27*, 2212–2223. [CrossRef]
12. Wang, S.; Chen, Z.; Li, J.; Wang, J.; Li, J.; Zhao, J. Flexible motion framework of the six wheel-legged robot: Experimental results. *IEEE/ASME Trans. Mechatron.* **2022**, *27*, 2246–2257. [CrossRef]
13. Chiu, C.H.; Peng, Y.F.; Sun, C.H. Intelligent decoupled controller for mobile inverted pendulum real-time implementation. *J. Intell. Fuzzy Syst.* **2017**, *32*, 3809–3820. [CrossRef]
14. Huang, J.; Ri, M.; Wu, D.; Ri, S. Interval type-2 fuzzy logic modeling and control of a mobile two-wheeled inverted pendulum. *IEEE Trans. Fuzzy Syst.* **2018**, *26*, 2030–2038. [CrossRef]
15. Su, Y.; Wang, T.; Zhang, K.; Yao, C.; Wang, Z. Adaptive nonlinear control algorithm for a self-balancing robot. *IEEE Access* **2020**, *8*, 3751–3760. [CrossRef]
16. Iwendi, C.; Alqarni, M.A.; Anajemba, J.H.; Alfakeeh, A.S.; Zhang, Z.; Bashir, A.K. Robust navigational control of a two-wheeled self-balancing robot in a sensed environment. *IEEE Access* **2019**, *7*, 82337–82348. [CrossRef]
17. Sekiguchi, S.; Yorozu, A.; Kuno, K.; Okada, M.; Watanabe, Y.; Takahashi, M. Human-friendly control system design for two-wheeled service robot with optimal control approach. *Robot. Auton. Syst.* **2020**, *131*, 103562. [CrossRef]
18. Li, C.H.G.; Zhou, L.P.; Chao, Y.H. Self-balancing two-wheeled robot featuring intelligent end-to-end deep visual-steering. *IEEE/ASME Trans. Mechatron.* **2021**, *26*, 2263–2273. [CrossRef]
19. Li, X.; Zhou, H.; Feng, H.; Zhang, S.; Fu, Y. Design and experiments of a novel hydraulic wheel-legged robot (WLR). In Proceedings of the IEEE/RSJ International Conference on Intelligent Robots and Systems, Madrid, Spain, 1–5 October 2018; pp. 3292–3297.
20. Li, X.; Zhou, H.; Zhang, S.; Feng, H.; Fu, Y. WLR-II, a hose-less hydraulic wheel-legged robot. In Proceedings of the IEEE/RSJ International Conference on Intelligent Robots and Systems, Macau, China, 3–8 November 2019; pp. 4339–4346.
21. Klemm, V.; Morra, A.; Salzmann, C.; Tschopp, F.; Bodie, K.; Gulich, L.; Küng, N.; Mannhart, D.; Pfister, C.; Vierneisel, M.; et al. Ascento: A two-wheeled jumping robot. In Proceedings of the International Conference on Robotics and Automation, Montreal, QC, Canada, 20–24 May 2019; pp. 7515–7521.
22. Zhang, C.; Liu, T.; Song, S.; Meng, M.Q.H. System design and balance control of a bipedal leg-wheeled robot. In Proceedings of the IEEE International Conference on Robotics and Biomimetics, Dali, China, 6–8 December 2019; pp. 1869–1874.
23. Zhao, L.; Yu, Z.; Chen, X.; Huang, G.; Wang, W.; Han, L.; Qiu, X.; Zhang, X.; Huang, Q. System design and balance control of a novel electrically-driven wheel-legged humanoid robot. In Proceedings of the IEEE International Conference on Unmanned Systems, Beijing, China, 15–17 October 2021; pp. 742–747.
24. Raza, F.; Hayashibe, M. Towards robust wheel-legged biped robot system: Combining feedforward and feedback control. In Proceedings of the IEEE/SICE International Symposium on System Integrations, Iwaki, Fukushima, Japan, 11–14 January 2021; pp. 606–612.
25. Dong, J.; Liu, R.; Lu, B.; Guo, X.; Liu, H. LQR-based balance control of two-wheeled legged robot. In Proceedings of the 41st Chinese Control Conference, Hefei, China, 25–27 July 2022; pp. 450–455.
26. Zhou, H.; Yu, H.; Li, X.; Feng, H.; Zhang, S.; Fu, Y. Configuration transformation of the wheel-legged robot using inverse dynamics control. In Proceedings of the IEEE International Conference on Robotics and Automation, Xi'an, China, 30 May–5 June 2021; pp. 3091–3096.
27. Hao, Y.; Lu, B.; Cao, H.; Dong, J.; Liu, R. Run-and-jump planning and control of a compact two-wheeled legged robot. In Proceedings of the 7th Asia-Pacific Conference on Intelligent Robot Systems, Tianjin, China, 1–3 July 2022; pp. 1–6.

28. Xin, Y.; Chai, H.; Li, Y.; Rong, X.; Li, B.; Li, Y. Speed and acceleration control for a two wheel-leg robot based on distributed dynamic model and whole-body control. *IEEE Access* **2019**, *7*, 180630–180639. [CrossRef]
29. Klemm, V.; Morra, A.; Gulich, L.; Mannhart, D.; Rohr, D.; Kamel, M.; Viragh, Y.; Siegwart, R. LQR-assisted whole-body control of a wheeled bipedal robot with kinematic loops. *IEEE Robot. Autom. Lett.* **2020**, *5*, 3745–3752. [CrossRef]
30. Xin, Y.; Vijayakumar, S. Online dynamic motion planning and control for wheeled biped robots. In Proceedings of the IEEE/RSJ International Conference on Intelligent Robots and Systems, Las Vegas, NV, USA, 24 October–24 January 2020; pp. 3892–3899.
31. Wang, Y.; Xin, Y.; Rong, X.; Li, Y. Whole-body motion planning and control for underactuated wheeled-bipdal robots. In Proceedings of the IEEE International Conference on Robotics and Biomimetics, Sanya, China, 27–31 December 2021; pp. 1071–1076.
32. Chen, H.; Wang, B.; Hong, Z.; Shen, C.; Wensing, P.M.; Zhang, W. Underactuated motion planning and control for jumping with wheeled-bipedal robots. *IEEE Robot. Autom. Lett.* **2021**, *6*, 747–754. [CrossRef]
33. Zhu, W.; Raza, F.; Hayashibe, M. Reinforcement learning based hierarchical control for path tracking of a wheeled bipedal robot with sim-to-real framework. In Proceedings of the IEEE/SICE International Symposium on System Integration, Narvik, Norway, 9–12 January 2022; pp. 40–46.
34. Cui, L.; Wang, S.; Zhang, J.; Zhang, D.; Lai, J.; Zheng, Y.; Zhang, Z.; Jiang, Z.P. Learning-based balance control of wheel-legged robots. *IEEE Robot. Autom. Lett.* **2021**, *6*, 7667–7674. [CrossRef]
35. Chang, H.T.; Tai, F.C.; Tsai, C.C. Adaptive fuzzy-basis-function-network trajectory tracking for a terrain-adaptive self-balancing two-wheeled mobile robot. In Proceedings of the International Conference on Fuzzy Theory and Its Applications, Taitung, Taiwan, 5–8 October 2021; pp. 42–47.
36. Wu, C.F.; Chen, B.S.; Zhang, W. Multiobjective investment policy for a nonlinear stochastic financial system: A fuzzy approach. *IEEE Trans. Fuzzy Syst.* **2017**, *25*, 460–474. [CrossRef]
37. Wu, L.F.; Li, T.H.S. Fuzzy dynamic gait pattern generation for real-time push recovery control of a teen-sized humanoid robot. *IEEE Access* **2020**, *8*, 36441–36453. [CrossRef]
38. Zhong, Z.; Zhu, Y.; Lin, C.M.; Huang, T. A fuzzy control framework for interconnected nonlinear power networks under TDS attack: Estimation and compensation. *J. Frankl. Inst.* **2021**, *358*, 74–88. [CrossRef]
39. Benzaouia, S.; K.M'Sirdi, N.; Rabhi, A.; Zouggar, S. Signed-distance fuzzy-logic sliding-mode control strategy for floating interleaved boost converter. In Proceedings of the 9th International Conference on Systems and Control, Caen, France, 24–26 November 2021; pp. 417–422.
40. Choi, B.J.; Kwak, S.W.; Kim, B.K. Design of a single-input fuzzy logic controller and its properties. *Fuzzy Sets Syst.* **1999**, *106*, 299–308. [CrossRef]
41. Londhe, P.S.; Santhakumar, M.; Patre, B.M.; Waghmare, L.M. Task space control of an autonomous underwater vehicle manipulator system by robust single-input fuzzy logic control scheme. *IEEE J. Ocean. Eng.* **2017**, *42*, 13–28. [CrossRef]
42. Li, I.H. Design for a fluidic muscle active suspension using parallel-type interval type-2 fuzzy sliding control to improve ride comfort. *Int. J. Fuzzy Syst.* **2022**, *24*, 1719–1734. [CrossRef]
43. Hsu, C.F.; Kao, W.F. Double-loop fuzzy motion control with CoG supervisor for two-wheeled self-balancing assistant robots. *Int. J. Dyn. Control* **2020**, *8*, 851–866. [CrossRef]
44. Nafia, N.; Kari, A.E.; Ayad, H.; Mjahed, M. Robust interval type-2 fuzzy sliding mode control design for robot manipulators. *Robotics* **2018**, *7*, 40. [CrossRef]
45. Hsu, C.F.; Kao, W.F. Perturbation wavelet-neural sliding-mode position control for a voice coil motor driver. *Neural Comput. Appl.* **2019**, *31*, 5975–5988. [CrossRef]
46. Lin, C.M.; Nguyen, H.B.; Huynh, T.T. A new self-organizing double function-link brain emotional learning controller for MIMO nonlinear systems using sliding surface. *IEEE Access* **2021**, *9*, 73826–73842. [CrossRef]
47. Ye, W.; Li, Z.; Yang, C.; Sun, J.; Su, C.Y.; Lu, R. Vision-based human tracking control of a wheeled inverted pendulum robot. *IEEE Trans. Cybern.* **2016**, *46*, 2423–2434. [CrossRef] [PubMed]
48. Li, J.; Qin, H.; Wang, J.; Li, J. OpenStreetMap-based autonomous navigation for the four wheel-legged robot via 3D-lidar and CCD camera. *IEEE Trans. Ind. Electron.* **2022**, *69*, 2708–2717. [CrossRef]

Article

Force Characteristics of Centrifugal Pump as Turbine during Start-Up Process under Gas–Liquid Two-Phase Conditions

Baodui Chai [1,2], Junhu Yang [1,*] and Xiaohui Wang [1]

1   School of Energy and Power Engineering, Lanzhou University of Technology, Lanzhou 730050, China
2   School of Chemistry and Chemical Engineering, Lanzhou Jiaotong University, Lanzhou 730070, China
*   Correspondence: lzyangjh@lut.cn

Abstract: The start-up process of a centrifugal pump as turbine (PAT) under gas–liquid two-phase conditions was simulated based on Fluent, and the evolution mechanism of the internal flow field and the variation law of force characteristics were studied in its start-up process under gas–liquid two-phase conditions. The results show that the area with high gas phase concentration corresponds to a strong vortex at the beginning of the start-up. The vortex intensity in the impeller gradually decreases with an increase in rotational speed. The gas volume fraction of the blade suction surface is more significant than that of the blade pressure surface. The higher the inlet gas volume fraction (IGVF) is, the more severely the blade load will fluctuate during the start-up process. As the rotational speed increases, the fluctuation of the blade load gradually weakens, and the maximum load is distributed near the inner edge of the blade after the rotational speed is stable. The periodic unbalanced radial force is produced in the start-up process. From the pure liquid conditions to the gas–liquid two-phase conditions with increasing IGVF, the dominant frequency amplitude of radial force shows a similar trend of decreasing first but then increasing. After the rotational speed tends to be stable, the dominant frequency of radial force is equal to the rotational frequency of the blade. With the increase in rotational speed, the dominant frequency amplitude of axial force decreases gradually. The higher the IGVF, the greater the dominant frequency amplitude of axial force at the same time.

Keywords: PAT; start-up process; gas–liquid two-phase; radial force; axial force

Citation: Chai, B.; Yang, J.; Wang, X. Force Characteristics of Centrifugal Pump as Turbine during Start-Up Process under Gas–Liquid Two-Phase Conditions. *Actuators* **2022**, *11*, 370. https://doi.org/10.3390/act11120370

Academic Editor: Luigi de Luca

Received: 8 November 2022
Accepted: 6 December 2022
Published: 8 December 2022

Publisher's Note: MDPI stays neutral with regard to jurisdictional claims in published maps and institutional affiliations.

1. Introduction

A centrifugal pump as turbine (PAT) is generally applied to residual pressure energy recovery of various devices in the petrochemical industry. The recovered medium of PAT usually contains a certain amount of gas. Gas–liquid two-phase flow can not only affect the external characteristics of PAT but also cause instability phenomena such as vibration and noise. Under gas–liquid two-phase conditions, the magnitude and direction of force acting on the PAT impeller can change with time during its start-up process. These unsteady forces directly affect the stability of the PAT start-up process [1–3]. Therefore, it is particularly essential to analyze the force characteristics of the PAT start-up process under gas–liquid two-phase conditions.

Some scholars have studied the force characteristics of a centrifugal pump under gas–liquid two-phase conditions. Barrios [4] and Gamboa [5] used visual experiments to observe that the bubble size in the impeller of the centrifugal pump increases with the increase in inlet gas volume fraction (IGVF) and decrease in rotational speed. Li et al. [6] obtained the variation law of a centrifugal pump blade load under different IGVFs. They pointed out that the imbalance of radial force on the impeller intensified, and the torque on the impeller reduced with the increase in IGVF. Yuan et al. [7] proposed that the gas volume fraction in the front shroud near the trading edge increased significantly with the rise in IGVF than in the back shroud. Luo et al. [8] found that the radial force of the centrifugal pump impeller and its pulsation amplitude increased first and then decreased with the increase in IGVF. The frequency corresponding to the peak value of radial force pulsation under each working condition was a multiple of the blade rotational frequency. Michael et al. [9] found that increasing the rotational speed would be beneficial to gas–liquid two-phase mixing. Due to the increase in flow rate, the occurrence of buzz and cavitation would increase with the growth of rotational speed. Yan et al. [10] proposed that a large number of gas blockages would appear in the centrifugal pump impeller with the rise in IGVF at a low flow rate, resulting in uneven pressure load on the blade surface and some large vortices in the flow channel. Zhang et al. [11] found that the gas in the centrifugal pump impeller mainly aggregated near the suction surface of the outlet area. The gas inhomogeneity and the interphase force in the impeller increased with the rise in IGVF.

Some papers have introduced force characteristics of PAT under stable working conditions. Anthony et al. [12] pointed out that the radial force of PAT at a low flow rate was a rotating force centered on the impeller, and the average amplitude of radial force increased with decreasing flow rate. Shi et al. [13] found that adding guide vanes could reduce the radial force of PAT and make the radial force distribution more uniform. Qu et al. [14] proposed that the shroud force of the PAT impeller at each level became smaller and smaller with the increase in stages at the best efficiency point, and the axial force at each level increased with the rise in the incoming flow head. Yang et al. [15] pointed out that the radial force of the double-volute turbine was higher than that of the single-volute turbine overall, while its balance was much poorer. Dai et al. [16] found that the radial force on the volute reduced and moved to the fourth quadrant as the blade warp angle increased. Shi et al. [17,18] pointed out that with the increase in IGVF, the relative average static pressure and pressure pulsation in the volute and impeller of PAT gradually decreased. In contrast, the hydraulic loss increased in the flow process. Shi et al. [19] found that with the increase in IGVF, the efficiency and power of PAT gradually decreased while the head steadily increased. The larger the flow rate, the greater the radial force, and the fluctuation of radial force under gas–liquid two-phase conditions was more significant than that under the pure liquid phase. However, the computational analysis of forces during the start-up process at different IGVFs is also crucial to the PAT.

Overall, the research on the force characteristics under gas–liquid two-phase conditions mainly focuses on centrifugal pumps, while the research on PAT mainly focuses on steady conditions and pure liquid phase conditions. There are few studies on the transient characteristics of the PAT start-up process, especially the force characteristics under gas–liquid two-phase conditions. Therefore, the PAT start-up process under gas–liquid two-phase conditions was studied using numerical simulation and experimental investigation. In this study, the combination of water and ideal gas was used as the medium to investigate the pressure distribution, the velocity streamline distribution, and the gas volume fraction distribution during the PAT start-up process. The main objective of the present study was to develop an in-depth understanding of the gas–liquid phase force characteristics and analyze the variation law of blade load, radial force, and axial force during the PAT start-up process at different IGVFs, which will help to improve the hydraulic performance of the PAT.

2. Numerical Simulation and Verification

2.1. Control Equation

The Mixture model can be used to simulate multiphase flow with different and same velocities of each phase. When the dispersed phase has a wide distribution and the interphase drag law is unknown, the Mixture model is the most desirable. The working medium of this paper is water and ideal gas. The gas–liquid two-phase is assumed to be homogeneous and consecutive. Water is the primary phase, and perfect gas is the secondary phase. There is no phase transition and mass transfer between the two phases. Considering the stability and economy of calculation, the Mixture model is used to study the PAT start-up process under gas–liquid two-phase conditions. The continuity equation of the Mixture model is [18]:

$$\frac{\partial}{\partial t}(\rho_m) + \nabla \cdot \left(\rho_m \vec{v}_m\right) = 0 \tag{1}$$

where $\vec{v}_m$ is the mass-averaged velocity and can be calculated as:

$$\vec{v}_m = \frac{\sum\limits_{k=1}^{n} \alpha_k \rho_k \vec{v}_k}{\rho_m} \tag{2}$$

α_k, ρ_k and $\vec{v}_k$ represent the volume fraction, the density, and the averaged velocity of the kth phase component, respectively. ρ_m is the mixture density and can be written as:

$$\rho_m = \sum_{k=1}^{n} \alpha_k \rho_k \tag{3}$$

By summing the momentum equations for all phases, the momentum equation of the Mixture model can be obtained and expressed as [18]:

$$\frac{\partial}{\partial t}\left(\rho_m \vec{v}_m\right) + \nabla \cdot \left(\rho_m \vec{v}_m \vec{v}_m\right) = -\nabla p + \nabla \cdot \left[\mu_m\left(\nabla \vec{v}_m + \vec{v}_m^T\right)\right] + \rho_m \vec{g} + \vec{F} + \nabla \cdot \left(\sum_{k=1}^{n} \alpha_k \rho_k \vec{v}_{dr,k} \vec{v}_{dr,k}\right) \tag{4}$$

where n is the phase number, and $\vec{F}$ is the body force. μ_m is the mixture viscosity and can be expressed as:

$$\mu_m = \sum_{k=1}^{n} \alpha_k \mu_k \tag{5}$$

μ_k represents the viscosity of the kth phase component. $\vec{v}_{dr,\,k}$ is the slip velocity of the second phase and can be expressed as:

$$\vec{v}_{dr,k} = \vec{v}_k - \vec{v}_m \tag{6}$$

2.2. Calculation Model

To analyze the force characteristics of the PAT start-up process under gas–liquid two-phase conditions, the IS80-50-315 centrifugal pump reversed as the turbine is selected as the study subject. The main design parameters of PAT are the rated speed $n_t = 1450$ r/min, the rated head $H_t = 50$ m, and the rated flow rate $Q_t = 50$ m^3/h. Figure 1 shows the structure of PAT, and its main structure parameters are the inlet diameter of impeller $D_1 = 315$ mm, the outlet diameter of impeller $D_2 = 80$ mm, the inlet width of blade $b_1 = 10$ mm, the inlet offset angle of blade $\beta_1 = 32°$, the wrap angle of blade $\theta = 150°$, the number of blades $z = 6$, the base circle diameter $D_0 = 320$ mm, the inlet diameter of volute $D_{in} = 50$ mm, and the outlet width of volute $b_0 = 24$ mm.

The entire fluid domain of PAT is modeled in CREO software. The hexahedral structured grids of the impeller, volute, and outlet pipe are generated by ICEM CFD, as shown in Figure 2. Fine boundary layer meshes are generated on the blade surface and the flow surface inside the volute, and mesh refinement is performed at the inlet and outlet sides of the blade and the volute tongue. The grid independence at IGVF of 0.15 has been verified, as shown in Figure 3. The X-axis represents the grid number, and the left Y-axis and the right Y-axis represent head and efficiency, respectively. As the number of grids increases, the head and efficiency of PAT gradually tend to be stable. When the number of grids reaches 1.9 million, the relative changes in head and efficiency are less than 1%. So it is reasonable that the grid numbers of the fluid domain are determined to be 1.9 million.

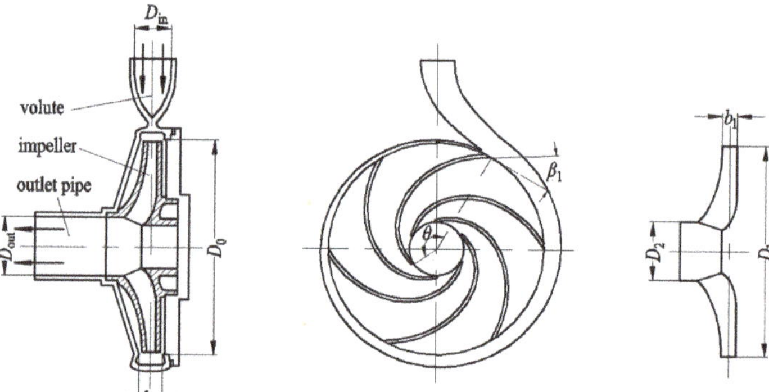

Figure 1. Structure of PAT.

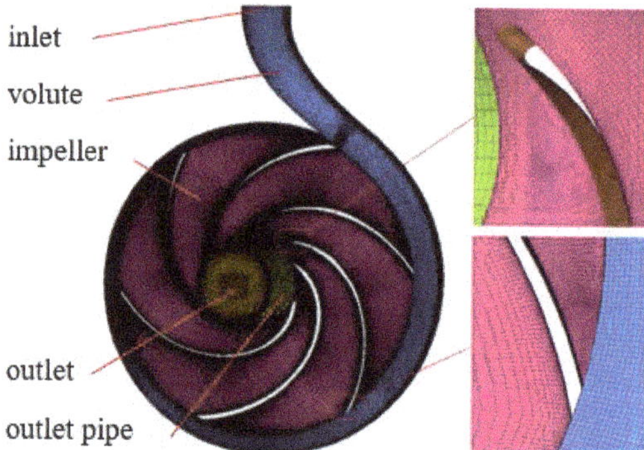

Figure 2. Fluid domain grid of PAT.

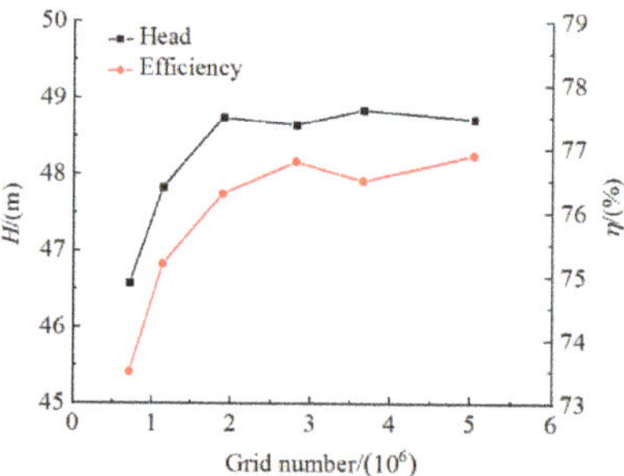

Figure 3. Check of grid independence at IGVF of 0.15.

2.3. Calculation Method

The rotational speed of PAT is accelerated in its start-up process under the action of the incoming flow. In addition to the torque provided by the incoming flow, the rotating parts are also subjected to the load torque and the friction resistance torque. The rotation equation of the rotor is based on the d'Alembert principle [20]:

$$J\frac{d\omega}{dt} = M_\text{t} - M_\text{l} - M_\text{f} \tag{7}$$

where J is the rotational inertia of the rotor, ω is the angular velocity of the rotor, t is the time, M_t, M_l, and M_f are the fluid flow torque, the load torque, and the frictional resistance torque acting on the rotor, respectively. M_t can be calculated as:

$$M_t = \frac{1}{\omega}\rho g H_t Q_t \eta_t \tag{8}$$

where ρ is the fluid density, g is the acceleration of gravity, H_t is the head of PAT, Q_t is the flow rate of PAT, and η_t is the efficiency of PAT.

The variation of angular velocity with time during PAT start-up is obtained by integrating Equation (7) as follows:

$$\omega = \left[\frac{2}{J}\int_0^t (\rho g H_t Q_t \eta_t - N_l - N_f)dt\right]^{\frac{1}{2}} \tag{9}$$

where N_l and N_f are the load consumption power and the frictional resistance loss power, respectively.

Based on the rotation equation of the rotor and the sliding mesh technique, the UDF program for calculating angular velocity is written and compiled in Fluent [21]. The angular velocity variation with time is transferred into the flow field solver, and the impeller domain is accelerated rotationally. The resultant torque on the impeller decreases with the increase in iteration times, and then the rotational speed gradually increases until the rated speed.

2.4. Boundary Condition

The Mixture model, RNG k-ε turbulence model, and SIMPLEC algorithm are employed to simulate the start-up process of PAT under gas–liquid two-phase conditions using Fluent software. The walls of the impeller, volute, and outlet pipe adopt adiabatic non-slip boundary conditions. The standard wall function method is used to modify the turbulence model for the near wall region. The residual convergence target is set at 10^{-5}.

It is assumed that the velocity and void fraction distribution of gas–liquid two-phase flow at the inlet is uniform and equal. The velocity inlet boundary condition is set at the inlet of the domain. The IGVF is set as 0, 0.05, 0.10, and 0.15, respectively. The IGVF is defined as:

$$\alpha_g + \alpha_l = 1 \tag{10}$$

$$\alpha_g = \frac{Q_g}{Q_g + Q_l} \tag{11}$$

where α_g is the IGVF, α_l is the inlet liquid volume fraction, Q_g is the gas volume flow, and Q_l is the liquid volume flow.

The outlet boundary condition is set to 0.5 MPa static pressure to avoid cavitation in industrial processes [17]. The rotational speeds during the start-up process under different time steps are compared to verify their independence [22]. When the time step is 0.0005 s, it has little effect on rotational speed in the start-up process. So the time step of the transient calculation is set at 0.0005 s. Steady results are used as initial conditions for transient calculation. Starting from the stationary state, the rotational speed of PAT increases with the iteration of the time step. When the resultant torque on the impeller reduces to nearly 0, the rotational speed no longer increases and keeps stable near the rated speed. It indicates that the transient calculation of the PAT start-up process is completed.

2.5. Experimental Verification

The test device for PAT characteristics is shown in Figure 4. The test device and equipment meet the II-level accuracy requirements in GB/T 3216-2005. Based on ensuring that the liquid flow rate is at a constant value, the gas flow rate is controlled by the gas regulating valve to obtain the gas–liquid mixture with different gas content so as to carry out the gas–liquid two-phase test of PAT. The water in the water storage tank is mixed with the air generated by the compressor in the gas–liquid mixing device after passing through the electromagnetic flowmeter, and then they are transported to the PAT. Taking α_g = 0.15 as an example, the head and efficiency of PAT after the end of the start-up are tested under different flow rates and compared with the simulated results, as shown in Figure 5.

The rotational speeds are acquired by testing once every 0.1 s during the start-up process. Figure 6 shows the comparison of rotational speed during the start-up process between simulated and experimental results at $\alpha_g = 0.15$. The simulated results are consistent with the experimental results; this indicates that the simulation method in this study is reliable. Since the inlet flow rate of PAT decreases with the increase in IGVF, the start-up time under gas–liquid two-phase conditions is longer than that under pure water conditions.

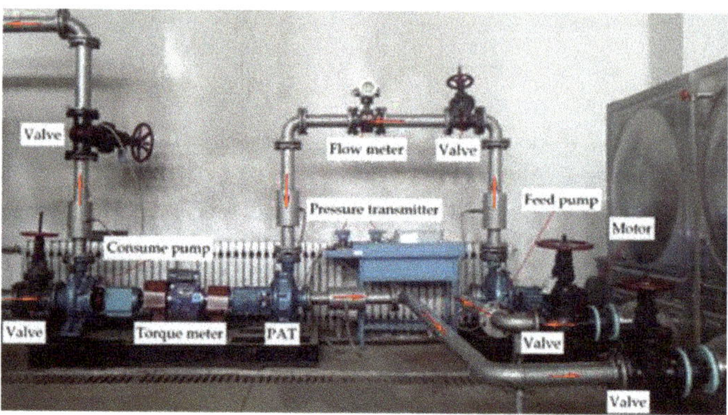

Figure 4. Test device for PAT characteristics.

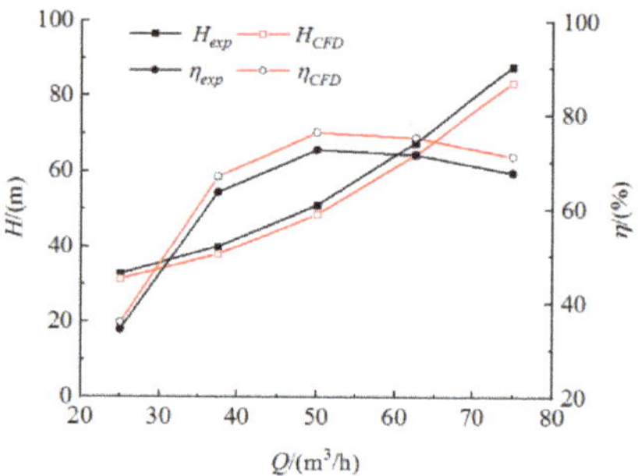

Figure 5. Comparison of performance between simulated and experimental results at $\alpha_g = 0.15$.

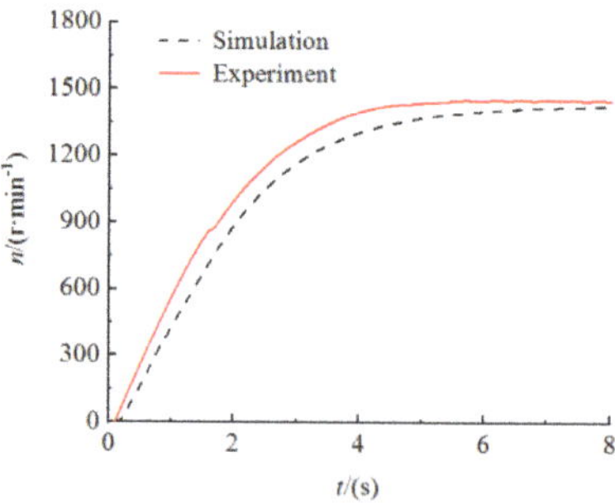

Figure 6. Comparison of rotational speed during the start-up process between simulated and experimental results at $\alpha_g = 0.15$.

3. Results and Analysis

3.1. Internal Flow Field Distribution

Whether the pressure distribution is uniform in the impeller channel directly affects the force of the impeller. The pressure distribution in the impeller is related to the change in rotational speed. The starting process is an accelerated motion with smaller and smaller acceleration. At the beginning of starting, the acceleration is rather big, and the pressure distributes more unevenly in the impeller. With the decrease in acceleration, the pressure distributes more uniformly and gradually becomes a gradient distribution in the impeller. From the most uneven at the beginning to the gradual gradient distribution afterward, the variation trend of pressure distribution during the start-up process is similar at different IGVFs. Thus the IGVF $\alpha_g = 0.15$ is taken as an example for analysis. Figure 7 shows the static pressure distribution of the axial vertical plane in the flow field domain during the start-up process. From the diagram, it is observed that a lot of low-pressure areas appear in the impeller flow channel at $t = 1$ s. At $t = 2$ s, with the increase in rotational speed, the low-pressure area in the impeller gradually decreases and converges to the inner edge of the impeller. After $t = 3$ s, the rotational speed tends to be stable, and the pressure decreases gradually from the volute inlet to the impeller outlet in gradient distribution along the channel direction. The low-pressure area is mainly concentrated on the inner edge of the impeller. The pressure on the suction surface of the blade is more significant than that on the pressure surface of the blade.

Figure 7. Pressure distribution of axial vertical plane at $\alpha_g = 0.15$. (**a**) $t = 1$ s; (**b**) $t = 2$ s; (**c**) $t = 3$ s; (**d**) $t = 4$ s.

Figure 8 shows the velocity streamline distribution of the axial vertical surface in the flow field during the start-up process. From the diagram, it is observed that large-scale high-intensity vortices are formed in the impeller flow channel at $t = 1$ s. At $t = 2$ s, with the increase in rotational speed, the vortex intensity in the impeller decreases, and the vortices aggregate near the outer edge of the impeller. After $t = 3$ s, the rotational speed tends to be stable, the vortex intensity in the impeller decreases further, and the vortices show a stability trend. They are mainly concentrated on the outer edge of the impeller.

Figure 8. Velocity streamlines distribution of the axial vertical plane at $\alpha_g = 0.15$. (**a**) $t = 1$ s; (**b**) $t = 2$ s; (**c**) $t = 3$ s; (**d**) $t = 4$ s.

Figure 9 shows the gas volume fraction distribution of the axial vertical plane in the flow field during the start-up process. From the diagram, it is observed that the gas phase distribution in the whole flow field is not uniform at $t = 1$ s. The gas volume fraction in the middle area of the impeller flow channel is high, and the distribution range is wide. Compared with the velocity streamline in Figure 8, it is found that the area with high gas phase concentration corresponds to a strong vortex at the beginning of start-up, which shows that the generation of vortex in the impeller flow channel has a great relationship with the accumulation of gas. At $t = 2$ s, the area of high gas phase concentration decreases significantly, and the gas phase gradually gathers to the central region of the impeller. After $t = 3$ s, the gas volume fraction increases gradually from the volute inlet to the draft tube outlet. The gas phase will gather more in the inner edge of the impeller, and the gas volume fraction near the blade suction surface is more significant than that near the blade pressure surface. It is indicated that the gas phase tends to move along the blade suction toward the impeller outlet. After entering the impeller, the liquid phase is subjected to a sizeable inertial force and centrifugal force, deviating from the typical streamlined trajectory and moving to the blade pressure surface. In comparison, the inertial and centrifugal force of the gas phase subjected is relatively less. Under the squeezing action of the liquid phase, the gas phase is forced to shift to the suction surface, so the gas volume fraction of the suction surface is higher.

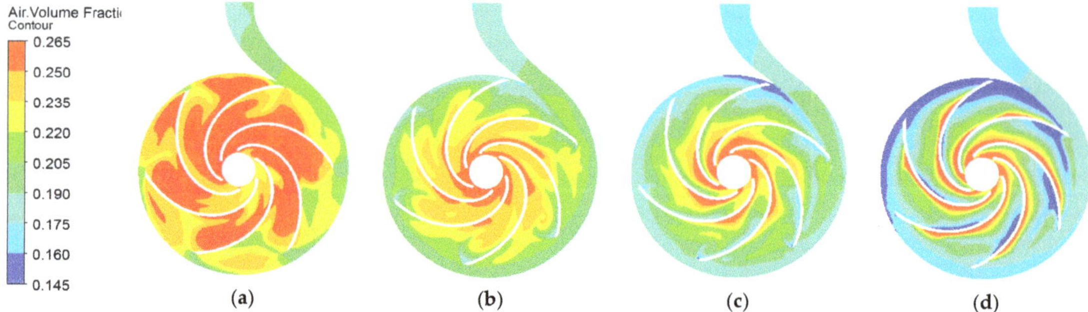

Figure 9. Gas volume fraction distribution of axial vertical plane at $\alpha_g = 0.15$. (**a**) $t = 1$ s; (**b**) $t = 2$ s; (**c**) $t = 3$ s; (**d**) $t = 4$ s.

3.2. Blade Load Distribution

The size and distribution characteristics of the PAT blade load directly affect the conversion efficiency between fluid and mechanical energy and the stable operation of the equipment. The blade load is obtained by calculating the pressure difference between the blade pressure surface and the blade suction surface. Figure 10 shows the distribution of blade load during the start-up process under different inlet volume fractions. The ordinate p in the figure is the blade load, and the abscissa L^* is the dimensionless parameter representing the blade position. The 0 position defines the blade inlet, and the 1 position defines the blade outlet. From the diagram, it is observed that the blade load will oscillate violently from blade inlet to blade outlet under pure water and gas–liquid two-phase conditions at $t = 1$ s. The maximum positive load is generated at the position of 0.24~0.34. At this moment, the pressure on the pressure surface is more significant than on the suction surface, resulting in the maximum dynamic torque. The blade load becomes 0 at the position of 0.48~0.53. The maximum negative load is generated at the position of 0.63~0.78. At this moment, the pressure on the pressure surface is less than that on the suction surface, resulting in the maximum resistance torque. At $t = 2$ s, the oscillation of the blade load distribution curve obviously weakens. The amplitude of the blade load is the largest under the condition of $\alpha_g = 0.15$. The wave peak of the blade load curve is at 0.27 position, and its wave trough is at 0.59 position. The maximum amplitude of the blade load is reduced to 0.59 times that at $t = 1$ s. At $t = 3$ s, the blade load distribution is relatively flat under gas–liquid two-phase conditions. The blade load is the largest under the condition of $\alpha_g = 0.15$, and the maximum blade load is 0.47 times that at $t = 1$ s. At $t = 4$ s, the distribution of blade load shows a similar trend of steady condition. The blade load increases rapidly along the flow direction of the fluid firstly and then changes gently to the maximum load at the position of 0.6~0.8. The maximum blade load is 0.36 times that at $t = 1$ s, and then the blade load decreases gradually to 0. It is proposed that the maximum load at the initial start-up moment is distributed near the outer edge of the blade. With the increase in the rotational speed, the oscillation of the blade load gradually weakens. After the rotational speed is stable, the maximum load is distributed near the inner edge of the blade. At the initial start-up moment, the greater the IGVF, the greater the amplitude of the blade load. The crowding effect of the gas phase on the liquid phase increases with the increase in the IGVF, which results in the uneven distribution of the liquid pressure in the impeller flow channel and weakens the workability of the liquid phase to the impeller. Therefore, decreasing the IGVF is helpful in reducing the oscillation of the blade load during the start-up process, and thus the stability of the start-up process will be improved.

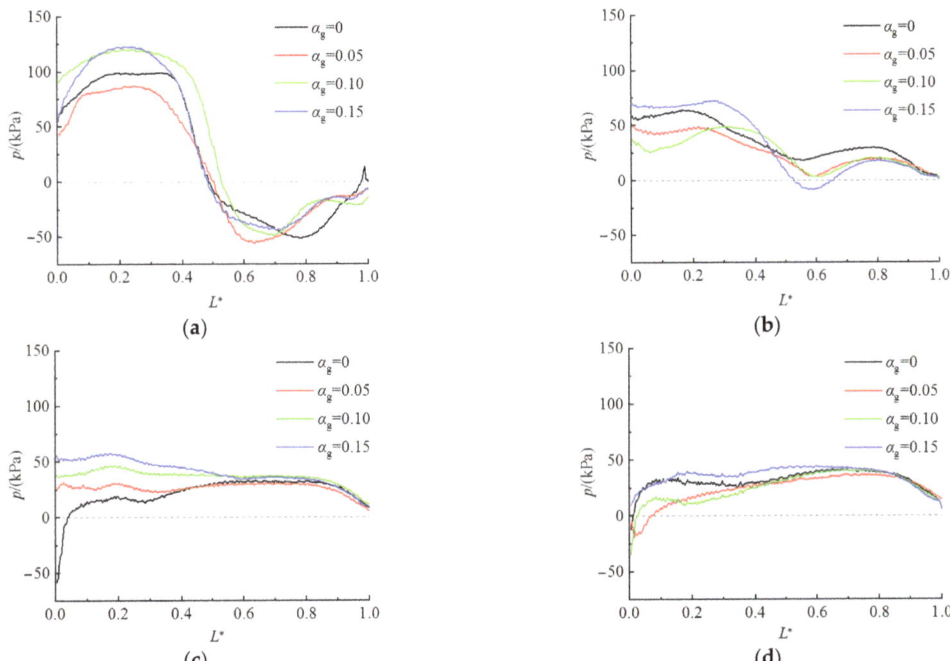

Figure 10. Blade load of PAT during the start-up process. (**a**) $t = 1$ s; (**b**) $t = 2$ s; (**c**) $t = 3$ s; (**d**) $t = 4$ s.

3.3. Radial Force Analysis

When the fluid enters the PAT impeller through the spiral volute, the fluid pressure distribution is not uniform around the impeller. Thus, the radial force acting on the impeller will be produced. The vibration is caused by dynamic and static interference between the rotating and stationary parts in the start-up process [23]. By monitoring the x-direction force F_x and y-direction force F_y of the entire impeller during the start-up process, their resultant force is the radial force F_r. Figure 11 shows the radial force vector distribution of a rotation period during the start-up process at $\alpha_g = 0.15$. The magnitude and direction of the radial force on the impeller change during the start-up process. At $t = 1$ s, the amplitude of radial force fluctuation is the largest. Due to the uneven distribution of the local vortex formed by the liquid flow in the impeller flow channel, the unbalanced radial force is produced on the impeller, and the vector trajectory of radial force is in the fourth quadrant.

At t = 2 s, the vector trajectory of radial force reduces and is still in the fourth quadrant, but its center shifts to the lower left side compared to t = 1 s. When $t \geq 3$ s, the vector trajectory of radial force is slightly enlarged, and its shape is approximately elliptical. This is because the PAT IGVF reaches a certain level, and the bubble destroys the symmetry of the internal flow of the impeller, resulting in uneven distribution of the liquid flow in each flow channel, and thus causing periodic unbalanced radial force, which will affect the stability of the start-up process.

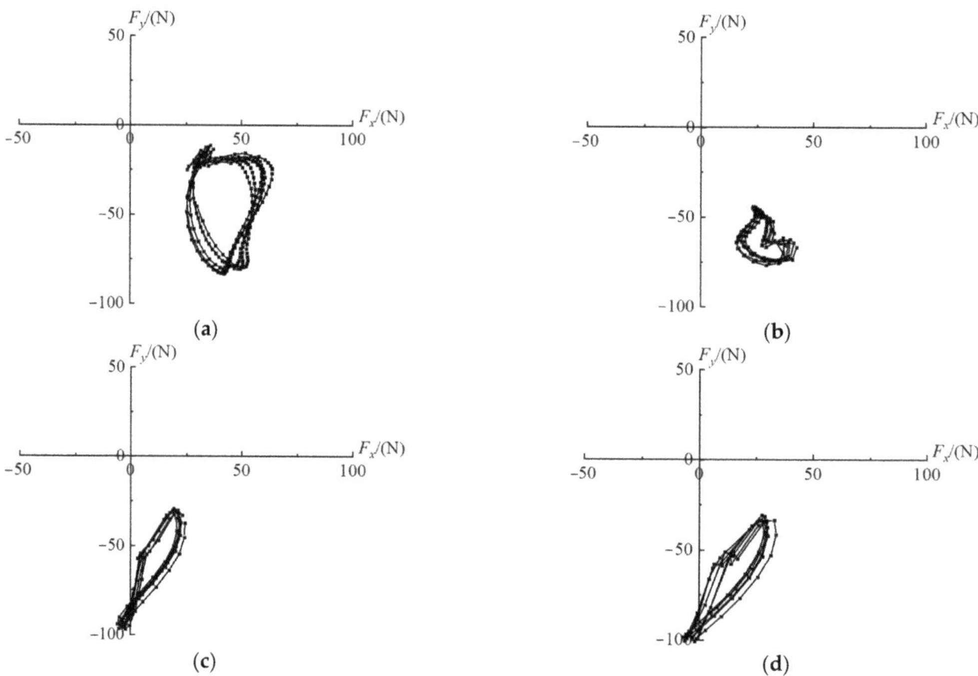

Figure 11. Vector distribution of radial force during the start-up process at α_g = 0.15. (**a**) t = 1 s; (**b**) t = 2 s; (**c**) t = 3 s; (**d**) t = 4 s.

To compare the effect of different IGVFs on the radial force time domain of the PAT start-up process, the start-up process has been respectively calculated under four conditions of α_g = 0, α_g = 0.05, α_g = 0.10, and α_g = 0.15. The time domain characteristic of radial force during the start-up process is shown in Figure 12. The abscissa is a rotation period when α_g = 0.15, and the ordinate is the radial force. Due to the lower IGVF, the greater the liquid inlet flow rate, the faster the PAT accelerates during the start-up process. At the same time, the more minor the IGVF, the more the number of radial force fluctuation. At t = 1 s, the time domain distribution of radial force is not uniform under different IGVFs. The amplitude of radial force is the largest under the condition of α_g = 0.10. At t = 2 s, the amplitude of radial force under α_g = 0 and α_g = 0.05 is much larger than that under α_g = 0.10 and α_g = 0.15. When $t \geq 3$ s, the time domain distribution of radial force tends to be stable and oscillates periodically. The amplitude of radial force at low IGVF is slightly larger than at high IGVF. The number of radial force fluctuations is consistent with the blade number in each change period. The impeller rotates one cycle, and the radial force fluctuates six times.

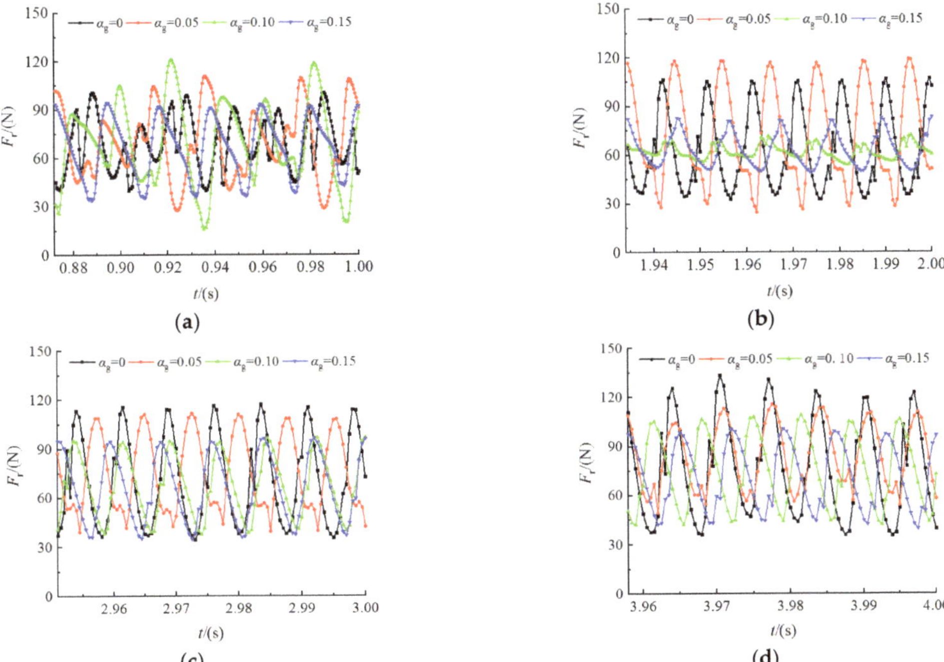

Figure 12. Time domain characteristic of radial force during the start-up process. (**a**) $t = 1$ s; (**b**) $t = 2$ s; (**c**) $t = 3$ s; (**d**) $t = 4$ s.

The frequency domain diagram of radial force is obtained by the Fast Fourier transform (FFT) for the time domain diagram of radial force in Figure 12, as shown in Figure 13. In the graph, the abscissa represents the multiple between the radial force frequency (f) and the impeller rotation frequency (f_n), and the ordinate represents the amplitude of radial force pulsation (A_r). At $t = 1$ s, the dominant frequency of radial force is $2\,f_n$, and its amplitude under gas–liquid two-phase conditions is more significant than that under pure water conditions. At $t = 2$ s, the dominant frequency of radial force is $4\,f_n$. At $t = 3$ s, the dominant frequency of radial force is $5\,f_n$. Its amplitude is the largest under the pure water conditions and the smallest under the condition of $\alpha_g = 0.10$. At $t = 4$ s, the dominant frequency of radial force is $6\,f_n$, that is, the blade rotation frequency and the secondary frequency is two times the blade rotation frequency. With the increase in IGVF, the dominant frequency amplitude of radial force increases slightly. From the pure liquid conditions to the gas–liquid two-phase conditions with increasing IGVF, at the same time, the dominant frequency amplitude shows a similar trend of decreasing firstly but then increasing. This is because a small number of bubbles will increase the non-uniformity of the flow rate, which prompts the mainstream direction to change, and causes the radial velocity of the internal fluid particles in the impeller to change. The amplitude of radial force pulsation becomes lower under gas–liquid two-phase conditions than pure water conditions. When the IGVF increases, the tiny bubbles rise and merge to form an unstable bubble flow. These bubbles continue to rupture or merge, which will exacerbate the oscillation of radial force.

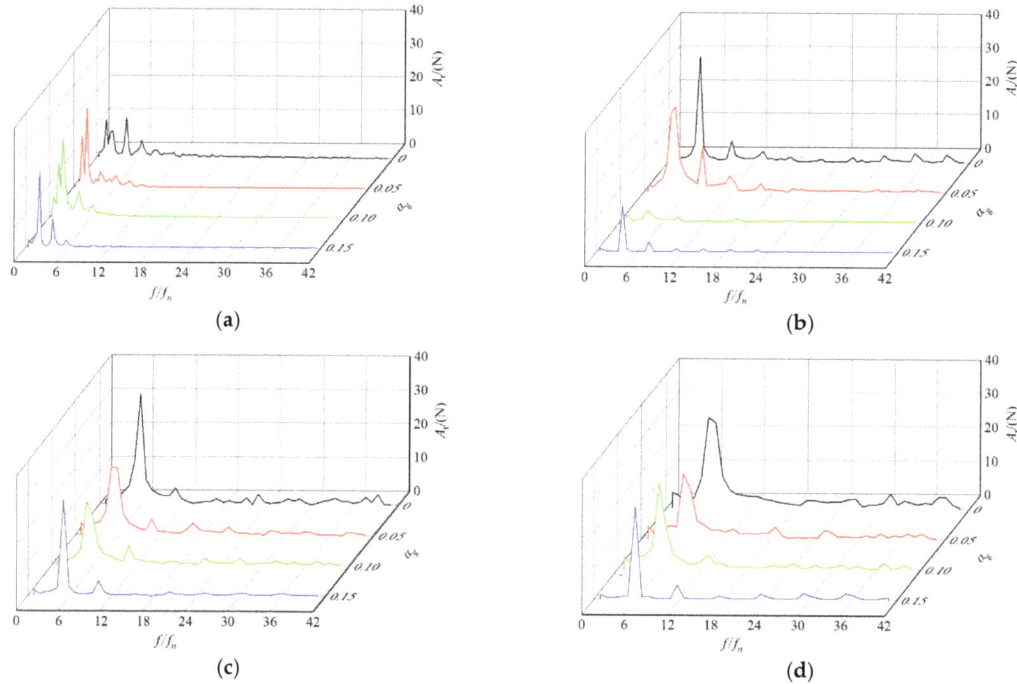

Figure 13. Frequency domain characteristic of radial force pulsation during the start-up process. (**a**) $t = 1$ s; (**b**) $t = 2$ s; (**c**) $t = 3$ s; (**d**) $t = 4$ s.

3.4. Axial Force Analysis

Due to the asymmetry between the front and the back shroud of the impeller and the difference in compression area, the axial force acting on the PAT impeller is produced in its start-up process. The axial force is mainly composed of the internal force acting on the front and the back shroud of the impeller, the external force acting on them, and the force acting on the blade [14]. The start-up process has been respectively calculated under four conditions of $\alpha_g = 0$, $\alpha_g = 0.05$, $\alpha_g = 0.10$, and $\alpha_g = 0.15$. The time domain characteristic of axial force during the start-up process is shown in Figure 14. At $t = 1$ s, the axial force shows the maximum amplitude and fluctuates twice in a fluctuation cycle under gas–liquid two-phase conditions. The magnitude and amplitude of the axial force under pure water conditions are greater than those under gas–liquid two-phase conditions. At $t = 2$ s, the axial force at $\alpha_g = 0$ and $\alpha_g = 0.05$ has a secondary fluctuation in a fluctuation period, and the axial force at $\alpha_g = 0.1$ and $\alpha_g = 0.15$ has a periodic oscillation. When $t \geq 3$ s, the axial force oscillates periodically under different IGVFs, and the amplitude of the axial force decreases gradually. The magnitude of axial force under low IGVF is more significant than those under high IGVF. The number of axial force fluctuations in each change cycle is consistent with the blade number.

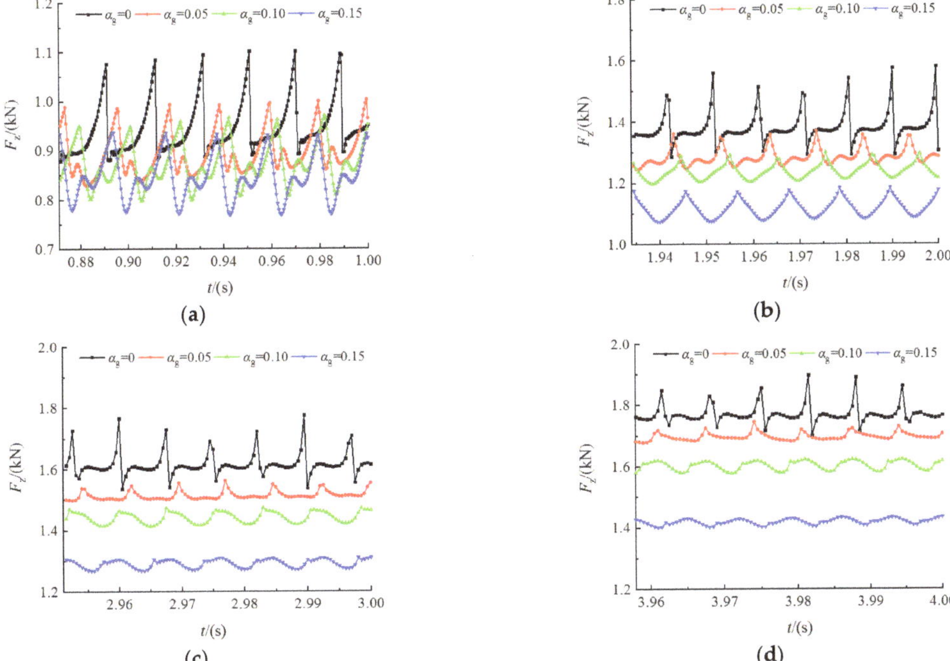

Figure 14. Time domain characteristic of axial force during the start-up process. (**a**) $t = 1$ s; (**b**) $t = 2$ s; (**c**) $t = 3$ s; (**d**) $t = 4$ s.

The frequency domain diagram of axial force is obtained by the FFT for the time domain diagram of axial force in Figure 14, as shown in Figure 15. At $t = 1$ s, the dominant frequency of axial force is $2 f_n$, and its amplitude under the gas–liquid two-phase conditions is more significant than that under the pure water conditions. At $t = 2$ s, the dominant frequency of axial force is $4 f_n$, and its amplitude is the largest at $\alpha_g = 0.15$. The frequency domain amplitude of axial force under pure water conditions attenuates slowly. The axial force still has a large frequency domain amplitude at the high-order frequency multiplication of the blade rotation frequency. At $t = 3$ s, the dominant frequency of axial force is $5 f_n$. At $t = 4$ s, the dominant frequency of axial force is $6 f_n$, that is, the blade rotation frequency, and the secondary frequency is equal to 2 times the blade rotation frequency. At the beginning of start-up, the dominant frequency amplitude of axial force is the largest, gradually decreasing with the increase in rotational speed. The higher the IGVF, the greater the dominant frequency amplitude of axial force at the same time.

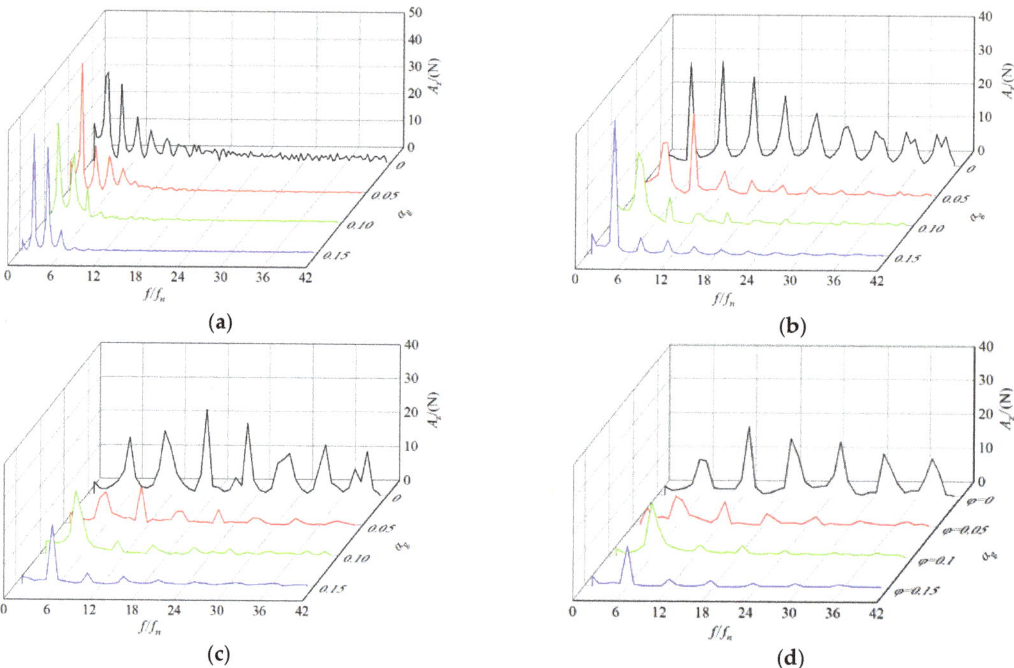

Figure 15. Frequency domain characteristic of axial force pulsation during the start-up process. (**a**) $t = 1$ s; (**b**) $t = 2$ s; (**c**) $t = 3$ s; (**d**) $t = 4$ s.

4. Conclusions

In this study, the evolution mechanism of the internal flow field and the force characteristics during the PAT start-up process under gas–liquid two-phase conditions were proposed using numerical simulation and experiments. The major findings are as follows:

(1) At the beginning of start-up, the gas volume fraction is high, its distribution is uneven in the impeller, and the area with high gas phase concentration corresponds to a strong vortex. With the increase in rotational speed, the low-pressure area, the high-concentration gas phase region, and the vortex intensity in the impeller gradually decrease. The gas phase gradually converges to the inner edge of the impeller, and the gas volume fraction of the blade suction surface is more significant than that of the blade pressure surface.

(2) At the beginning of start-up, the blade load oscillates violently, and the maximum load is distributed near the outer edge of the blade. As the rotational speed increases, the oscillation of the blade load gradually weakens, and the maximum load is distributed near the inner edge of the blade after the rotational speed is stable. The larger the IGVF, the more severe the blade load oscillation during the start-up process. Therefore, reducing the IGVF is helpful in improving the stability of the PAT start-up process.

(3) The periodic unbalanced radial force is produced in the start-up process. From the pure liquid conditions to the gas–liquid two-phase conditions with increasing IGVF, the dominant frequency amplitude of radial force shows a similar trend of decreasing first but then increasing. After the rotational speed tends to be stable, the dominant frequency of radial force is equal to the rotational frequency of the blade. With the increase in IGVF, the dominant frequency amplitude of radial force decreases slightly at the same time.

(4) At the beginning of start-up, the axial force shows the maximum amplitude and fluctuates twice in a fluctuation cycle under gas–liquid two-phase conditions. With the

increase in rotational speed, the dominant frequency amplitude of axial force decreases gradually. The higher the IGVF, the greater the dominant frequency amplitude of axial force at the same time.

Author Contributions: Conceptualization, B.C. and J.Y.; methodology, B.C. and J.Y.; software, B.C.; validation, B.C. and X.W.; formal analysis, B.C.; investigation, B.C. and X.W.; writing—original draft preparation, B.C.; writing—review and editing, B.C.; project administration, J.Y.; funding acquisition, J.Y. All authors have read and agreed to the published version of the manuscript.

Funding: This research was funded by the National Natural Science Foundation of China (grant number: 52169019) and the University Industry Support Project of Gansu Province (grant number: 2020C-20).

Data Availability Statement: The data presented in this study are available on request from the corresponding author.

Acknowledgments: The authors would like to express sincere appreciation to the editor and the anonymous reviewers for their valuable comments and suggestions for improving the presentation of the manuscript.

Conflicts of Interest: The authors declare no conflict of interest.

References

1. Jain, S.; Patel, R. Investigations on pump running in turbine mode: A review of the state-of-the-art. *Renew. Sustain. Energy Rev.* **2014**, *30*, 841–868. [CrossRef]
2. Wang, X.; Yang, J.; Shi, F. Status and prospect of study on energy recovery hydraulic turbines. *J. Drain. Irrig. Mach. Eng.* **2014**, *32*, 742–747.
3. Maxime, B.; Su, W.; Li, X.; Li, F.; Wei, X.; An, S. Investigation on pump as turbine (PAT) technical aspects for micro hydropower schemes: A state of the art review. *Renew. Sustain. Energy Rev.* **2017**, *79*, 148–179.
4. Barrios, L. Visualization and Modeling of Multiphase Performance Inside an Electrical Submersible Pump. Ph.D. Thesis, The University of Tulsa, Tulsa, OK, USA, 2007.
5. Gamboa, J. Prediction of the Transition in Two-Phase Performance of an Electrical Submersible Pump. Ph.D. Thesis, The University of Tulsa, Tulsa, OK, USA, 2008.
6. Li, G.; Wang, Y.; Zheng, Y.; Ma, X.; Liang, L.; Hu, R. Unsteady internal flow and thrust analysis of centrifugal pump under gas-liquid two-phase flow conditions. *J. Drain. Irrig. Mach. Eng.* **2016**, *34*, 369–374.
7. Yuan, J.; Zhang, K.; Si, Q.; Zhou, B.; Tang, Y.; Jin, Z. Numerical investigation of gas-liquid two-phase flow in centrifugal pumps based on inhomogeneous model. *Trans. Chin. Soc. Agric. Mach.* **2017**, *48*, 89–95.
8. Luo, X.; Yan, S.; Feng, J.; Zhu, G.; Sun, S.; Chen, S. Force characteristics of gas-liquid two-phase centrifugal pump. *Trans. Chin. Soc. Agric. Eng.* **2019**, *35*, 66–72.
9. Michael, M.; Saketh, K.; Dominique, T. Investigations on the effect of rotational speed on the transport of air-water two-phase flows by centrifugal pumps. *Int. J. Heat Fluid Flow* **2022**, *94*, 108939.
10. Yan, S.; Luo, X.; Feng, J.; Zhu, G.; Zhang, L.; Chen, S. Numerical simulation of a gas-liquid centrifugal pump under different inlet gas volume fraction conditions. *Int. J. Fluid Mach. Syst.* **2019**, *12*, 56–63.
11. Zhang, W.; Yu, Z.; Zahid, M.; Li, Y. Study of the gas distribution in a multiphase rotodynamic pump based on interphase force analysis. *Energies* **2018**, *11*, 1069. [CrossRef]
12. Anthony, C.; Laurent, G.; Daniel, P. Characteristics of centrifugal pumps working in direct or reverse mode: Focus on the unsteady radial thrust. *Int. J. Rotating Mach.* **2013**, *2013*, 279049.
13. Shi, F.; Yang, J.; Wang, X. Numerical prediction of radial force in hydraulic turbine based on fluent. *Adv. Mater. Res.* **2013**, *716*, 717–720. [CrossRef]
14. Qu, X.; Kong, F.; Chen, H. A numerical simulation and analysis of impeller axial force of pumps as turbines. *China Rural. Water Hydropower* **2013**, *7*, 96–102.
15. Yang, J.; Li, T. Influence of volute configurations on radial force of hydraulic turbine. *J. Drain. Irrig. Mach. Eng.* **2015**, *33*, 651–656.
16. Dai, C.; Kong, F.; Dong, L.; Zhang, H.; Feng, Z. Effect of blade wrap angle on the radial force of centrifugal pump as turbine. *J. Vib. Shock* **2015**, *34*, 69–72.
17. Shi, F.; Yang, J.; Wang, X. Unsteady analysis of on effect hydraulic turbine under variable gas content. *J. Aerosp. Power* **2017**, *32*, 2265–2272.
18. Shi, F.; Yang, J.; Miao, S.; Wang, X. Investigation on the power loss and radial force characteristics of pump as turbine under gas-liquid two-phase condition. *Adv. Mech. Eng.* **2019**, *11*, 1687814019843732.
19. Shi, G.; Liu, Y.; Luo, K. Analysis of gas-liquid two-phase flow field in hydraulic turbine considering gas compressibility. *J. Eng. Therm. Energy Power* **2018**, *33*, 40–45.

20. Sun, K.; Li, Y.; Zhao, J. Transient starting performance analysis of vertical axis tidal turbine. *Huazhong Univ. Sci. Tech. Nat. Sci. Ed.* **2017**, *45*, 51–56.
21. Chen, W.; Liu, Y.; Chen, L.W. Study on hydrodynamic performance of horizontal tidal turbine rotating passively based on UDF. *Ocean Eng.* **2018**, *36*, 119–126.
22. Untaroiu, A.; Wood, H.; Allaire, P.; Ribando, R. Investigation of self-starting capability of vertical axis wind turbines using a computational fluid dynamics approach. *J. Sol. Energy Eng.* **2011**, *133*, 125–134. [CrossRef]
23. Barrio, R.; Fernandez, J.; Blanco, E.; Parrondo, J. Estimation of radial load in centrifugal pumps using computational fluid dynamics. *Eur. J. Mech.-B/Fluids* **2011**, *3*, 316–324. [CrossRef]

Article

A Novel High-Speed Third-Order Sliding Mode Observer for Fault-Tolerant Control Problem of Robot Manipulators

Van-Cuong Nguyen [1,†], **Xuan-Toa Tran** [2,†] **and Hee-Jun Kang** [1,*,†]

1 Department of Electrical, Electronic and Computer Engineering, University of Ulsan, 93 Daehak-ro, Ulsan 44610, Korea
2 NTT Hi-Tech Institute, Nguyen Tat Thanh University, 300A Nguyen Tat Thanh Street, Ho Chi Minh City 70000, Vietnam
* Correspondence: hjkang@ulsan.ac.kr
† These authors contributed equally to this work.

Abstract: In this paper, a novel fault-tolerant control tactic for robot manipulator systems using only position measurements is proposed. The proposed algorithm is constructed based on a combination of a nonsingular fast terminal sliding mode control (NFTSMC) and a novel high-speed third-order sliding mode observer (TOSMO). In the first step, the high-speed TOSMO is proposed for the first time to approximate both the system velocity and the lumped unknown input with a faster convergence time compared to the TOSMO. The faster convergence speed is obtained thanks to the linear characteristic of the added elements. In the second step, the NFTSMC is designed based on a nonsingular fast terminal sliding (NFTS) surface and the information obtained from the proposed high-speed TOSMO. Thanks to the combination, the proposed controller–observer tactic provides excellent features, such as a fast convergence time, high tracking precision, chattering phenomenon reduction, robustness against the effects of the lumped unknown input and velocity requirement elimination. Especially, the proposed observer does not only improve the convergence speed of the estimated signals, but also increases the system dynamic response. The system's finite-time stability is proved using the Lyapunov theory. Finally, to validate the efficiency of the proposed strategy, simulations on a PUMA560 robot manipulator are performed.

Keywords: high-speed third-order sliding mode observer; nonsingular fast terminal sliding mode control; controller–observer strategy; faster convergence; fault-tolerant control

Citation: Nguyen, V.-C.; Tran, X.-T.; Kang, H.-J. A Novel High-Speed Third-Order Sliding Mode Observer for Fault-Tolerant Control Problem of Robot Manipulators. *Actuators* **2022**, *11*, 259. https://doi.org/10.3390/act11090259

Academic Editor: Zhuming Bi

Received: 9 August 2022
Accepted: 6 September 2022
Published: 8 September 2022

Publisher's Note: MDPI stays neutral with regard to jurisdictional claims in published maps and institutional affiliations.

1. Introduction

In the industry, robot manipulators are employed widely in various applications, such as material handling, milling, painting, welding and roughing. Along with the growth of robot manipulator applications, interest in robotic control has been increased [1–4]. In some research, the robot's end-effector position is required to track the desired trajectory. For this purpose, the kinematics control approach is preferred [5,6]. Another approach is to use the dynamics control when joint angles are preferred [7,8]. Generally, robot manipulators are difficult to control in both theoretical and practical aspects due to some of their characteristics. First, robot manipulator systems have a highly nonlinear and very complex dynamic in coupling terms. Additionally, payload changes, friction, external disturbances, etc., lead to robot dynamic uncertainty. Therefore, obtaining the robot's correct dynamics is arduous or even impossible. In some special cases, in long-term operation, unknown faults can occur when the robot is operating, which includes actuator faults or sensor faults. Further, to reduce the weight/size and save costs, in some cases, manufacturers remove the velocity sensors in the robot. These are big problems that have been challenged by many researchers. To simplify the presentation and avoid duplication, in this article, the dynamic uncertainty and the unknown fault are treated and abbreviated as a lumped unknown input.

To deal with the aforementioned lumped unknown input, many control methods have been developed, such as proportional–integral–derivative (PID) control [9,10], adaptive control [11], fuzzy control [12], neural network (NN) control [13] and sliding mode control (SMC) [14,15]. PID control is well-known as a simple and monotonic controller, which does not require the dynamic model of the robot system. However, this controller cannot achieve high tracking performance. Adaptive control is an effective method to deal with matched uncertainties; however, it is not appropriate for the problem of mismatched uncertainties. Intelligent control schemes are widely employed, such as NN control and fuzzy control. The learning ability and good approximation of nonlinear function with the arbitrary accuracy of NN controllers make them a good choice for modelling complex processes and compensating for mismatched uncertainties. However, transient performance in the presence of a disturbance can be degraded due to the required online learning procedure. The fuzzy logic control method was developed based on expert knowledge and experience; however, its main disadvantages are difficulties in the stability analysis and comprehensive knowledge of the requirements of a system. SMC is one of the most powerful robust controllers that has been widely utilized in the fault-tolerant control problem of robot manipulators due to its effectiveness in rejecting the effects of the lumped unknown input [16–18]. Moreover, the design procedure of SMC is not complex and quite popular in the literature [19–21]. Unfortunately, conventional SMC uses a linear sliding surface that means the finite-time convergence cannot be guaranteed. In order to achieve a finite-time convergence, a nonlinear sliding surface is utilized instead of a linear one in the design process of the controller; this technique is well-known as terminal SMC (TSMC) [22,23]. Compared to conventional SMC, TSMC extends two outstanding properties, which are finite-time convergence and achieving higher accuracy when parameters are carefully designed. Unfortunately, TSMC only obtains a faster convergence when system states are near the equilibrium point, but slower when the system states are far from the equilibrium point. In addition, TSMC suffers from the singularity problem. The two problems have been handled separately using fast TSMC (FTSMC) [24,25] and nonsingular TSMC (NTSMC) [26–28]. In order to solve both problems at the same time, nonsingular fast TSMC (NFTSMC) was developed [29–31]. Due to its excellent control features, such as providing finite-time convergence, eliminating the singularity problem, achieving high-position tracking precision and robustness against the lumped unknown input, NFTSMC has been applied widely in the literature. However, same as SMC and TSMC, utilizing a discontinuous switching element with a big and fixed sliding gain to handle the effects of the lumped unknown input in the designing process of NFTSMC is the root of high-frequency oscillations, the so-called chattering [32]. This phenomenon harms the system; therefore, it decreases the practical applicability of SMC. In addition, the design procedure of NFTSMC involves real velocity information, which is sometimes unavailable in practical systems.

To resolve the chattering problem, the elementary idea is to reduce the sliding gain in the switching control component. In this approach, the lumped unknown input is first completely or partially estimated. After that, the estimated unknown input is applied in a controller design as a compensator to reduce the lumped unknown input effects. Therefore, the switching control component is now utilized to carry out the impacts of the estimation error instead of the lumped unknown input as in the original controller. As a result, the sliding gain is chosen with a smaller value, thus, the chattering phenomenon can be reduced. In the literature, various techniques for fault diagnosis have been proposed to approximate the lumped unknown input, such as the NN observer [33,34], adaptive observer [35,36], time-delay estimation [37,38], linear extended state observer (LESO) [39,40], second-order sliding mode observer (SOSMO) [41] and third-order sliding mode observer (TOSMO) [42,43]. Among them, the SOSMO and the TOSMO stand out with the capability to estimate not only the lumped unknown input, but also the velocity information; therefore, the requirement of the tachometer is eliminated without the need of an additional state observer. Concerning the comparison of the above two observers, the main advantage of the SOSMO is its ability to provide a faster approximation speed. In contrast, the TOSMO can provide estimation signals with a higher estimation accuracy and less chattering, without any filtration. Unfortunately, as a trade-off, its convergence speed becomes slower than that of the SOSMO. Therefore, it is necessary to design an observer that can combine the wonderful properties of both the SOSMO and the TOSMO.

Motivated by all the above discussions, this paper first proposes a novel high-speed TOSMO for the robot manipulator system by adding additional terms to the original TOSMO. This observer can not only maintain the remarkable benefits of the original TOSMO, but also obtain a faster convergence time. The estimated velocity and unknown input are then applied to design a fault-tolerant control based on NFTSMC. The major contributions of this paper are summarized as follows:

1. The proposal of a novel high-speed TOSMO that can obtain a faster convergence speed while maintaining the high estimation accuracy of the TOSMO;
2. The proposal of a fault-tolerant control law based on NFTSMC and the proposed high-speed TOSMO that handles the effects of the lumped unknown input to achieve a higher tracking accuracy and low chattering phenomenon;
3. The provision of proof of the system finite-time stability when combining a controller and observer.

This paper is organized into six parts. Following the introduction, the mathematical dynamics model of robot manipulators and problem formulation are presented in Section 2. In Section 3, the design of the high-speed TOSMO is presented, followed by the design procedure of the fault-tolerant control law based on NFTSMC for the robot manipulators shown in Section 4. To confirm the efficiency of the proposed method, computer simulations on a PUMA560 robot manipulator are shown in Section 5. Finally, Section 6 gives some conclusions.

2. Mathematical Dynamics Model of Robot Manipulators and Problem Formulation

2.1. Robot Dynamics

Let us consider a serial n-link robot manipulator under the effects of dynamic uncertainty and unknown fault as follows:

$$\ddot{q} = M(q)^{-1}\left[\tau(t) - C(q,\dot{q})\dot{q} - G(q) - F_r(\dot{q}) - \tau_d(t)\right] + \Omega(q,\dot{q},t) \tag{1}$$

where $q,\dot{q},\ddot{q} \in \Re^n$ correspondingly represent the robot's joint positions, velocity and acceleration vectors; $M(q) \in \Re^{n \times n}$ represents the inertia matrix, which is symmetric and positively definite, making it invertible; $C(q,\dot{q}) \in \Re^{n \times n}$, $G(q) \in \Re^n$ and $F_r(\dot{q}) \in \Re^n$ denote the Coriolis and centripetal forces, gravitational vector and friction vector, respectively; $\tau_d(t) \in \Re^n$ denotes the disturbance vector; $\tau(t) \in \Re^n$ denotes the control input signal; and $\Omega(q,\dot{q},t) = \omega\left(t - T_f\right)\Phi(q,\dot{q})$ denotes the unknown but bounded fault with the time profile $\omega\left(t - T_f\right) = diag\left\{\omega_1\left(t - T_f\right), \omega_2\left(t - T_f\right), \ldots, \omega_n\left(t - T_f\right)\right\}$, where $\omega_i\left(t - T_f\right) = \begin{cases} 0 & if\ t \le T_f \\ 1 - e^{-\zeta_i(t - T_f)} & if\ t \ge T_f \end{cases}$. The unknown fault function $\Phi(q,\dot{q})$ occurs at time T_f with evolution rate $\zeta_i > 0$, $(i = 1, 2, \ldots, n)$.

By defining $u = \tau(t)$ and $x = \begin{bmatrix} x_1^T & x_2^T \end{bmatrix}^T$ with $x_1 = q$, $x_2 = \dot{q}$, we could transfer system (1) into the state space form as

$$\begin{aligned} \dot{x}_1 &= x_2 \\ \dot{x}_2 &= \Psi(x) + M(x_1)^{-1}u + \Delta(x,t) \end{aligned} \tag{2}$$

where $\Psi(x) = M(q)^{-1}\left[-C(q,\dot{q}) - G(q)\right]$ represents the nominal model of robot manipulators and $\Delta(x,t) = M(q)^{-1}\left[-F_r(\dot{q}) - \tau_d\right] + \Omega(q,\dot{q},t)$ denotes the lumped unknown but bounded dynamic uncertainty and unknown fault.

Remark 1. *In this paper, the unknown fault was considered as an additional dynamic uncertainty; therefore, their total effects on the system were carried out. We simply named them the lumped unknown input.*

2.2. Problem Formulation

Let $x_d \in \Re^n$ be an expected trajectory of a robot's joint, where the tracking error is defined as

$$e = x_1 - x_d \tag{3}$$

The central purpose of this paper was divided into two parts. First, a novel high-speed TOSMO was proposed for the first time to estimate both a system's states and the lumped unknown input with high precision and a fast response time. Second, based on the achieved information from the proposed observer, a fault-tolerant control approach using NFTSMC was then designed for system (2) to ensure that the joint position x_1 could track the desired trajectory x_d with high accuracy, even in the presence of the lumped unknown input and the absence of the velocity measurement. In addition, the controller further demonstrated the effectiveness of the proposed high-speed TOSMO. The proposed algorithm was constructed based on assumptions as follows:

Assumption 1. *The desired trajectory x_d was a twice continuously differentiable function in respect to time t.*

Assumption 2. *The lumped unknown input $\Delta(x, t)$ was bounded by a positive constant Δ_D as*

$$|\Delta(x, t)| \leq \Delta_D \tag{4}$$

Assumption 3. *The derivative of the lumped unknown input $\Delta(x, t)$ in respect to time existed and was bounded by a positive constant $\Delta_{\dot{D}}$ as*

$$\left| \frac{d}{dt} \Delta(x, t) \right| \leq \Delta_{\dot{D}} \tag{5}$$

Note that Assumption 3 is realistic and has been used in much research [44–46].

3. Design of Observer

3.1. High-Speed Third-Order Sliding Mode Observer

In this section, a novel high-speed TOSMO was proposed by adding additional terms to the original TOSMO. Thanks to the linear characteristic of these terms, which can strongly deal with perturbations that are very far away from the origin, the slow convergence problem of the TOSMO was improved.

Based on system (2), the novel high-speed TOSMO was proposed as follows:

$$\begin{aligned}
\dot{\hat{x}}_1 &= k_1 \left| x_1 - \hat{x}_1 \right|^{2/3} sign\left(x_1 - \hat{x}_1 \right) + \hat{x}_2 + \boldsymbol{\Gamma}\left(\boldsymbol{x_1 - \hat{x}_1} \right) \\
\dot{\hat{x}}_2 &= \Psi(\hat{x}) + M\left(x_1 \right)^{-1} u + k_2 \left| x_1 - \hat{x}_1 \right|^{1/3} sign\left(x_1 - \hat{x}_1 \right) + \boldsymbol{\Gamma}\left(\dot{\boldsymbol{x}}_1 - \hat{\boldsymbol{x}}_2 \right) - \hat{z} \\
\dot{\hat{z}} &= -k_3 sign\left(x_1 - \hat{x}_1 \right)
\end{aligned} \tag{6}$$

where $\hat{x}$ is the estimation of x, $k_i (i = 1,\ 2,\ 3)$ denotes the sliding gains and Γ is a positive constant. In the proposed observer, the additional terms were bolded to distinguish them from the original TOSMO.

Theorem 1. *For the robot manipulator system given in (2) with the high-speed TOSMO (6), if the sliding gains of the observer were chosen as Remark 3, then the proposed observer was stable and the estimation states $(\hat{x}_1,\ \hat{x}_2)$ would achieve the real states $(x_1,\ x_2)$ in finite time.*

Proof of Theorem 1. By subtracting (6) from (2), we obtained the estimation errors as

$$
\begin{aligned}
\dot{\tilde{x}}_1 &= -k_1|\tilde{x}_1|^{2/3}sign(\tilde{x}_1) + \tilde{x}_2 - \Gamma\tilde{x}_1 \\
\dot{\tilde{x}}_2 &= -k_2|\tilde{x}_1|^{1/3}sign(\tilde{x}_1) - \Gamma\left(\dot{\hat{x}}_1 - \hat{x}_2\right) + \Delta(x,t) - \delta(x,\tilde{x}) + \hat{z} \\
\dot{\hat{z}} &= -k_3 sign(\tilde{x}_1)
\end{aligned}
\tag{7}
$$

where $\tilde{x}_1 = x_1 - \hat{x}_1$ and $\tilde{x}_2 = x_2 - \hat{x}_2$ represent the position and velocity estimation errors, respectively. The estimation error $\delta(x,\tilde{x}) = \Psi(\hat{x}) - \Psi(x)$. To facilitate the next design approach, in this phase, we assumed that $|\delta(x,\tilde{x})| \leq \Delta_\delta$ and $\left|\dot{\delta}(x,\tilde{x})\right| \leq \Delta_{\dot{\delta}}$, where Δ_δ and $\Delta_{\dot{\delta}}$, were positive constants. Note that this assumption was the same as Assumption 3; thus, it could be supported in both theoretical and practical applications.

Substituting the first term of (6) into (7), we obtained

$$
\begin{aligned}
\dot{\tilde{x}}_1 &= -k_1|\tilde{x}_1|^{2/3}sign(\tilde{x}_1) + \tilde{x}_2 - \Gamma\tilde{x}_1 \\
\dot{\tilde{x}}_2 &= -k_2|\tilde{x}_1|^{1/3}sign(\tilde{x}_1) + \Delta(x,t) - \delta(x,\tilde{x}) + \hat{z} - \Gamma k_1|\tilde{x}_1|^{2/3}sign(\tilde{x}_1) - \Gamma^2\tilde{x}_1 \\
\dot{\hat{z}} &= -k_3 sign(\tilde{x}_1)
\end{aligned}
\tag{8}
$$

By defining the new variable $\bar{x} = \tilde{x}_2 - \Gamma\tilde{x}_1$, the estimation errors in (8) were rewritten as

$$
\begin{aligned}
\dot{\tilde{x}}_1 &= -k_1|\tilde{x}_1|^{2/3}sign(\tilde{x}_1) + \bar{x} \\
\dot{\bar{x}} &= -k_2|\tilde{x}_1|^{1/3}sign(\tilde{x}_1) + \Delta(x,t) - \delta(x,\tilde{x}) + \hat{z} \\
&\quad -\Gamma k_1|\tilde{x}_1|^{2/3}sign(\tilde{x}_1) - \Gamma^2\tilde{x}_1 \\
&\quad \underbrace{+\Gamma k_1|\tilde{x}_1|^{2/3}sign(\tilde{x}_1) + \Gamma^2\tilde{x}_1 - \Gamma\tilde{x}_2}_{-\Gamma\dot{\tilde{x}}_1} \\
\dot{\hat{z}} &= -k_3 sign(\tilde{x}_1)
\end{aligned}
\tag{9}
$$

Letting $\hat{\Delta}(x,t) = \Delta(x,t) - \delta(x,\tilde{x}) + L$, the system (9) became

$$
\begin{aligned}
\dot{\tilde{x}}_1 &= -k_1|\tilde{x}_1|^{2/3}sign(\tilde{x}_1) + \bar{x} \\
\dot{\bar{x}} &= -k_2|\tilde{x}_1|^{1/3}sign(\tilde{x}_1) + \hat{\Delta}(x,t) + \hat{z} \\
\dot{\hat{z}} &= -k_3 sign(\tilde{x}_1)
\end{aligned}
\tag{10}
$$

where $L = -\Gamma\tilde{x}_2$ with the assumptions were that $|L| \leq \Delta_L$ and $\left|\frac{d}{dt}L\right| \leq \Delta_{\dot{L}}$.

By defining $\tilde{x}_3 = \hat{z} + \hat{\Delta}(x,t)$, the system in (10) could be rewritten in the same form of the second-order sliding mode differentiator [47] as

$$
\begin{aligned}
\dot{\tilde{x}}_1 &= -k_1|\tilde{x}_1|^{2/3}sign(\tilde{x}_1) + \bar{x} \\
\dot{\bar{x}} &= -k_2|\tilde{x}_1|^{1/3}sign(\tilde{x}_1) + \tilde{x}_3 \\
\dot{\tilde{x}}_3 &= -k_3 sign(\tilde{x}_1) + \dot{\hat{\Delta}}(x,t)
\end{aligned}
\tag{11}
$$

Equation (11) is also as well-known as the TOSMO [48]. By selecting the candidate Lyapunov function V_0 and using the same demonstrating process as in [48], it could be concluded that system (11) was stable and the differentiators $\tilde{x}_1$, $\bar{x}$ and $\tilde{x}_3$ converged to zero in finite time. Thus, system (7) was stable and the estimation errors $\tilde{x}_1$, $\tilde{x}_2$ converged to zero in finite time. $\square$

Remark 2. *The proposed high-speed TOSMO in (6) was designed based on the original TOSMO in [42]. The linear characteristics of the added terms were utilized to increase the convergence speed of the estimated signals.*

Remark 3. *The observer gains of (6) were selected according to [47] as $k_1 = \alpha_1 \overline{\Delta}^{1/3}$, $k_2 = \alpha_2 \overline{\Delta}^{2/3}$ and $k_3 = \alpha_3 \overline{\Delta}$, where $\alpha_1 = 2$, $\alpha_2 = 2.12$ and $\alpha_3 = 1.1$ with $\overline{\Delta} = \Delta_{\dot{D}} + \Delta_{\dot{\delta}} + \Delta_{\dot{L}}$.*

3.2. Unknown Input Identification

After the convergence time, the estimated velocity reached the real velocity, $\hat{x}_2 = x_2$; thus, the term $L = -\Gamma \tilde{x}_2$ converged to zero. The third term of system (11) then became

$$\dot{\tilde{x}}_3 = -k_3 sign(\tilde{x}_1) + \dot{\hat{\Delta}}(x,t) \equiv 0 \tag{12}$$

Notably, because the velocity estimation error $\tilde{x}_2$ converged to zero, the auxiliary unknown input term $\hat{\Delta}(x,t) = \Delta(x,t) - \delta(x,\tilde{x}) + L$ converged to $\Delta(x,t)$.

The lumped unknown input terms could be rebuilt as

$$\hat{\Delta}(x,t) = \int k_3 sign(\tilde{x}_1) \tag{13}$$

Since the estimated unknown input in (13) included an integral operation, the lumped unknown input terms could be rebuilt directly from the output injection term, and the chattering of the obtained function was partially eliminated without the need for a lowpass filter. In addition, the proposed observer in (6) not only maintained the advantages of the conventional TOSMO such as the finite-time convergence and high estimation accuracy for both velocity and the lumped unknown input, but also provided a faster convergence time than that of the TOSMO. The outstanding features of the proposed high-speed TOSMO were verified in the simulation part.

Remark 4. *The obtained lumped unknown input could be used for fault detection and fault isolation and could also be applied to the fault-tolerant control to eliminate its effect on the system. The estimated velocity could be employed in the controller design process instead of the real measured velocity.*

4. Design of Control Algorithm

In this section, a fault-tolerant control tactic using the NFTSMC algorithm was proposed to carry out the effects of the lumped unknown input of system (2). In addition, in some special cases, the tachometers in robots would be cut off by manufacturers to save cost and reduce weight. For that reason, this paper assumed that the tachometers did not exist. The estimated velocity, $\hat{x}_2$, which was obtained from the proposed observer in Section 3, was utilized; therefore, the requirement of the real measured velocity was eliminated.

4.1. Design of Nonsingular Fast Terminal Sliding Surface

Let us define the estimated velocity error as

$$\hat{e} = \hat{x}_2 - \dot{x}_d \tag{14}$$

where $\dot{x}_d$ describes the desired velocity.

A NFTS surface was selected as in [49]

$$\hat{s} = \hat{e} + \int \left[\beta_1 |e|^{\gamma_1} sign(e) + \beta_2 \left| \hat{e} \right|^{\gamma_2} sign\left(\hat{e} \right) + \beta_3 e + \beta_4 e^3 \right] dt \tag{15}$$

where the parameters β_1, β_2, β_3, β_4 are positive constants and the parameters γ_1, γ_2 could be selected as $0 < \gamma_1 < 1$, $\gamma_2 = 2\gamma_1/(1 + \gamma_1)$.

According to the SMC theory, the following conditions were satisfied when the robot system reached the sliding mode:

$$\hat{s} = 0$$

$$\dot{\hat{s}} = 0 \tag{16}$$

Thus, the sliding mode dynamics could be acquired as

$$\hat{e} = -\int \left[\beta_1 |e|^{\gamma_1} sign(e) + \beta_2 \left|\hat{e}\right|^{\gamma_2} sign\left(\hat{e}\right) + \beta_3 e + \beta_4 e^3 \right] dt \tag{17}$$

Theorem 2. *For the sliding mode dynamics in (17), the origin was defined as the stable equilibrium point and the state trajectories converged to zero in finite time.*

Proof of Theorem 2. Taking the derivative of the tracking error in (3) in respect to time yielded

$$\begin{aligned} \dot{e} &= \dot{x}_1 - \dot{x}_d \\ &= x_2 - \dot{x}_d \end{aligned} \tag{18}$$

Based on the definition of the estimation errors in Section 3, the velocity error (14) was rewritten as

$$\begin{aligned} \hat{\dot{e}} &= \hat{x}_2 - \dot{x}_d \\ &= x_2 - \dot{x}_d - \tilde{x}_2 \end{aligned} \tag{19}$$

After the convergence time of the estimation errors (7), the estimated velocity, $\hat{x}_2$, reached the true velocity, x_2. Hence, the velocity error (19) became

$$\hat{\dot{e}} = x_2 - \dot{x}_d = \dot{e} \tag{20}$$

The sliding mode dynamics in (17) became

$$\dot{e} = -\int \left[\beta_1 |e|^{\gamma_1} sign(e) + \beta_2 |\dot{e}|^{\gamma_2} sign(\dot{e}) + \beta_3 e + \beta_4 e^3 \right] dt \tag{21}$$

Then, the following sliding mode dynamics could be obtained

$$\ddot{e} = -\beta_1 |e|^{\gamma_1} sign(e) - \beta_2 |\dot{e}|^{\gamma_2} sign(\dot{e}) - \beta_3 e - \beta_4 e^3 \tag{22}$$

A Lyapunov function candidate was selected as

$$V_1 = \frac{\beta_1}{\gamma_1 + 1} |e|^{\gamma_1 + 1} + \frac{1}{2}\dot{e}^2 + \frac{\beta_3}{2}e^2 + \frac{\beta_4}{4}e^4 \tag{23}$$

Taking the time derivative of the Lyapunov function candidate (23) and substituting the result from (22) yielded

$$\begin{aligned} \dot{V}_1 &= \beta_1 |e|^{\gamma_1} sign(e)\dot{e} + \dot{e}\ddot{e} + \beta_3 e\dot{e} + \beta_4 e^3 \dot{e} \\ &= \dot{e}\left(-\beta_1 |e|^{\gamma_1} sign(e) - \beta_2 |\dot{e}|^{\gamma_2} sign(\dot{e}) - \beta_3 e - \beta_4 e^3\right) \\ &\quad + \beta_1 |e|^{\gamma_1} sign(e)\dot{e} + \beta_3 e\dot{e} + \beta_4 e^3 \dot{e} \\ &= -\beta_2 |\dot{e}|^{\gamma_2 + 1} \end{aligned} \tag{24}$$

From (23) and (24), it could be concluded that $V_1 > 0$ and $\dot{V}_1 < 0$, therefore, the origin of the sliding mode dynamics (17) was a stable equilibrium point and the state trajectories e and $\hat{e}$ converged to zero in finite time. Thus, Theorem 2 was successfully proven. $\square$

4.2. Observer-Based NFTSMC Design

To obtain the control law for the robot manipulator system (2), an observer-based NFTSMC, as shown in Figure 1, was proposed. The control input signal was design as follows:

$$u = -M(x_1)\left(u_{eq} + u_{sw}\right) \tag{25}$$

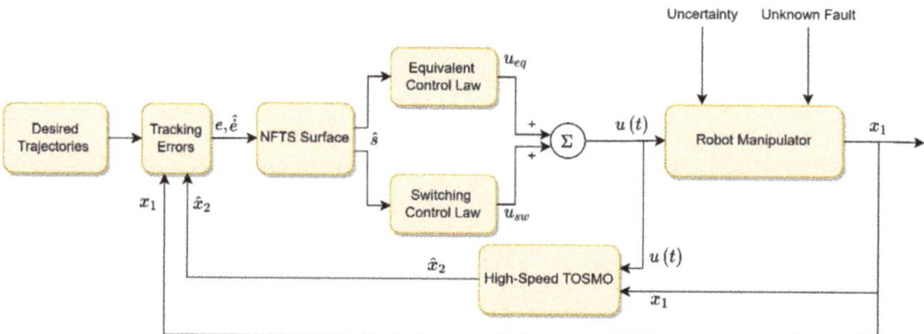

Figure 1. Overall structure of the proposed fault-tolerant control approach.

Here, the equivalent control law, u_{eq}, played the role of holding the error state variables on the sliding surface, and was designed as follows:

$$u_{eq} = \Psi(x) + k_2|\tilde{x}_1|^{1/3}sign(\tilde{x}_1) + \Gamma\left(\dot{\hat{x}}_1 - \hat{x}_2\right) + \int k_3 sign(\tilde{x}_1) + A - \ddot{x}_d \qquad (26)$$

where $A = \beta_1|e|^{\gamma_1}sign(e) + \beta_2\left|\hat{\dot{e}}\right|^{\gamma_2}sign\left(\hat{\dot{e}}\right) + \beta_3 e + \beta_4 e^3$.

The switching control law, u_{sw}, played the role of driving the state variables to the sliding surface, and was designed as follows:

$$u_{sw} = (\Delta_\delta + \mu)sign(\hat{s}) \qquad (27)$$

where μ is a small positive constant.

The proposed control input was presented in the following Theorem 3:

Theorem 3. *Consider the robot manipulator system given by (2); if NFTSMC was designed as (25)–(27), then system (2) was stable. Additionally, the finite-time convergence of the tracking errors was guaranteed.*

Proof of Theorem 3. Taking the derivative of both the sliding surface (15) and the tracking velocity error (14) in respect to time, we obtained

$$\dot{\hat{s}} = \frac{d}{dt}\hat{\dot{e}} + A \qquad (28)$$

$$\frac{d}{dt}\hat{\dot{e}} = \dot{\hat{x}}_2 - \ddot{x}_d \qquad (29)$$

Substituting the second term of the proposed observer (6) into (29) yielded

$$\frac{d}{dt}\hat{\dot{e}} = -\ddot{x}_d + \Psi(\hat{x}) + M(x_1)^{-1}u + k_2|\tilde{x}_1|^{1/3}sign(\tilde{x}_1) + \Gamma\left(\dot{\hat{x}}_1 - \hat{x}_2\right) + \hat{z}$$
$$\dot{\hat{z}} = k_3 sign(\tilde{x}_1) \qquad (30)$$

Substituting (30) into (28), we obtained

$$\dot{\hat{s}} = -\ddot{x}_d + \Psi(\hat{x}) + M(x_1)^{-1}u + k_2|\tilde{x}_1|^{1/3}sign(\tilde{x}_1) + \Gamma\left(\dot{\hat{x}}_1 - \hat{x}_2\right) + \hat{z} + A$$
$$= -\ddot{x}_d + \Psi(x) + \delta(x, \tilde{x}) + M(x_1)^{-1}u + k_2|\tilde{x}_1|^{1/3}sign(\tilde{x}_1) + \Gamma\left(\dot{\hat{x}}_1 - \hat{x}_2\right) + \hat{z} + A \qquad (31)$$
$$\dot{\hat{z}} = k_3 sign(\tilde{x}_1)$$

Applying control input in (25)–(27) to (31) gave

$$\dot{\hat{s}} = -(\Delta_\delta + \mu)sign(\hat{s}) + \delta(x, \tilde{x}) \qquad (32)$$

Let us define the Lyapunov function candidate as follows:

$$V_2 = \frac{1}{2}\hat{s}^T\hat{s} \tag{33}$$

With the result of (32), the time derivative of the Lyapunov function candidate (33) yielded

$$
\begin{aligned}
\dot{V}_2 &= \hat{s}^T\dot{\hat{s}} \\
&= \hat{s}^T\left(-(\Delta_\delta + \mu)sign(\hat{s}) + \delta(x,\tilde{x})\right) \\
&= -(\Delta_\delta + \mu)\sum_{i=1}^{n}|\hat{s}_i| + \delta(x,\tilde{x})^T\hat{s} \leq -\mu\sum_{i=1}^{n}|\hat{s}_i| \\
&\leq -\mu\|\hat{s}\| = -\sqrt{2}\mu V_2^{1/2} < 0, \; \forall \hat{s} \neq 0
\end{aligned}
\tag{34}
$$

From (33) and (34), it could be concluded that system (2) was stable, and the finite-time convergence of the tracking errors was guaranteed. Thus, Theorem 3 was successfully proven. $\square$

Remark 5. *In the equivalent control law (26), we could see that the estimated lumped unknown input, $\int k_3 sign(\tilde{x}_1)$, which was obtained from the high-speed TOSMO in Equation (6), was included. Consequently, the switching control law now was utilized to handle the effects of the estimation errors; therefore, a small value of sliding gain could be chosen. By this way, the high-frequency chattering phenomenon would be significantly decreased in the control input signal.*

Remark 6. *In combining the observer and controller, the convergence speed of the controller was dependent on the convergence speed of the designed observer. Therefore, the proposed high-speed TOSMO not only supported early fault detection, but also helped the controller to achieve a faster convergence speed than when combined with the TOSMO.*

Remark 7. *Although the NFTS surface was selected according to [49], the proposed equivalent control law in (26) was different. Therefore, it could be considered as a contribution of this paper.*

5. Numerical Simulations

To validate the efficiency of the suggested algorithm, in this section, we used the PUMA560 robot manipulator with the last three joints blocked for a computer simulation. The structure of the PUMA560 robot is shown in Figure 2. The specific parameter values of the PUMA560 robot dynamic model were provided in [50]. In the simulation analysis, the MATLAB/Simulink program was used with a sampling time of 10^{-3} s.

In the simulation, the desired trajectories of the three were assumed as

$$
q_d = \begin{bmatrix} q_{d_1} \\ q_{d_2} \\ q_{d_3} \end{bmatrix} = \begin{bmatrix} \cos(\pi t/5) - 1 \\ \sin(\pi t/5 + \pi/2) - 1 \\ \sin(\pi t/5 + \pi/2) - 1 \end{bmatrix}
$$

The initial states of the robot were selected as $q_1(0) = q_2(0) = q_3(0) = -0.5$ and $\dot{q}_1(0) = \dot{q}_2(0) = \dot{q}_3(0) = 0$.

The friction vector and external disturbance vector were assumed as

$$
F_r = \begin{bmatrix} F_{r1} \\ F_{r2} \\ F_{r3} \end{bmatrix} = \begin{bmatrix} 1.9\sin(\dot{q}_1) \\ 2.03\sin(\dot{q}_2) \\ 1.76\sin(\dot{q}_3) \end{bmatrix} \quad \tau_d = \begin{bmatrix} \tau_{d_1} \\ \tau_{d_2} \\ \tau_{d_3} \end{bmatrix} = \begin{bmatrix} 1.1\dot{q}_1 + 1.2\sin(3q_1) \\ 1.65\dot{q}_2 + 2.14\sin(2q_2) \\ -3.01\dot{q}_3 + 1.3\sin(q_3) \end{bmatrix}
$$

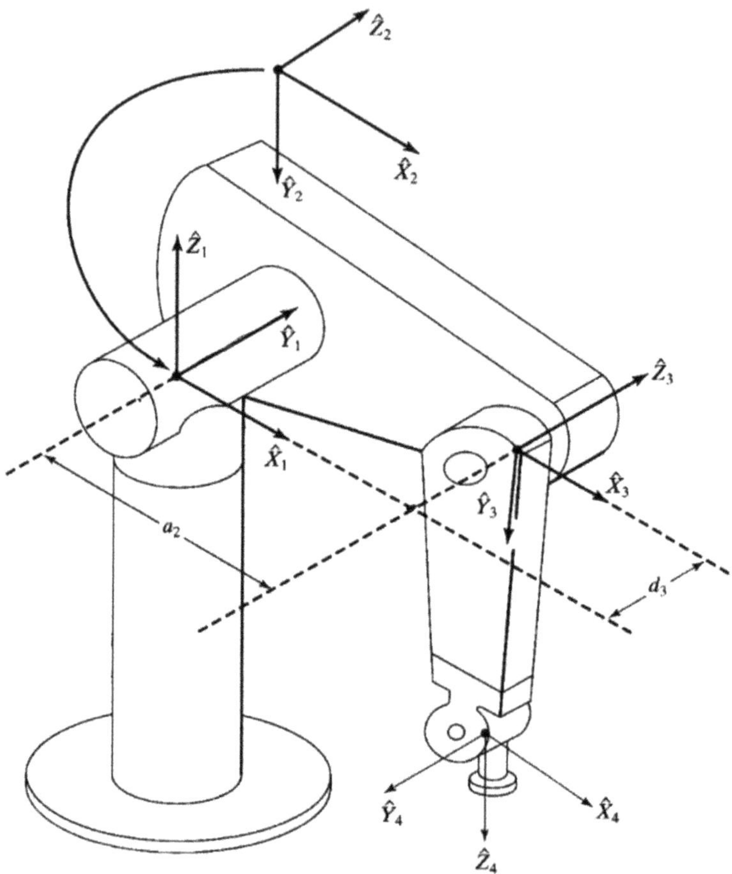

Figure 2. Structure of the PUMA560 robot manipulator.

The fault was assumed to occur at $T_f = 10$ s, with the fault signal as follows:

$$\Phi = \begin{bmatrix} \Phi_1 \\ \Phi_2 \\ \Phi_3 \end{bmatrix} = \begin{bmatrix} 10\sin(q_1 q_2) + 3.7\cos(\dot{q}_1 q_2) + 5.2\cos(\dot{q}_1 \dot{q}_2) \\ 15\sin(q_1 q_2) + 3.6\cos(\dot{q}_1 q_2) + 2.7\cos(\dot{q}_1 \dot{q}_2) \\ 0 \end{bmatrix}$$

The parameters of the controllers and observers were selected as follows: $\gamma_1 = 1/2$, $\gamma_2 = 2/3, \beta_1 = diag(15, 15, 15), \beta_2 = diag(10, 10, 10), \beta_3 = diag(10, 10, 10), \beta_4 = diag(5, 5, 5)$, $\Delta_\delta = 0.5, \overline{\Delta} = 22, \mu = 0.01$ and $\Gamma = 5$.

The simulation consisted of two parts: First, the estimation results of the proposed high-speed TOSMO were compared with that of the TOSMO and the SOSMO, which were designed as in [42]. Second, the proposed fault-tolerant technique was compared with three controllers: (1) NFTSMC without compensation; (2) NFTSMC with the SOSMO compensation (NFTSMC-SOSMO); (3) NFTSMC with the TOSMO compensation (NFTSMC-TOSMO).

In the first part, the comparison results among three observers are shown in Figures 3–5. The obtained velocity estimation errors are shown in Figure 3. As in the results, the SOSMO (green solid line) provided a faster estimation of velocity compared to the TOSMO (blue solid line). In contrast, the TOSMO provided a velocity estimation with higher precision and less chattering compared to the SOSMO. The red solid line in the figure shows the estimation results of the high-speed TOSMO. It was easy to see that the estimated information of the high-speed TOSMO maintained the higher precision and lesser chattering characteristic of the TOSMO, while the convergence speed was significantly increased and matched the speed of the SOSMO. The faster velocity estimation would help the controller reach a faster convergence time. In terms of the lumped unknown input estimation, the results are shown in Figures 4 and 5. As shown in the results, the SOSMO provided estimation information with a lower accuracy compared to the TOSMO and the proposed high-speed TOSMO due to the time delay when using a lowpass filter to reconstruct the estimation signal. On the contrary, the TOSMO and the high-speed TOSMO could reconstruct the lumped unknown input directly without any filtration. However, the convergence time of the TOSMO was a little slower. The same applied to the velocity estimation results, where the lumped unknown input estimation results of the high-speed TOSMO could maintain the high estimation performance of the TOSMO and reached the convergence speed of the SOSMO. It is worth mentioning that the faster estimation speed helped in early fault detection, thus, reducing the robot's failure rate.

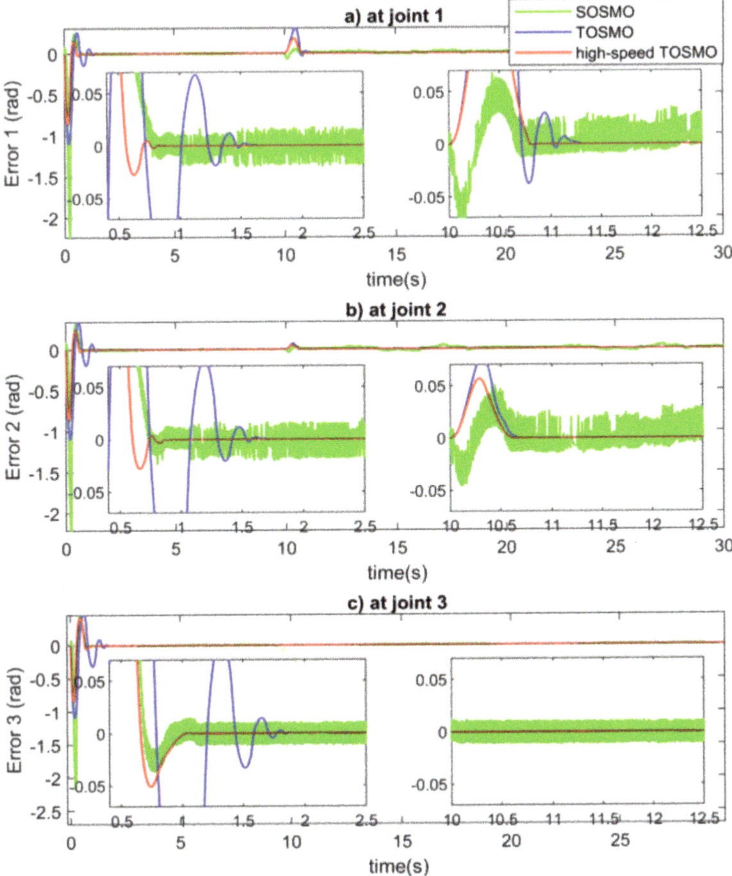

Figure 3. Velocity estimation errors at (**a**) joint 1, (**b**) joint 2 and (**c**) joint 3.

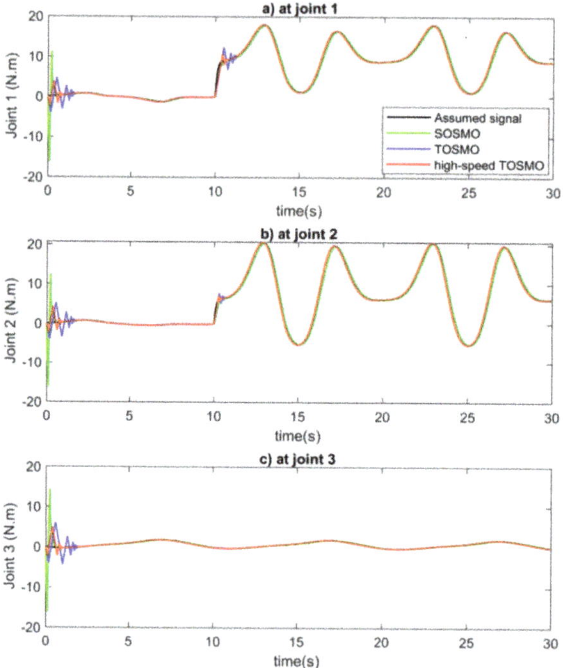

Figure 4. Lumped unknown input estimation at (**a**) joint 1, (**b**) joint 2 and (**c**) joint 3.

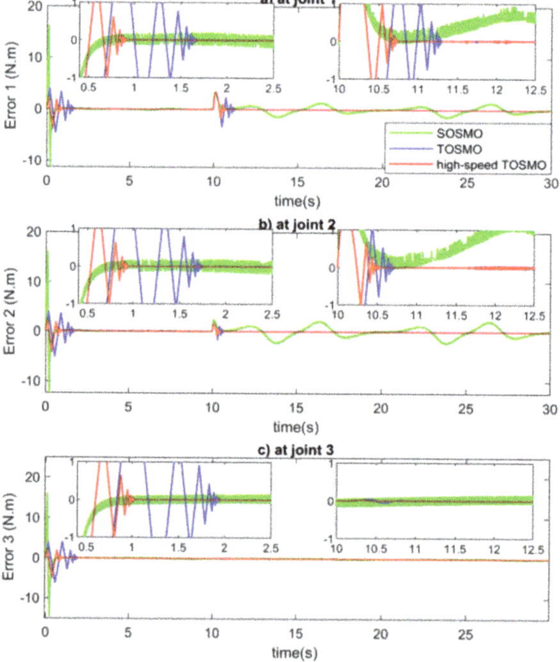

Figure 5. Lumped unknown input estimation errors at (**a**) joint 1, (**b**) joint 2 and (**c**) joint 3.

In the second part, the comparison results among four controllers were shown in Figures 6 and 7. Figure 6 shows the results of the tracking error at each joint. As in the figure, in terms of the tracking performance, NFTSMC without compensation (green solid line) and the NFTSMC-SOSMO (black dash line) provided quite good tracking precision. However, with better approximation information, the NFTSMC-TOSMO (blue solid line) and the proposed fault-tolerant control strategy (red solid line) provided higher control performance. The two controllers provided almost the same tracking accuracy due to the estimation accuracy of the TOSMO and the high-speed TOSMO being not much different. In terms of the convergence speed, NFTSMC without compensation and the NFTSMC-TOSMO converged simultaneously because they used the same velocity signal in the design process. According to the effect of the velocity signal, the proposed fault-tolerant control strategy converged faster compared to the above two controllers, and almost the same as the NFTSMC-SOSMO. Therefore, it could be concluded that the proposed high-speed TOSMO not only obtained estimation information faster, but also helped the designed controller achieve better control performance. The comparison of control input torque is shown in Figure 7. As in the figure, the control input of NFTSMC without compensation was under the effect of the chattering phenomenon because of using a large sliding gain. After compensation, the sliding gain could be chosen with a smaller value; therefore, the chattering phenomenon in the control inputs of the NFTSMC-SOSMO, the NFTSMC-TOSMO and the proposed fault-tolerant control were significantly reduced. The time response of the proposed NFTS surface is shown in Figure 8.

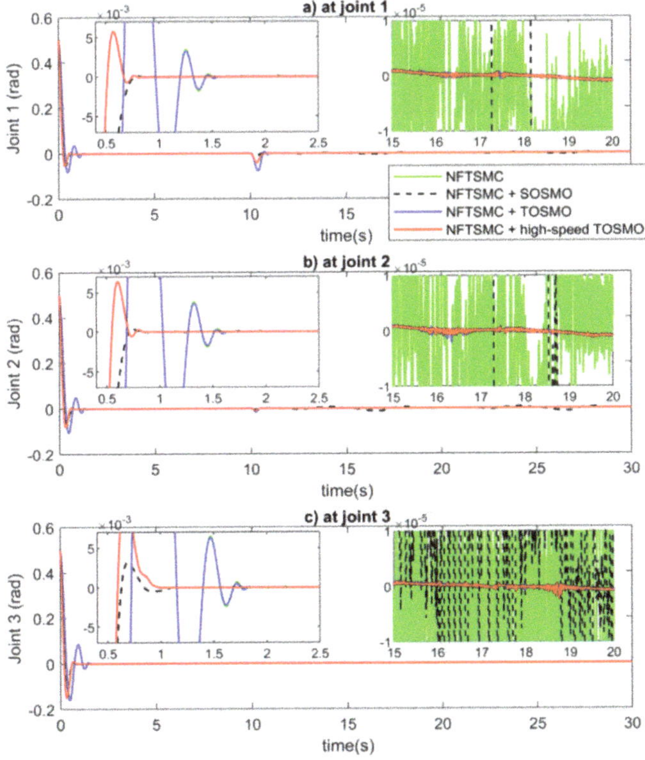

Figure 6. Comparison of tracking errors among controllers at (**a**) joint 1, (**b**) joint 2 and (**c**) joint 3.

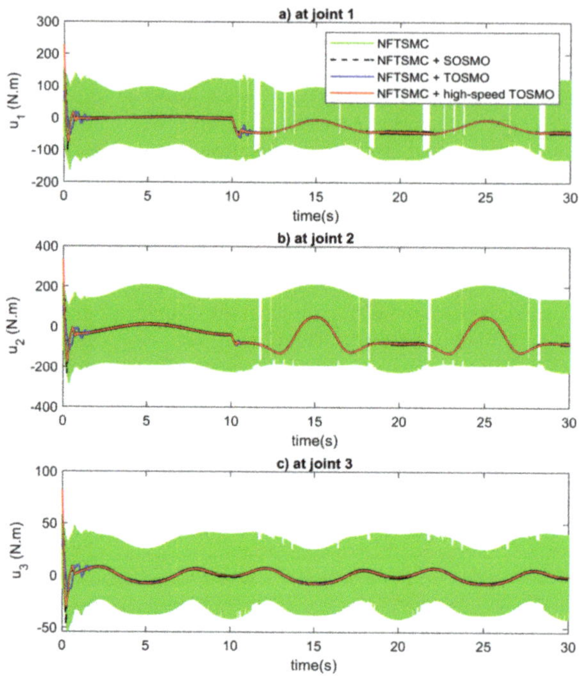

Figure 7. Comparison of control input torque among controllers at (**a**) joint 1, (**b**) joint 2 and (**c**) joint 3.

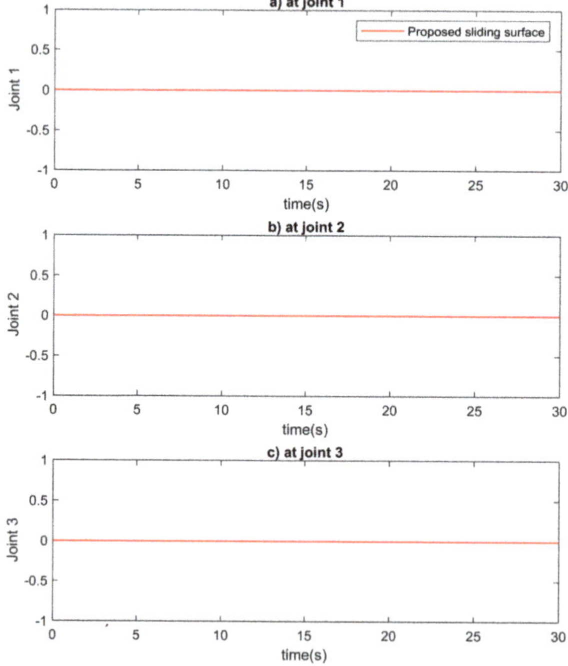

Figure 8. Time response of the proposed sliding surface at (**a**) joint 1, (**b**) joint 2 and (**c**) joint 3.

6. Conclusions

This paper proposed a novel fault-tolerant control strategy for robot manipulator systems using only position measurements. Thanks to the linear characteristic of the added elements, the proposed high-speed TOSMO could estimate both the velocity signal and the lumped unknown input with a faster convergence time compared to the TOSMO. The obtained information from the observer was combined with NFTSMC in designing the fault-tolerant controller. The proposed controller–observer tactic provided excellent properties, such as a fast convergence time, high-position tracking precision, finite-time convergence, chattering phenomenon reduction, robustness against the effects of the lumped unknown input and velocity requirement elimination. The faster convergence characteristic of the observer also improved the convergence speed of the designed controller. The system stability and finite-time convergence were proved using the Lyapunov stability theory. Finally, the efficiency of the proposed algorithm was validated with simulations on the PUMA560 robot manipulator. Due to the efficiency of the proposed algorithm, it would be possible to implement it in real robot system in the future. In addition, designing a fixed-time observer based on the proposed high-speed TOSMO is a promising idea.

Author Contributions: Conceptualization, V.-C.N. and X.-T.T.; methodology, V.-C.N. and X.-T.T.; software, V.-C.N. and H.-J.K.; validation, X.-T.T.; formal analysis, V.-C.N. and H.-J.K.; investigation, H.-J.K.; resources, X.-T.T. and H.-J.K.; data curation, V.-C.N. and X.-T.T.; writing—original draft preparation, V.-C.N. and X.-T.T.; writing—review and editing, V.-C.N., X.-T.T. and H.-J.K.; visualization, X.-T.T.; supervision, H.-J.K.; project administration, H.-J.K.; funding acquisition, H.-J.K. All authors have read and agreed to the published version of the manuscript.

Funding: This research was supported by the Basic Science Research Program through the National Research Foundation of Korea (NRF) funded by the Ministry of Education (2019R1D1A3A03103528).

Institutional Review Board Statement: Not applicable.

Informed Consent Statement: Not applicable.

Data Availability Statement: Not applicable.

Conflicts of Interest: The authors declare no conflict of interest.

References

1. Jin, L.; Li, S.; Yu, J.; He, J. Robot manipulator control using neural networks: A survey. *Neurocomputing* **2018**, *285*, 23–34. [CrossRef]
2. Zhang, S.; Dong, Y.; Ouyang, Y.; Yin, Z.; Peng, K. Adaptive neural control for robotic manipulators with output constraints and uncertainties. *IEEE Trans. Neural Netw. Learn. Syst.* **2018**, *29*, 5554–5564. [CrossRef] [PubMed]
3. Madsen, E.; Rosenlund, O.S.; Brandt, D.; Zhang, X. Adaptive feedforward control of a collaborative industrial robot manipulator using a novel extension of the Generalized Maxwell-Slip friction model. *Mech. Mach. Theory* **2021**, *155*, 104109. [CrossRef]
4. Nubert, J.; Köhler, J.; Berenz, V.; Allgöwer, F.; Trimpe, S. Safe and fast tracking on a robot manipulator: Robust mpc and neural network control. *IEEE Robot. Autom. Lett.* **2020**, *5*, 3050–3057. [CrossRef]
5. Xie, Z.; Jin, L.; Luo, X.; Hu, B.; Li, S. An Acceleration-Level Data-Driven Repetitive Motion Planning Scheme for Kinematic Control of Robots With Unknown Structure. *IEEE Trans. Syst. Man Cybern. Syst.* **2021**, *52*, 5679–5691. [CrossRef]
6. Fan, J.; Jin, L.; Xie, Z.; Li, S.; Zheng, Y. Data-driven motion-force control scheme for redundant manipulators: A kinematic perspective. *IEEE Trans. Ind. Inform.* **2021**, *18*, 5338–5347. [CrossRef]
7. Van, M.; Do, X.P.; Mavrovouniotis, M. Self-tuning fuzzy PID-nonsingular fast terminal sliding mode control for robust fault tolerant control of robot manipulators. *ISA Trans.* **2020**, *96*, 60–68. [CrossRef]
8. Vo, A.T.; Kang, H.-J.; Nguyen, V.-C. An output feedback tracking control based on neural sliding mode and high order sliding mode observer. In Proceedings of the 2017 10th International Conference on Human System Interactions (HSI), Ulsan, Korea, 17–19 July 2017; pp. 161–165.
9. Song, Y.; Huang, X.; Wen, C. Robust adaptive fault-tolerant PID control of MIMO nonlinear systems with unknown control direction. *IEEE Trans. Ind. Electron.* **2017**, *64*, 4876–4884. [CrossRef]
10. Alibeji, N.; Sharma, N. A PID-Type Robust Input Delay Compensation Method for Uncertain Euler—Lagrange Systems. *IEEE Trans. Control Syst. Technol.* **2017**, *25*, 2235–2242. [CrossRef]
11. Tutsoy, O.; Barkana, D.E. Model free adaptive control of the under-actuated robot manipulator with the chaotic dynamics. *ISA Trans.* **2021**, *118*, 106–115. [CrossRef]
12. Muñoz-Vázquez, A.J.; Treesatayapun, C. Model-free discrete-time fractional fuzzy control of robotic manipulators. *J. Frankl. Inst.* **2022**, *359*, 952–966. [CrossRef]

13. Song, Y.; Guo, J. Neuro-adaptive fault-tolerant tracking control of Lagrange systems pursuing targets with unknown trajectory. *IEEE Trans. Ind. Electron.* **2017**, *64*, 3913–3920. [CrossRef]
14. Nguyen, V.-C.; Vo, A.-T.; Kang, H.-J. Continuous PID Sliding Mode Control Based on Neural Third Order Sliding Mode Observer for Robotic Manipulators. In Proceedings of the International Conference on Intelligent Computing, Nanchang, China, 3–6 August 2019; pp. 167–178.
15. Cheng, X.; Liu, H.; Lu, W. Chattering-Suppressed Sliding Mode Control for Flexible-Joint Robot Manipulators. *Actuators* **2021**, *10*, 288. [CrossRef]
16. Zhou, W.; Wang, Y.; Liang, Y. Sliding mode control for networked control systems: A brief survey. *ISA Trans.* **2021**, *124*, 249–259. [CrossRef]
17. Alwi, H.; Edwards, C. Fault detection and fault-tolerant control of a civil aircraft using a sliding-mode-based scheme. *IEEE Trans. Control Syst. Technol.* **2008**, *16*, 499–510. [CrossRef]
18. Utkin, V.I. *Sliding Modes in Control and Optimization*; Springer Science & Business Media: New York, NY, USA, 2013.
19. Yu, S.; Yu, X.; Shirinzadeh, B.; Man, Z. Continuous finite-time control for robotic manipulators with terminal sliding mode. *Automatica* **2005**, *41*, 1957–1964. [CrossRef]
20. Zhihong, M.; Paplinski, A.P.; Wu, H.R. A robust MIMO terminal sliding mode control scheme for rigid robotic manipulators. *IEEE Trans. Automat. Control* **1994**, *39*, 2464–2469. [CrossRef]
21. Islam, S.; Liu, X.P. Robust sliding mode control for robot manipulators. *IEEE Trans. Ind. Electron.* **2010**, *58*, 2444–2453. [CrossRef]
22. Nguyen, V.-C.; Le, P.-N.; Kang, H.-J. Model-Free Continuous Fuzzy Terminal Sliding Mode Control for Second-Order Nonlinear Systems. In Proceedings of the International Conference on Intelligent Computing, Shenzhen, China, 12–15 August 2021; pp. 245–258.
23. Xu, Q. Piezoelectric nanopositioning control using second-order discrete-time terminal sliding-mode strategy. *IEEE Trans. Ind. Electron.* **2015**, *62*, 7738–7748. [CrossRef]
24. Truong, T.N.; Vo, A.T.; Kang, H.-J. Implementation of an Adaptive Neural Terminal Sliding Mode for Tracking Control of Magnetic Levitation Systems. *IEEE Access* **2020**, *8*, 206931–206941. [CrossRef]
25. Mobayen, S. Fast terminal sliding mode controller design for nonlinear second-order systems with time-varying uncertainties. *Complexity* **2015**, *21*, 239–244. [CrossRef]
26. Cruz-Ortiz, D.; Chairez, I.; Poznyak, A. Non-singular terminal sliding-mode control for a manipulator robot using a barrier Lyapunov function. *ISA Trans.* **2022**, *121*, 268–283. [CrossRef] [PubMed]
27. Shao, X.; Sun, G.; Xue, C.; Li, X. Nonsingular terminal sliding mode control for free-floating space manipulator with disturbance. *Acta Astronaut.* **2021**, *181*, 396–404. [CrossRef]
28. Zaare, S.; Soltanpour, M.R. Continuous fuzzy nonsingular terminal sliding mode control of flexible joints robot manipulators based on nonlinear finite time observer in the presence of matched and mismatched uncertainties. *J. Frankl. Inst.* **2020**, *357*, 6539–6570. [CrossRef]
29. Nguyen, V.-C.; Kang, H.-J. A Fault Tolerant Control for Robotic Manipulators Using Adaptive Non-singular Fast Terminal Sliding Mode Control Based on Neural Third Order Sliding Mode Observer. In Proceedings of the International Conference on Intelligent Computing, Sanya, China, 4–6 December 2020; pp. 202–212.
30. Vo, A.T.; Kang, H.-J. A novel fault-tolerant control method for robot manipulators based on non-singular fast terminal sliding mode control and disturbance observer. *IEEE Access* **2020**, *8*, 109388–109400. [CrossRef]
31. Nguyen, V.-C.; Le, P.-N.; Kang, H.-J. An Active Fault-Tolerant Control for Robotic Manipulators Using Adaptive Non-Singular Fast Terminal Sliding Mode Control and Disturbance Observer. *Actuators* **2021**, *10*, 332. [CrossRef]
32. Utkin, V.; Guldner, J.; Shi, J. *Sliding Mode Control in Electro-Mechanical Systems*; CRC Press: Boca Raton, FL, USA, 2009.
33. Zhou, Q.; Shi, P.; Xu, S.; Li, H. Observer-based adaptive neural network control for nonlinear stochastic systems with time delay. *IEEE Trans. Neural Netw. Learn. Syst.* **2012**, *24*, 71–80. [CrossRef]
34. Abdollahi, F.; Talebi, H.A.; Patel, R. V A stable neural network-based observer with application to flexible-joint manipulators. *IEEE Trans. Neural Netw.* **2006**, *17*, 118–129. [CrossRef]
35. Ya-Li, D.; Sheng-Wei, M.E.I. Adaptive observer for a class of nonlinear systems. *Acta Autom. Sin.* **2007**, *33*, 1081–1084.
36. Jiang, B.; Staroswiecki, M.; Cocquempot, V. Fault diagnosis based on adaptive observer for a class of non-linear systems with unknown parameters. *Int. J. Control* **2004**, *77*, 367–383. [CrossRef]
37. Wang, Y.; Leibold, M.; Lee, J.; Ye, W.; Xie, J.; Buss, M. Incremental Model Predictive Control Exploiting Time-Delay Estimation for a Robot Manipulator. *IEEE Trans. Control Syst. Technol.* **2022**, 1–16, (Early Access). [CrossRef]
38. Van, M.; Ge, S.S.; Ren, H. Finite Time Fault Tolerant Control for Robot Manipulators Using Time Delay Estimation and Continuous Nonsingular Fast Terminal Sliding Mode Control. *IEEE Trans. Cybern.* **2017**, *47*, 1681–1693. [CrossRef] [PubMed]
39. Wang, X.; Liao, R.; Shi, C.; Wang, S. Linear extended state observer-based motion synchronization control for hybrid actuation system of more electric aircraft. *Sensors* **2017**, *17*, 2444. [CrossRef] [PubMed]
40. Saleki, A.; Fateh, M.M. Model-free control of electrically driven robot manipulators using an extended state observer. *Comput. Electr. Eng.* **2020**, *87*, 106768. [CrossRef]
41. Van, M.; Kang, H.-J.; Suh, Y.-S. A novel neural second-order sliding mode observer for robust fault diagnosis in robot manipulators. *Int. J. Precis. Eng. Manuf.* **2013**, *14*, 397–406. [CrossRef]

42. Nguyen, V.-C.; Vo, A.-T.; Kang, H.-J. A Non-singular Fast Terminal Sliding Mode Control Based on Third-Order Sliding Mode Observer for a Class of Second-Order Uncertain Nonlinear Systems and Its Application to Robot Manipulators. *IEEE Access* **2020**, *8*, 78109–78120. [CrossRef]
43. Nguyen, V.-C.; Vo, A.-T.; Kang, H.-J. A Finite-Time Fault-Tolerant Control Using Non-Singular Fast Terminal Sliding Mode Control and Third-Order Sliding Mode Observer for Robotic Manipulators. *IEEE Access* **2021**, *9*, 31225–31235. [CrossRef]
44. Feng, Y.; Han, F.; Yu, X. Chattering free full-order sliding-mode control. *Automatica* **2014**, *50*, 1310–1314. [CrossRef]
45. Van, M.; Ge, S.S.; Ren, H. Robust fault-tolerant control for a class of second-order nonlinear systems using an adaptive third-order sliding mode control. *IEEE Trans. Syst. Man Cybern. Syst.* **2016**, *47*, 221–228. [CrossRef]
46. Tran, X.-T.; Kang, H.-J. Continuous adaptive finite-time modified function projective lag synchronization of uncertain hyperchaotic systems. *Trans. Inst. Meas. Control* **2018**, *40*, 853–860. [CrossRef]
47. Levant, A. Higher-order sliding modes, differentiation and output-feedback control. *Int. J. Control* **2003**, *76*, 924–941. [CrossRef]
48. Ortiz-Ricardez, F.A.; Sánchez, T.; Moreno, J.A. Smooth Lyapunov function and gain design for a second order differentiator. In Proceedings of the 2015 54th IEEE Conference on Decision and Control (CDC), Osaka, Japan, 15–18 December 2015; pp. 5402–5407.
49. Tran, X.-T.; Kang, H.-J. A Novel Adaptive Finite-Time Control Method for a Class of Uncertain Nonlinear Systems. *Int. J. Precis. Eng. Manuf.* **2015**, *16*, 2647–2654. [CrossRef]
50. Armstrong, B.; Khatib, O.; Burdick, J. The explicit dynamic model and inertial parameters of the PUMA 560 arm. In Proceedings of the 1986 IEEE International Conference on Robotics and Automation, San Francisco, CA, USA, 7–10 April 1986; Volume 3, pp. 510–518.

 *actuators*

Article

Image Servo Tracking of a Flexible Manipulator Prototype with Connected Continuum Kinematic Modules

Ming-Hong Hsu [1], Phuc Thanh-Thien Nguyen [1], Dai-Dong Nguyen [1] and Chung-Hsien Kuo [2,*]

[1]   Department of Electrical Engineering, National Taiwan University of Science and Technology, Taipei 106, Taiwan
[2]   Department of Mechanical Engineering, National Taiwan University, Taipei 106, Taiwan
*   Correspondence: chunghsien@ntu.edu.tw

Abstract: This paper presents the design and implementation of a flexible manipulator formed of connected continuum kinematic modules (CKMs) to ease the fabrication of a continuum robot with multiple degrees of freedom. The CKM consists of five sequentially arranged circular plates, four universal joints intermediately connecting five circular plates, three individual actuated tension cables, and compression springs surrounding the tension cables. The base and movable circular plates are used to connect the robot platform or the neighboring CKM. All tension cables are controlled via linear actuators at a distal site. To demonstrate the function and feasibility of the proposed CKM, the kinematics of the continuum manipulator were verified through a kinematic simulation at different end velocities. The correctness of the manipulator posture was confirmed through the kinematic simulation. Then, a continuum robot formed with three CKMs is fabricated to perform Jacobian-based image servo tracking tasks. For the eye-to-hand (ETH) experiment, a heart shape trajectory was tracked to verify the precision of the kinematics, which achieved an endpoint error of 4.03 in Root Mean Square Error (RMSE). For the eye-in-hand (EIH) plugging-in/unplugging experiment, the accuracy of the image servo tracking system was demonstrated in extensive tolerance conditions, with processing times as low as 58 ± 2.12 s and 83 ± 6.87 s at the 90% confidence level in unplugging and plugging-in tasks, respectively. Finally, quantitative tracking error analyses are provided to evaluate the overall performance.

Keywords: continuum robot; image servo tracking; image jacobian; autonomous manipulation

Citation: Hsu, M.-H.; Nguyen, P.T.-T.; Nguyen, D.-D.; Kuo, C.-H. Image Servo Tracking of a Flexible Manipulator Prototype with Connected Continuum Kinematic Modules. *Actuators* **2022**, *11*, 360. https://doi.org/10.3390/act11120360

Academic Editor: Gary M. Bone

Received: 28 October 2022
Accepted: 30 November 2022
Published: 2 December 2022

Publisher's Note: MDPI stays neutral with regard to jurisdictional claims in published maps and institutional affiliations.

1. Introduction

1.1. Motivation

Recently, automation-related equipment, especially robotic manipulators, has been successfully used widely in industrial applications [1,2], medical services [3], and disaster-based narrow-space exploration [4]. In addition to the conventional articulated manipulators, soft and flexible manipulators are becoming more attractive to robotics researchers [5,6]. Continuum robots are typical examples of flexible manipulators, which have better agility than conventional articulated manipulators. In general, a continuum robot is practical for performing soft grasping and manipulating tasks, in the same vein as the manipulations of the octopus tentacle or the elephant trunk. Hence, continuum robots are often operated in narrow spaces because of their hyper degrees of freedom (DOF) motions [7,8]. Although providing flexibility and agility in manipulating tasks, the deployment of multi-segment continuum manipulators that exhibit non-constant curvature is still a formidable challenge because of the complexity of deriving continuum kinematics.

1.2. Related Works

Continuum robots are typical examples of flexible manipulators capable of performing with better agility than conventional articulated manipulators. The designs of continuum robots can be divided into two types: tension cable-driven and pneumatic-driven. For tension cable-driven robots, Hannan et al. [9] designed a continuum robot formed by 32 coupled joints and a total of eight actuatable cables, making it capable of performing a motion similar to that of an elephant trunk. Jones et al. [10] further proposed synthesizing the kinematic relationship between a general continuum skeleton and a continuum robot. The coordinates are related to the input of the controller (e.g., air pressure and tendon length) to realize real-time tasks and shape control of a continuum robot. As Yoon et al. [11] proposed, springs are also used as the backbone of the plate connection. Such a mechanism design is mainly used to enhance the elasticity and flexibility of the overall continuum structure to ensure the safety of the continuum robot when it collides with the human body. Another design proposed by Tokunaga et al. [12] used an elastic center column and springs, where the springs are installed outside the center column to avoid unexpected shapes such as twisting, which is beneficial for handling weights and loads on the end effector. For pneumatic-driven robots, the use of pneumatic artificial muscles is available for developing continuum robots. Liu et al. [13] presented a light soft manipulator, where thin McKibben pneumatic artificial muscles were utilized for continuously controllable stiffness actuation. Their study successfully overcomes problems such as that most continuum robots cannot continuously change their stiffness at a fixed end position. Dalvand et al. [14] proposed an analytical loading model by considering the number of tendons and the load distribution of the tendons to avoid the tendon relaxation problem that leads to inaccurate motion control. Pneumatic-type continuum robots need a pneumatic source, and, therefore, they are not convenient for installation, mobility, and maintenance. However, the tension cable-driven type usually uses motors or linear actuators as the source of kinematic motion control. Consequently, tension cable-driven continuum robots have advantages in terms of installation, mobility, and maintenance. However, the kinematics of the tension cable-driven continuum robot is much more complicated than those of conventional articulated robots.

In addition, bioinspired continuum robot designs have been studied. For example, Hassan et al. [15] developed an active-braid design to fabricate a bioinspired continuum manipulator. Their study adopted flexible cross-linked spiral array structures to form the continuum structure. Inspired by the biological structure of snakes, Zhang et al. [16] proposed a compound continuum robot combining concentric tubes and a notched continuum robot to achieve a smaller diameter and a larger central cavity.

Because continuum robots can navigate and manipulate in a narrow space, they have attracted considerable attention in designing surgical robots. Safety is an essential aspect of robotic surgery. Comin et al. [17] presented a solution of combining a pneumatic soft continuum robot and a rigid robot arm in terms of series connection. The rigid part maintains a safe tool contact force, while the soft part follows the required cutting path. Their solution can demonstrate teleoperated diathermic tissue-cutting tasks from a safety consideration; although, the proposed system still was not friendly-using and only worked on designed scenarios. In addition, Zhao et al. [18] proposed a variable stiffness design for a continuum manipulator. Such a redundant continuum structure is formed with an elastic backbone that exhibits a continuously constrained bending curvature. Agility and miniaturization are design considerations in minimally invasive surgery (MIS) operations. Qu et al. [19] presented a continuum manipulator design for MIS operation, with a bioinspired wire-driven multi-backbone structure design being applied in this study. A super-elastic backbone was utilized to connect a series of thin disks to form a continuum structure with a 12 mm outer diameter and a total length of 180 mm. Another study was proposed for flexible endoscopic surgery purposes. Hwang et al. [20] designed a novel constrained continuum manipulator that used several auxiliary links attached to the primary continuum for payload capability improvement. Similar to the design of three 6-DOF continuum robots formed with cable-driven concentric tube mechanisms [21], the design of a snake-like surgical continuum robot [22] and a continuum manipulator with parallel and shifted-routing cable-driven control mechanisms for robotic surgeries of endometrial regeneration [23], provide the design paradigms for novel continuum manipulators. Although making some achievements, these methods had to consider finding a balance between the manipulator's workspace and stiffness without increasing the system's complexity.

Furthermore, to survey the structural designs of continuum manipulators, the control designs of continuum manipulators are also discussed. Shen et al. [24] presented a study on a flexible backbone cable-driven continuum manipulator to improve accuracy to overcome such factors as gravity or mechanism effects. The authors proposed a method consisting of a kinematic model and data-driven Gaussian process regression (GPR) based on experiments applied to hardware platforms to reduce the position and orientation errors by 68.72% and 51.74%, respectively. Other control systems applied to continuum robots also provide helpful information in designing continuum robot control systems, such as pose planning of a multi-section continuum manipulator in terms of an imitation learning-based approach [25] and absolute positioning accuracy improvement of a continuum surgical manipulator by utilizing the closed-loop control approach [26].

Images-based visual servoing (IBVS) studies are also discussed to demonstrate path tracking and autonomous manipulation abilities. Lai et al. [27] proposed a vision-based adaptive control scheme based on a soft continuum manipulator with a bidirectional two-segment configuration. The proposed controller was realized with the IBVS approach. The key point positioning error could reach within 6.5% based on the manipulator length, and the robot will find the best fit to the desired shape if the goal positions are not reachable. Yang et al. [28] presented the stereo tracking of a continuum surgical manipulator. A wrist marker was designed to realize the closed-loop image-based servoing control scheme. The feature points extracted from the stereoscopic images were evaluated to align with the actual pose, including the position and direction of the target. Although the tip positioning error was reduced to 25.23% during trajectory tracking, the control cycle is longer than the preferred, making the manipulator overshoot when it changed motion direction. Finally, many studies were proposed to realize the visual servoing ability of continuum robots for a wide variety of applications, such as visual servoing and compliant control [29], image-based laser beam steering control [30], model-free visual servoing combined with singularity avoidance to enhance safety [31], and optical coherence tomography (OCT)-guided visual servoing for micromotion manipulations [32,33], vision-based shape control for cable-driven manipulator [34], hybrid EIH and ETH for visual servoing [35], and OCT path scanning [36].

1.3. Contribution

This paper proposes an ideal constant curvature assumption to overcome the problem of deriving continuum kinematics. With the well-defined continuum kinematics in [9] and [10], this paper presents the kinematics of a continuum kinematic module (CKM) via the parameters κ, φ, and l (the arc length, curvature, and rotation angle in the x-y plane of CKM, respectively). Therefore, a complete continuum robot formed of connected CKMs can be accumulated. It is noted that three CKMs are utilized to produce a continuum robot for image-based servo tracking practices. Moreover, image-based path tracking and autonomous manipulation are typical demonstrations for investigating the applicability of hyper DOF continuum manipulators for possible deployment of service robots to provide manipulation compliance and ensure safety in operation when performing human-robot interaction (HRI). The main contribution of this study is summarized as follows:

- This paper presents a three-segment CKM-based continuum robot design that is convenient for the fabrication of a continuum robot and efficient at obtaining the overall kinematic model for control purposes.
- For the proposed design and control validation, experiments of image-based servoing path tracking and autonomous manipulation are established, utilizing Jacobian images to track the desired image targets by controlling the continuum robot. In the eye-to-hand (ETH) experiment, a heart shape trajectory was tracked to verify the precision of the kinematics with acceptably low endpoint errors. In the eye-in-hand (EIH)

experiment, a stereo vision-based object detection algorithm was developed for the power socket grasping task with high accuracy and efficient operation in real-time.

1.4. Limitation

Although addressing the problem of deriving continuum kinematics, the wire-driven CKM causes posture deviation of the proposed three-segment continuum robot due to the influence of gravity. Therefore, the mentioned limit may lead to the performance of the visual servoing system in the EIH plugging/unplugging experiment being imperfect under low-tolerance conditions.

The remainder of this study is organized as follows: In Section 2, the detail of the multi-segment CKM-based continuum robot design, the architecture of a single CKM, and the implementation of the image-based servo tracking systems are carefully described. Next, Section 3 presents the experimental materials and methods, followed by the results and corresponding subsequent analyses. Finally, the conclusions and future work are summarized in Section 4.

2. Proposed Method

In this study, the proposed flexible manipulator with visual servoing, as shown in Figure 1, comprises three continuum kinematic modules (CKMs). Each CKM includes five circular plates, four universal joints connecting the circular plate, three independent tension cables, and compression springs surrounding the tension cables. The architecture of a single CKM is extended to the proposed flexible manipulator (i.e., continuum robot) composed of three CKMs.

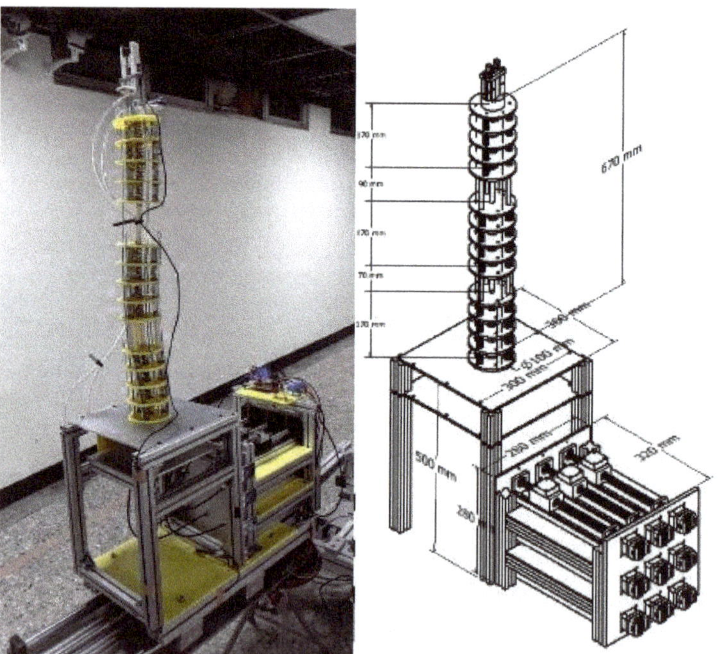

Figure 1. The proposed flexible manipulator with three connected CKMs.

The visual servo control element deals with image-related information, identification and selection of tracking points, Image Jacobian, and velocity kinematics calculations. For the visual servo control to be able to achieve real-time processing effects, this system uses a

Win10 computer, based around an i7-9750H CPU and NVIDIA GTX1660 GPU, and an Intel Realsense D435i camera.

The continuum-manipulator platform is controlled based on the microcontroller Teensy 4.0, which is mainly responsible for calculating forward and inverse kinematics. The control of the sliding table motor and air pressure-related components, the signal processing of the sensor, and sending of the control signal to each sliding table, allow the obtaining of the status of each sliding table. Finally, serial communication is used as the communication protocol between the two systems. Figure 2 presents the overall structure of the whole proposed system.

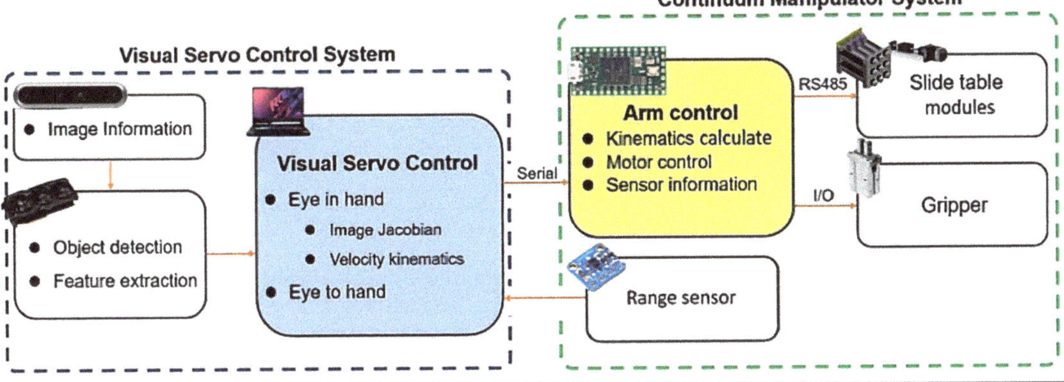

Figure 2. The overall architecture of the proposed visual servo continuum manipulator.

The detailed CKM architecture, the design and kinematics of the continuum robot, and image servo tracking will be introduced in the following subsections.

2.1. Design and Kinematics of A Continuum Kinematic Module (CKM)

Each CKM is composed of five circular plates and three tension cables, as shown in Figure 3. The backbone of the CKM is formed with universal joints and compressive springs. Tension cables are then used to control the curvature of the CKM. The compressive springs and tension cables are intermediately arranged at the outer ring of the circular plates to stabilize CKM backbone deformation. Three tension cables are then symmetrically located at L1, L2, and L3, and three compressive springs are symmetrically located at L4, L5, and L6.

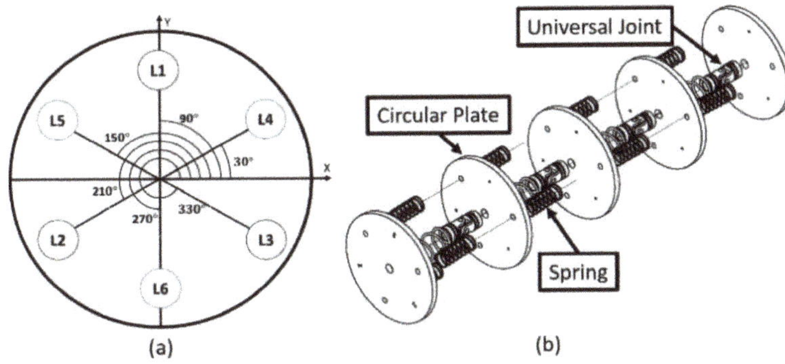

Figure 3. (**a**) Position of the compression springs and tension cables; (**b**) composition of a CKM.

The forward kinematics of the proposed CKM is obtained by referring to [9]. First, assume that the CKM bends as an ideal curvature shape. Then, for the convenience of calculation, assume the arc centerline length is l, the arc radius is r, the angle of the arc is θ, the arc curvature is κ, the center of the arc is o, and the angle that the CKM rotates on the x-y plane is φ. The illustration of the arc parameter (φ, κ, l) is shown in Figure 4.

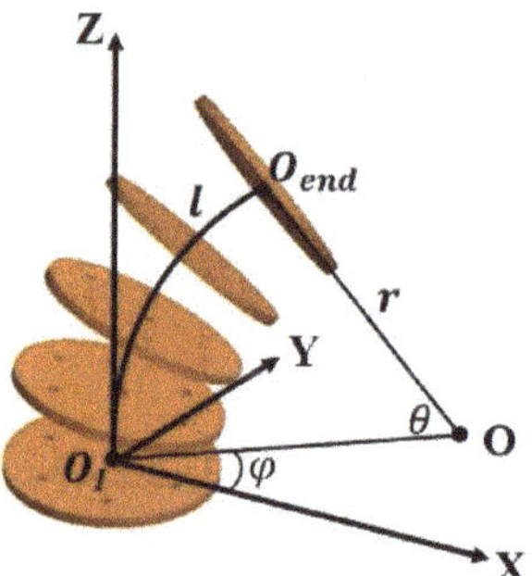

Figure 4. Illustration of the CKM kinematics parameters.

As in [37], the rotation of the arm in space is divided into five groups of Denavit–Hartenburg parameters. As illustrated in Figure 4, in the first turn, θ rotates to an angle of φ along the z-axis and α rotates to an angle of $\pi/2$ along the x-axis in space. In the second turn, θ is rotated to an angle of $\kappa l/2$ along the z-axis while α is rotated to an angle of $-\pi/2$ along the x-axis, which means that the z-axis is aligned with the endpoint. In the third turn, the linear distance between the bottom and the end d_3 is extended along the z-axis, and α is rotated at an angle of $\pi/2$ along the x-axis. In the fourth turn, θ is rotated along the z-axis and α is rotated at an angle of $-\pi/2$ along the x-axis so that the coordinate axis is positioned in the positive direction toward the endpoint of the CKM. Finally, θ is rotated at an angle of $-\varphi$ to offset the first set of rotations in the last turn. In summary, the D–H matrix of a single CKM can be obtained through the abovementioned rotation relationship, as in Table 1.

Table 1. Single CKM D–H matrix.

Transform Turn	θ	d	a	α
1	φ	0	0	$\pi/2$
2	$\kappa l/2$	0	0	$-\pi/2$
3	0	$2/\kappa \times \sin(\kappa l/2)$	0	$\pi/2$
4	$\kappa l/2$	0	0	$-\pi/2$
5	$-\varphi$	0	0	0

Through the D–H matrix in Table 1, and by substituting it into the calculation, the transformation matrix from the center coordinates of the circular bottom plate to the center coordinates of the end circular plate can be obtained:

$$T_1^5 = \begin{bmatrix} \cos^2\Phi(\cos(kl)-1)+1 & \sin\Phi\cos\Phi(\cos(kl)-1) & \cos\Phi\sin(kl) & \frac{\cos\Phi(1-\sin(kl))}{k} \\ \sin\Phi\cos\Phi(\cos(kl)-1) & \cos^2\Phi(1-\cos(kl))+\cos(kl) & \sin\Phi\sin(kl) & \frac{\sin\Phi(1-\cos(kl))}{k} \\ -\cos\Phi\sin(kl) & -\sin\Phi\sin(kl) & \cos(kl) & \frac{\sin(kl)}{k} \\ 0 & 0 & 0 & 1 \end{bmatrix} \quad (1)$$

The points (x, y, z) in the working plane must be converted into arc parameters (φ, κ, l) with inverse kinematics using the inverse kinematics in [2]. The conversion shows its bending geometric relationship, as illustrated in Figure 5. The target point can be projected from three-dimensional space to two-dimensional space:

$$\varphi = \tan^{-1}\frac{y}{x} \tag{2}$$

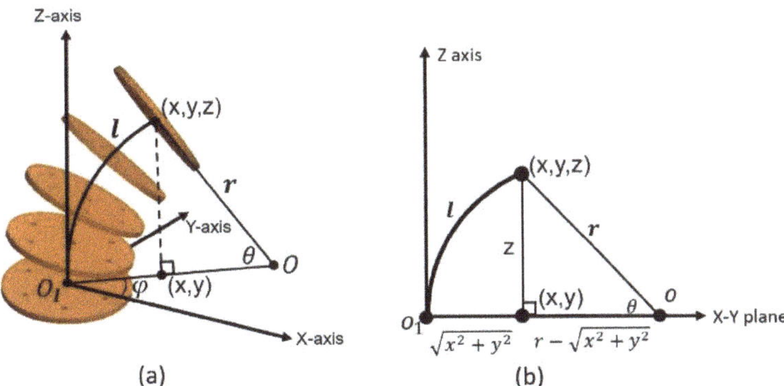

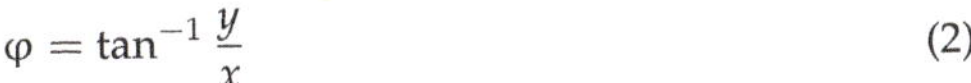

Figure 5. The posture of a single CKM reaching the target point: (**a**) three-dimensional and (**b**) two-dimensional.

A single CKM bends in an ideal arc shape, so the intersection of the end circular plate and the bottom circular plate extension line is the arc center of the CKM centerline. Therefore, the connecting line from the circular bottom plate to the center of arc o passes through the projected point to the x-y plane (x, y). Through geometric relations, the arc angle θ and arc radius r can be obtained:

$$r = \frac{x^2 + y^2 + z^2}{2\sqrt{x^2 + y^2}} \tag{3}$$

$$\theta = \cos^{-1}\left(\frac{r - \sqrt{x^2 + y^2}}{r}\right) \tag{4}$$

The arc curvature κ is the reciprocal of the arc radius r. Then, the arc length l can be expressed as:

$$l = r\theta = \frac{\cos^{-1}(1 - \kappa\sqrt{x^2 + y^2})}{\kappa}; \ \kappa = \frac{1}{r} \tag{5}$$

The arc parameters (φ, κ, l) are used in kinematics. However, linear actuators mainly control tension cables to manipulate the CKM. Therefore, the arc parameters (φ, κ, l) need to be converted into the length of each tension cable, which controls the CKM. The position of the tension cable is the three vertices of the equilateral triangle, so the arc length l is the average length of the three control cables (i.e., l_{C1}, l_{C2}, l_{C3}), as.

$$l = \frac{l_{C1} + l_{C2} + l_{C3}}{3} \tag{6}$$

Figure 6a explains the relation between the tension cables and the arc parameters in space. From Figure 6b, it is determined that $\varphi_{C1} = 90° - \varphi$, and φ_{C2} and φ_{C3} can be deduced through a geometric relation:

$$\begin{cases} \varphi_{C2} = 210° - \varphi \\ \varphi_{C3} = 330° - \varphi \end{cases} \tag{7}$$

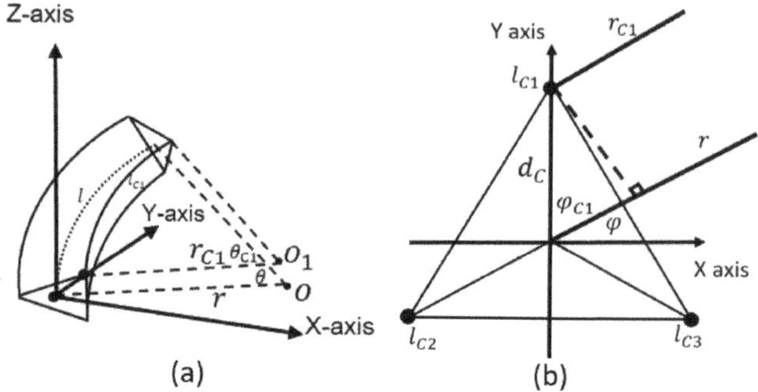

Figure 6. Relation between the tension cables and arc parameters: (**a**) three-dimensional and (**b**) two-dimensional.

Then, the arc radius of the tension cable r_{C1} and the central arc radius r are two parallel lines; the relation between r_{C1} and r is expressed through the trigonometric function:

$$r = r_{C1} + d_C * \cos \varphi_{C1} \tag{8}$$

where d_C is the distance between each cable and the center of the circular plate. Then, each arc radius of the tension cables r_{Ci} can be derived from the relation with the central arc radius r:

$$l_{Ci} = l - \theta d_C \cos \varphi_i; i = \{1, 2, 3\} \tag{9}$$

The arc length of each tension cable l_{Ci} can be deduced as:

$$r_{Ci} = r - d_C \cos \varphi_i; \; i = \{1, 2, 3\} \tag{10}$$

2.2. Design and Kinematics of the Flexible Manipulator Formed with Three CKMs

In this paper, the flexible manipulator consists of three CKMs. As Figure 7a shows, the tendon cables of each segment are controlled via linear actuators. The first circular plate segment passes through nine tendon cables, the second through six segments, and the third through three. Therefore, the design of each CKM circular plate is slightly different, as shown in Figure 7b.

The forward kinematics of the continuum robot is calculated. According to the mechanism configuration, the relationship of each coordinate system is shown in Figure 8.

The coordinate system of the first and third CKMs is the same, while the coordinate system of the second CKM is 180° inversed from the others. Because each segment is calculated independently, the CKM transformation matrix is obtained from the forward kinematic Equation (1). Therefore, each segment and fixed-segment transformation matrix, T_{si} and T_{di} are stated as the following:

$$T_{si} = \begin{bmatrix} \cos^2 \varphi_i(\cos \kappa_i l_i - 1) + 1 & \sin \varphi_i \cos \varphi_i(\cos \kappa_i l_i - 1) & \cos \varphi_i \sin \kappa_i l_i & \frac{\cos \varphi_i(1 - \cos \kappa_i l_i)}{\kappa_i} \\ \sin \varphi_i \cos \varphi_i(\cos \kappa_i l_i - 1) & \cos^2 \varphi_i(1 - \cos \kappa_i l_i) + \cos \kappa_i l_i & \sin \varphi_i \sin \kappa_i l_i & \frac{\sin \varphi_i(1 - \cos \kappa_i l_i)}{\kappa_i} \\ -\cos \varphi_i \sin \kappa_i l_i & -\sin \varphi_i \sin \kappa_i l_i & \cos \kappa_i l_i & \frac{\sin \kappa_i l_i}{\kappa_i} \\ 0 & 0 & 0 & 1 \end{bmatrix} \tag{11}$$

$$T_{di} = \begin{bmatrix} -1 & 0 & 0 & 0 \\ 0 & -1 & 0 & 0 \\ 0 & 0 & 1 & d_i \\ 0 & 0 & 0 & 1 \end{bmatrix} \tag{12}$$

Finally, the transformation matrix of the continuum robot can be obtained by multiplying T_{si} and T_{di}:

$$T_m = T_{s_1} T_{d_1} T_{s_2} T_{d_2} T_{s_3} = \begin{bmatrix} a_{11} & a_{12} & a_{13} & a_{14} \\ a_{21} & a_{22} & a_{23} & a_{24} \\ a_{31} & a_{32} & a_{33} & a_{34} \\ a_{41} & a_{42} & a_{43} & a_{44} \end{bmatrix} \tag{13}$$

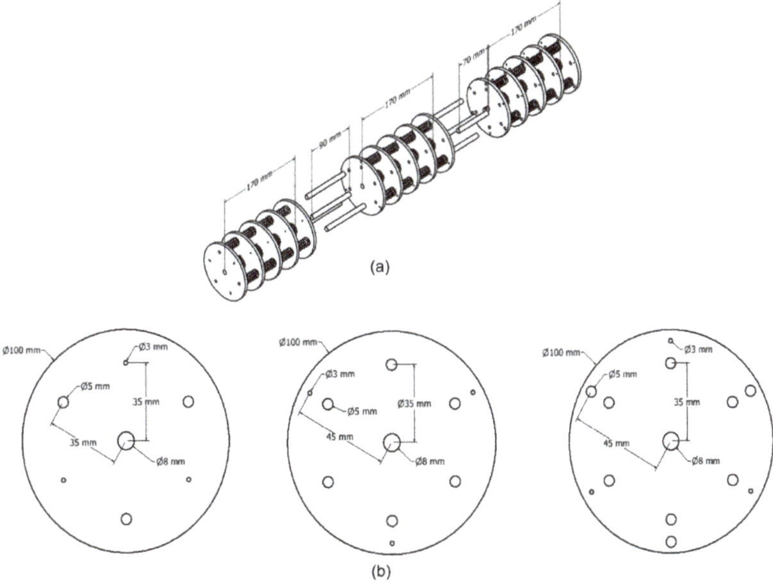

Figure 7. Relation between the tension cables and arc parameters: (**a**) three-dimensional and (**b**) two-dimensional.

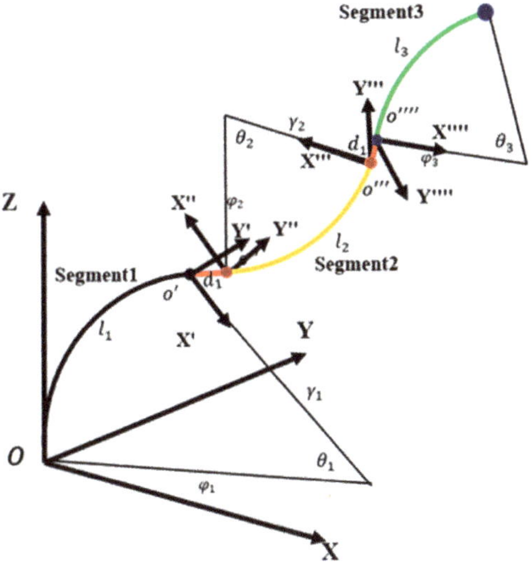

Figure 8. The coordinate system of the three-segment continuum robot.

According to [1], the velocity kinematics $v_{jacobian}$ can be written as Equation (14), with x as a vector in the task space, including the position or both the position and direction; $\dot{x}$ implies the x differential with time or velocity, $\dot{q}$ is the change of the arm arc parameters of each axis as Equation (15), and J is the Jacobian matrix:

$$v_{jacobian} = \begin{bmatrix} v_x & v_y & v_z & \omega_x & \omega_y & \omega_z \end{bmatrix}^T = \dot{x} = J\dot{q} \tag{14}$$

$$q = \begin{bmatrix} \varphi_1 & k_1 & l_1 & \varphi_2 & k_2 & l_2 & \varphi_3 & k_3 & l_3 \end{bmatrix} \tag{15}$$

Finally, the transformation matrix of the continuum robot can be differentiated by the chain rule:

$$\dot{T}_{s_1}^{s_3} = \dot{T}_{s_1} T_{d_1}^{s_3} + T_{s_1} \dot{T}_{d_1} T_{s_2}^{s_3} + T_{s_1}^{d_1} \dot{T}_{s_2} T_{d_2}^{s_3} + T_{s_1}^{s_2} \dot{T}_{d_2} T_{s_3} + T_{s_1}^{d_2} \dot{T}_{s_3} = \begin{bmatrix} \alpha_{11} & \alpha_{12} & \alpha_{13} & \alpha_{14} \\ \alpha_{21} & \alpha_{22} & \alpha_{23} & \alpha_{24} \\ \alpha_{31} & \alpha_{32} & \alpha_{33} & \alpha_{34} \\ \alpha_{41} & \alpha_{42} & \alpha_{43} & \alpha_{44} \end{bmatrix} \tag{16}$$

$$J\dot{q} = \begin{bmatrix} \alpha_{14} & \alpha_{24} & \alpha_{34} \end{bmatrix}^T \tag{17}$$

As shown in (14), $v_{jacobian}$ includes a linear and angular velocity, so $x = \begin{bmatrix} x & y & z \end{bmatrix}^T$ will be rewritten as $x = \begin{bmatrix} x & y & z & \theta_x & \theta_y \end{bmatrix}^T$. Because of the structure limitation, it cannot rotate on the z-axis, so θ_z is removed. The tangent vector $t = \begin{bmatrix} t_x & t_y & t_z \end{bmatrix}$ that can be defined by $\begin{bmatrix} \alpha_{13} & \alpha_{23} & \alpha_{33} \end{bmatrix}$ from Equation (16), and θ_t can be defined as Equation (18), with its differentiation version as Equation (19).

$$\theta_t = \begin{bmatrix} \arctan\frac{t_y}{t_z} \\ \arctan\frac{t_x}{t_z} \end{bmatrix} \tag{18}$$

$$\dot{\theta}_t = \begin{bmatrix} \frac{\dot{t}_y t_z - t_y \dot{t}_z}{t_z^2 + t_y^2} \\ \frac{\dot{t}_x t_z - t_x \dot{t}_z}{t_z^2 + t_x^2} \end{bmatrix} \tag{19}$$

The time derivative of the tangent vector t and derived angular Jacobian vector can be written as:

$$\dot{t} = J_t \dot{q} = \begin{bmatrix} \alpha_{13} & \alpha_{23} & \alpha_{33} \end{bmatrix}^T, J_t = \begin{bmatrix} J_{t1} & J_{t2} & J_{t3} \end{bmatrix}^T \tag{20}$$

$$J_\theta = \begin{bmatrix} J_{\theta_1} \\ J_{\theta_2} \end{bmatrix} = \begin{bmatrix} \frac{1}{t_z^2 + t_y^2}(t_z J_{t2} - t_y J_{t3}) \\ \frac{1}{t_z^2 + t_x^2}(t_z J_{t1} - t_x J_{t3}) \end{bmatrix} \tag{21}$$

The velocity kinematics is rewritten as Equation (22), and the Jacobian matrix can be factored out from it

$$v_{jacobian} = \begin{bmatrix} v_x & v_y & v_z & \omega_x & \omega_y \end{bmatrix}^T = J\dot{q} = \begin{bmatrix} \alpha_{14} & \alpha_{24} & \alpha_{34} & J_{\theta_1 \dot{q}} & J_{\theta_2 \dot{q}} \end{bmatrix} \tag{22}$$

The Jacobian matrix combined with forward kinematics can control the posture of the manipulator through the velocity and direction of the end. When W is the identity matrix, the solution obtained by $J(q)^+\dot{x}$ is the least square solution, with $J(q)^+$ is the pseudoinverse of the Jacobian matrix and $I - J(q)^+J(q)$ is the zero-space projection matrix:

$$\dot{q} = J(q)^+ \dot{x} + \{I - J(q)^+ J(q)\}\varepsilon \tag{23}$$

$$J^+ = W^{-1}J'\left(JW^{-1}J'\right)^{-1} \tag{24}$$

Next, the arc parameters of each segment are calculated through Equation (23) and the end-effector velocity. Finally, the length of the tension cables of each segment can be calculated through Equations (7)–(10). Then, control of the manipulator posture is achieved.

Finally, the proposed continuum robot's working space and limitations are elaborated, as shown in Table 2 and Figure 9. The manipulator's current end-effector position can be obtained by substituting arc parameters (φ, κ, l) into the forward kinematics. Then the operating range of the flexible manipulator is in an ideal situation. However, the actual length of the compression springs is limited by ± 20 mm, so after the restriction is substituted in and recalculated, the actual working range of the arm is spherical.

Table 2. Limitations of the arc parameters.

$\theta_1, \theta_2, \theta_3$	$\varphi_1, \varphi_2, \varphi_3$	$\kappa_1, \kappa_2, \kappa_3$
$0^o \sim 90^o$	$0^o \sim 360^o$	$\in R^+$

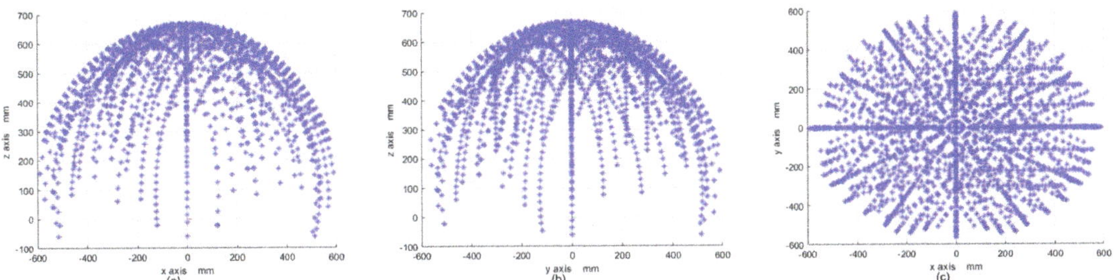

Figure 9. Working space analysis of the proposed continuum robot: (**a**) *x*-axis, (**b**) *y*-axis, (**c**) *z*-axis.

2.3. Implementation of the Image Servo Tracking Systems

In image-based servo tracking, it is necessary to convert the velocities of the camera and the feature point. The transformation matrix of this conversion is called an Image Jacobian and represents the velocity conversion relation between the feature point and the camera. The camera's location defines the camera's location coordinates, the feature points are the pixel coordinates in the image plane, and the conversion of the two points is considered the projection model of the pinhole camera. In the perspective projection model, the relation between the coordinates of the feature point (x_f, y_f, z_f) in the working space and the projection point (u, v) on the image plane can be written as:

$$\begin{bmatrix} u & v \end{bmatrix}^T = \frac{\lambda}{z_f} \begin{bmatrix} x_f & y_f \end{bmatrix}^T; \quad \lambda \text{ is the focal length} \tag{25}$$

If the camera is placed in a static environment and the camera moves at a translation velocity $V = \begin{bmatrix} V_x & V_y & V_z \end{bmatrix}^T$ and a rotation velocity $\omega = \begin{bmatrix} \omega_x & \omega_y & \omega_z \end{bmatrix}^T$, then the velocity of the feature point P in the camera coordinates can be expressed as:

$$\frac{dP}{dt} = -\omega \times P - V \tag{26}$$

$$\begin{cases} \dot{x}_f = -z_f \omega_y + z_f v \omega_z / \lambda - V_x \\ \dot{y}_f = -z_f u \omega_z / \lambda + z_f \omega_x - V_y \\ \dot{z}_f = -z_f v \omega_x / \lambda - z_f u \omega_y / \lambda - V_z \end{cases} \tag{27}$$

Finally, after differentiating Equation (25) to time and substituting Equation (27), Equation (28) can be factored out where J_{image} is the Image Jacobian, and V_{camera} is the camera's velocity in the camera coordinate system:

$$\begin{bmatrix} \dot{u} \\ \dot{v} \end{bmatrix} = J_{image} V_{camera} = \begin{bmatrix} -\frac{\lambda}{z} & 0 & \frac{u}{z} & \frac{uv}{\lambda} & -\frac{\lambda^2+u^2}{\lambda} & v \\ 0 & -\frac{\lambda}{z} & \frac{v}{z} & \frac{\lambda^2+v^2}{\lambda} & -\frac{uv}{\lambda} & -u \end{bmatrix} \begin{bmatrix} v_x \\ v_y \\ v_z \\ \omega_x \\ \omega_y \\ \omega_z \end{bmatrix} \tag{28}$$

where (u, v) is in length units, but the image obtained in the camera is a series of pixels, so the current (u, v) unit needs to be converted into a pixel unit:

$$\begin{bmatrix} u & v \end{bmatrix}^T = \frac{\lambda}{z} \begin{bmatrix} \frac{x}{S_x} & \frac{y}{S_y} \end{bmatrix}^T; \lambda_x = \frac{\lambda}{S_x}, \lambda_y = \frac{\lambda}{S_y} \tag{29}$$

where S_x, S_y and λ are the scale and focal length camera intrinsic parameters. Therefore, Equation (28) can be converted as:

$$J_{image} = \begin{bmatrix} -\frac{\lambda_x}{z} & 0 & \frac{u}{z} & \frac{uv}{\lambda_y} & -\frac{\lambda_x{}^2+u^2}{\lambda_x} & \frac{\lambda_x}{\lambda_y}v \\ 0 & -\frac{\lambda_y}{z} & \frac{v}{z} & \frac{\lambda_y{}^2+v^2}{\lambda_y} & -\frac{uv}{\lambda_x} & -\frac{\lambda_y}{\lambda_x}u \end{bmatrix} \tag{30}$$

2.3.1. Eye-To-Hand (ETH) Visual Servoing Configuration

In the ETH configuration, the camera is in a fixed position, so the depth is fixed in the calculation. The camera is obtained as the pixel coordinate system in the image plane, and the manipulator is the robot coordinate system, so it is necessary to multiply the transformation matrix to convert the pixel coordinate system to the spatial coordinate system. The relation can then be formulated as follows:

$$\begin{bmatrix} x_{robot} \\ y_{robot} \end{bmatrix} = [R] \begin{bmatrix} u \\ v \end{bmatrix} + [T] \tag{31}$$

Through the conversion relation, the points in the image plane can be transferred to the robot coordinate system, and through kinematics Equations (14)–(24), the manipulator can perform tasks.

2.3.2. Eye-In-Hand (EIH) Visual Servoing Configuration

The velocity of the camera in the camera coordinate system v_{camera} and the velocity of the camera in the robot coordinate system $v_{jacobian}$ are in different references. Therefore, to combine the two values, a coordinate conversion is required. Thus, Hutchinson et al. [37] proposed the conversion as follows:

$$v_{jacobian} = \begin{bmatrix} R_m v_{tcamera} - R_m v_{wcamera} \times r_m \\ R_m v_{wcamera} \end{bmatrix} = \begin{bmatrix} R_m v_{tcamera} + r_m \times R_m v_{wcamera} \\ R_m v_{wcamera} \end{bmatrix} = \begin{bmatrix} R_m & sk(r_m)R_m \\ 0 & R_m \end{bmatrix} \begin{bmatrix} v_{tcamera} \\ v_{wcamera} \end{bmatrix} \tag{32}$$

$$R_m = \begin{bmatrix} T_{m11} & T_{m12} & T_{m13} \\ T_{m21} & T_{m22} & T_{m23} \\ T_{m31} & T_{m32} & T_{m33} \end{bmatrix}, r_m = \begin{bmatrix} T_{m14} \\ T_{m24} \\ T_{m34} \end{bmatrix} \tag{33}$$

$$v_{jacobian} = \begin{bmatrix} R_m & sk(r_m)R_m \\ 0 & R_m \end{bmatrix} V_{camera} \tag{34}$$

Because the camera is installed at the tail of the manipulator, the rotation vector R_m and the translation vector r_m of the camera coordinate system's origin in the robot coordinate system can be obtained as Equation (33). After the calculation, the conversion relationship between the velocity in the camera coordinate system and the camera's velocity in the robot coordinate system can be obtained as in Equation (34).

With the dramatic development of deep learning, deep learning-based computer vision techniques are emerging in various engineering fields. For object detection tasks, the YOLO algorithm [38] and its variants [39,40] provide a reliable detection tool to detect objects in real-time. Furthermore, by integrating with the robot system, some deep vision-based methods were proposed to do the picking task with high accuracy [41,42]. In this study, for plugging and unplugging tasks, we applied the well-known object detection open-source framework, YOLOv4 [40], to detect the plugging area before extracting the target tracking point for action accomplishment.

3. Experimental Results

3.1. CKM-Based Continuum Robot Implementation

In the proposed three-segment CKM continuum robot, the three CKMs are independent, each controlled separately by three independently actuated tension cables. Therefore, the first CKM circular plates need to pass nine steel cables, the second CKM has six cables, and the third CKM has three cables. In each CKM, non-control tension cables are covered with a tension spring layer, allowing the non-control tension cables to change their length without affecting the posture of the current CKM.

When a tension cable is pulled to the tension-cable control module, it needs to bend because of the different directions, so using a brake housing helps to change direction. Controlling the cable through the brake housing can change the direction but will not affect the force of the cable. Due to the proposed continuum manipulator with three CKMs and nine tension cables, control requires a total of nine sliding tables. Therefore, the nine linear sliding tables are integrated into a rectangular tension-cable control module to facilitate construction. To reduce the total manipulator height, the tension-cable control module is localized horizontally in 3×3 square layers with three sliding tables in each layer controlling a CKM. After all the control cables pass through the outer brake housings fixed on the aluminum plate, they will be fixed on the slider. Finally, each CKM can be controlled by a group of sliders. The break housings and the sliding tables are illustrated in Figure 10.

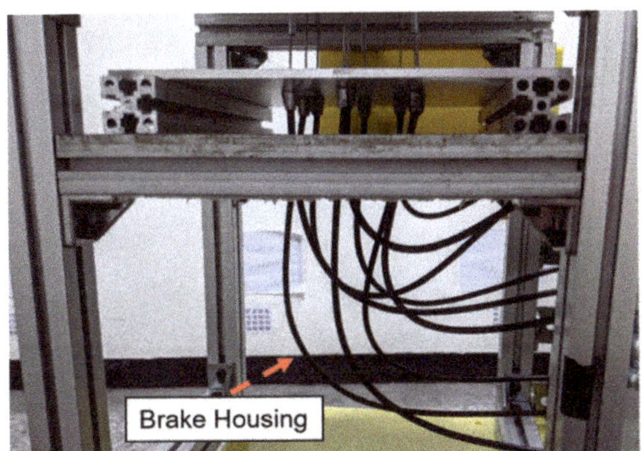

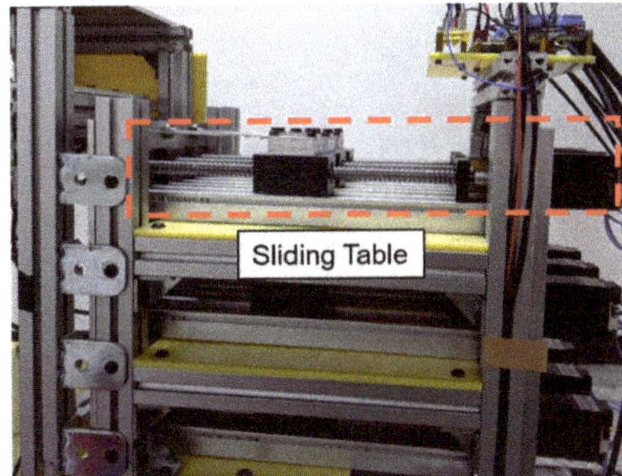

Figure 10. The brake housings (**left**) and a single sliding table (**right**).

3.2. Kinematic Simulations

Simulation of each kinematics is performed in MATLAB to verify the correctness of the kinematics and posture of the manipulator. First, the arc parameters (κ, φ, l) are substituted in the kinematics of a single CKM, κ is set to 0.003, and l is set to 170. Then, as shown in Figure 11, the simulation of the rotation angle of a single CKM is achieved. Finally, the arc parameters (φ, κ, l) of each segment are substituted in the kinematics of the flexible manipulator.

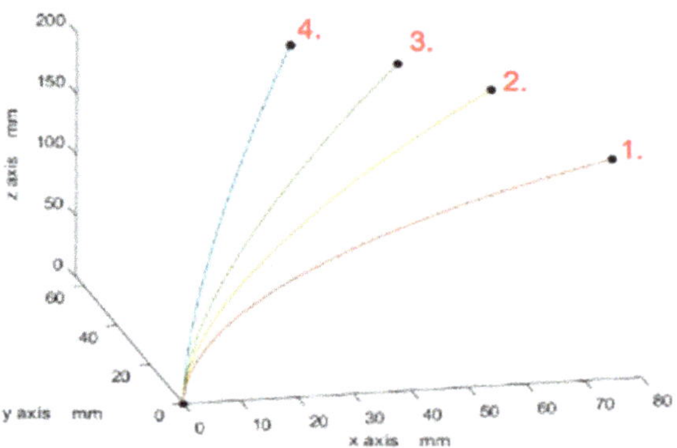

Figure 11. Simulation of the rotation angle φ of single CKM kinematic: **1.** $\varphi = 0$; **2.** $\varphi = \pi/2$; **3.** $\varphi = \pi/4$; **4.** $\varphi = \pi/3$.

In Figure 12a, substituting the same rotation angle φ in the three segments, the rotation angle φ is $(0 - 2\pi$, scale of $\pi/4)$, and the curvatures of the three segments κ are set as 0.003, 0.002, and 0.001, respectively. In Figure 12b, the rotation angle φ is substituted in the first and third segments, the rotation angle $\varphi + \pi$ is substituted in the second segment, and the rotation angle φ is $(0 - 2\pi$, scale of $\pi/4)$, where the curvatures κ of three segments are 0.003, 0.002, and 0.001, respectively.

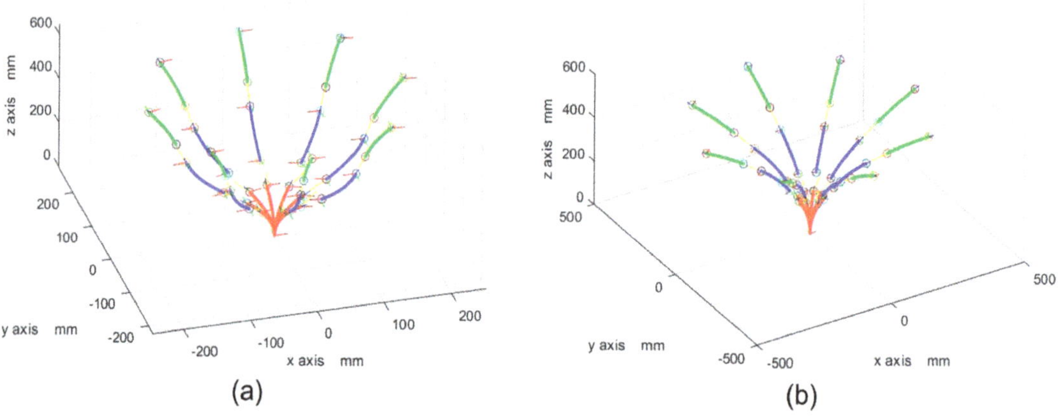

(a)　　　　(b)

Figure 12. Kinematic simulations of the flexible manipulator: (**a**) same rotation angle φ in three segments, (**b**) rotation angle φ in the first segment (red) and third segment (green), rotation angle $\varphi + \pi$ in the second segment (blue).

Next, the change in the curvature κ is simulated. The rotation angles φ for the three segments are fixed, and the order is $\pi/3$, $4\pi/3$, and $\pi/3$. Figure 13a shows that the arc curvature κ of the second segment changes from 0.002 to 0.007, each time increasing by 0.001, and the curvature κ of the first and third segments are fixed sequentially to 0.003 and 0.001. Figure 13b shows that the arc curvature κ of the third segment changes from 0.001 to 0.021, each time increasing by 0.005, and the curvature κ of the first and third segments are fixed sequentially to 0.003 and 0.002, respectively.

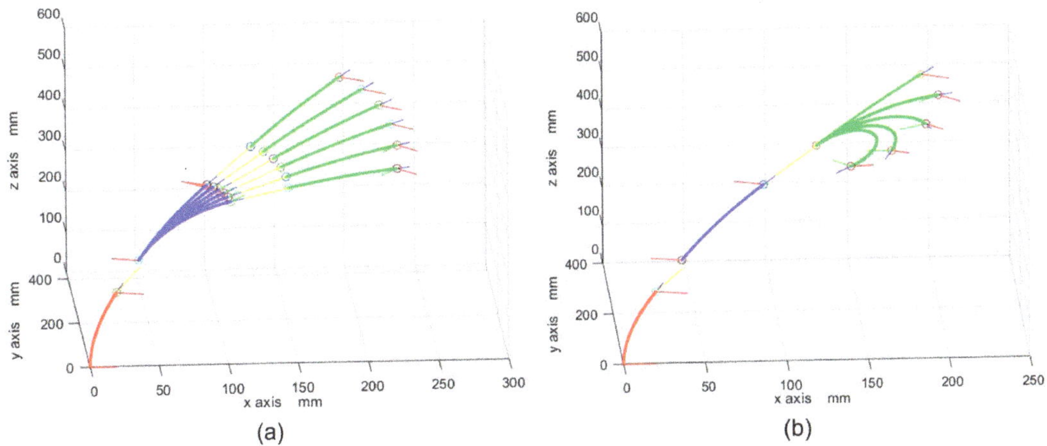

Figure 13. Kinematic simulation of the flexible manipulator: (**a**) change in the arc curvature κ of the second segment, (**b**) change in the arc curvature κ of the third segment.

The last is the simulation of velocity kinematics, as shown in Figure 14. Given an initial posture of the flexible manipulator, the first segment arc parameter (κ, φ, l) is $(0.003, 0, 170)$, the second segment arc parameter (κ, φ, l) is $(0.002, \pi/2, 170)$, and the third segment arc parameter (κ, φ, l) is $(0.001, \pi/2, 170)$. The kinematics are simulated by substituting different end velocities and directions in the kinematics. Finally, the kinematics have been verified for feasibility.

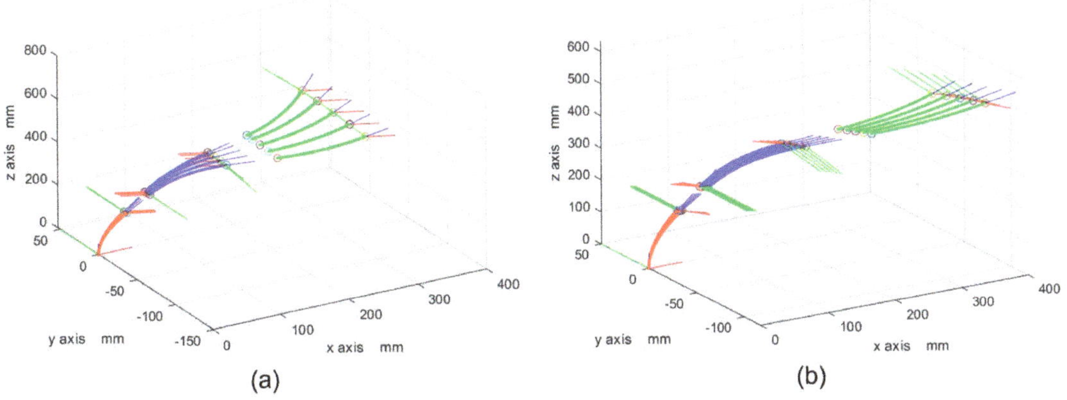

Figure 14. Velocity kinematic simulation of the flexible manipulator: (**a**) change in the end velocity $[0\ 20\ 0\ 0\ 0\ 0]$ (**b**) change in the end velocity $[20\ 0\ 0\ 0\ 0\ 0]$.

The posture of the arm movement over 4 s is simulated, as shown in Figure 14. The endpoint records of the two simulation results of each second are shown in Table 3. The end velocity vector $(0, 20, 0, 0, 0, 0)$, which means that it moves 20 mm per second along the positive y-axis during the end velocity vector $(20, 0, 0, 0, 0, 0)$, which means that it moves 20 mm per second in the positive x-axis. Table 3 shows that the error was within 10 mm, and the maximum error in the z-axis was within 50 mm. The feasibility of kinematics can be verified through the above simulations, whether using single CKM kinematics, flexible manipulator kinematics, or velocity kinematics.

Table 3. Velocity kinematics simulations-Endpoint error of the manipulator posture.

Speed Variation	(0, 20, 0, 0, 0, 0)					
	X	**Y**	**Z**	**X_error**	**Y_error**	**Z_error**
Start point	279.6	−101.5	586.8	-	-	-
1	281.6	−79.38	591.0	−1.8	−2.12	−4.2
2	282.7	−57.43	594.4	−3.1	−4.07	−7.6
3	283.5	−36.03	596.8	−3.9	−5.47	−10.0
4	283.6	−16.42	598.4	−4.0	−5.08	−11.5
Speed variation	**(20, 0, 0, 0, 0, 0)**					
	X	**Y**	**Z**	**X_error**	**Y_error**	**Z_error**
Start point	279.6	−101.5	586.8	-	-	-
1	298.4	−103.5	575.5	1.2	−2.0	11.3
2	316.7	−105.3	563.3	2.9	−3.8	23.5
3	334.3	−106.7	551.2	5.3	−1.4	35.6
4	351.2	−107.7	538.4	8.4	−2.4	48.4

3.3. Heart Shape Trajectory Tracking

In this experiment, the trajectory tracking task of eye-to-hand (ETH) is used to verify the effectiveness of the Jacobian control method. The experimental environment is shown in Figure 15. The camera position was fixed in the experiment, and the distance from the laser point imaging screen was 0.23 m. The exact size heart shape with different numbers of tracking points is generated in the image plane. Finally, the manipulator achieves trajectory tracking through velocity kinematics. The reference trajectory is the heart shape trajectory point defined in the image plane, with the generated heart shape trajectory point calculated using the following formula:

$$\begin{cases} y(x) = 60 - \sqrt{\left(1600 - (x - 40)^2\right)} \\ y(x) = 60 - 20(arcos(1 - (x/40)) - \pi) \end{cases} \tag{35}$$

Figure 15. The environment of the trajectory tracking.

According to the number of different heart shape trajectory points, five experiments were performed, and the results were averaged to verify the stability and error of the tracking. As a result, the tracking trajectory is shown in Figure 16. The error is the distance between the current laser and target points.

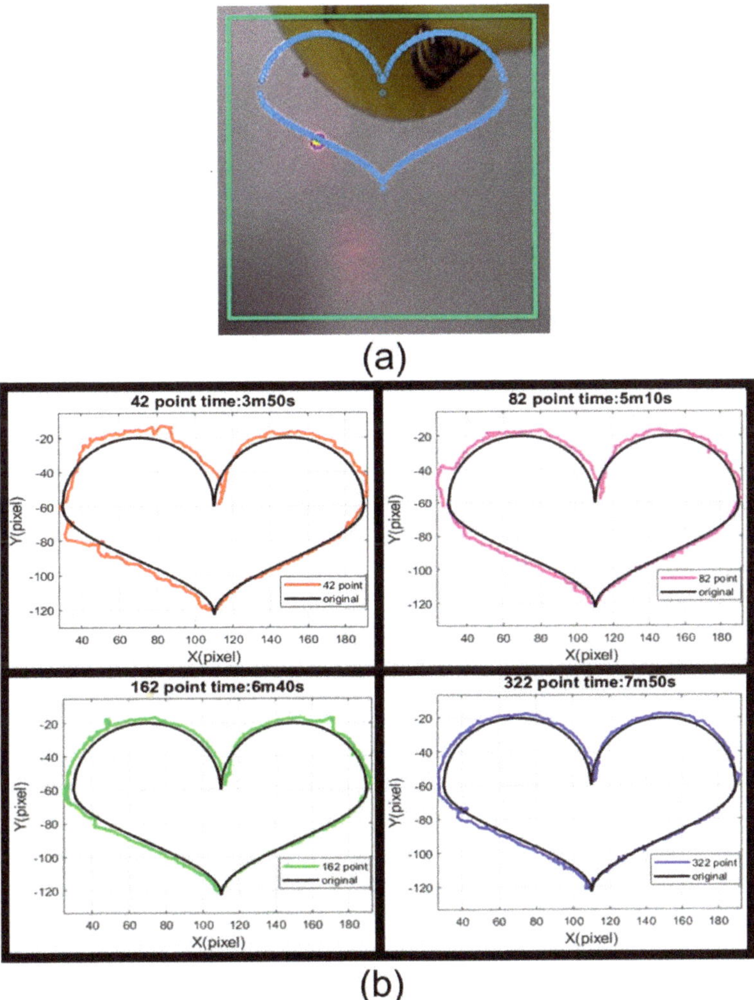

Figure 16. Heart shape trajectory tracking (**a**) Image obtained during trajectory tracking (**b**) Tracking trajectory with different numbers of tracking points.

The experimental results are shown in Table 4. The experiment shows that the proposed control method can achieve accurate path tracking.

3.4. Image-based Servo Tracking Experiment

The effectiveness and feasibility of image-based servo tracking are verified by the plugging/unplugging task of eye-in-hand (EIH), with a stereo camera and pneumatic grippers installed at the manipulator's end. The experimental environment and the processes of autonomous "plug-in" and "unplug" of electric power sockets are shown in Figures 17 and 18, respectively.

Table 4. Velocity kinematics simulations-Endpoint error of the manipulator posture.

Number of Points	Test	Max Error (pixel)		Min Error (pixel)		Average Error (pixel)		RMSE
		X	Y	X	Y	X	Y	
42	1	14.43	27.00	0.10	0.04	3.70	5.21	7.80
	2	15.23	32.30	0.02	0.41	3.89	5.34	7.67
	3	16.34	35.05	0.04	0.10	3.64	5.17	7.63
	4	15.69	28.31	0.03	0.39	3.77	5.26	7.74
	5	16.89	33.41	0.21	0.19	3.98	5.31	7.87
	Average	15.72	31.21	0.08	0.22	3.80	5.26	7.74
82	1	40.00	19.24	0.00	0.33	2.82	4.08	6.05
	2	38.20	20.45	0.00	0.33	2.79	3.99	6.08
	3	42.32	18.78	0.08	0.07	2.67	4.12	6.13
	4	39.64	19.67	0.02	0.09	2.37	4.05	6.09
	5	37.25	20.26	0.07	0.05	2.88	3.98	6.03
	Average	39.50	19.68	0.03	0.17	2.71	4.04	6.08
162	1	36.00	14.42	0.00	0.06	1.81	3.09	4.26
	2	37.23	14.36	0.02	0.02	1.86	3.07	4.32
	3	34.05	13.96	0.00	0.03	1.82	3.08	4.23
	4	32.35	14.85	0.00	0.04	1.83	3.02	4.36
	5	35.87	13.56	0.02	0.01	1.79	3.08	4.33
	Average	35.10	14.23	0.01	0.03	1.82	3.07	4.30
322	1	6.02	11.09	0.02	0.00	1.79	3.10	3.97
	2	6.04	11.05	0.01	0.02	1.80	2.97	4.07
	3	5.98	11.09	0.00	0.03	1.73	2.99	4.05
	4	6.06	11.12	0.00	0.01	1.75	3.12	4.01
	5	6.03	10.99	0.00	0.01	1.76	3.08	4.03
	Average	6.03	11.07	0.01	0.02	1.77	3.05	4.03

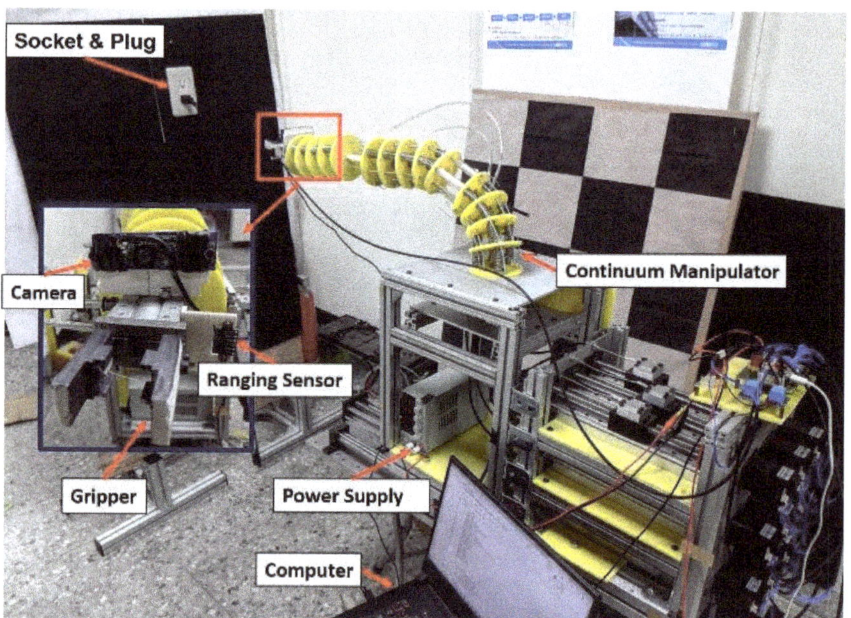

Figure 17. Environment for the plugging/unplugging experiment.

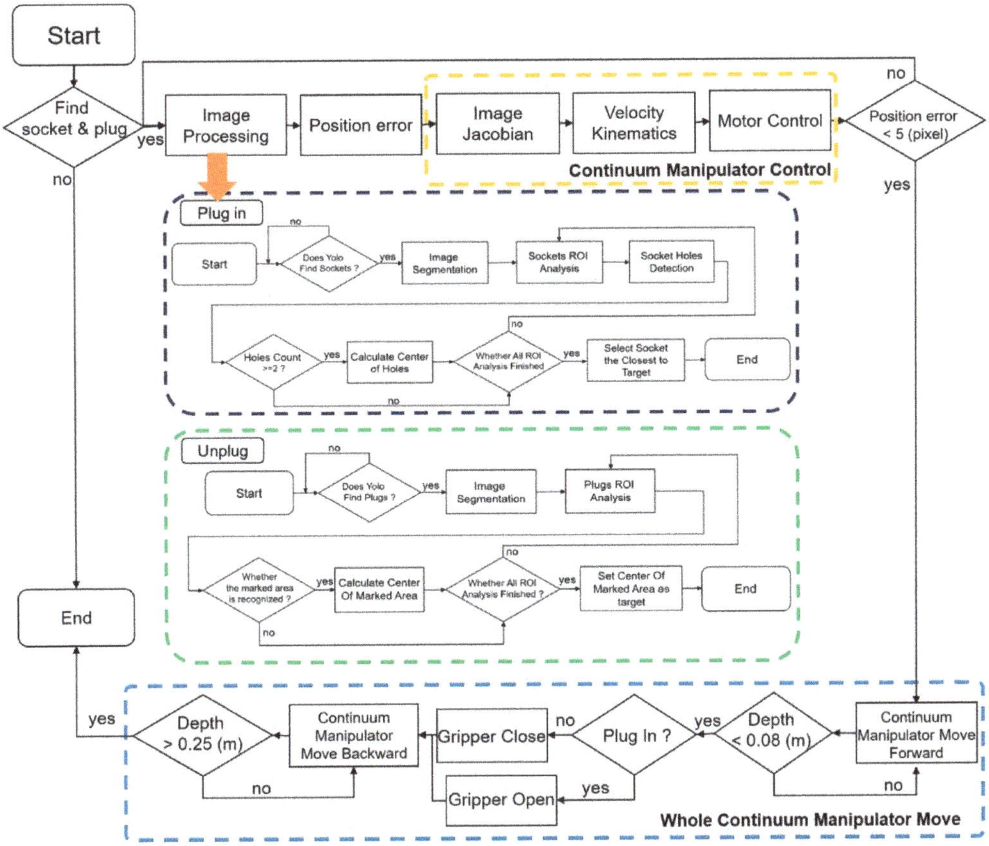

Figure 18. The algorithm flowchart of the Plugging/Unplugging task.

The experiment has two main parts: plugging and pulling the plug. In both experiments, the socket and plug specifications are in Figure 19. A widely used object detection open-source framework, YOLOv4 [39], was used to identify the target area first, and then the target tracking point was analyzed through image processing. Figure 20 shows the results of the detected electric socket and plug.

After the target tracking point is obtained, the error can be calculated between the current position and the tracking point. The arc parameters (φ, κ, l) of each segment of the manipulator can be obtained by substituting the error into the Image Jacobian and velocity kinematics. Then, the manipulator can perform the task of plugging/unplugging through arc parameters (φ, κ, l).

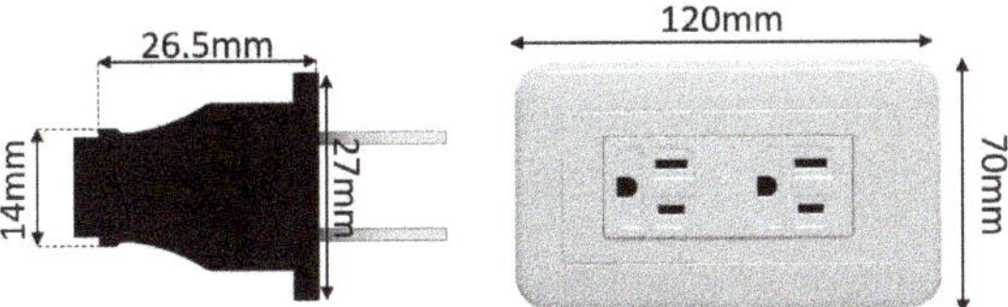

Figure 19. Specifications of the socket and plug.

Figure 20. Object detection results using YOLO for the electric power socket (left-hand-side) and the electric power plug (right-hand-side). It is noted that the detected object areas are indicated as rectangle boxes.

Finally, Figures 21 and 22 show the results of the plugging and unplugging experiments, respectively. Both parts of the experiment (plugging-in and plugging-out) were performed five times, and the experimental results are shown in Table 5. We set the start points for each experimental trial as the final points, around 50 cm in front of the target. For five trials of unplugging, the success rate is 100%. However, the plugging-in experiments show a success rate of only 60%. The visual servo system could achieve lows of 58 ± 2.12 s and 83 ± 6.87 s in processing time at the 90% confidence level in the unplugging and plugging-in tasks, respectively. The main reason is that the power plugging-in into the socket needs a higher accuracy because of the small holes in the socket. Therefore, the control accuracy and rigidity of the continuum manipulator should be further improved in the future.

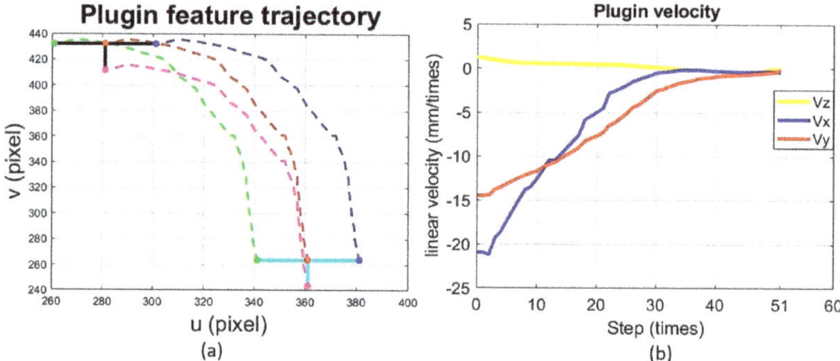

Figure 21. Plugging-in experiment: (**a**) trajectory of the feature points, (**b**) end-effector linear velocity v_x, v_y and v_z.

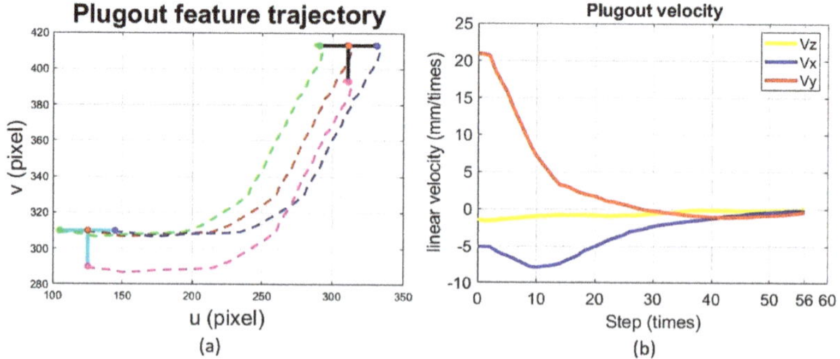

Figure 22. Unplugging experiment: (**a**) trajectory of the feature points, (**b**) end-effector linear velocity v_x, v_y and v_z.

Table 5. Plugging/unplugging experimental results for each test.

	Unplugging Task		Plugging Task	
	Time (s)	Mission Completion	Time (s)	Mission Completion
1	61	Success	88	success
2	55	Success	80	success
3	62	Success	-	fail
4	58	Success	75	success
5	57	Success	-	fail

4. Conclusions

In this paper, a continuum kinematic module (CKM) was proposed. The function and feasibility of the CKM were demonstrated using a continuum manipulator composed of three CKMs. First, the continuum manipulator's kinematics were verified through a kinematic simulation and analysis, and the correctness of the manipulator posture was confirmed through the simulation. Then, an ETH heart shape tracking experiment was established to verify the correctness of the kinematics and the accuracy of control of the continuum manipulator in practical applications. Finally, the accuracy and feasibility of image servo tracking in an EIH plugging-in/unplugging experiment were demonstrated in extensive tolerance conditions. At the same time, the results show that the accuracy needs to be improved in low-tolerance conditions. Furthermore, the wire-driven CKM easily causes posture deviation due to the influence of gravity.

In future work, an inertial measurement unit (IMU) could be installed to correct posture deviation through feedback. To increase the accuracy of the image-based servo tracking, the current IBVS can be changed to PBVS because the IBVS method does not involve manipulator posture considerations. The PBVS will instead estimate the current pose of the target relative to the camera. Because the continuum manipulator has a greater degree of freedom, the wire-driven manipulator is susceptible to the influence of gravity, which causes posture deviation. Therefore, if PBVS-based image-based servo tracking is used, the accuracy of the image-based servo tracking may be improved. In addition, the continuum manipulator is limited in movement. Therefore, it is proposed that the mounting of the robot on an AGV can remove the restriction in the movement of the manipulator, and indoor positioning will allow the manipulator to perform more diversified tasks. Finally, the continuum manipulator can be applied to collaborative or service robots.

Author Contributions: Methodology, M.-H.H.; Project administration, C.-H.K.; Software, M.-H.H.; Supervision, C.-H.K.; Validation, M.-H.H.; Visualization, M.-H.H.; Writing—original draft, M.-H.H.; Writing—review & editing, M.-H.H., P.T.-T.N. and D.-D.N. All authors have read and agreed to the published version of the manuscript.

Funding: This research received no external funding.

Institutional Review Board Statement: Not applicable.

Informed Consent Statement: Not applicable.

Data Availability Statement: Not applicable.

Conflicts of Interest: The authors declare no conflict of interest.

References

1. Angrisani, L.; Grazioso, S.; Gironimo, G.D.; Panariello, D.; Tedesco, A. On the use of soft continuum robots for remote measurement tasks in constrained environments: A brief overview of applications. In Proceedings of the 2019 IEEE International Symposium on Measurements & Networking (M&N), Catania, Italy, 8–10 July 2019; pp. 1–5.
2. Boonchai, P.; Tuchinda, K. Design and Control of Continuum Robot for Using with Solar Cell System. In Proceedings of the 3rd International Conference on Vision, Image and Signal Processing (ICVISP 2019), Vancouver, BC, Canada, 26–28 August 2019.
3. Burgner-Kahrs, J.; Rucker, D.C.; Choset, H. Continuum Robots for Medical Applications: A Survey. *IEEE Trans. Robot.* **2015**, *31*, 1261–1280. [CrossRef]

4. Yamauchi, Y.; Ambe, Y.; Nagano, H.; Konyo, M.; Bando, Y.; Ito, E.; Arnold, S.; Yamazaki, K.; Itoyama, K.; Okatani, T.; et al. Development of a continuum robot enhanced with distributed sensors for search and rescue. *ROBOMECH J.* **2022**, *9*, 8. [CrossRef]

5. Zhang, J.; Sheng, J.; O'Neill, C.T.; Walsh, C.J.; Wood, R.J.; Ryu, J.H.; Desai, J.P.; Yip, M.C. Robotic Artificial Muscles: Current Progress and Future Perspectives. *IEEE Trans. Robot.* **2019**, *35*, 761–781. [CrossRef]

6. Li, S.; Hao, G. Current Trends and Prospects in Compliant Continuum Robots: A Survey. *Actuators* **2021**, *10*, 145. [CrossRef]

7. Dong, X.; Wang, M.; Mohammad, A.; Ba, W.; Russo, M.; Norton, A.; Kell, J.; Axinte, D. Continuum Robots Collaborate for Safe Manipulation of High-Temperature Flame to Enable Repairs in Challenging Environments. *IEEE/ASME Trans. Mechatron.* **2022**, *27*, 4217–4220. [CrossRef]

8. Wooten, M.; Frazelle, C.; Walker, I.D.; Kapadia, A.; Lee, J.H. Exploration and Inspection with Vine-Inspired Continuum Robots. In Proceedings of the 2018 IEEE International Conference on Robotics and Automation (ICRA), Brisbane, QLD, Australia, 21–25 May 2018; pp. 5526–5533.

9. Hannan, M.W.; Walker, I.D. Kinematics and the implementation of an elephant's trunk manipulator and other continuum style robots. *J. Robot. Syst.* **2003**, *20*, 45–63. [CrossRef]

10. Jones, B.A.; Walker, I.D. Kinematics for multi-section continuum robots. *IEEE Trans. Robot.* **2006**, *22*, 43–55. [CrossRef]

11. Yoon, H.S.; Yi, B.J. A 4-DOF flexible continuum robot using a spring backbone. In Proceedings of the International Conference on Mechatronics and Automation, Changchun, China, 9–12 August 2009; pp. 1249–1254.

12. Tokunaga, T.; Oka, K.; Harada, A. 1segment continuum manipulator for automatic harvesting robot-prototype and modeling. In Proceedings of the IEEE International Conference on Mechatronics and Automation (ICMA), Takamatsu, Japan, 6–9 August 2017; pp. 1655–1659.

13. Liu, Y.; Yang, Y.; Peng, Y.; Zhong, S.; Liu, N.; Pu, H. A Light Soft Manipulator With Continuously Controllable Stiffness Actuated by a Thin McKibben Pneumatic Artificial Muscle. *IEEE/ASME Trans. Mechatron.* **2020**, *25*, 1944–1952. [CrossRef]

14. Dalvand, M.M.; Nahavandi, S.; Howe, R.D. An Analytical Loading Model for n-Tendon Continuum Robots. *IEEE Trans. Robot.* **2018**, *34*, 1215–1225. [CrossRef]

15. Hassan, T.; Cianchetti, M.; Mazzolai, B.; Laschi, C.; Dario, P. Active-Braid, a Bioinspired Continuum Manipulator. *IEEE Robot. Autom. Lett.* **2017**, *2*, 2104–2110. [CrossRef]

16. Zhang, G.; Du, F.; Xue, S.; Cheng, H.; Zhang, X.; Song, R.; Li, Y. Design and Modeling of a Bio-Inspired Compound Continuum Robot for Minimally Invasive Surgery. *Machines* **2022**, *10*, 468. [CrossRef]

17. Comin, F.J.; Saaj, C.M.; Mustaza, S.M.; Saaj, R. Safe Testing of Electrical Diathermy Cutting Using a New Generation Soft Manipulator. *IEEE Trans. Robot.* **2018**, *34*, 1659–1666. [CrossRef]

18. Zhao, B.; Zhang, W.; Zhang, Z.; Zhu, X.; Xu, K. Continuum Manipulator with Redundant Backbones and Constrained Bending Curvature for Continuously Variable Stiffness. In Proceedings of the IEEE/RSJ International Conference on Intelligent Robots and Systems (IROS), Madrid, Spain, 1–5 October 2018; pp. 7492–7499.

19. Qu, T.; Chen, J.; Shen, S.; Xiao, Z.; Yue, Z.; Lau, H.Y.K. Motion control of a bio-inspired wire-driven multi-backbone continuum minimally invasive surgical manipulator. In Proceedings of the IEEE International Conference on Robotics and Biomimetics (ROBIO), Qingdao, China, 3–7 December 2016; pp. 1989–1995.

20. Hwang, M.; Kwon, D. Strong Continuum Manipulator for Flexible Endoscopic Surgery. *IEEE/ASME Trans. Mechatron.* **2019**, *24*, 2193–2203. [CrossRef]

21. Zhang, S.; Li, Q.; Yang, H.; Zhao, J.; Xu, K. Configuration Transition Control of a Continuum Surgical Manipulator for Improved Kinematic Performance. *IEEE Robot. Autom. Lett.* **2019**, *4*, 3750–3757. [CrossRef]

22. Wang, T.; Wang, Z.; Wu, G.; Lei, L.; Zhao, B.; Zhang, P.; Shang, P. Design and Analysis of a Snake-like Surgical Robot with Continuum Joints. In Proceedings of the 5th International Conference on Advanced Robotics and Mechatronics (ICARM), Shenzhen, China, 18–21 December 2020; pp. 178–183.

23. Li, J.; Zhou, Y.; Tan, J.; Wang, Z.; Liu, H. Design and Modeling of a Parallel Shifted-Routing Cable-Driven Continuum Manipulator for Endometrial Regeneration Surgery. In Proceedings of the IEEE/RSJ International Conference on Intelligent Robots and Systems (IROS), Las Vegas, NV, USA, 24 October 2020–24 January 2021; pp. 3178–3183.

24. Shen, W.; Yang, G.; Zheng, T.; Wang, Y.; Yang, K.; Fang, Z. An Accuracy Enhancement Method for a Cable-Driven Continuum Robot With a Flexible Backbone. *IEEE Access* **2020**, *8*, 37474–37481. [CrossRef]

25. Seleem, I.A.; El-Hussieny, H.; Assal, S.F.M.; Ishii, H. Development and Stability Analysis of an Imitation Learning-Based Pose Planning Approach for Multi-Section Continuum Robot. *IEEE Access* **2020**, *8*, 99366–99379. [CrossRef]

26. Li, Q.; Yang, H.; Chen, Y.; Xu, K. Closed Loop Control of a Continuum Surgical Manipulator for Improved Absolute Positioning Accuracy. In Proceedings of the IEEE International Conference on Robotics and Biomimetics (ROBIO), Dali, China, 6–8 December 2019; pp. 1551–1556.

27. Lai, J.; Huang, K.; Lu, B.; Chu, H.K. Toward Vision-based Adaptive Configuring of A Bidirectional Two-Segment Soft Continuum Manipulator. In Proceedings of the IEEE/ASME International Conference on Advanced Intelligent Mechatronics (AIM), Boston, MA, USA, 6–9 July 2020; pp. 934–939.

28. Yang, H.; Wu, B.; Liu, X.; Xu, K. A Closed-Loop Controller for a Continuum Surgical Manipulator Based on a Specially Designed Wrist Marker and Stereo Tracking. In Proceedings of the IEEE/ASME International Conference on Advanced Intelligent Mechatronics (AIM), Boston, MA, USA, 6–9 July 2020; pp. 335–340.

29. Song, K.; Tsai, H. Visual Servoing and Compliant Motion Control of a Continuum Robot. In Proceedings of the 18th International Conference on Control, Automation and Systems (ICCAS), PyeongChang, Republic of Korea, 17–20 October 2018; pp. 734–739.
30. Mo, H.; Wei, R.; Ouyang, B.; Xing, L.; Shan, Y.; Liu, Y.; Sun, D. Control of a Flexible Continuum Manipulator for Laser Beam Steering. *IEEE Robot. Autom. Lett.* **2021**, *6*, 1074–1081. [CrossRef]
31. Wu, K.; Zhu, G.; Wu, L.; Gao, W.; Song, S.; Lim, C.M.; Ren, H. Safety-Enhanced Model-Free Visual Servoing for Continuum Tubular Robots Through Singularity Avoidance in Confined Environments. *IEEE Access* **2019**, *7*, 21539–21558. [CrossRef]
32. del Giudice, G.; Orekhov, A.; Shen, J.; Joos, K.; Simaan, N. Investigation of Micro-motion Kinematics of Continuum Robots for Volumetric OCT and OCT-guided Visual Servoing. *IEEE/ASME Trans. Mechatron.* **2020**, *26*, 2604–2615. [CrossRef]
33. Lindenroth, L.; Merlin, J.; Bano, S.; Manjaly, J.G.; Mehta, N.; Stoyanov, D. Intrinsic Force Sensing for Motion Estimation in a Parallel, Fluidic Soft Robot for Endoluminal Interventions. *IEEE Robot. Autom. Lett.* **2022**, *7*, 10581–10588. [CrossRef]
34. Xu, F.; Wang, H.; Chen, W.; Miao, Y. Visual Servoing of a Cable-Driven Soft Robot Manipulator With Shape Feature. *IEEE Robot. Autom. Lett.* **2021**, *6*, 4281–4288. [CrossRef]
35. AlBeladi, A.; Ripperger, E.; Hutchinson, S.; Krishnan, G. Hybrid Eye-in-Hand/Eye-to-Hand Image Based Visual Servoing for Soft Continuum Arms. *IEEE Robot. Autom. Lett.* **2022**, *7*, 11298–11305. [CrossRef]
36. Zhang, Z.; Rosa, B.; Caravaca-Mora, O.; Zanne, P.; Gora, M.J.; Nageotte, F. Image-Guided Control of an Endoscopic Robot for OCT Path Scanning. *IEEE Robot. Autom. Lett.* **2021**, *6*, 5881–5888. [CrossRef]
37. Webster, R.J., III; Jones, B.A. Design and Kinematic Modeling of Constant Curvature Continuum Robots: A Review. *Int. J. Robot. Res.* **2010**, *29*, 1661–1683. [CrossRef]
38. Redmon, J.; Divvala, S.; Girshick, R.; Farhadi, A. You Only Look Once: Unified, Real-Time Object Detection. *arXiv* **2016**, arXiv:1506.02640.
39. Redmon, J.; Farhadi, A. YOLOv3: An Incremental Improvement. *arXiv* **2018**, arXiv:1804.02767.
40. Bochkovskiy, A.; Wang, C.Y.; Liao, H.Y.M. YOLOv4: Optimal Speed and Accuracy of Object Detection. *arXiv* **2020**, arXiv:2004.10934.
41. Wang, H.; Lin, Y.; Xu, X.; Chen, Z.; Wu, Z.; Tang, Y. A Study on Long-Close Distance Coordination Control Strategy for Litchi Picking. *Agronomy* **2022**, *12*, 1520. [CrossRef]
42. Tang, Y.; Zhou, H.; Wang, H.; Zhang, Y. Fruit detection and positioning technology for a Camellia oleifera C. Abel orchard based on improved YOLOv4-tiny model and binocular stereo vision. *Expert Syst. Appl.* **2023**, *211*, 118573. [CrossRef]

MDPI

St. Alban-Anlage 66

4052 Basel

Switzerland

Tel. +41 61 683 77 34

Fax +41 61 302 89 18

www.mdpi.com

Actuators Editorial Office

E-mail: actuators@mdpi.com

www.mdpi.com/journal/actuators